DYNAMIC AGRICULTURE YEARS 11–12 3RD EDITION

LISLE BROWN
ROBERT HINDMARSH
ROSS MCGREGOR

Australia • Brazil • Mexico • Singapore • United Kingdom • United States

Dynamic Agriculture Years 11–12
3rd Edition
Lisle Brown
Robert Hindmarsh
Ross McGregor

Senior publishing editor: Eleanor Gregory
Editor: Nadine Anderson-Conklin
Copyeditor: Marcia Bascombe
Proofreader: Jane Fitzpatrick
Indexer: Bruce Gillespie
Permissions researcher: Flora Smith
Text designer: Miranda Costa
Cover designer: Olga Lavecchia
Cover image: Getty Images/Jamie Evans
Production controllers: Emily Moore and Emma Roberts
Typesetter: Q2AMedia
Reprint: Alice Kane

National Library of Australia Cataloguing-in-Publication Data
Brown, Lisle, author.
Dynamic agriculture : years 11-12 /
Lisle Brown, Robert Hindmarsh, Ross McGregor.

3rd edition.
9780170265560 (paperback)
Includes index.
For secondary school age.

Agriculture--Australia--Textbooks.
Agriculture--Study and teaching (Secondary)--Australia.

Hindmarsh, Robert, author.
McGregor, Ross, author.

630.994

Cengage Learning Australia
Level 7, 80 Dorcas Street
South Melbourne, Victoria Australia 3205

Cengage Learning New Zealand
Unit 4B Rosedale Office Park
331 Rosedale Road, Albany, North Shore 0632, NZ

For learning solutions, visit **cengage.com.au**

Printed in China by 1010 Printing International Limited
10 11 12 25 24

CONTENTS

ISBN 9780170265560

Introduction

This appealing new edition of *Dynamic Agriculture Years 11–12* addresses effectively the entire revised NSW Years 11–12 Agriculture Stage 6 syllabus.

As with previous editions of *Dynamic Agriculture*, this new text has been written with the understanding that students' enjoyment of agriculture is enhanced by hands-on experience. The book is packed with a broad range of practical skills, activities and questions to cater for a range of abilities. Margin questions throughout the text enable students to review the content. Agricultural terms and key words are highlighted at the beginning of each chapter and throughout the text. They also appear in an end-of-book glossary so students can search for terms alphabetically. The chapter review sections contain extension, research and assessment material, including ICT activities. Questions and activities within this section are classified according to the cognitive demand on the student (according to Bloom's revised taxonomy).

Resources for each chapter are provided on the accompanying NelsonNet website. The worksheets provide students with revision, consolidation, practical and skill-based activities for a range of abilities.

This lively, accessible book draws senior agriculture students into the subject and enables them to build on their knowledge, understanding and skills to participate in the dynamic world of modern agriculture.

Shutterstock.com/Im Perfect Lazybones

How to use this book

The **outcomes** page at the beginning of each chapter lists the key knowledge and skills that students will learn throughout the chapter. The outcomes are classified according to year level, with outcomes in orange being Preliminary outcomes, and outcomes in blue being HSC outcomes. Chapter outcomes in black assist students in understanding course outcomes overall.

The **words to know** list at the beginning of each chapter presents the student with a glossary of bolded new terms introduced within the chapter. These terms also appear in an alphabetical glossary at the end of the book.

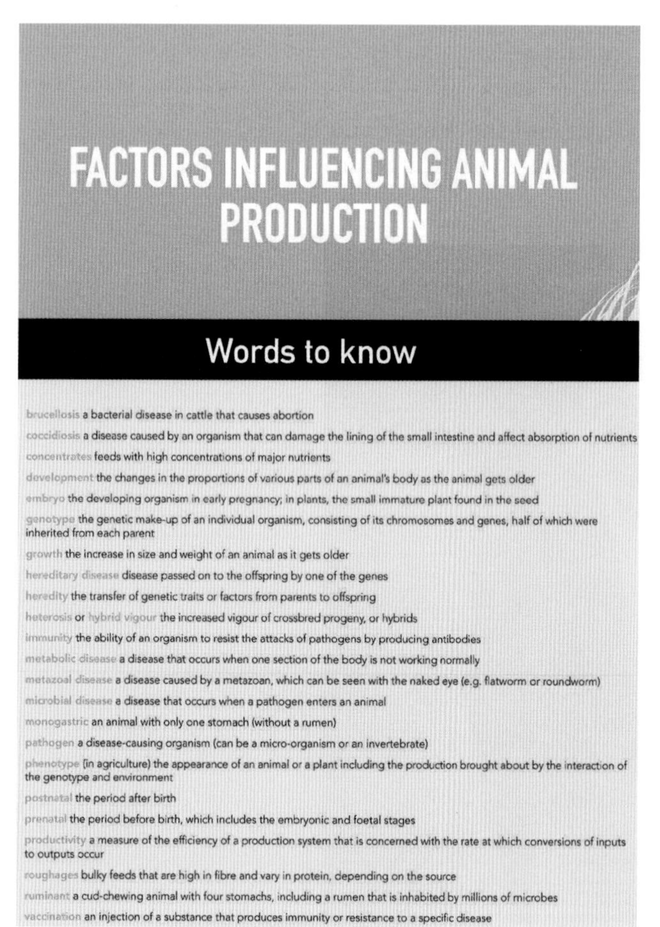

FACTORS INFLUENCING ANIMAL PRODUCTION

Words to know

brucellosis a bacterial disease in cattle that causes abortion
coccidiosis a disease caused by an organism that can damage the lining of the small intestine and affect absorption of nutrients
concentrates feeds with high concentrations of major nutrients
development the changes in the proportions of various parts of an animal's body as the animal gets older
embryo the developing organism in early pregnancy; in plants, the small immature plant found in the seed
genotype the genetic make-up of an individual organism, consisting of its chromosomes and genes, half of which were inherited from each parent
growth the increase in size and weight of an animal as it gets older
hereditary disease disease passed on to the offspring by one of the genes
heredity the transfer of genetic traits or factors from parents to offspring
heterosis or hybrid vigour the increased vigour of crossbred progeny, or hybrids
immunity the ability of an organism to resist the attacks of pathogens by producing antibodies
metabolic disease a disease that occurs when one section of the body is not working normally
metazoal disease a disease caused by a metazoan, which can be seen with the naked eye (e.g. flatworm or roundworm)
microbial disease a disease that occurs when a pathogen enters an animal
monogastric an animal with only one stomach (without a rumen)
pathogen a disease-causing organism (can be a micro-organism or an invertebrate)
phenotype (in agriculture) the appearance of an animal or a plant including the production brought about by the interaction of the genotype and environment
postnatal the period after birth
prenatal the period before birth, which includes the embryonic and foetal stages
productivity a measure of the efficiency of a production system that is concerned with the rate at which conversions of inputs to outputs occur
roughages bulky feeds that are high in fibre and vary in protein, depending on the source
ruminant a cud-chewing animal with four stomachs, including a rumen that is inhabited by millions of microbes
vaccination an injection of a substance that produces immunity or resistance to a specific disease

Margin question sets contain comprehension-style questions to enable students to revise what they have read and learnt. Each question set appears directly adjacent to the relevant piece of text.

4 Define the terms 'environment', 'ecosystem' and 'community'.

Connect boxes enable students to find out extra information beyond their textbook. Connect boxes link to relevant sites, such as the NSW Farmers' Federation, Department of Primary Industry and CSIRO.

connect

Australia's trade agreements – DFAT

Students and teachers can also link directly to the external websites referred to in the Connect boxes in *Dynamic Agriculture Years 11–12* via the free, unprotected weblinks site located at http://dynagsenior.nelsonnet.com.au.

The **NelsonNet** teacher website provides students with extra worksheets.

Disclaimer: Please note that complimentary access to NelsonNet is only available to teachers who use the accompanying student textbook as a core educational resource in their classroom. Contact your sales representative for information about access codes and conditions.

ISBN 9780170265560

The **end-of-chapter review** contains lots of different ideas for activities, research and extension. The shading on the question number indicates the level of thinking required. Light shaded numbers indicate lower order thinking (Understanding) with shading increasing up to the highest order thinking (Evaluating and Creating).

Material for the **mandatory practicals** (Chapters 15, 17 and 18) and the **mandatory analysis of a research study** (Chapters 31, 32 and 33) is also provided in this section.

Things to do provide students with a range of interesting activities to undertake, which include tasks such as drawing, constructing, observing, experimenting and surveying.

Things to find out challenge students to research information for themselves by visiting farms or other establishments, websites and other sources of information.

Extended response questions require students to answer questions in detail, to practise the style of question they will encounter on their end of year HSC examination.

Chapter review

Things to do

1. Describe how temperature, humidity levels and lighting conditions are regulated in an intensive animal production shed for either poultry or pigs.
2. Using data loggers and various probes to measure environmental factors, assess the physical aspects of the environment of a selected animal; for example, for chickens, slightly vary the temperature of the brooder and record behavioural changes, then restore brooder to optimal operation.
3. Debate the usefulness of the five principles that define good animal welfare practice.

Things to find out

1. For a named animal husbandry practice (e.g. lamb marking, drenching, castration, dehorning, hoof trimming or pairing, or crutching), list how the following factors can be managed to reduce stress and minimise the risks to farm animal welfare:
   - use of appropriate equipment
   - skill of the operator
   - timing of the animal practice
   - management of the animals after completion of the practice.
2. For a named animal species, identify two physical and two behavioural characteristics for which a farmer's knowledge can assist with the management of the animal.
3. Outline two examples showing how an understanding of an animal's physical and behavioural characteristics can assist in the management of the animal. Some examples include panoramic vision of animals, flight zones, and herd or mob instinct.

Extended response questions

1. Investigate animal welfare legislation for a specific farm animal and discuss the implications of the legislation for the relevant production system.
2. Discuss the ethical or moral issues associated with ONE of the following: mulesing of lambs; live sheep, beef cattle or goat export; battery cage egg production; or the use of farrowing crates in intensive piggeries.
3. Evaluate the role of legislation in the development of animal management systems.

About the authors

Lisle Brown

BScAgr, DipEd, MEd

Lisle Brown has 36 years of agricultural teaching experience, including 29 years as Head Teacher of Agriculture at the selective James Ruse Agricultural High School. Lisle has extensive experience in syllabus development as an active member of junior and senior agriculture syllabus committees and in the setting and marking of the Higher School Certificate (HSC) examination.

Lisle's teaching career was recognised by the NSW Association of Agriculture Teachers (NSWAAT) with the 2008 J.A. Sutherland Award for significant contribution to the teaching of agriculture in NSW.

Robert Hindmarsh

BScAgr, DipEd, MEd

Bob Hindmarsh has had a 33-year teaching career in more than 15 schools, ranging from selective agricultural high schools to comprehensive schools. In this time he introduced agriculture into the curriculum of two schools and worked from scratch to establish viable farms in both schools. He has been Head Teacher of Agriculture at a large selective agricultural high school and a developing comprehensive high school, as well as a deputy principal and principal in comprehensive schools.

He has published several books on agriculture and been an active member of both the junior and senior agriculture syllabus committees. He has been part of the examination committees for the HSC and School Certificate-level for Agriculture as well as an established marker for the HSC exam in previous years.

As a past president of the NSW Agriculture Teachers' Association, he has provided support to a range of teachers in various locations across the state in previous years as well as being part of an effective team of dedicated agriculture teachers.

Ross McGregor

BScAgr

Ross McGregor has had over 35 years of agricultural teaching experience, including the last 5 years as Head Teacher of Agriculture at Hurlstone Agricultural High School. Ross has extensive experience in teaching, HSC exam setting, HSC marking and syllabus development. He has a keen interest in cattle showing.

Acknowledgements

The authors and publisher would like to thank the following people for reviewing the *Dynamic Agriculture Years 11–12* manuscript throughout the writing process.

Grant Jackson, Gosford High School, NSW
Samantha Jarrett, Mount View High School, NSW
Dianne Toynton, Broken Hill High School, NSW
Melissa Willcocks, Ashford Central School, NSW

ISBN 9780170265560

UNIT 1
THE WORLD OF AGRICULTURE

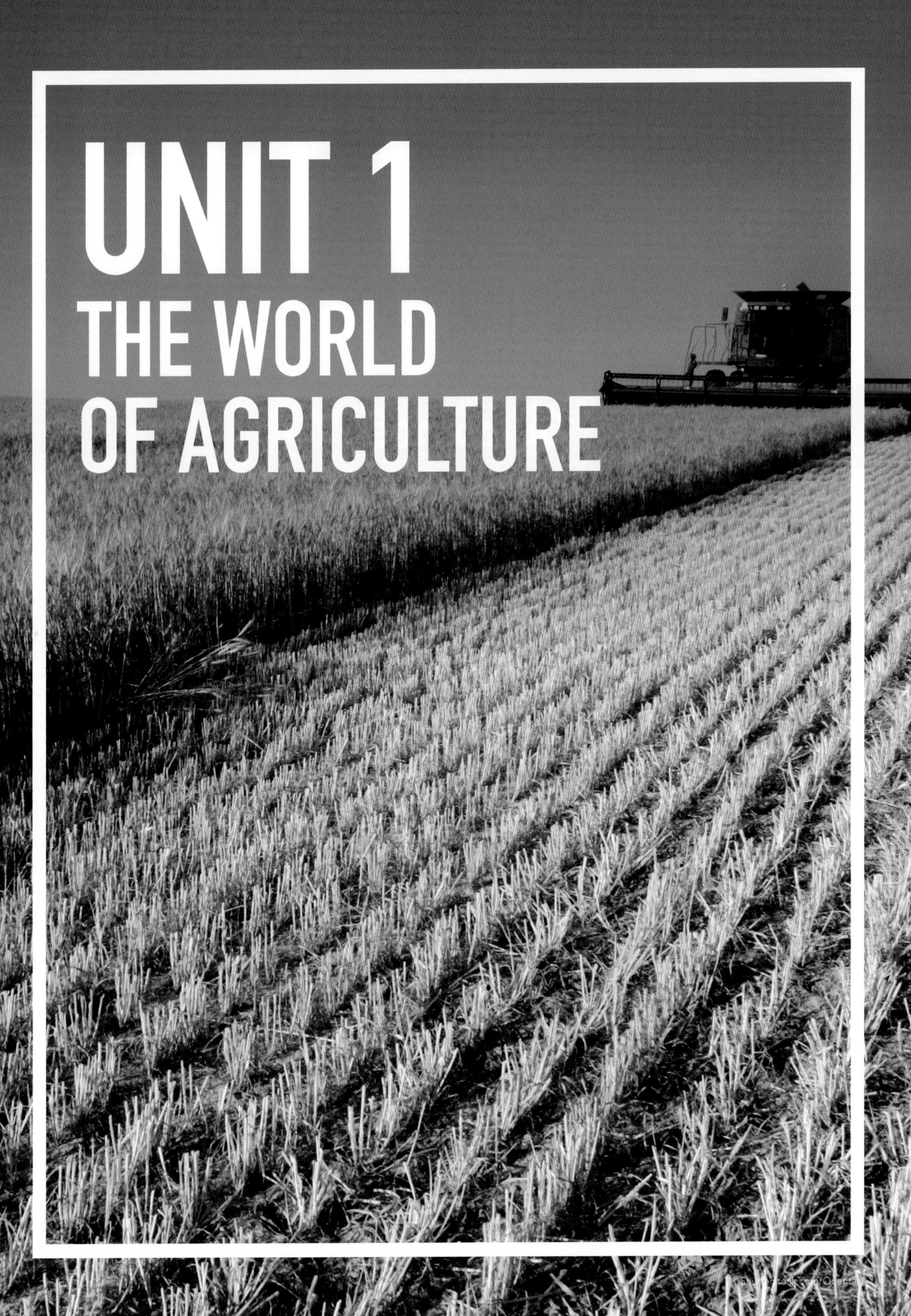

CHAPTER 1

Outcomes

Students will learn about the following topics.

1 Agricultural systems and the interactions between farm subsystems.
2 The differences between natural and managed farm systems.
3 The impact of physical, biological, social, historical and economic factors on agricultural systems.
4 Limiting factors to agriculture production systems.
5 Agricultural careers.

Students will be able to demonstrate their learning by carrying out these actions.

1 Define the nature of agriculture.
2 Distinguish between a system and a subsystem.
3 Describe the role of a farm and the farmer in an agricultural production system.
4 Distinguish between extensive and intensive agricultural systems.
5 Discuss factors that influence the intensity of agricultural production.
6 Discuss the importance of agriculture in the development of Australia.
7 Analyse a farm system, in terms of a model, illustrating system components.
8 Distinguish between a managed system and a natural system.
9 Assess how energy is cycled through a system.
10 Assess how material is cycled through a system.
11 Describe the main features of managed and natural systems.
12 Compare and contrast the movement of energy and materials through managed and natural systems.
13 Discuss the influence of limiting factors on agricultural production systems.

ISBN 9780170265560

MANAGED AND NATURAL SYSTEMS

Words to know

agribusiness the many industries that directly or indirectly support the production, processing (value adding), distribution and retailing of agricultural products

biological efficiency how well living processes function

biotic factor related to living things; biotic factors in the environment are the effects of living organisms

climate the average weather conditions over a long period of time (at least 30 years)

community a group of organisms that live together, sharing the same environment

disease any condition that produces an adverse change in the normal functioning of an organism

economic efficiency how much money is returned by a particular system in relation to the cost of establishing, running and maintaining the system

ecosystem the relationship between an interacting community of organisms and its physical environment

efficiency the extent and rate of conversion of inputs to outputs

environment any non-genetic factors that affect plant growth, including climate, soil, other plants, pests, diseases and management practices

extensive farming the production of plants or animals over large areas of land

farm system a group of parts (subsystems) that interact to achieve a purpose

feedback the information received by the farmer on the performance of the system

input the items or materials that go into a (farming) system (e.g. fertiliser)

intensive farming the production of a large number of plants or animals on a small area of land

limiting factor any factor that lowers the production potential of a system

output the items or materials produced by a system (e.g. farming) and removed from it (e.g. wheat)

pasture a balanced community of plants (generally grasses and legumes) that provides grazing animals with their food requirements

pest any organism that injures, irritates or damages livestock and livestock products and can adversely affect productivity

process an action that changes input to output

rate a measure of production over time

subsystem a system that forms part of a larger system (e.g. a sheep subsystem)

succession a sequence of different communities in a particular area that form over a period of time

value adding processing a product in some manner, which enables the product to be sold at a higher price

ISBN 9780170265560

Introduction

Agriculture is concerned with the raising of crops and animals to provide food, fibre and other materials needed by society. Every day we are all connected directly to farming because of our need for food. Much of our clothing has its origins in material produced on the farm, such as cotton and wool. Timber products are used for building construction, paper production and in the manufacture of furniture. Perfumes, medicines, ornamental plants, tea, coffee and flowers all have a place in our lives. Production on a farm involves the management of an animal's or a plant's growth pattern, reproductive capacity and lifespan by the farmer. In **extensive farming** or production situations, such as the management of grazing animals (Fig. 1.1), a farmer must manipulate the interactions that occur between the grazing animal and the plant **community**. In **intensive farming** or production situations (Fig. 1.2), the farmer must develop and manipulate several environmental factors to maintain high and consistent levels of production from the farm. Table 1.1 indicates some of the main features of these two systems.

Table 1.1 Characteristics of intensive and extensive agricultural systems

Feature	Intensive system	Extensive system
Size and location of farms	On high value land close to urban centres and near transport links to enable quick and cheap transport of product to market. Because of high land values the farms tend to be small, a half to 10 hectares in general.	Land is further away from townships and cities; consequently land values are lower. Farms are large in scale, usually 50 hectares and larger.
Nature of enterprise	Intensive with high numbers of plants or animals per square metre of available space, e.g. polyhouses, greenhouses, market gardens, piggeries, aquaculture, poultry enterprises.	Grazing animals with low numbers per hectare, usually ruminants, e.g. cattle, goats, sheep, although alpacas are often found in these farming areas as well.
Digestive requirements	Monogastric animals that have efficient digestive systems and are able to tolerate high-protein diets.	Ruminants in general, pasture based relying on grasses and legumes, hay and silage. Ability to tolerate fibre or cellulose in their diets.
Disease	Diseases can rapidly infect large numbers of animals or plants in intensive systems due to contact, dispersion by wind and water or body fluids from one animal or plant to another.	Lower stocking rates, good vaccination and drenching programs lower the incidence of rapid spread of diseases.
Reproductive management	Use of artificial insemination or hand pollination methods. Often plants or animals are only kept for production.	Need to replenish plants or animals in this system to maintain sustainability or link production levels to the management of reproductive cycles, e.g. in the dairy industry.
Labour and capital requirements	Very high due to the investment in technology, housing and numbers of plants and animals requiring management.	Costs and labour needs are much lower and often additional labour is only needed at particular times of the year, e.g. shearing or fruit picking.
Markets	Require fast and reliable access to markets.	Transport costs can be a problem with distance. Costs are associated with the need to build handling and loading yards.

ISBN 9780170265560

Figure 1.1 An extensive system – beef cattle grazing

Figure 1.2 An intensive system – battery system for laying hens

The nature of agriculture

The farm is the basic unit of production in agriculture. It can be thought of as a collection of plants, animals and non-living components that interact to achieve a purpose, or goal. The farmer sets these goals and is a manager and manipulator in such a system. A **farm system** is made up of many interconnected parts, which all function together to produce materials for people to use. The interacting parts – animals, plants, soil and **climate**, micro-organisms, invertebrates and the management style of the farmer – function together to achieve a purpose or goal (Fig. 1.3).

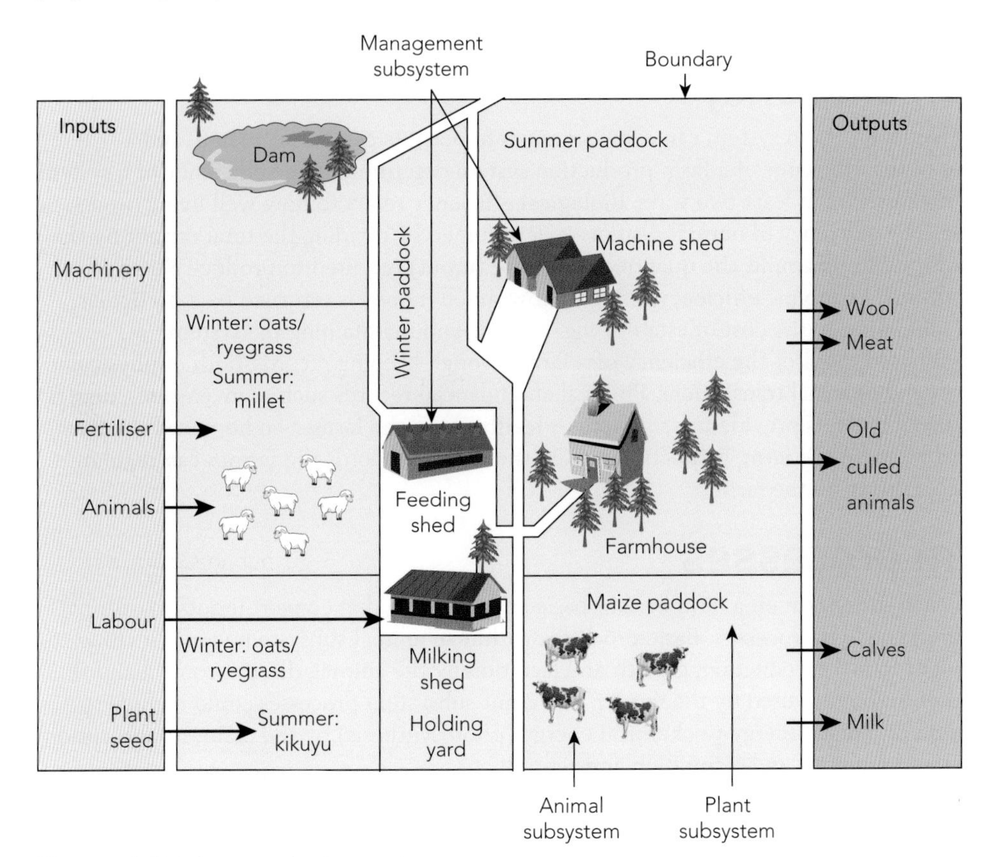

Figure 1.3 A farm production system

ISBN 9780170265560

The farm can be considered a system, and each individual part a **subsystem**. For example, a mixed farm might contain a dairy subsystem, a beef cattle subsystem, a **pasture** subsystem and a farm management subsystem. These subsystems all function together to produce materials for others to use, while the farm as a whole is responsive to the management decisions of the farmer.

On a farm, production occurs within a definite area, separated from surrounding areas or systems by a definite boundary. The boundary might be natural, such as mountains, or artificial, such as a fence line.

Factors such as how much money a farmer has available and the level of farming skills a farmer possesses, as well as natural land formations, all contribute to the positioning of the boundary to a farm system. Boundaries can take a number of forms depending upon the system being described. When considering a whole animal as a system, the skin can be thought of as its boundary.

Farm inputs

In order to produce material a farm requires the addition of particular materials, or **inputs**. The farmer must decide on the type and amount of each input to use and when to add the input. The farmer must make decisions based upon knowledge, skill, experience, and economic and personal factors. Common decisions about inputs include the type and amount of plants or animals managed, fertiliser levels, water requirements and the types of machinery used.

Farm outputs

Materials produced on a farm are called **outputs**. For example, these can be animals, hay, grain, wool, milk or eggs.

Farmers need to know exactly what the output levels are compared to input levels. **Rates** are used to quantify the efficiency of output levels. Consequently, farmers talk of litres of milk produced per cow per day, piglets born per sow per year or the weight gain in kilograms per day for beef cattle. Rates include a reference to a length of time, for example, per day, per month or per year. Rates are useful in determining and comparing productivity on farms and productivity of individual subsystems.

Farm efficiency

How effectively a farm system can convert inputs to outputs can be calculated, and a measure of the **efficiency** of a farm production system determined. The efficiency of a system can be measured in at least two ways. **Biological efficiency** refers to how well living processes function. The efficiency of natural systems is determined by dividing the total output by the level of input; for example, the quantity of pasture (input) required to produce a litre of milk (output). **Economic efficiency** refers to how much money is returned by a particular system in relation to the cost of establishing, running and maintaining the system.

Farmers can monitor the efficiency of a farm through keeping records, both of physical resources and financial transactions. Physical and financial records such as inventories, diaries and financial budgets provide information, or **feedback**, to the farmer on how well the farm, or a subsystem on the farm, is functioning. On the basis of records the farmer can regulate what is occurring on the farm.

Farm processes

Within each subsystem on a farm certain basic actions occur that convert inputs to outputs. These are known as **processes**. Basic processes within an animal subsystem would include digestion of food, reproduction, growth and lactation by the animal, **disease** control and the method of husbandry used by the farmer. In a plant subsystem processes could include the conversion of radiant energy to chemical energy (photosynthesis) by the plant, its absorption of water and minerals, and respiration and reproduction.

In various countries across the world many forms of agriculture are practised, depending on the social and cultural needs of the people (Fig. 1.4).

 ISBN 9780170265560

Shutterstock.com/Jirapolphoto

Figure 1.4 A farming system – paddy rice

Each of these types of agriculture – for example, raising pigs in intensive sheds, primitive slash and burn methods of crop production, range land farming and nomadic farming – can be thought of as a system. The term 'system' can apply to a farm of any particular size or type as well as to a level of organisation, such as the type of agriculture practised in a country.

The level of production that can be obtained from farms will depend on the efficiency by which solar energy or energy contained in food is converted to energy within the animal or plant. This energy is used for growth, repair of cells and production of materials such as milk, eggs, wool, fruit, wheat, cotton and fibre.

The intensity of agricultural production will depend upon:

- the influence of physical and biological factors such as landform, soil and climate, which can often determine the type and number of plants or animals found in an area
- cultural, social and religious beliefs, or historical factors, which determine the scale and type of farming operations
- economic factors such as land values, commodity prices, marketing costs and the availability of cheap transport, which influence the success of any farming activity.

Agriculture is important in Australia because people need products made from plants and animals. Many people in different business activities depend on farmers to buy products such as machinery, fertiliser or chemicals, while other members of the community service the farm sector with advice. Numerous country towns and cities owe their existence to the concentration of local farming activities. Dairy farms along the coastal areas of New South Wales led to the formation of focal points for transport and manufacturing of the raw materials. These centres later developed into towns and cities. Wheat was responsible for the establishment of many towns and the opening up of New South Wales by rail lines.

Global population estimates for the year 2050 are heading towards 10 billion. To feed a population of this size, global food production needs to increase by approximately 70–75%. To achieve this level of increased productivity, agricultural production systems need to be economically viable and sustainable within their environmental settings while increasing output. High establishment costs for farm businesses, the effects of climate and disease, unsustainable practices, changes in government policy both local and international, poor infrastructure and an ageing workforce will all affect how successfully Australia and other countries can increase farm productivity to feed this growing world population.

Managed systems

On a single farm several activities often occur, such as raising sheep for wool and growing grain crops. Although the whole farm is a system, each unit that can be identified on the farm, such as a sheep enterprise or a wheat enterprise, is called a subsystem (Fig. 1.5). The subsystems interact but are small functional parts of a larger arrangement.

ISBN 9780170265560

Fotolia.com/Alexandr Ozerov

Figure 1.5 Cattle and sheep subsystems

People direct many of the interactions on farms. Simple examples of this direction include the selection of crops and animals to raise, the selective breeding of living organisms, the application of fertilisers and methods of climatic manipulation, such as the building of dams and glasshouses (Fig. 1.6) or the use of irrigation. Agricultural systems are managed systems.

When a change is made to a farm system, all of the subsystems are affected because of their interactions with one another. In addition, as many farmers and communities are now finding out, management decisions made for one agricultural system can influence adjoining systems, both rural and urban (Fig. 1.7).

Alamy/OJO Images Ltd

Figure 1.6 Controlling disease in a glasshouse

Shutterstock.com/Loco

Figure 1.7 As a result of excessive land clearing there are no shade trees for stock

Because of the drain of plant and animal products from any farming system, minerals are often not effectively recycled, and so available mineral reserves might decrease. The farmer usually adds minerals, feed concentrates and other material into a farm system to maintain productivity and to sustain the farming system over time. A rainforest is an example of a natural system that is self-sustaining, whereas the farm system is an artificial system requiring constant management of inputs (Fig. 1.8).

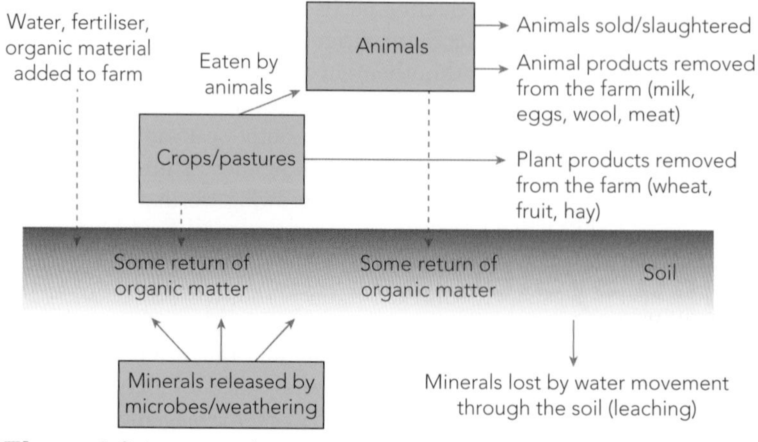

Figure 1.8 A managed system

1 Define the terms 'system' and 'subsystem'.
2 List three ways in which farmers can manipulate interactions between subsystems.
3 List three examples of how the management of one system might influence other adjoining systems.

ISBN 9780170265560

A lack of understanding of the natural environment, along with unexpected outcomes of management decisions and economic necessity, has often produced unfavourable ecological effects in some farming situations. Examples of these effects include excessive soil acidity, soil salinity problems, erosion of top soil (Fig. 1.9) and the build-up of destructive pests and diseases over time.

Figure 1.9 The results of wind erosion

Natural ecosystems

An organism's immediate surroundings are known as its **environment**. An environment is composed of two parts: a non-living part concerned with resources such as soil and climate; and a living, or **biotic**, part concerned with living organisms such as plants, animals, invertebrates and micro-organisms. The study of living organisms and their environment is called ecology.

All organisms that live together in an area form a community. Where communities live and how successfully they reproduce depend upon the suitability of factors such as climate, resources (food and shelter) and the influence of organisms on one another. Communities change over time, eventually reaching an equilibrium that benefits all members of the community. The stages that a community passes through to reach the final, or climax, community are called **successions**.

The relationship that develops between an interacting community of organisms and their physical environment is described as an **ecosystem**. An ecosystem forms the main functional unit in the study of natural systems by ecologists. It has a definite structure in which living and non-living components interact. Within an ecosystem, energy flow and cycling of materials can be identified. Figure 1.10 illustrates the main components of an ecosystem.

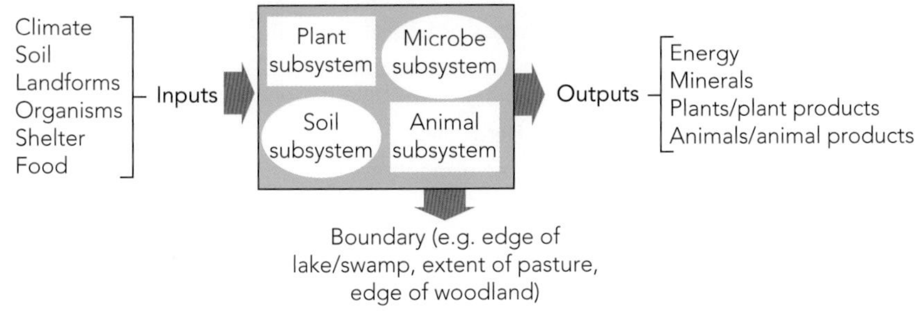

Figure 1.10 A model of an ecosystem

Scientists studying these systems group organisms into three main types (Fig. 1.11, page 10).

1 Producers (or autotrophs) are organisms that make their own food, and include plants and some bacteria.
2 Consumers (or heterotrophs) are animals that eat:
   - plants (herbivores)
   - other animals (carnivores) or
   - plants and animals (omnivores).

3 Decomposers are invertebrates and micro-organisms that break down organic matter, thus gaining energy and mineral resources for themselves and releasing these materials back into the environment.

Within a natural ecosystem, two things are evident.

1 All organisms are supplied with energy.
2 Minerals are recycled from the dead remains of plants and animals back into living communities.

In most ecosystems the source of energy is the Sun. Plants, as the main type of producer organism, convert this radiant energy into chemical energy.

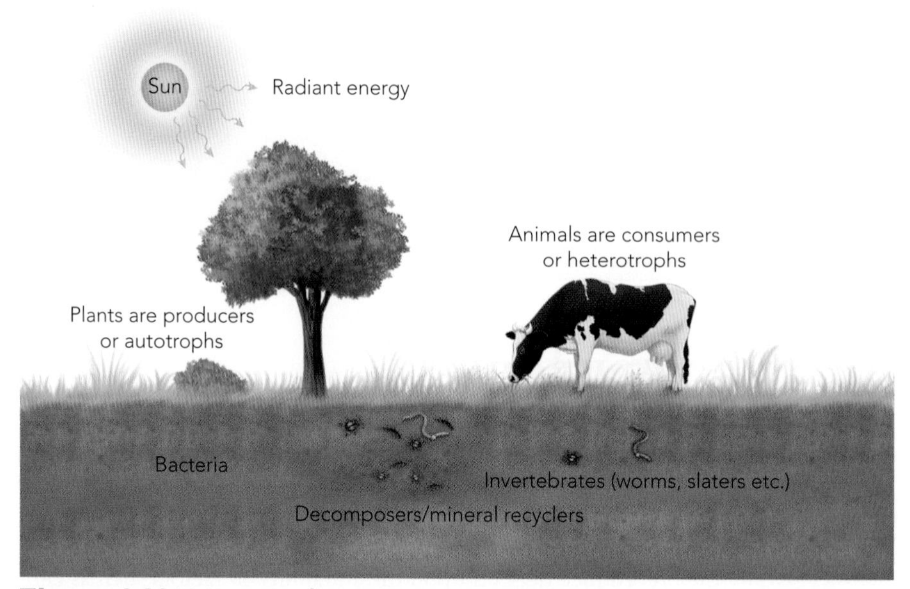

Figure. 1.11 Main types of organisms in an ecosystem

Herbivores are animals that eat plants to obtain their energy requirements. Animals that live off other animals obtain their energy indirectly from plants. The energy stored in plants is passed through the ecosystem via a series of food chains, defined by which organism eats what (Fig. 1.12).

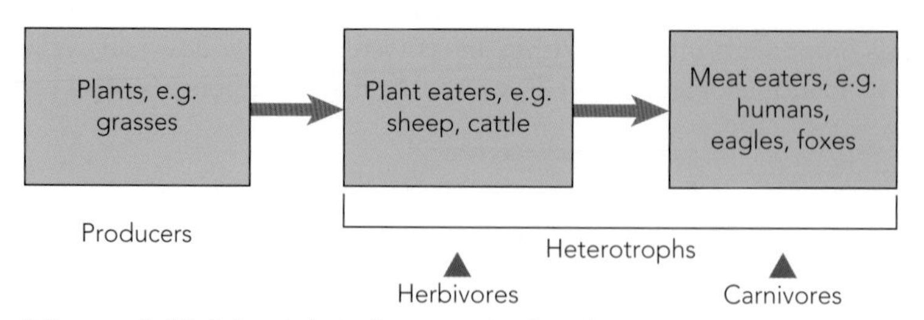

Figure 1.12 A food chain for an agricultural system

4 Define the terms 'environment', 'ecosystem' and 'community'.
5 Explain how energy is transferred through an ecosystem.
6 Describe how minerals are cycled in an ecosystem.
7 List the main components of an ecosystem and describe the role of each.

Animals within an ecosystem use the energy gained from their food for maintenance of body functions; for example, body cell repair, movement, the maintenance of body temperature and production. Energy used for production results in the manufacture of fur, fibre, milk and eggs, among many other products (Fig. 1.13).

Nutrients are cycled between the living and non-living parts of the environment by the activities of many organisms. Bacteria, fungi and many invertebrates decompose the remains of plants and animals, and thereby release minerals, organic compounds and gases. The minerals are absorbed by plant roots and used by microbes. The cycling of minerals in this fashion forms a closed, or cyclic, system (Fig. 1.14).

 ISBN 9780170265560

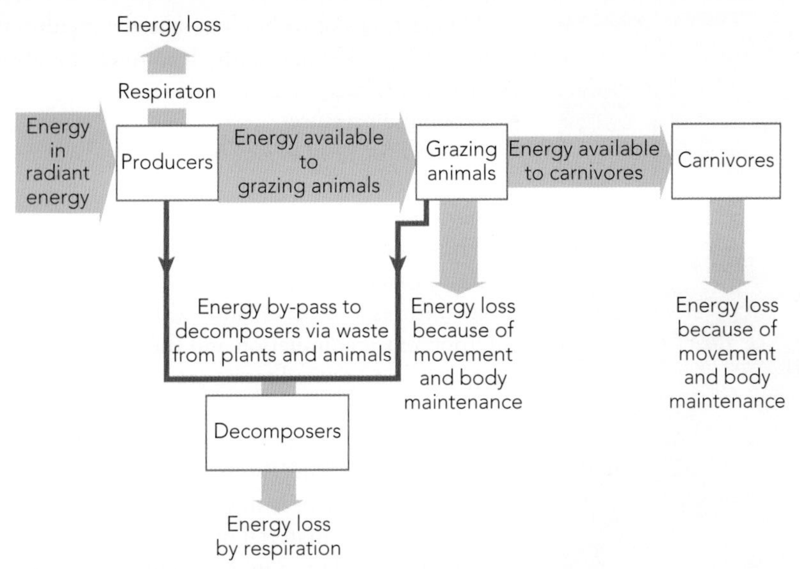

Figure 1.13 Energy pathways in a system

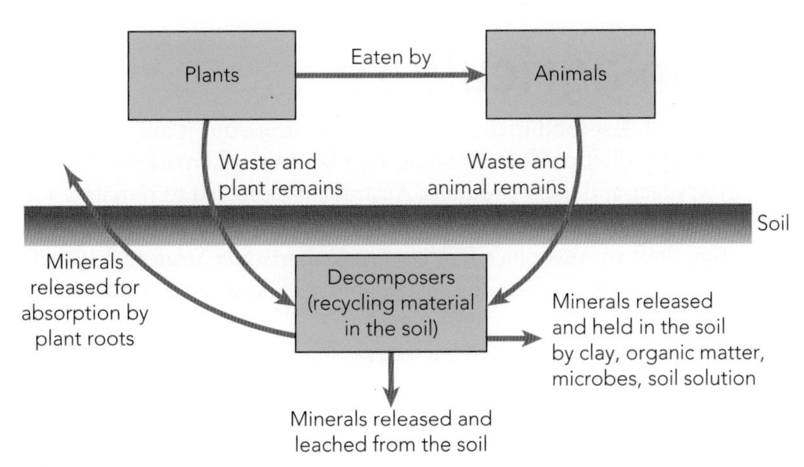

Figure 1.14 Mineral cycling in an ecosystem

Agricultural systems

Agricultural systems require constant management as products are removed from the system for human use. Agricultural systems are managed systems because inputs such as water, fertiliser, seeds, plants and animals must be constantly added to replace those products that are marketed.

Agricultural systems managed by many of the world's original cultures, such as the Australian Aborigines and the Indian people of the Deccan, feature sustainable development. However, care should be taken in the general use of such statements, as several indigenous cultures have not used sustainable development and have ruined their environments, altering both the soil and animal populations in some instances.

Getty images/SPL Creative

Figure 1.15 Lot feeding cattle in a high-tech agricultural system

Management of technology-based agricultural systems (Fig. 1.15) may easily lead to exploitation of the environment, because of:

- economic pressures
- demand for products of specific standards by consumers
- lack of available labour.

These forces spawned the evolution of a more efficient, machine- and market-dominated system. Monocultures, extensive land clearance operations, environment manipulation and the development of large corporate-owned farms are some features of agriculture today.

The rapid discovery of knowledge and its application to production systems (both agricultural and industrial) have meant that people now live longer and choose from a greater range of material options. On the other hand, the non-renewable resource system (the atmosphere, and soil and water systems) is being heavily exploited and ultimately run down.

Australian agriculture

Australia produces much less food than many other countries, but is able to export well over half of its agricultural production. It is a leading supplier to world markets of beef, sheep meat, wheat, barley, sugar and dairy products. Australia is expected to remain a surplus producer of agricultural products but agricultural production faces challenges.

The consultation draft of Australia's *Biodiversity Conservation Strategy 2010–2020* has identified many factors that will affect the sustainability of Australian agricultural systems. These include:

- climate change resulting in conditions such as prolonged drought
- available and adequate water resources
- invasive species of plants, animals and insect pests
- breach of quarantine by diseases and pests
- loss, fragmentation and degradation of habitat
- unsustainable use of natural resources
- changes to the aquatic environment and water flows
- inappropriate fire regimes.

Due to advances in technology and research there have been many positive developments for Australian agricultural systems as well. These successes include the following achievements.

- Australian primary industries have continued to reduce greenhouse gas emissions; however, drought has been a major factor in reducing crop and stock levels (Fig. 1.16). Emissions have reduced but are projected to increase into the future.

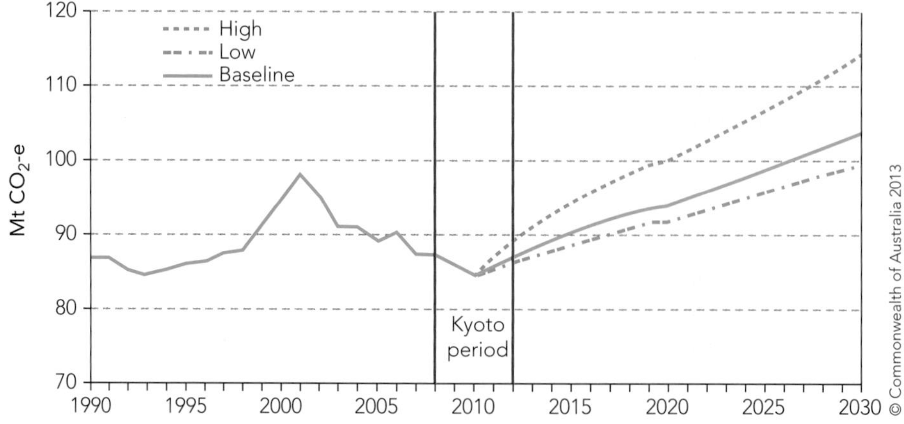

Figure 1.16 Baseline agriculture emissions, 1990–2030

ISBN 9780170265560

- Natural Resource Management (NRM) is a fundamental activity on Australian farms. In fact, 94% of Australian farms actively undertake natural resource management. (Australian Bureau of Statistics, *Natural Resource Management on Australian Farms 2006–07*)
- Of the 417.3 million hectares of land managed by agricultural businesses, 9.2 million hectares have been set aside specifically for conservation/protection purposes by 36% of the land managers. (Australian Bureau of Statistics, *Year Book Australia 2009–10*)
- In 2009–10, 65% of agricultural businesses reported native vegetation on their property, with 55% actively preserving native vegetation for conservation purposes. (Australian Bureau of Statistics, *Land Management and Farming in Australia 2009–10*, Catalogue No. 4627.0)
- In 2009–10, 52% of agricultural businesses undertook activities to protect native vegetation, 45% wetland protection and 49% river or creek bank protection. (Australian Bureau of Statistics, *Year Book Australia 2009–10*)
- Estimates suggest that the management of soil resources, water resources and biodiversity costs $3.5 billion annually, or 10% of agriculture's GDP. For every government dollar invested, farmers are estimated to have invested $2.60 in NRM and environmental protection. (OECD (2008), *Environmental Performance of Agriculture in OECD Countries Since 1990*: Australia Country section)

8 List some examples of the effects of exploitative agricultural management systems on the environment.

9 Describe three examples of beneficial outcomes from the use of technology and innovative thinking to develop sustainable farming systems.

Agribusiness

The term **agribusiness** arose from the work of Davis and Goldberg in 1957 to indicate that farms did not just produce food and fibre products but used inputs provided by local industries to run their farms. They also recognised that the final product derived from raw farm outputs often underwent further processing (known as **value adding**) along with a particular pattern of marketing and retailing. Agribusiness refers to activities in the production, processing and distribution of food and fibre products. Consequently, the farm is only one part of the system in this situation. The nature of farm products developed for particular markets and the range of processing, distribution and marketing arrangements are also part of the agribusiness system. We can adjust the boundaries of the agribusiness system even further and consider how farmers gain the skills and knowledge to develop local and overseas markets and the various mechanisms of government assistance. Figure 1.17 on page 14 illustrates the agribusiness value chain. Agribusiness is an interconnected chain of industries that either directly or indirectly support the production, processing (value adding), distribution and marketing of agricultural products.

Agribusinesses in Australia contribute 12% of our gross domestic product and farm-related industries employ approximately 1.6 million people or 17% of the total work force. Australian farms export 60% of what they grow and produce and are able to supply 93% of Australia's daily domestic food supply (Year of the Farmer, 2012).

The growth of the farm sector is slowing from an average rate of 2.8% to 1% per annum, illustrating the need for a greater focus on research into the factors that limit productivity and career training.

The challenge in the future is to achieve production levels to satisfy demand in a manner that can be sustained and that will blend with the natural environment and allow cycling of materials (Fig. 1.18, page 14).

This achievement will require:

- an appreciation of how to assess situations
- a knowledge of problem-solving skills
- the application of technology in obtaining and analysing data
- the presentation of information to various groups of people in the community.

Innovation is important to the growth, profitability and sustainability of Australia's rural industries. Various agencies are involved in the development and promotion of innovative practices at the Commonwealth, state and territory government levels. Examples of organisations involved in research include the 15 rural research development corporations (RDC), cooperative research centres (CRC), universities funded through the Australian Research Council and the CSIRO. Visit the Council of Rural Research and Development Corporations website for more information.

10 Define the term 'agribusiness'.

11 Explain the information shown in Figure 1.17 on page 14.

12 List three factors to be considered to achieve levels of production on farms that are both sustainable and able to supply a growing demand for agricultural products.

13 Name three organisations that assist with the development of innovative technologies and practices to enhance sustainability in farming.

14 List three agribusinesses in your district that employ local people.

connect

Council of Rural Research and Development Corporations

ISBN 9780170265560

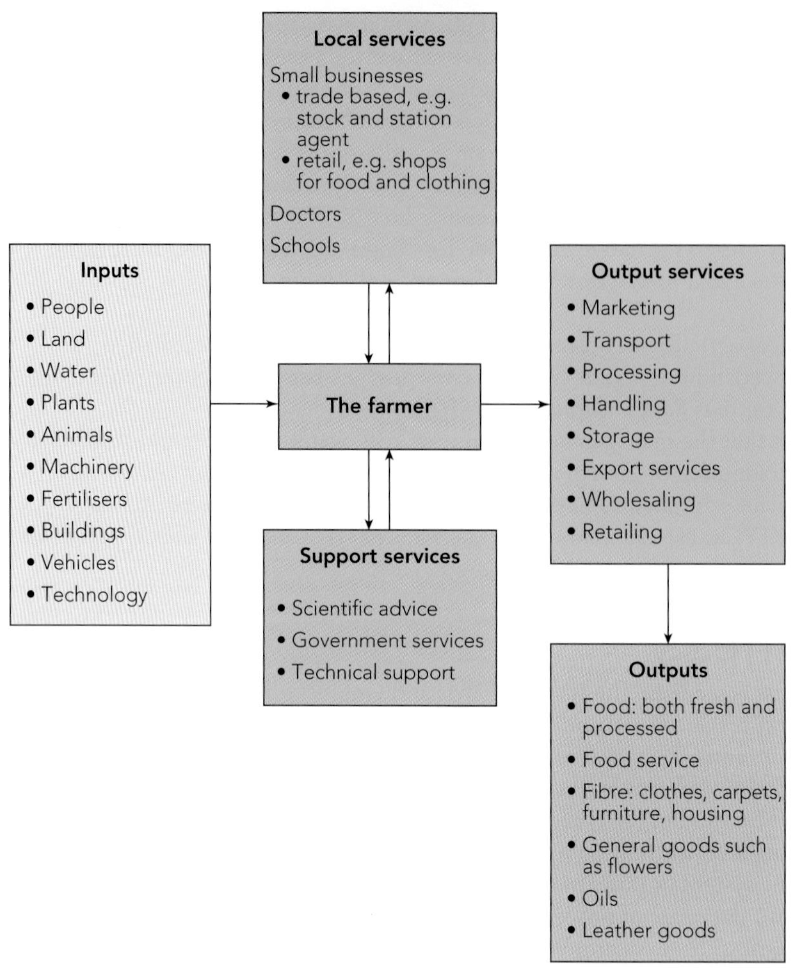

Figure 1.17 The agribusiness value chain
Source: Year of the Farmer, 2012

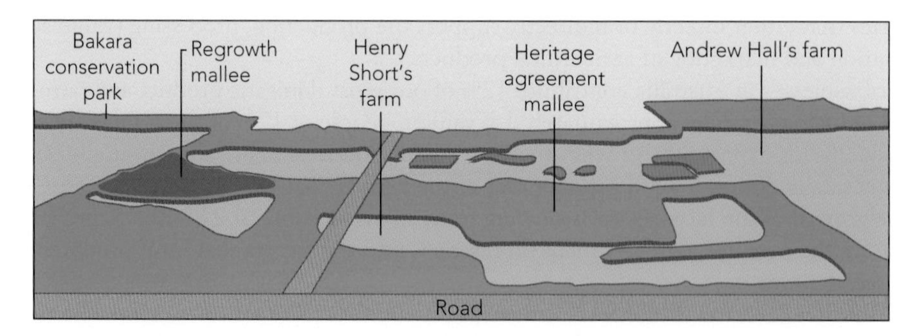

Figure 1.18 Sustainable system management: wildlife corridors between conservation areas and neighbouring farms blend into a working farm landscape

Limiting factors

Limiting factors operate in a manner that reduces the productive capacity of a system. Insect or weed infestations will lower production levels. Decreased levels of production could be due to competition for light, space, mineral nutrients or water in the case of plants or the effect of disease-causing organisms or poor nutrition on an animal's metabolism.

The upper limit to any production system is set by the genetic makeup of the plant or animal within the system. In several industries genetic development has slowed, and in some, such as the broiler meat industry, it has reached a plateau.

ISBN 9780170265560

The extent to which an organism's genetic potential is reached is determined by its environment. This covers the interaction of factors such as:

- nutrition – including the form and type of food used to supply the energy and protein requirements of the organism
- climate – including temperature, amount of sunlight and availability of water
- disease – any condition that adversely affects a plant or animal
- land degradation – the impact and control of weeds costs Australian agriculture more than $4 billion per year. Farmers consider weed control as one of their highest priorities in preventing long-term land degradation. (Department of Agriculture, Fisheries and Forestry, *At a glance*, 2010)
- predators and vermin – wild dogs, rabbits, foxes, wild pigs, pest birds and mice are estimated to cost Australian agricultural industries approximately $745 million per annum (Department of Agriculture, *Australia's Agriculture, Fisheries and Forestry at a glance 2010*)
- husbandry skills of the farmer – including the manipulation of reproductive processes, the degree of crowding of animals and the design of yards and sheds
- natural disasters – such as droughts, floods and bushfires.

15 Explain the concept of a limiting factor to farm production.

connect

Why do scientists work with farmers?

Watch the presentation and then answer the questions.

Agricultural careers

Agriculture is a knowledge-intensive sector with a strong demand for skilled professionals. Estimates indicate a potential demand for 6000 tertiary-qualified graduates per year in the sector. However, the sector faces a significant undersupply of graduates, with Australian universities graduating fewer than 800 graduates per year in agriculture. This trend is beginning to change but the demand for trained people to meet the challenges of agricultural production and the development of innovative technologies and practices will increase as the demands for reliable food supplies for a rising global population increase (Pratley and Hay (2010), *The job market in agriculture in Australia*, Australian Council of Deans of Agriculture, Charles Sturt University).

Chapter review

Things to do

1. Select a farm area (or use the school farm) that you can observe over a period of 4 weeks.
   a. List all of the living organisms that you observe, and classify them into producers and consumers.
   b. Examine the area for evidence of recycling organisms. Record your observations in a diagram. Use arrows to establish links between these living organisms. What patterns emerge?
2. a. Repeat the activity in question 1 in a natural area.
   b. Compare the diversity and complexity of relationships that you discover in the two areas.
3. Visit a local farm and undertake the following activities.
   a. Describe the activities that occur on the farm. The information you record should emphasise the essential aspects of farm management and reflect any unique features of the farm.
   b. Describe the same farm using the terms 'boundary', 'inputs', 'subsystems', 'outputs', 'processes' and 'limiting factors'.
   c. Attempt to model the farm using one of the models presented in the chapter.
4. Establish a small-scale living system such as a terrarium or worm farm. How is matter and energy cycled in the system you have developed? Will you need to manage the system? Why or why not?
5. Talk to a farmer and collect or write down examples of records (both on physical and financial aspects of the farm) that are kept by the farmer. Why does the farmer keep these types of records?
6. Visit an intensive animal production system. Use the following formulae to calculate the efficiency of the system:

$$\text{Biological efficiency} = \frac{\text{outputs}}{\text{inputs}}$$

$$= \frac{\text{kilograms of product produced}}{\text{kilograms of feed consumed by the animals}}$$

$$\text{Economic efficiency} = \frac{\text{value of outputs}}{\text{cost of inputs}}$$

Once these figures are determined, compare them to industry averages.

Things to find out

1. Obtain a rural newspaper or magazine. Read an article that describes the activities that occur on a farm. Such an article must describe the various enterprises found on the farm, the purpose of production and the main inputs. Construct a simple systems model of the farm and justify the basis of your model.
2. For a local farm, list the various ways the farmer has manipulated and managed the natural environment. Assess if any major cycles have been disrupted. Discuss whether or not the farmer has been successful with the management of the natural system.
3. For your local area, determine a major environmental problem caused by agricultural activities. How have agricultural practices contributed to this problem? Describe the effects of these agricultural practices. Suggest alternatives.
4. What social or legal implications or regulations exist for the use of certain inputs such as growth hormones and genetically modified materials?

ISBN 9780170265560

Extended response questions

1. 'Farming is less of a business and more of a way of life; therefore, there is little need for a form of system analysis on farms.' Discuss this statement.
2. By using examples, show how a farmer might modify the natural system to improve farm productivity.
3. Some components of a farm system pass through it in a non-cyclic manner, while others are cycled.
   a. By discussion and illustration, show that you appreciate the difference between cyclic and non-cyclic movements within a system.
   b. How do such movements influence farm management?
4. For a major animal production system in your area:
   a. outline the main limiting factors to production
   b. discuss the reasons for these factors limiting production
   c. outline some means of overcoming these problems.
5. A farm is made up of a number of complex interacting components. Describe the approach you would adopt if you were analysing a farm's operation.
6. 'Agriculture is a system that exploits the environment with little consideration for long-term consequences.' Critically comment on this statement.

ISBN 9780170265560

CHAPTER 2

Outcomes

Students will learn about the following topics.

1. Farm modelling.
2. Systems theory.
3. The impact of physical, biological, social, historical and economic factors on agricultural systems.
4. The principles of experimental design and research.
5. Collection and analysis of data.
6. Methods of presenting data.
7. Report writing.

Students will be able to demonstrate their learning by carrying out these actions.

1. Outline the main steps involved in the scientific method of inquiry.
2. Describe the concepts of 'hypothesis', 'theory' and 'bias'.
3. Evaluate methods of data presentation.
4. Clearly set out results and present data.
5. Describe the systems method of inquiry.
6. Illustrate the concept of a system and its component parts.
7. Draw or model a system clearly.
8. Compare and contrast various methods of solving problems.
9. Write up a practical report in a logical manner.
10. Present data.
11. Interpret data.
12. Evaluate practical reports.
13. Consider design aspects of experiments in applied situations.
14. Discuss reasons for experiments being performed.
15. Distinguish between the terms 'population', 'sample', 'parameter' and 'statistic'.
16. Discuss aspects of experimental design.
17. Evaluate experimental design.
18. Define the terms 'randomisation', 'replication', 'control' and 'standardisation'.
19. Analyse data.
20. Define the terms 'mean', 'mode', 'median', 'range', 'variance' and 'coefficient of variation'.
21. Calculate the mean and standard deviation of trial data.
22. Measure variability through the use of standard deviation.
23. Explain the need for a test of significance to be carried out.
24. Determine if a result is significant by use of standard error calculations.
25. Discuss aspects of a normal distribution.
26. Describe the term 'significance' as applied to experimental results.
27. Assess the significance of experimental results.
28. Use appropriate methods of presenting data.

ISBN 9780170265560

METHODS OF SOLVING PROBLEMS AND EXPERIMENTAL DESIGN

Words to know

analyse to examine information relating to a problem posed by an experimenter

bias a form of prejudice, or slant, in obtaining a result

coefficient of variation an absolute measure of dispersion of the data

control the standard with which a new technique or variety is compared, or a part of an experiment that does not receive any treatment

conclusion a series of judgements or inferences made on information gained through an experiment

data information gained through measurement or the collection of observations

discussion a detailed examination of the findings of an experiment

experimental error a deviation from the true measurement expressed as a percentage

germination the process whereby a seed changes from a dormant state to active growth

hypothesis a concept or idea to be assessed or tested

mean the average value

median the middle value in a group of values

mode the most commonly occurring value

parameter the measurement gained from total populations

population (in biology) a group of one kind of organism living in a particular place at any one time; (in statistics) the entire pool from which a statistical sample is drawn

randomisation method used to ensure that all members of a population have an equal chance of being involved in an experiment

range a measure of spread of the values

replication repeating the same experiment a number of times

sample a representative section of a population

scientific method a method of problem-solving that involves the testing of an idea through the use of experiments and the analysis of data

standard deviation (s) the square root of the variance

standardisation to make all conditions in the different trials of an experiment (e.g. climate, soil, slope) as similar as possible

statistics the measurements gained from population samples

theory a generally accepted explanation of a principle or observation

variable a particular feature that is being measured

variance (s^2) a measure of how closely values cluster around the mean

varieties types of plants or animals

weather daily changes in Earth's atmosphere in precipitation, temperature, wind, pressure, cloud cover and other factors

ISBN 9780170265560

Introduction

Agriculture is an applied science, and farmers often need to find answers to specific 'on-farm' problems, such as the type and amount of fertiliser to apply or the correct method of budgeting. Consequently, there is a need to develop skills in understanding concepts, gathering information and applying knowledge and **theories** gained through research to practical situations. In this chapter some of these skills will be examined and developed using two approaches to problem-solving in agriculture. These approaches are often complementary rather than mutually exclusive.

Experiments are carried out for many reasons. Some experiments are designed to test if one variety of plant or animal has any significant advantages over more accepted **varieties**. Farmers often require advice on which is the best variety of crop or breed of animal to use for their particular locality. On the basis of statistically proven evidence gained through experiments, farmers invest both time and money in developing a management system to suit recommendations.

Experiments are also designed to show farmers differences between management techniques at field days. A variety of material can be tested, from machinery performance, cultivation techniques, fertiliser types or rates of application, to grazing strategies and total farm management systems. All experiments share a common link – they all compare one method with another, or one method with an accepted standard, to allow valid and easy comparisons between strategies for farmers.

Many experiments are not performed by the farmer because of lack of time or expertise in experimental design or analysis. The farmer relies on experts from various government departments and private industry to conduct these experiments and advise or display their findings. However, it is still wise for a farmer to be able to judge the merits of each experiment.

The scientific approach

Everyone has a natural curiosity about what is happening around them. Scientists are simply people who have specialised in studying what is occurring, or has occurred, in a vast array of areas. Some study living systems, some study physical or chemical systems, while others study social systems.

Scientists attempt to use existing knowledge to solve problems or apply theories to practical situations. In studying events or attempting to solve problems scientists generate a number of new ideas, facts and even more problems to be researched. The method scientists use to examine events is called the **scientific method.** This can be broken down to a set number of steps that all scientists follow.

Outline the problem

You need to develop a very clear statement of the problem or investigation you are about to make. This is so that everyone reading about your experiment knows what you are trying to do. Others may wish to repeat your experiment to see if your results were correct.

Propose the problem

In scientific terms, when you have formulated a guess as to what is the cause of the event you wish to investigate, you have developed a **hypothesis.** The statement 'I think it works this way' assists in the types of observations or facts that you collect and the method you use to collect these facts or observations.

ISBN 9780170265560

Gather facts or data

Once you have discovered something you wish to investigate and worked out a hypothesis, you need to gather facts, **data** or make a series of careful observations. Facts are gathered by performing a series of experiments and recording the results.

Experiments test your ideas, confirm your observations and clarify your deductions (if you are lucky). Experiments need to be repeated several times and the results compared. They must be carried out according to well-defined rules outlined later in this chapter.

Make observations and measurements

Critical to any experiment are accurate observations or measurements of experimental results. It is important that what you record actually reflects, or is a measure of, what is really occurring. Measurements must be objective and free from observer **bias**.

Unfortunately, many factors might confuse the observer or confound the measurements being made. For instance, if you are measuring the effects of applying nitrogen to wheat by recording variations in plant height, how can you make sure that your measurements reflect the effects of the fertiliser rather than the effects of soil type, light intensity or temperature variation? If you are observing a minute organism under an electron microscope for the first time, how can you be sure that you are seeing an organism and not a foreign particle, such as the embedding material?

You must repeat experiments under identical conditions to test the validity of your observations. On the basis of your observations or measurements, the hypothesis or idea might be accepted, rejected or modified.

1. Suggest reasons for the importance of knowing how to use and access areas of knowledge, rather than attempting to know everything about a topic.
2. List and describe the main steps involved in the investigation of a problem based on a scientific method of analysis.
3. What does the term 'hypothesis' mean?
4. Why is it important to remove all sources of bias in an experiment?
5. What are some important points to consider when presenting results?
6. Suggest three ways of presenting data, and for each method outline what should be considered when presenting data in this manner.

Present the results

Once a series of measurements has been made or observations recorded, you must clearly outline and present them. The reader must be able to interpret without any confusion the results that you have obtained. There are many ways of presenting results, such as diagrams, photos, scientific papers, graphs (Fig. 2.1, page 22) and tables. It is important to match the way you present your data with the intended audience. A scientific paper, detailing all of your ideas and measurements, might be necessary for a gathering of scientists, while a precise summary of your experimental findings might be all that is needed for publication in a general newspaper.

Diagrams should be clear and not cluttered by too many labels. Measurements should be well presented and not overwhelming to the reader. Extract the important results that you wish to talk about and leave the others in an appendix to your experiment. Where possible, you should estimate the extent of **experimental error** when measurements are used.

Make a conclusion

Conclusions are reached on the basis of experimental evidence, and eventually lead to an acceptance, modification or rejection of the original hypothesis. In reaching a valid conclusion – that is, one accepted by the majority because it seems to accurately model or reflect what has happened – you have to account for all bias. Experimental design is an important factor that will allow you to eliminate factors that could lead to wrong conclusions. Many of your conclusions will lead to further experimentation and study.

Through the application of the scientific method, over time, scientists have added to the growing body of knowledge and understanding of the real world.

ISBN 9780170265560

a Line graphs
Line graphs connect or show trends in data

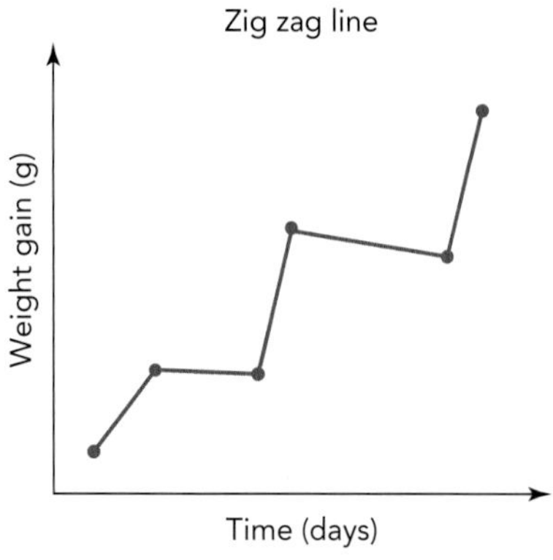

Straight lines connect data in this graph

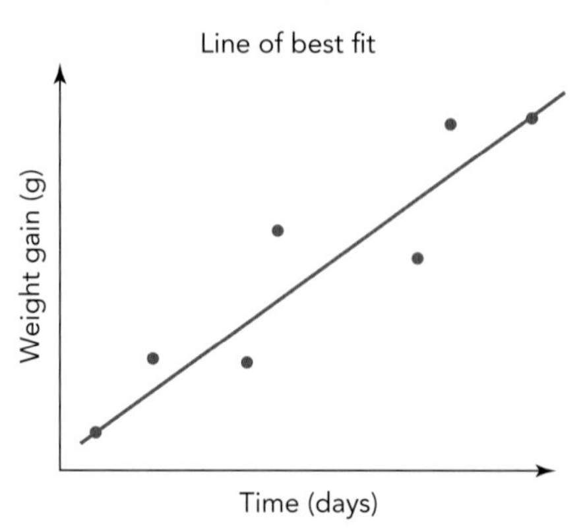

This graph shows a trend across data

b Pie graphs
Pie graphs illustrate data in percentage or absolute number terms that form part of a whole

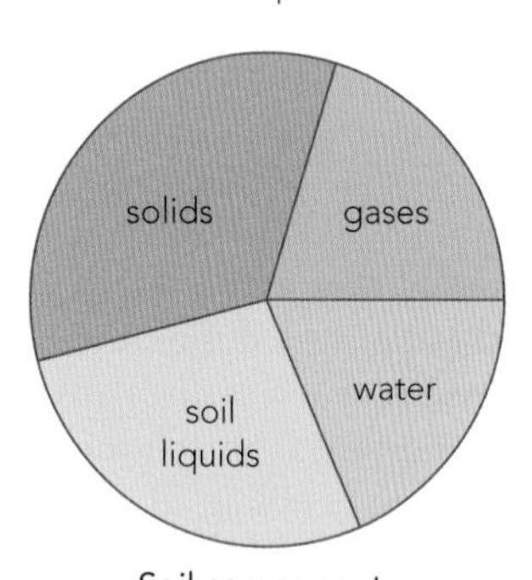

Soil components
percentage or absolute number terms that form part of a whole

c Column graphs
Column graphs are used to show how different things compare with each other

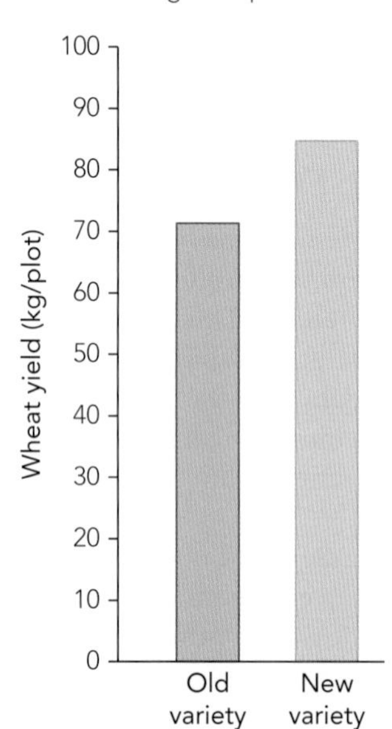

d Bar graphs
Bar graphs are similar to pie graphs in that they may be used to illustrate data in percentage or absolute number terms, forming part of a whole

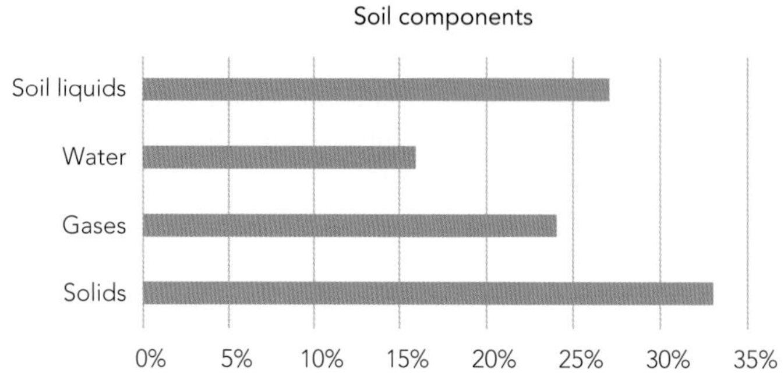

Figure 2.1 Graphing results

ISBN 9780170265560

The systems approach

The systems approach is another approach that can be used in a problem-solving situation. This approach concentrates on the way in which various parts work together as a whole. Once you know how an entire system operates, you can determine the effects of changes on the system. Consequently, with a systems approach you need to examine the effects of physical, social and economic factors on the operation of a farm (Fig. 2.2).

Agriculture, in fact, could be considered the study of a complex of interactions between living, economic and social aspects of our environment (Fig. 2.3).

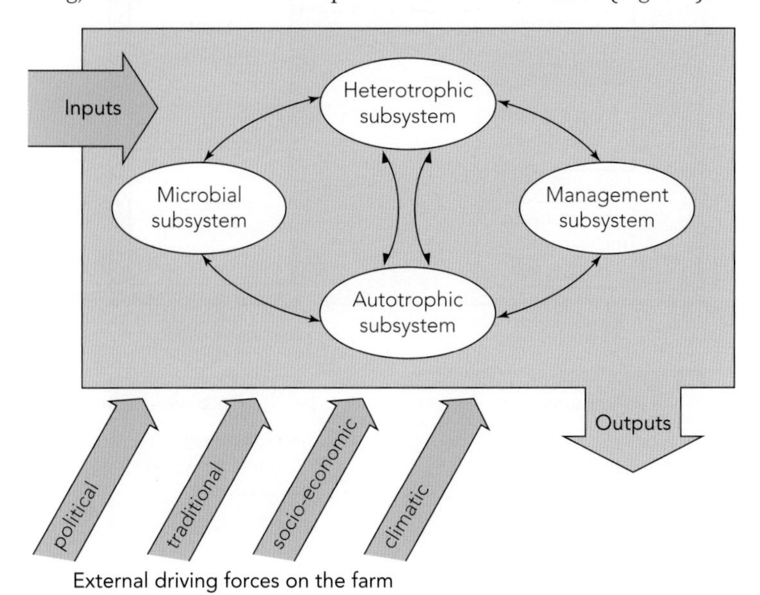

Figure 2.2 A systems model of a farm

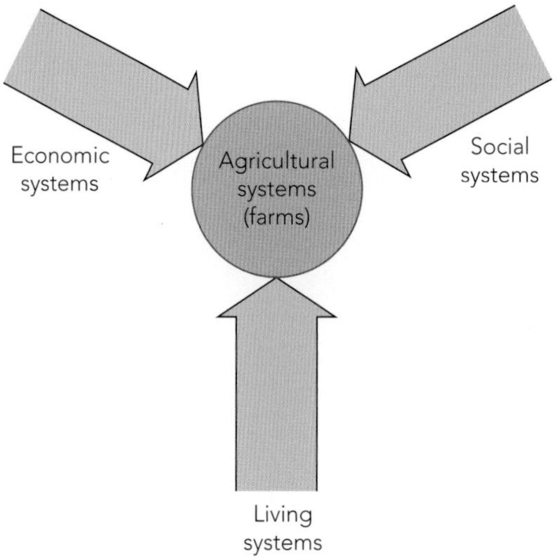

Economic systems
Agricultural systems (farms)
Social systems
Living systems

Figure 2.3 External forces acting on farm systems

By studying all of the components and their interactions, it is possible to suggest ways of improving any given situation.

With a systems approach to a problem situation, the problem is not reduced to a single aspect to be exclusively investigated; rather, the problem is described in terms of the system in which it is located. Such an approach requires a definition of the system for study by reference to its boundaries, inputs, major subsystems, processes, interactions and outputs. Models can then be constructed for any part of a farm as illustrated in Figure 2.4.

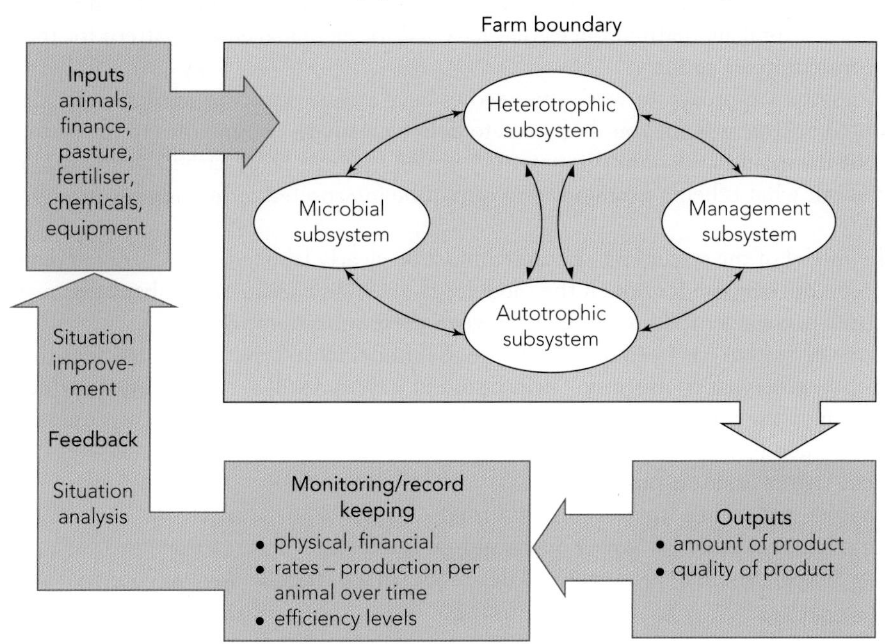

Figure 2.4 A model of a whole farm

Farms can be compared in terms of their system models, rather than the type of agriculture practised on them, their method of management or location in the world. In fact, it is possible, using systems terminology, to construct computer models of agricultural systems that allow people to predict what would happen under various conditions. Figure 2.5 illustrates how a researcher may determine what factors limit output levels in a dairy farm system.

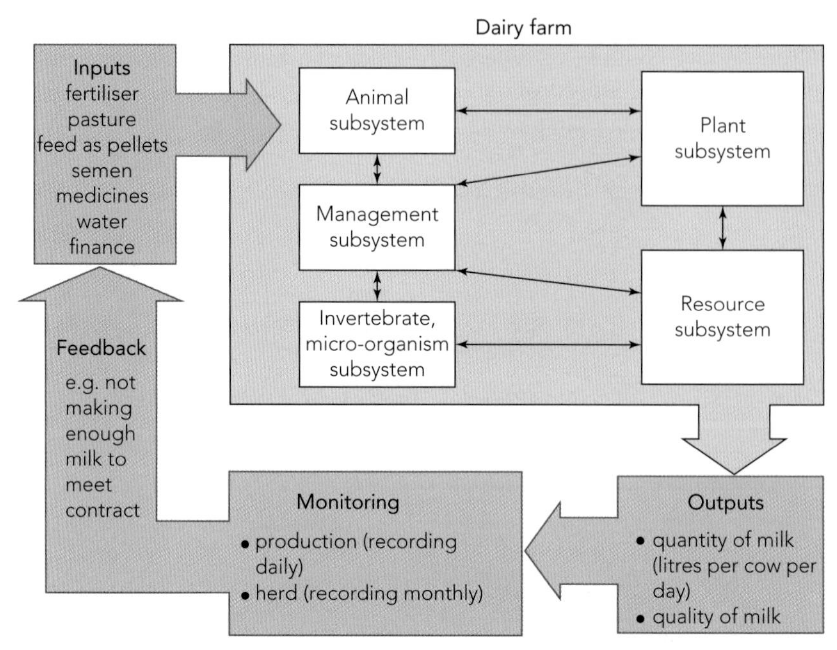

Figure 2.5 A dairy farm system model

7 Examine and redraw Figure 2.5 into your books. How does this diagram differ from Figure 2.4?

8 Can you think of additional inputs into the system illustrated in Figure 2.5?

9 What is the boundary to the system in Figure 2.5?

10 What problem is being investigated with this system model in Figure 2.5?

11 What alterations would you make in the system shown in Figure 2.5 to improve it? Give reasons for your decisions.

There are many problems in agriculture that can be stated in terms of specific hypothetical statements, such as 'How much water is required to grow cotton in a certain area?' or 'What is the optimum level of fertiliser required by oats?'. However, in reducing observations down to collecting evidence in specific areas, solutions to problems can be overlooked. For example, if every farmer continually added nitrogen fertiliser to the soil, over time soil acidification would probably occur. Lakes could also become overgrown with blue-green algae because of the leaching of excess fertiliser into the waterways, or other undesirable environmental or social effects might result. Since agricultural production systems are basically human activity systems, it often makes sense to consider the social, economic, cultural and natural forces that have led to the management pattern for the whole farm, community or country.

For a system approach to work you need to have a clear purpose in order to decide what is essential in your system. You then need to define clearly the boundaries to the system and work out the main inputs into the system. A model that can be described clearly removes doubt about what is being examined and cannot be interpreted differently by different people.

The model of the system being studied must be clearly labelled. The degree of detail will depend on the purpose for which the model was made. In particular, if subsystems need to be identified, consider the interactions between these subsystems (Fig. 2.6).

Once the system has been described, use the model to determine how efficient the system is in achieving its purpose. The purpose of a system will vary from one in which you are examining the physical or economic levels of output to one used to answer questions relating to such problems as these.

- What factors limit output?
- What happens when input levels are varied?
- What happens when an aspect of the system is altered?
- How feasible is the development of this model in a particular area, or country?

ISBN 9780170265560

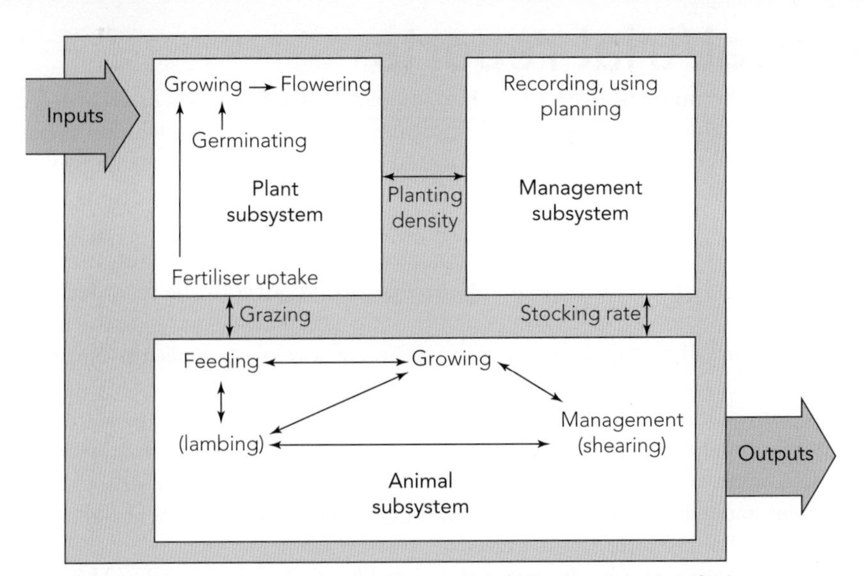

Within each subsystem basic processes are shown and interactions identified. Between each subsystem, interactions are shown relating to an aspect of animal and plant management on the farm

Figure 2.6 Farm subsystems and interactions

12 Redraw Figure 2.6 and then use the basic principle of this diagram to construct a system diagram of one farm enterprise that you have seen.

13 Why is it important to examine broader aspects of a problem involving economic, social or historical factors? Give some examples.

Problem-solving in agriculture

During your agricultural course you will be in a position to question certain practices and, hopefully, present intelligent solutions to problems. In order to gain the necessary skills to question and test concepts, you will need to develop the ability to design experiments, collect and record data, and then use this material to produce a written report in which results are interpreted and conclusions are drawn and presented. Graphing is an essential skill along with the ability to interpret both graphs and figures in an agricultural context. Farmers require answers to questions that directly relate to the management of the farm. Many of the trials and experiments performed involve the collection of measurements or data from large field sites in different locations. Experimenting in the field is a very different proposition from experimenting in a laboratory. In the laboratory many conditions can be accounted for or controlled, such as water levels, and light and temperature conditions. Even the soil can be made uniform by the selection of particular mixes for use in all pots. In the field the soil can vary in its physical and chemical makeup across a paddock in which a field trial is growing, while daily **weather** patterns can significantly influence an experimental result.

Experiments involving animals are even more difficult to set up, **analyse** and develop conclusions from. A common problem is insufficient animals being available on which to base valid conclusions for the experiment. Sex, breed or age differences between animals involved in experiments can also cause complications. Other considerations in any experiment involving animals include what to feed the animals and how to keep environments identical, along with how to look after the welfare of the animals during the experiments.

The most important thing to keep in mind is whether the design of the experiment and the steps followed by the experimenter allow you to make a valid conclusion on the basis of the data collected and analysed. You must consider whether factors other than those considered in the experiment could have affected the final result. Also, you must decide whether or not the experimenter made the right measurements and analysed them correctly.

ISBN 9780170265560

Practical/trial reports

Practical reports should be neatly written into a separate practical book or recorded using your computer or tablet. Diagrams, graphs and calculations should be included as part of any practical activity. Reports should be presented under the following headings, although students in consultation with their teachers should develop the ability to decide which headings are appropriate for particular experiments.

- *Title.* This should be a brief, specific statement of what the work is about.
- *Date.* The date(s) on which the experiment was carried out must be recorded at the top of the report.
- *Students involved.* If the work involved the efforts of several students, this must be acknowledged.
- *Aim and introduction.* The specific aim, purpose or hypothesis should be stated. An introduction might be warranted where any relevant theoretical information is needed. You might want to include a synopsis here as well. This provides a snapshot of your experiment and allows the reader to know a little about the experiment before reading through the whole report.
- *Apparatus and material.* Specific materials needed for the experiment should be mentioned. Standard pieces of scientific apparatus, such as test tubes, need not be mentioned here.
- *Method.* Steps taken in performing the experiment should be briefly and clearly explained. When doing this, use the third-person past tense, for example, 'The soil was sieved prior to use'. For all experiments involving numerical data collection and analysis, methods used to cope with variation should be outlined. Nature is characterised by its variability.
- *Diagram.* These are useful when describing specific pieces of apparatus, observations, or experimental designs. In all cases use an HB pencil when using a book. This works well in the field, especially if wet conditions are encountered, and allows mistakes to be corrected easily. Keep diagrams as clean lines; don't colour them in. Figure 2.7 shows some features of good diagrams.

14 What headings are required in a good practical report?

15 Why do experiments in field situations require more care in their design and data analysis than those in a laboratory?

16 List three problems confronting people using animals for research purposes.

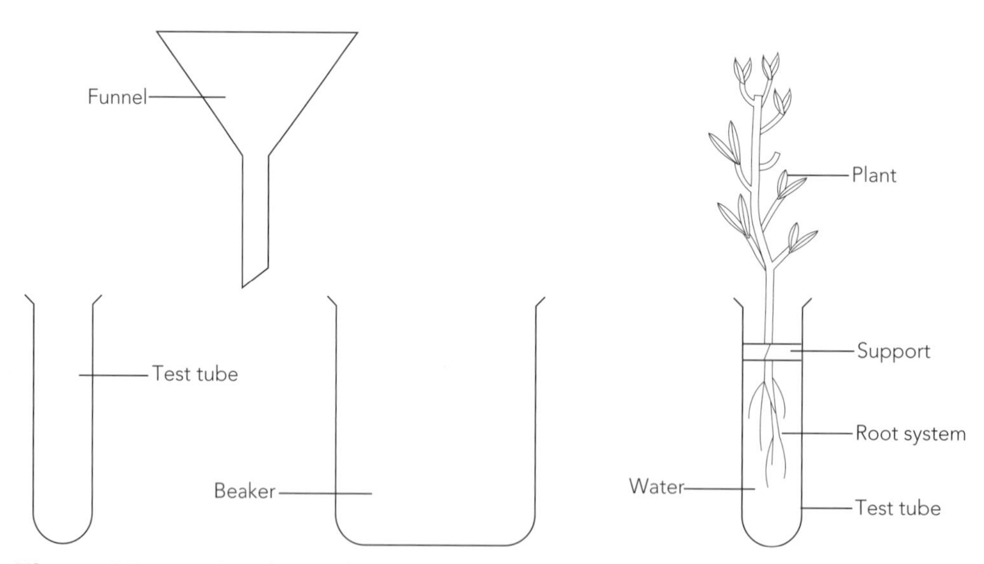

Figure 2.7 Examples of scientific diagrams

- *Results.* Observations and measurements should be recorded in table form during the experiment. Units of measurement should be shown, along with an estimate of error. Calculations must be shown along with relevant formulae and working out.

  Graphs need to be drawn on graph paper and in pencil. They should be large enough to show all details. The axes must be labelled and suitable units of measurement indicated. Each graph should be titled, and should include keys to identify features if needed. Figure 2.8 illustrates important aspects of a graph.

  Some analysis of the data might be needed. Biometric calculations showing means and other statistical measurements might be relevant here.

ISBN 9780170265560

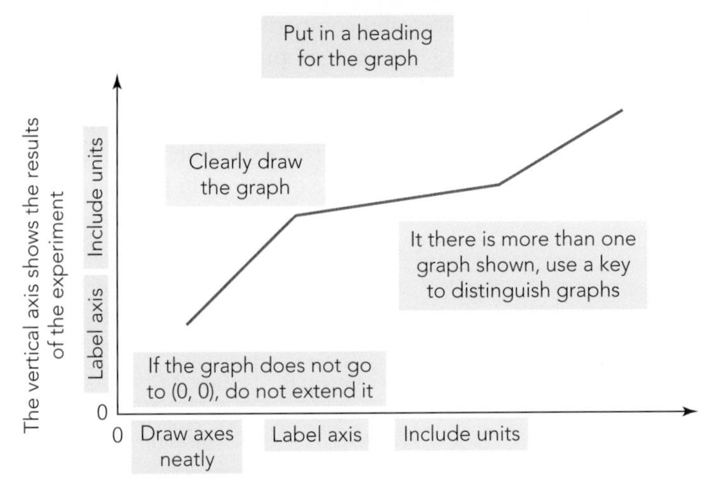

Figure 2.8 Graphing data

- *Analysis*. Here the general trends shown in the results are described and summarised. Reference is made to any trends in the observations or data, and any patterns seen in the graph(s) of the data collected.
- *Conclusions*. This section should contain statements of the actual findings, which should be directly related to the original aim stated in the introduction of the report. You should attempt to interpret data from tables, graphs or diagrams.
- *Discussion*. The **discussion** section should contain relevant points arising from the practical/experiment/trial, including an outline of problems encountered, comparisons with class averages to establish accuracy of observations and suggestions of further issues for investigation. Discussion of the role of experimental design factors and an analysis of the design of the experiment are important at this stage. In other words, the steps taken in order to develop the conclusions should be analysed critically.
- *Questions*. Questions relating to practical work are often included at this point to link the practical aspect of the investigations to theoretical course aspects. These might be provided by the teacher.
- If there are no questions, in your conclusion you should attempt to synthesise the practical results with the theoretical concepts developed during the course. There should be provision for a bibliography, as all references should be noted and other author's work acknowledged.

17 Distinguish between a discussion and a conclusion.

18 List and describe three methods of presenting results.

Good experimental design

Experiments are performed to test or compare ideas. A series of plots of ground of known area, or a number of animals, might be subjected to a series of treatments or tests. How the test subjects respond is recorded by a series of measurements or observations. A measurable feature of a **population** is called a **variable** and, as the name implies, not many of the measurements will be the same.

In order to ensure total accuracy when trying to determine how a population of organisms has responded to a treatment (for example, the response of a worm population in a herd of Hereford cattle to drenching rates) every organism should be examined, weighed or measured. This is an impossible task in many instances. To record changes in height of every oat plant in a hectare of land that has been treated with varying levels of nitrogen fertiliser would be a daunting task. Therefore, to obtain an accurate measure of population responses to treatments, a **sample** is taken and analysed. A sample must be truly representative of the population from which it is drawn. Estimates about a population are obtained by the use of random samples. In such a sample each member of the population has an equal chance of appearing.

Measurements that characterise actual populations are called **parameters**. Measurements that are obtained from samples of a population are called **statistics**. These will vary according to the nature of the samples obtained.

19 Outline reasons for experiments being carried out in agriculture.

20 Define the terms 'variable', 'population', 'sample', 'parameter' and 'statistic'.

ISBN 9780170265560

21 Is a realistic estimate of the weight of animals, such as chickens, in a population obtained by simply catching and weighing those easiest to corner? Why or why not?

22 Examine the shed design in Figure 2.9, which is being used to test different treatments of Groups A–D. What problems can you list that could influence the final result?

23 The aim of the experiment shown in Figure 2.9 was to weigh the various groups of birds and compare their growth rates. Group A contained all male birds; group B, all female birds; and groups C and D, mixtures of unsexed birds. What do you think the experimenter was trying to establish? What factors would have to be controlled so that an accurate conclusion could be reached?

Good experimental design allows for valid comparisons between treatments and assists in determining what is occurring in a living population. Good design tests what the experimenter wants to examine, not the interaction of many factors. An important part of good experimental work is the manner in which measurements are made and analysed. An experiment is considered to be valid if the correct experimental procedures have been followed. Examine Figure 2.9 and then answer the questions in the margin on design.

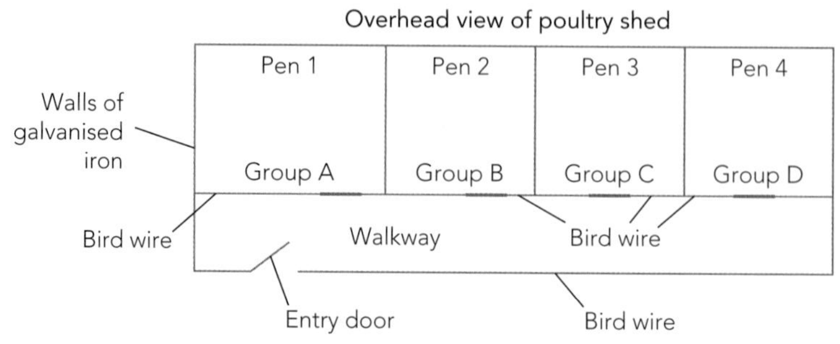

Figure 2.9 Experiment setup

Principles of experimental design

In order to minimise variation due to factors outside the experimental treatment, a careful approach to the experiment is necessary. Careful thought should be given to the design of the experiment.

A sound knowledge of the management techniques required for the establishment and growth of the organism being tested is essential. Otherwise, the result of the experiment could be confused with poor management technique. The experimental site must be carefully chosen to avoid errors arising from drainage problems and variation in soil quality, and to prevent environmental effects, such as draughts, influencing animal health. Shade cast by trees on the perimeter of experimental areas, the effects of weeds or contamination effects from neighbouring herds of animals should also be considered. Border effects have significant influences on the results of experiments conducted in field situations, and buffer areas might need to be established.

Each treatment should be repeated several times within the experiment to provide an estimate of experimental error. **Replication** attempts to overcome problems associated with atypical responses exerting an influence, due to very good or very poor environmental conditions. Replication also increases the accuracy of the results.

Treatments should be distributed in such a way that all the plots or animals have an equal chance of receiving the treatment. This **randomisation** process assists in preventing bias during sampling and treatment allocation.

The use of a **control**, or a part of the experiment that does not receive any treatment, should be standard so that variations between treatments can be compared. This is an important feature of any experiment. As far as possible all plots or animals in the experiment should be managed in exactly the same way so that the only thing different is the variable or treatment being tested. This is known as **standardisation**. In a fertiliser trial the only thing that varies from plot to plot is the amount of fertiliser applied. Everything else is the same for each plot, such as plot size and shape, irrigation and planting rate.

Care should also be taken in the design phase to make the number of animals or the plot area large enough so that useful measurements can be obtained. The size of the experiment should allow for effective use of money and time.

Several experimental designs exist to account for variation within or between experimental treatments. One simple design is known as a completely random design. As shown in Figure 2.10, where three varieties of oats labelled as A, B and C are compared, the treatments are set out side by side (see Things to do Question 5, page 34).

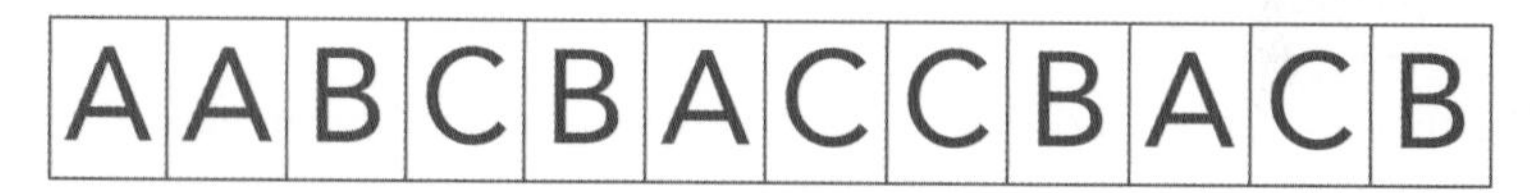

Figure 2.10 A completely random design

ISBN 9780170265560

The completely random design is the easiest to lay out, but it may be influenced by a number of factors not directly related to the treatments being investigated, such as variations in soil type, border effects or microclimate changes. The design is only suitable for small numbers of treatments and requires the use of uniform experimental material.

A randomised complete block design is shown in Figure 2.11. This design is used where greater numbers of treatments are involved. Treatments are grouped together in blocks, which are arranged in a particular pattern. In Figure 2.11a, five treatments are contained within each block. This design gives greater accuracy.

24 List and describe the factors you have to consider when planning an experiment.

25 Define the purpose of randomisation, replication and the use of a control in an experiment.

26 Identify two common experimental designs and explain the difference between them.

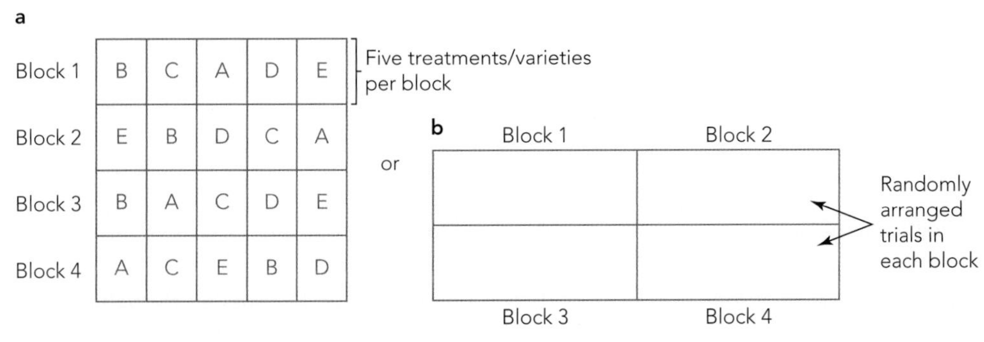

Figure 2.11a, b Two possible arrangements of a randomised complete block design

Analysing data

There are several ways of determining the uniformity and, consequently, the accuracy of the measurements. The first measurements (the mean and the mode) indicate central tendency.

1 The **mean**, or average value, can be calculated for a population using this formula:

$$\text{Mean} = \frac{\text{sum of (variables} \times \text{frequency)}}{\text{total frequency}}$$

or

$$\text{Mean} = \frac{\Sigma fx}{\Sigma f}$$

where f = frequency
x = variable
Σ means 'sum of'

The mean for a sample is calculated as:

$$\text{Mean} = \frac{\text{sum of all variables}}{\text{total number of variables}}$$

or

$$\bar{x} = \frac{\sum_{i=1}^{n} x_i}{n}$$

where n = number of variables
x_i = measurement value
$\sum_{i=1}^{n}$ means 'sum of the variables'

2 The **mode**, or most commonly occurring value, is found by examining the frequency of occurrence of each measure.

3 The **median** is the halfway value when variables are arranged in order of magnitude.

Yields over 6 years recorded for two varieties of cereal crop						
Variety	Year 1	Year 2	Year 3	Year 4	Year 5	Year 6
A	25	28	30	27	20	34
B	23	34	30	28	23	20

ISBN 9780170265560

Measures of spread or dispersion of variables are obtained by calculating the following values: range, variance and coefficient of variation. (Widely spread values indicate greater variation.)

1 The **range** is the difference between the highest and lowest of the variables. This does not consider the frequency of the variables.

2 **Variance** (s^2) is a measure of how closely values cluster around the mean. The **standard deviation** (s) is defined as the square root of the variance. Variance can be calculated as:

$$s^2 = \frac{\text{sum of each value squared} - \left[\frac{(\text{total sum of all variables})^2}{\text{number of variables}}\right]}{\text{number of variables } -1}$$

$$s^2 = \frac{\sum_{i=1}^{n} x_i^{\,2} - \frac{(\Sigma x_i)^2}{n}}{n-1}$$

3 The **coefficient of variation** is calculated as:

$$\text{Coefficient of variation} = \left(\frac{\text{standard deviation}}{\text{sample mean}} \times \frac{100}{1}\right)\%$$

This is regarded as an absolute measure of dispersion of the data. The smaller the value of the coefficient of variation, the more reliable the results. Reasons for high values include effects due to border influences, soil interactions or insufficient replications to provide greater significance to the experimental results.

The normal distribution

As a sample from a population becomes larger, agreement between sample values and population values increases, and so the sample mean ($\bar{x}$) approaches the value of the population mean (μ). The variance of the sample (s^2) approaches the variance within the population (σ^2). With increasing numbers of measurements a trend is observed that, when graphed, takes the form of a bell-shaped curve (Fig. 2.12).

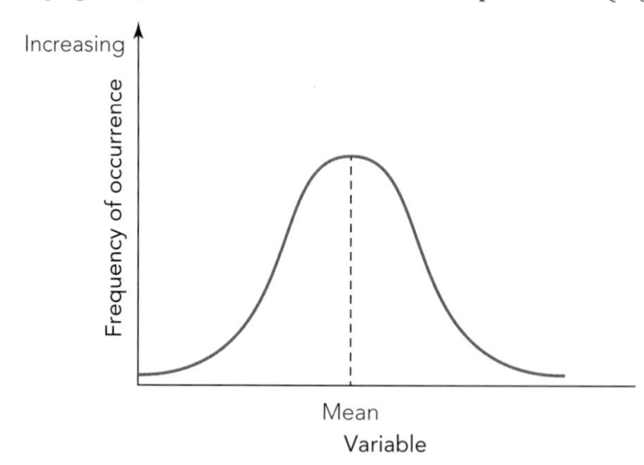

Figure 2.12 Normal distribution

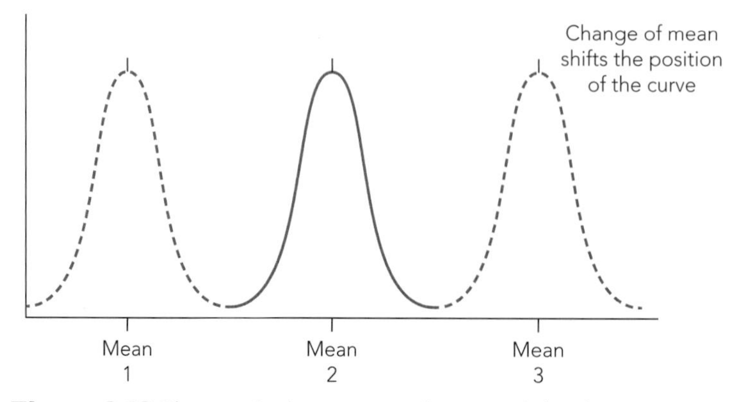

Figure 2.13 Changes in the mean on the normal distribution

27 The table at the bottom of page 29 shows the yields over 6 years recorded for two varieties of cereal crop.

a Calculate the mean for each variety.

b Calculate the difference between the two means.

28 Measurements were made of the heights of oat plants in a field. Their heights (in cm) were: 75, 100, 85, 65, 60, 75, 90, 75, 65, 100, 60, 70, 75, 80, 90, 65, 80, 75, 62, 67, 70, 70, 80, 90, 95, 85, 90, 95, 85, 75, 63, 75, 51, 58, 60, 75, 80, 90, 107, 85.

a Construct a frequency table of measurements by completing the following table for all heights.

Height of plant (cm)	Frequency
51	1
58	1
60	3

b Find the mode.

c Calculate the median value.

d Plot frequency against height. Choose a line of best fit for the graph. (Put frequency on vertical axis, height on horizontal axis.)

29 The heights of 20 recently germinated rice plants (in mm) are as follows: 5, 7, 7, 6, 8, 7, 5, 3, 10, 9, 7, 8, 9, 3, 5, 7, 8, 4, 6, 6.

a Calculate the sample range.

b Calculate the sample mean.

c Calculate the sample variance.

d Calculate the sample standard deviation.

e Calculate the sample coefficient of variation.

ISBN 9780170265560

This curve is symmetric and approaches the horizontal axis but doesn't touch it, and is called a normal distribution. The experimental design and analysis values of the mean, mode and median coincide and the area under the curve is defined as equal to 1. Figure 2.13 indicates what happens when the mean value changes; Figure 2.14 shows what happens when variance changes.

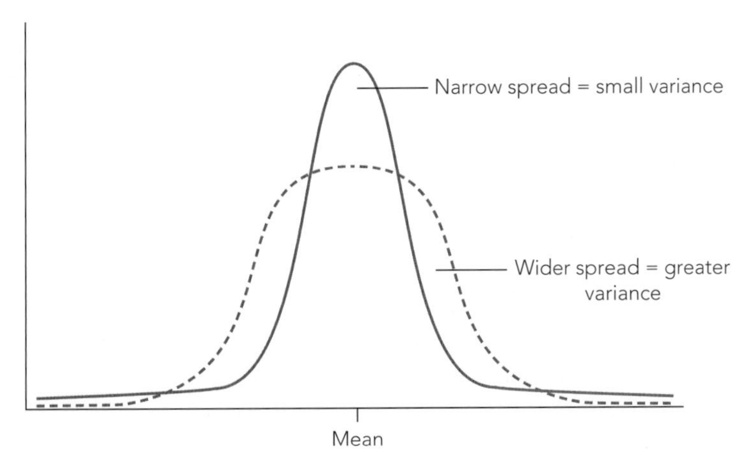

Figure 2.14 Changes in the variance on the normal distribution

Probability and the normal distribution

Refer to Figure 2.15, which indicates probability values under the normal curve. By definition, the area under the curve is unity (1). The deviation of z going from 0 to 1 is 0.3413 by definition, and for z to go from 0 to –1 is also 0.3413. Consequently, the area under the probability curve for $-1 < z < 1$ is 0.6826. The area beyond this is 1.0 – 0.6826 = 0.3174, or approximately one-third. Thus, the probability of a random value deviating by more than 1 unit from the mean in a normal distribution is less than one-third.

Consider the point 1.96, the area $\int_{oZ}^{1.96} = 0.4750$. The probability of a point falling into the interval $P(-1.96 < z < 1.96) = 0.95$, and so the probability of a random unit deviating by more than 1.96 from the mean = 0.05, or 5%. At this point you are 95% confident of your calculations and errors would only occur 5% of the time.

If you move to the point 2.58, the area outside the interval is only 0.01, or 1%. This gives a 99% confidence interval, and your results could only be wrong 1% of the time.

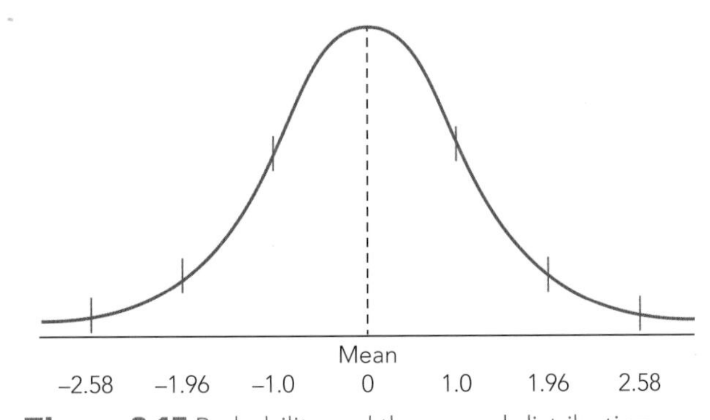

Figure 2.15 Probability and the normal distribution

The **significance** of an experimental result is the probability, at a chosen significance level, that the result will hold true. The normal significance level chosen for agricultural experiments is 5%, which means that the result obtained in the experiment should be obtainable 95% of the time.

30 Redraw Figures 2.12, 2.13 and 2.14 into your notebook and describe the importance of these graphs.

31 Distinguish between a 1% and a 5% level of error in an experiment.

32 What does the term 'significance' mean when applied to experimental results?

ISBN 9780170265560

Assessing experimental results

Once a result is achieved, knowing if it is significant is extremely important. This may be determined or assessed in several ways.

Using measures of central tendency

Look at the derived data in the following table. This shows the differences in yield between two varieties of triticale growing in the same area.

It is possible to conclude from this data that variety B gives a higher yield and is more consistent.

	Yield (kg/ha)	
	Variety A	Variety B
Mean	61.7	71.4
Mode	61.0	70.5
Median	61.0	70.5
Range	15.5	13.0

Comparing standard errors

Standard errors may be used to determine if the differences between two or more sets of data are significant; that is, whether or not the differences are due to chance (sample error or environmental effects) or to the factors being tested (for example, varietal differences or fertiliser levels).

To arrive at a decision the researcher needs to know the following:

1. the mean
2. the standard deviation
3. the significance of the data differences.

It has been generally agreed by statisticians that if the difference between the arithmetic mean of the various sets of data is more than twice the standard error, the results are significant; that is, they cannot be attributed to mere chance. In terms of the normal curve, this translates to a 5% probability that the difference is due to chance.

Examine the following data comparing the population of bacteria (*Escherichia coli*) in samples of milk collected in two dairies. One of the dairies washes the udders before milking, the other dry wipes the udders.

Farm A (washing udders)

Sample number	Microbes/sample	Deviation from mean (d)	d^2
1	160	5	25
3	154	1	1
2	155	0	0
4	151	4	16
	$\bar{x} = 155$		$\Sigma d^2 = 42$

Farm B (dry wiping udders)

Sample number	Microbes/sample	Deviation from mean (d)	d^2
1	140	4.5	20.5
2	132	3.5	12.25
3	125	10.5	110.25
4	145	9.5	90.25
	$\bar{x} = 135.5$		$\Sigma d^2 = 233.00$

ISBN 9780170265560

The standard deviation for farm A is calculated as:

$$\text{Standard deviation} = \sqrt{\frac{\sum d^2}{n}}$$
$$= \sqrt{\frac{42.4}{4}}$$
$$= \sqrt{10.5}$$
$$= 3.2$$

The standard deviation for farm B is calculated as:

$$\text{Standard deviation} = \sqrt{\frac{\sum d^2}{n}}$$
$$= \sqrt{\frac{233}{4}}$$
$$= \sqrt{58.25}$$
$$= 7.6$$

The standard error is determined as:

$$\text{Standard error} = \sqrt{\frac{(\text{sample standard deviation})^2}{n} + \frac{(\text{sample standard deviation})^2}{n} + \ldots}$$
$$= \sqrt{\frac{(3.2)^2}{4} + \frac{(7.6)^2}{4}}$$
$$= 4.12$$

The significance of the results can be worked out as follows.

- The difference between the two means is 155 – 135.5 = 19.5.
- The standard error × 2 = 4.12 × 2 = 8.24.

The difference between the two means is more than twice the standard error. You can conclude that the differences are significant. On the basis of such a result, you could make recommendations relating to farming practice.

ISBN 9780170265560

Chapter review

Things to do

1. Draw a systems model of one enterprise found on a local farm. Write this activity up as a practical report. The aim will be to develop a system diagram of a farm enterprise.
2. Perform a simple animal husbandry operation, such as drenching, vaccinating or crutching an animal. Write up a practical report on this activity.
3. Examine the following information and answer the questions, which will guide you through the process of recording and presenting information.

   A class decided to raise broiler chickens as a fundraising project. It purchased 150 broiler chickens and raised them to a marketable weight over a period of 8 weeks. The class finished the project with 143 broilers, with an average weight of 2.2 kg. This gave an average dressed weight of 1.6 kg. Weights throughout the period are shown in the table on the right.

Week	Weight (kg)
0	0.030
1	0.070
2	0.109
3	0.239
4	0.534
5	0.850
6	1.300
7	1.950
8	2.200

   During the exercise the birds consumed three bags of starter pellets and seven bags of finisher pellets. The bags weighed 40 kg each. The starter pellets cost $22.50 per bag while the finisher cost $20.00 per bag. The chickens cost $1.00 per bird. The cost of killing and processing was $0.95 per bird.

   a Make a labelled line graph of weight gain.
   b Calculate the total cost of day-old chickens.
   c Calculate the total cost of processing.
   d Calculate the cost of feeding the birds.
   e List three other aspects of production that would need to be costed before an accurate total cost for production could be calculated.
   f What was the mortality rate?
   g What was the dressing percentage (percentage carcase weight per live weight)?
   h What was the total weight of feed consumed and how much did this cost?
   i How would you determine the minimum selling price to cover costs?

   Present the information in a readable and logical form.
4. Select two varieties of seed, for example, two varieties of radish seed. Obtain 12 Petri dishes and arrange them into two groups of six. Place 25 seeds on each plate so that in the end you have six plates of each variety. Place both sets of plates in a similar environment and provide them with similar treatment. Count the number of seeds that germinate in each plate after 5 days.

   Determine the following:

   a the mean seed **germination** rate for each variety
   b the sample range
   c the sample variance
   d the standard error
   e the significance of the results.
5. Perform a field experiment by comparing the effect of different levels of nitrogen fertiliser on the growth of oats (through measuring plant height from stem base to top of flag leaf).

   Divide your plot up into 1 m × 1 m squares. Standardise the experimental site as much as possible. Sow the oats at a standard recommended rate. Apply fertiliser at a rate equivalent to the following field applications (kilograms per hectare): 0, 50, 100, 150, 200, 250, 300, 350, 400, 450, 500. Make sure you apply the principles of randomisation, replication and controls in this experiment.

   Collect and graph your results. Analyse the significance of your results. What recommendations can you make?

ISBN 9780170265560

6. A farmer wished to fertilise a vegetable garden to increase production. In terms of interactions, describe how the following factors influence fertiliser application: soil moisture levels, soil pH, bank balance, sowing rate, variety, fertiliser history, cost of fertiliser, expected yield, market price and sowing time.
7. a Examine your local farm area and write down a problem that affects one aspect of farm production. Think of ways to overcome this problem and propose a hypothesis.
   b Write out clearly how you would proceed to design and conduct an experiment to support or refute your hypothesis.

Things to find out

1. Why have scientists adopted a particular method to solve problems?
2. How can local farmers find out about the results of experiments in agriculture that could influence the way they manage their farms?
3. Examine a local farming problem, either on the school farm or in the community. Clearly state the nature of the problem and produce a model for investigation of this problem. Outline the steps involved in the investigation. What factors will you need to take into consideration when assessing the nature of the problem? What data will you need to collect? Describe how you will go about collecting the data. What problems do you expect to encounter?
4. How can experimental trials be used in field days?
5. How are experimental findings reported to the farming community?
6. When reading advertisements in farming publications that quote the results of experiments to justify the claims made in the advertisement, what other information should you check before accepting the results?

Extended response questions

1. For an experiment you have performed, outline the purpose of the experiment. Discuss the steps taken in doing the experiment. What conclusions were reached? How did you account for sample variation in your experimental design?
2. The scientific method involves defining the problem, developing a hypothesis, designing an experiment and drawing logical conclusions. Illustrate how the scientific method has been followed in the investigation of a major agricultural problem in your local area.
3. Compare and contrast various methods used to solve problem situations that develop on farms.
4. 'Many of the changes that occur on farms are driven not by the farmer but by external change agents.' Explain this comment and use examples of external change agents to illustrate your answer.
5. An experiment is conducted to test the effectiveness of a new drench preparation. Results indicate that the drench is very effective, achieving a 10% better kill than an existing type of drench.
   a How can the researcher be reasonably certain that this 10% is not due to random variation?
   b How certain should the experimenter be about the result before promoting it to farmers?
6. In many instances models are used as devices to achieve situation improvements on farms.
   a Use a model, or series of models, to illustrate situation improvements.
   b Explain how this method of problem analysis can assist a farmer in improving a situation.

CHAPTER 3

Outcomes

Students will learn about the following topics.

1. The social aspects related to agriculture.
2. The changing role of the family farm in Australian agriculture.
3. The role of agriculture in the Australian economy and the place of the farm in the wider agribusiness sector.
4. The role of agriculture in Australian world trade arrangements.
5. The problems of resource adjustment within the agriculture sector.

Students will be able to demonstrate their learning by carrying out these actions.

1. State the meaning of the term 'farming family'.
2. Describe the role of management in setting goals, deciding the enterprise mix of the farm and deciding day-to-day priorities.
3. Describe the various roles that the farm family plays in the local community.
4. Explain that agriculture is a significant sector of the Australian economy and has played a major part in its development.
5. Outline the changes in the number and size of farms in the last 30 years.
6. Describe the kinds of enterprises likely to be found on Australian farms.
7. Briefly describe the cost–price squeeze problem faced by the rural sector.
8. Outline the changes that have occurred to the number of people employed in the rural sector.
9. Explain agriculture's role as a consumer in the economy.
10. Explain that agriculture will continue to be a significant part of the Australian economy, and indicate where growth in exports might occur in the future.
11. Distinguish between the cost–price squeeze and the farm problem.
12. Discuss why people employed in farming find it difficult to move on to other occupations in the economy.
13. Discuss why lending institutions are reluctant to lend capital for investment in farming.
14. Outline what assistance is available to farmers.

ISBN 9780170265560

SOCIAL FACTORS IN AGRICULTURE AND THE ECONOMIC PRINCIPLES OF AUSTRALIAN AGRICULTURE

Words to know

bilateral trade trade between two countries

capital money that has been invested or is available for investment

collateral security property pledged as guarantee for the repayment of a loan (in agriculture, usually the farm)

economic viability the ability of a farm to produce sufficient profit to provide a comfortable living for the farming family

farming family a family operating a farm as a business and a lifestyle

finance money

gross domestic product (GDP) the monetary measure that assesses the value of all the finished goods and services that are produced within a country, usually determined on a yearly basis; often used as a measure of the standard of living for a country

hybrid the result of crossing different varieties of plants or different animals of the same species to create new varieties or breeds

jetting forcing an insecticide under pressure into a fleece to prevent flystrike

laser levelling earthmoving technique, guided by lasers, that enables large areas of land to be levelled and giving a gentle slope for irrigation

pesticide any chemical substance, usually dust or spray, used for the destruction of any pest; usually an insecticide

prime lamb lamb produced for consumption as meat

sheep off shears sheep that have just been shorn

squatter's run an area of land taken up for grazing without the consent of the government in the early days of colonisation of Australia

terms of trade for farms, the ratio of the cost of inputs to the price received for products

top making taking wool part-way through processing to the point where it is ready for spinning

ISBN 9780170265560

Introduction

Farming is one of the important industries that has helped to build the Australian economy and the Australian character. In this chapter one of the family farms that make up most of the farming businesses in Australia is described, giving an insight into the lives of the family that operates it.

The current position of agriculture in the Australian economy is still significant, although it can no longer be said that 'Australia rides on the sheep's back'. Agriculture's main contribution is to export earnings. The major problem facing farm businesses is the fact that prices received for farm products have not risen at the same rate as the cost of inputs. This has led to the necessity for changes to be made in farming, involving the reallocation of resources such as labour.

Case study: a farming family

Most (95%) of Australia's farms are family farms or family businesses. Within Australia 99% of agribusinesses and 89% of farming land are Australian owned. Typically farms are run by the members of one family. One such **farming family** is the Greens, who own and operate 'Cluster', a 570-hectare farm situated on the Lachlan River, 40 km west of Forbes, New South Wales. The Greens also lease 100 hectares of a neighbouring property. Neville and Jane Green have four school-aged children: a daughter Bessie and three sons Dyson, Eddie and Gerald. 'Cluster' has been owned by the Green family since it was first taken up as a land grant from the New South Wales Government in 1860 by Neville's great-great-grandfather, Mitchell Green. Neville took over the management of the farm when his father, Caleb, retired some years ago. Prior to 1860, the land was part of a very large **squatter's run**, 'Burrawong'. It is in Wiradjuri country.

Management

Neville and his parents Caleb and Nellie are partners in the farm business, and Neville provides the management for the farm. Their management operates at three levels:

1 setting goals for the family farm to maintain its **economic viability**
2 deciding the enterprise mix, and making adjustments that will achieve the goals that they have set
3 making day-to-day decisions about what priority should be given to the multitude of things that need to be done on the farm.

Setting goals

1 Assume that you have the managerial responsibility for 'Cluster'. Arrange the Greens' goals in the order of priority that you would give them. State why you gave each this priority.

The goals that Caleb, Nellie and Neville have set include:

- making enough money to provide a reasonable standard of living
- maintaining and improving the farm so that it will continue to sustainably provide a reasonable living for future generations of the family should they decide to take up farming.

Deciding the enterprise mix

The current enterprise mix on the farm is a combination of the enterprises inherited from Caleb and Caleb's father and new enterprises that have been added. The enterprises now on the farm are listed below.

1 *Prime lamb production*: 600 first-cross ewes are run, being crossed with Dorset Horn rams to produce **prime lambs**. This is a change from earlier days, when most of the ewes were Corriedales. Having the ewes lamb twice a year instead of the traditional once has increased output.
2 *Border Leicester stud*: 200 stud ewes are run, and 140 flock rams are produced each year for sale to breeders of first-cross ewes. This stud was started in 1931 by Caleb's father.
3 *Beef cattle*: 40 breeders are run, and steers and heifers are sold as yearlings after being supplementary-fed grain to finish them.

ISBN 9780170265560

4 *Cropping*: 150 hectares of crops are grown each year, comprising wheat, barley and oats. The barley and oats are stored on the farm and used for cattle and sheep feed in times of shortage. The wheat is sold through a grain company. Occasionally crops such as canola, field peas and lupins are grown and the grain sold.
5 *Lucerne seed*: 80 hectares of laser-levelled irrigated land is being used to grow lucerne for seed production under contract to a seed company.
6 *Lucerne*: is harvested for hay when the opportunity presents. It is used to feed sheep and cattle in times of shortage. Lucerne is also used for grazing. All the hay is made into large bales that are handled with a front-end loader.

2 List the enterprises on the Greens' farm. Indicate the enterprises that produce outputs that are sold from the farm and the enterprises that provide inputs for other enterprises.

3 Draw a diagram that shows the links between the various enterprises on the farm.

iStockphoto.com/kerriekerr

Figure 3.1 Sowing crops to produce feed for the cattle and sheep

Deciding day-to-day priorities

Deciding what is to be done from day to day depends on such things as the weather, both current and predicted, the time of year in relation to livestock and cropping management programs, and the unexpected, such as an outbreak of disease (e.g. blowfly strike), natural disaster (e.g. flood) or family matters.

General features of the farm

The general features of the farm impinge on decisions at all levels.

The land itself is very flat. Two-thirds of the farm is subject to flooding, the other one-third being protected by levy banks. Floods large enough to cover half the farm occur once every 3 years, and floods large enough to cover two-thirds once every 10 years (Fig. 3.2).

Shutterstock.com/Dan Schreiber

Figure 3.2 A once-in-10-year flood

ISBN 9780170265560

4 What factors have influenced goal setting for this farming family?
5 Explain what factors have influenced the choice of enterprises on the farm.
6 List three general features of the farm that influence decisions made on the farm.

Annual average rainfall is approximately 500 mm, with a slight predominance in the late autumn and winter. Frosts occur in winter, and summer temperatures can often exceed 40°C.

An area of 150 hectares is laid out for flood irrigation. The system was set up in the late 1940s and was designed for more labour than it is now possible to afford. Two-thirds of the system has been restructured using **laser levelling**, so that the land can be easily irrigated by just one person (Fig. 3.3).

Soils are red-brown earths and grey clays, with two of the highest points on the farm being loamy sands. There are still plenty of trees on the farm, and species include grey box, yellow box, river red gums and cypress pine.

Shutterstock.com/Phillip Minnis

Figure 3.3 A laser-levelled irrigation paddock with a good stand of lucerne established

Farm workforce and their roles

The farm workforce over a 12-month period comprises:

- Neville Green
- Caleb Green
- Nellie Green
- contract shearers
- casual labour
- other relatives who visit and contribute in their holidays.

Neville Green

Neville Green works, on average, a 10-hour day throughout the year. His work includes:

- mixing cattle and sheep feed
- tractor work (ploughing, cultivating, sowing, spraying) (Fig. 3.4)
- harvesting and haymaking
- irrigating
- sheep work (drenching, dipping, marking, ear tagging, tattooing, crutching, foot paring, **jetting**, blowfly treatment, mustering)
- transporting stock to market
- transporting supplies back to the farm (fertiliser, fencing material)
- fencing
- general repairs and maintenance
- accounts
- feeding sheep and cattle
- using the computer to track the farm business and obtain information on the internet on topics such as weather forecasts and commodity prices.

ISBN 9780170265560

iStockphoto.com/Chris_Westwood

Figure 3.4 Tractor work

Caleb Green

Caleb Green's work on the farm has been somewhat reduced since he lost two fingers and a thumb in a harvesting accident. He takes care of the farm when Neville and his family are away on holiday and through the year his work includes assisting with:

- mixing cattle and sheep feed
- tractor work (ploughing, cultivating, sowing, spraying)
- harvesting and haymaking
- sheep work (drenching, marking, ear tagging, tattooing, crutching, foot paring, jetting, blowfly treatment, mustering)
- classing the wool clip
- weed control
- transporting supplies back to the farm (fertiliser, fencing material)
- fencing
- general repairs and maintenance
- feeding sheep and cattle.

Nellie Green

Nellie Green works on the farm occasionally. Her work includes:

- sheep work (recording information at stud lamb marking)
- accounts.

As well as the farm work Nellie occasionally drives the school bus.

Jane Green

Jane has a supporting role and much of her time is directed toward the needs of the four children. She has part-time employment with a NSW government department in a nearby country town.

Contract shearers

The contract shearers are employed for general shearing and for lamb shearing, which would amount to 2.5 weeks work. They work an 8-hour day of four 2-hour runs, and are paid according to the number of sheep that they shear.

Casual labour

From time to time the farmer on the next door property is engaged to assist with transporting livestock to market and grain at harvest to storage and market. This would amount to nearly 3 weeks a year.

7 List the personnel in the family farm workforce. Rate each member on a scale of 1 to 10 to indicate their importance to the farm business (1 being least important, 10 being very important).

Other relatives

Relatives of the Greens (brothers and a sister), who live in the city or town, visit the farm for holidays. They bring their children with them, and together they assist with whatever is being undertaken at the time. Any sheep work can always do with extra people to assist with mustering and droving. Jobs such as fencing and clearing sticks from paddocks are things that the relatively unskilled city folks can assist with. These tasks can be arranged to involve some social activities (for example barbecues and picnics), which helps to make them enjoyable times for all (Figs 3.6 and 3.7). A taste of the outdoor life for a week or so is seen as most recreational.

Alamy/K Philip Harrison

Figure 3.5 Lamb marking with help from city relatives

Corbis/Matthias Ritzmann

Figure 3.6 The Green family

Corbis/Matthias Ritzmann

Figure 3.7 A picnic on the farm – a welcome break from work

The next generation

It is not yet clear if any of Neville and Jane's children is interested in becoming farmers in the future. Eddie shows signs of liking farm work. As a pre-schooler he took every opportunity to accompany Neville as he worked on the farm. What the children's preferences are for their vocations and whether or not this involves the farm will become clearer as they grow up and progress through their education.

After HSC Neville completed a degree in agriculture, farm business management and marketing, and returned to the farm to become an industrious manager of the business. He has encouraged the adoption of new enterprises such as lucerne seed production, use of modern technology and the development of the irrigation system.

ISBN 9780170265560

Tomorrow's successful farmers must be even more multiskilled than those of today, to be able to see and discern what the market wants and produce a product for that market, rather than just produce something and then try to sell it. Tomorrow's farmers will also have the same responsibility as today's farmers: that is, to maintain and improve the productive capacity of their farms so that the farms remain productive for future generations. They must practise sustainable agriculture.

8 In your opinion is a tertiary qualification in agriculture, such as Neville's, important for someone taking up farming? Support your answer with reasons.

9 Outline the responsibility of tomorrow's farmers.

10 There is quite a list of things in the community that the farm family is involved in. For each, suggest why the farm family is involved.

Role in the local community

The farm family does not live in isolation; its life is intertwined with that of the local community. There are a number of groups and organisations within the community to which the family contributes and from which they benefit. These groups include a community trust fund, which is responsible for the maintenance and improvement of local community facilities (the hall, church and cemetery); the volunteer bushfire brigade, which is an important, but hopefully little used, community organisation (a fire-fighting appliance is stored ready for action on 'Cluster'); groups for sport such as tennis and waterskiing; the annual show, which is held at the local showground; and the church, where the whole family are active members.

Agriculture in the economy

The Australian economy consists of several sectors, and agriculture is one important sector. Other sectors include manufacturing, mining and service industries. The agriculture sector has made a major contribution to the development of Australia in the last 200 years. At the farm gate the agricultural sector contributes 3% to Australia's total **gross domestic product (GDP)**. In 2010–11 the gross value of Australian farm production was $48.7 billion. However, when value-adding (processes that food and fibre products go through once these products leave the farm) and all the economic activities supporting farm production through farm inputs (such as fertilisers, fuel, seed and livestock feeds) are taken into account, the contribution of agriculture to GDP averages approximately 12% or $ 155 billion (National Farmers' Federation, *Farm Facts 2012*).

11 What is the place of agriculture in the Australian economy?

The contribution to GDP of the various sectors can be compared over the period 2000–2011 to establish growth areas and sectors that have deteriorated. Agriculture has only dropped marginally. The trends are shown in Figure 3.8.

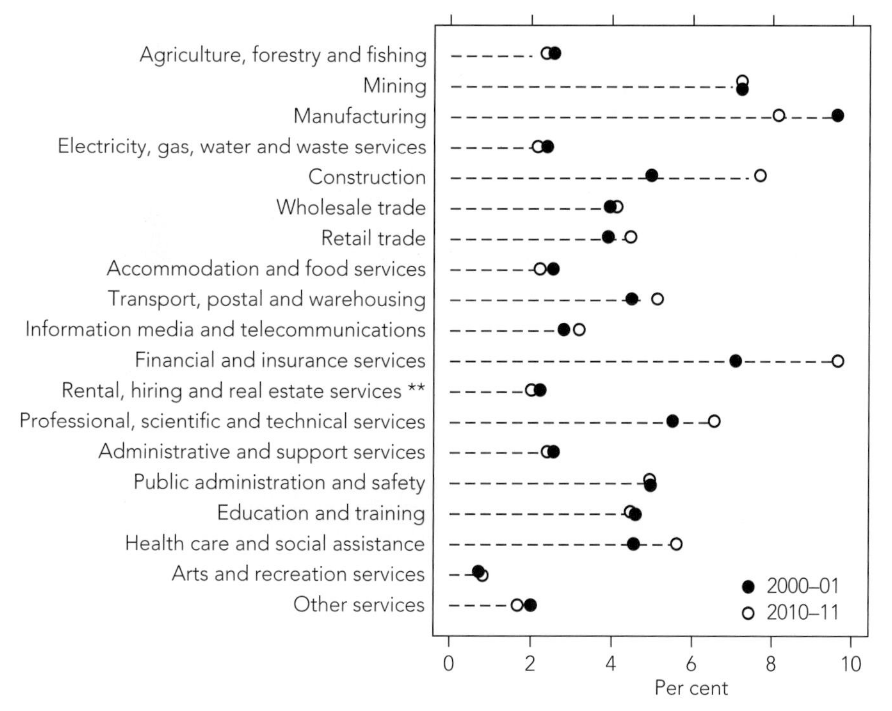

Source: ABS, Australian System of National Accounts, cat. no.5204.0.

Figure 3.8 The contribution of sectors in the Australian economy to gross domestic product 2000–2001 and 2010–2011 (industry gross value added as a proportion of gross domestic product)

The gross value of Australian farm production (at farm-gate) totals $48.7 billion a year. The top three agricultural commodities produced nationally (ranked by gross dollar value) are show in Table 3.1 below.

Table 3.1 Major agriculture commodities produced nationally

Cattle and calves	$7.3 billion
Wheat	$4.8 billion
Whole milk	$3.4 billion

Australia is a reliable global agricultural producer, though farm and fisheries production accounts for only approximately 2.4% of Australia's GDP, with a gross production value of $50.3 billion in Australia's financial year 2010–11 (Australian Bureau of Agricultural and Resource Economics and Sciences 2012). Agriculture is an important part of the Australian economy. Australia is a competitive net agricultural exporter, with approximately two-thirds of total production exported. In 2011 food and agricultural products accounted for 13.8% of Australian merchandise exports. In 2010 Australia accounted for 2.3% of all global agricultural exports (based on the World Trade Organization definition of agriculture, which excludes fisheries, forestry and rubber).

Agricultural production and yields vary widely across Australia, resulting from the different geographical and climatic conditions. Australia currently produces sufficient food to feed up to 60 million people but has a population of less than 23 million. The export of surplus agricultural products increases global food supplies.

Australia's share of total global agricultural food production is quite small overall (approximately 3%). However, for some important agricultural food products, particularly beef, sheep meat, wheat, barley, sugar, dairy, canola and pulses, Australia is a net exporter. In 2011 Australia ranked 17th among food exporting countries. The trend in agricultural exports for Australia for 2002–12 is shown in Figure 3.9.

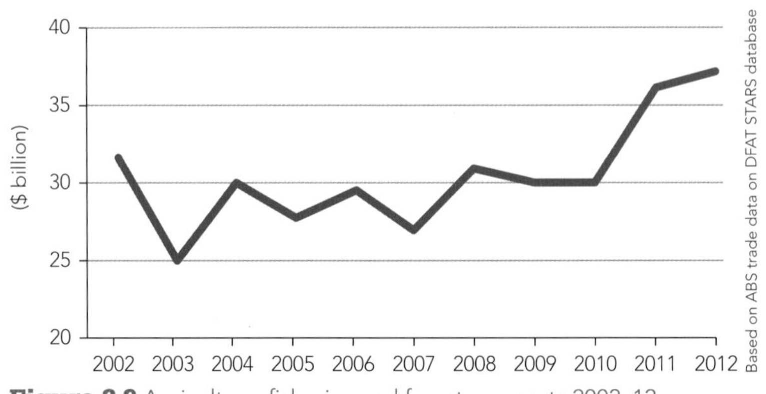

Figure 3.9 Agriculture, fisheries and forestry exports 2002–12

Table 3.2 outlines the main agricultural products exported in 2012, showing value of exports in millions of dollars and the percentage contribution to total exports.

Table 3.2 Australia's top agriculture, fisheries and forestry exports* 2012 by value and percentage

Rank	Commodity	Value ($ million)	Contribution (%)
1	Wheat	6531	17.6
2	Beef	4754	12.8
3	Cotton	2626	7.1
4	Wool and other animal hair (including **top making**)	2524	6.8
5	Meat (excluding beef)	2370	6.4
6	Wine	1891	5.1

continued

ISBN 9780170265560

Rank	Commodity	Value ($ million)	Contribution (%)
7	Oilseeds and oleaginous fruits, soft	1780	4.8
8	Barley	1317	3.5
9	Animal feed	1301	3.5
10	Vegetables	1249	3.4
11	Milk, cream, whey and yoghurt	1234	3.3
12	Live animals (excluding seafood)	1050	2.8
13	Hides and skins, raw (excluding fur skins)	819	2.2
14	Edible products and preparations	810	2.2
15	Cheese and curd	740	2.0
16	Fruit and nuts	670	1.8
17	Wood in chips or particles	666	1.8
18	Cereal preparations	664	1.8
19	Crustaceans	632	1.7
20	Rice	343	0.9
Total agriculture, fisheries and forestry exports		37158	100.0

* Based on the WTO definition of agriculture, fisheries and forestry

Source: DFAT STARS database, based on ABS trade data

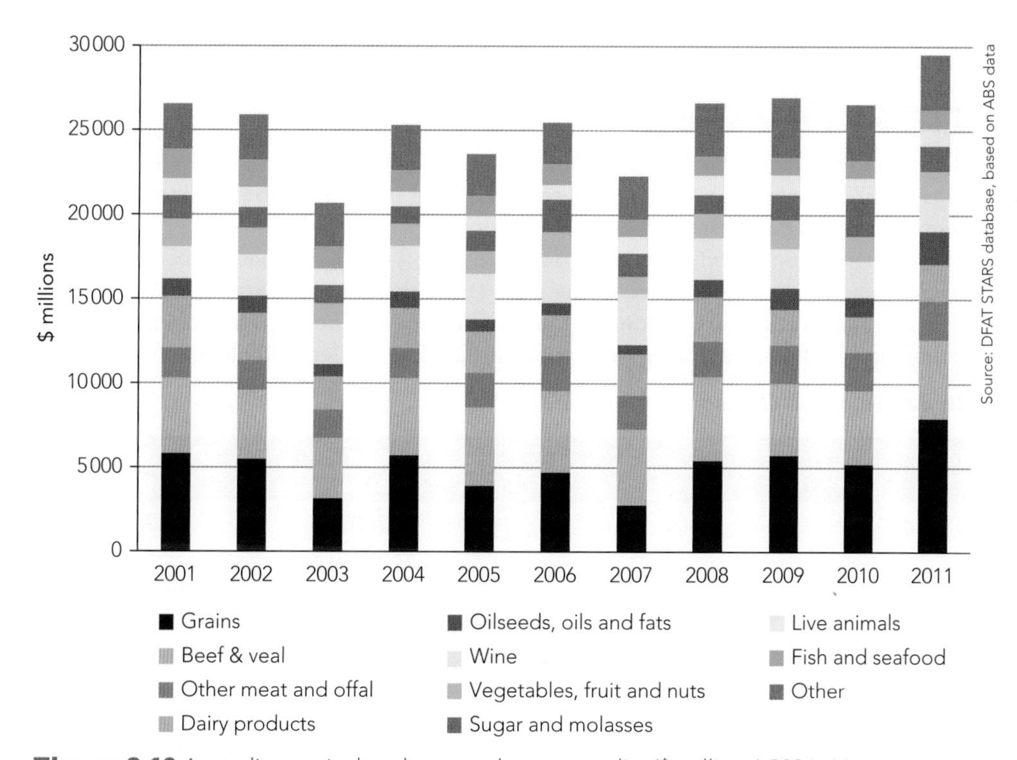

Figure 3.10 Australian agricultural exports by commodity ($ millions) 2001–11

Australia is expected to continue to produce a surplus even as domestic demand grows (Linehan *et al.* 2012). This growth in Australian demand will flow primarily from an increase in population: the Australian Bureau of Statistics (ABS) estimates Australia's population will be more than 35 million by 2056, on current trends (ABS 2010, cited by Department of Foreign Affairs and Trade (DFAT) Australia and China Joint Working Group December 2012).

Current trends in Australia's consumption profile are expected to continue: consumers are likely to continue to favour high-protein, highly processed, convenience-oriented food, with an emphasis on quality, safety and traceability. However, the increase in domestic food demand in Australia will be insignificant compared to the major expansion in food demand in other parts of the world, especially Asia. The main export markets for Australian agricultural foods are shown in Table 3.3.

Table 3.3 Australia's major agriculture export markets (by value): 2011–12 financial year

Major agriculture export market	Value (%)
China	18.3
Japan	12.0
EU27*	7.1
Indonesia	7.0
Republic of Korea	6.9
United States	6.4
New Zealand	4.0
Thailand	2.7
Malaysia	2.5
Vietnam	2.5

* The 27 members of the European Union (EU27) are Belgium, France, Austria, Bulgaria, Italy, Poland, Czech Republic, Cyprus, Portugal, Denmark, Latvia, Romania, Germany, Lithuania, Slovenia, Estonia, Luxembourg, Slovakia, Ireland, Hungary, Finland, Greece, Malta, Sweden, Spain, Netherlands and the United Kingdom.

Source: DFAT STARS Database, based on ABS cat. no. 5368.0, February 2013 data; ABS Special Data Service.

12 List the agricultural products that Australia exports in order of their importance.

13 Which component of Australian agricultural exports has significantly contributed to the increase in food exports from Australia in the last 10 years? Refer to Figure 3.10 on page 45.

14 Where are the main markets for Australia's agricultural exports?

Farms in Australia

Farms account for 51% of the Australian landmass (ABS 2010), and agriculture in Australia accounts for 52% of natural water use. There are 134 148 farms in Australia – including those for whom farming is not their primary business. However, there are 120 112 farms solely dedicated to agricultural production. In 1998 there were 144 800 establishments in Australia whose main activity was agricultural. In the early 1950s there were 205 000. They range in size from small horticultural holdings of just 1–2 hectares to huge grazing operations of many thousands of hectares. For the state totals for 2012–13, refer to Table 3.4.

Table 3.4 Distribution of farms in Australia 2012–13

State	Number of farms	Percentage of total
New South Wales	40 260	32.81
Victoria	29 317	23.89
Queensland	25 326	20.64
South Australia	12 396	10.10
Western Australia	11 050	9.00
Tasmania	3830	3.12
Northern Territory	439	0.36
Australian Capital Territory	71	0.06

Source: ABS Agricultural Commodities 2012–13, cat. no. 7121.0

Australian farms have a wide range of sizes, from a 5-hectare nursery on the outskirts of Sydney, to properties over 20 000 hectares in South Australia, south-western Queensland and Northern Territory. In general, while the number of farms has decreased, the resulting amalgamations have resulted in larger farms to be managed.

 ISBN 9780170265560

The total land area devoted to farming in Australia increased in 2010–11, by 3% over the previous year to 409.7 million hectares, reversing the trend of decline established in recent years as indicated in Figure 3.11. In 2011–12 the amount of land used for agriculture fell 1% to 405 million hectares. Approximately 88% of Australian agricultural land was mainly used for grazing.

The majority of Australian farms are designated as family farms. They are owned and operated by a family partnership (60%) or sole operator (29%). Less than 2% of farms are operated by non-farming enterprises, and 5% are corporate farms being run as public or private companies. On the family farm the whole family often contributes to its running. These family farms generate most of the production from Australian agriculture. Sheep grazing, cattle grazing and grain growing are the main enterprises on Australia's farms. Often they have a combination of two or three of these enterprises. Many farms are involved in dairying, horticulture and growing sugar cane. Figure 3.12 illustrates the percentage of farmers across Australia engaged in particular farming operations.

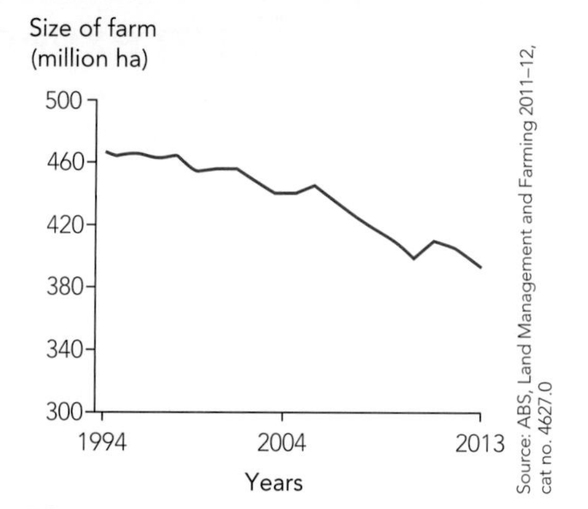

Figure 3.11 Area of Australian farms, 1994–2012

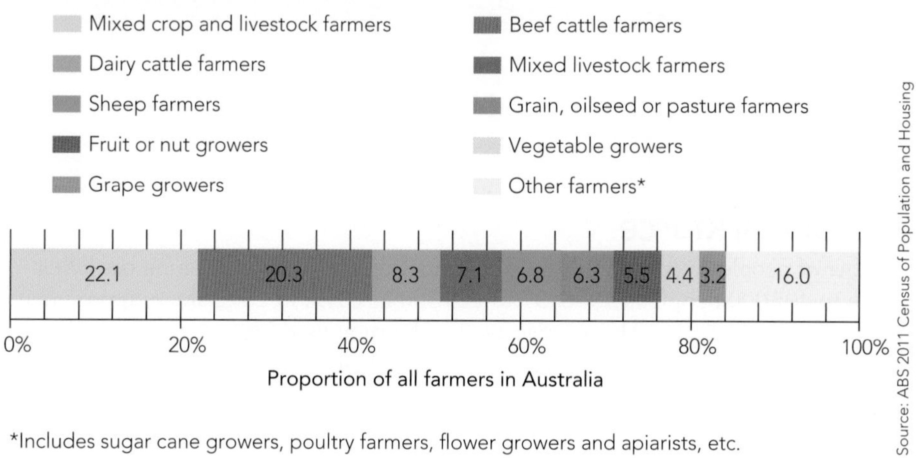

Figure 3.12 The proportion of farmers in Australia 2011

Review the value of livestock products from 2006 and 2007 to the current year by visiting the websites opposite.

View current statistical information on the structure of the agricultural industry in Australia and production levels for Australian crop and livestock sectors from the ABS websites listed opposite.

connect

Value of Australian agricultural commodities

connect

Industry structure

connect

Crops

connect

Livestock

The impact of agriculture on the local community

Farming activities impact upon surrounding communities in two main ways: economically and ecologically.

Economic impact

Farmers and their families buy goods and services from members of the local community. This flow of money allows the development of commercial centres and the growth of ancillary services and infrastructure improvements, such as better transport systems. If farming families leave a rural town, shops close, as do local banks, libraries and even schools. Road work and bridge repairs become financially more difficult.

Ecological impact

Ecological impacts include the effect of **pesticide** spray drifts on surrounding properties, possible pollution of waterways and the deterioration of soil quality due to increased erosion, soil salinity or acidification arising from poor farming practices.

ISBN 9780170265560

Farm expenditure

The inputs required by farmers generate a large amount of economic activity in the service and manufacturing industries. Table 3.5 shows typical expenditure on a farm.

Table 3.5 How farm expenditure is allocated to various inputs in an average year

Inputs	Expenditure (%)
Marketing	14
Machinery, capital items	13
Wages	11
Repairs and maintenance	10
Livestock purchases	10
Fodder and seed	7
Fertiliser	7
Fuel	7
Interest and land rent	7
Contractors	5
Chemicals	3
Veterinary	2
Miscellaneous	4

15 What has happened to the total land area in Australia devoted to farming activities over recent years?

16 From Table 3.5, what are the top five inputs in terms of expenditure on a farm?

17 What kinds of enterprises are most likely to be found on Australian farms?

The rural workforce

The number of people employed full-time in agriculture, forestry and fishing has fallen from close to 405 000 people in 1996–97 to 229 500 in 2013. In 2012 alone, approximately 18 000 people left the sector. The average age for the rural workforce currently is 53 years compared to the average age of members of the Australian workforce of 40 years (Fig. 3.13). Within the farming population, 23% are over the age of 65 and 13% are less than 35 years. More than half of all farmers work more than 50 hours per week compared to 16% of all employed persons. With an ageing workforce the ability to adopt newer technologies slows and the opportunity to pass knowledge and practices on to a younger generation diminishes (ABS *Labour Force, Australia*, cat. no. 6291.0).

Australian agriculture has important links with other sectors of the economy and, therefore, contributes to these flow-on industries. The farm-dependent economy – the agricultural sector plus the farm-output sector and farm-input sector – employs 1.6 million people, or 17.2% of the labour force (Department of Agriculture, Fisheries and Forestry, *Australian Food Statistics 2009–10*; Australian Farm Institute, *Australia's Farm-Dependent Economy: Analysis of the Role of Agriculture in the Australian Economy*; modelling undertaken by Econtech).

Table 3.6 People employed full-time in agriculture, forestry and fishing (2014)

State	Number of full-time labourers
NSW	64 300
QLD	40 600
VIC	54 500
WA	27 500
SA	31 200
TAS	9 000
NT	2 000
ACT	400

These numbers are rounded by the ABS, so they do not directly match the total figures in the text.
Source: ABS (2014), *Labour Force, Australia*, cat. no. 6291.0.55.003

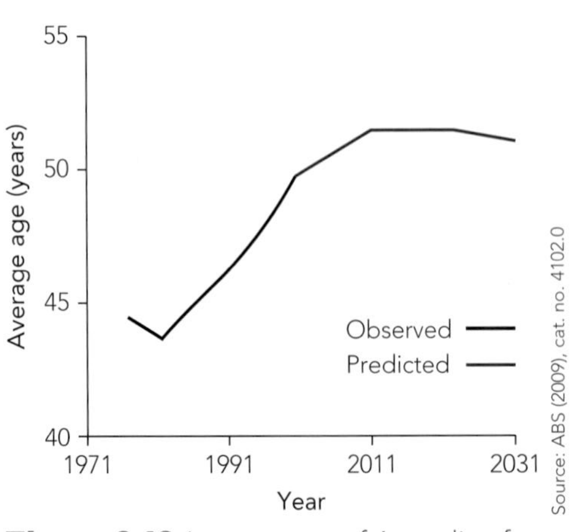

Figure 3.13 Average age of Australian farmers

In 1998, 394 000 people were employed in agricultural production systems; as of May 2011, 236 200 people were directly employed on-farm, full-time, in the Australian farm sector. Table 3.6 indicates distribution by state. Reasons for this decline include the increasing cost of labour, the impact of technological innovations that have improved farming efficiency and the desire to take up work opportunities in cities resulting in fewer family members continuing with a farming career.

There has also been a steady increase in the number of people employed in the industries servicing agriculture.

ISBN 9780170265560

This is because farming relies more and more on improved technology and services to increase the efficiency of production. People are employed in services such as those that supply chemicals, fertiliser, fuel, machinery, marketing services, insurance, banking and **finance** to farmers. They are also found in government agencies or departments that carry out research and conduct advisory services for the rural industries (for example CSIRO, Department of Primary Industries).

There has been a steady drift of people from rural areas to the large cities. Now there are (very slightly) more people moving from cities to rural areas. People have not gone inland but to country areas along the coast. They have gone there for the lifestyle, which is less stressful with fewer pressures, and to be near the beach. There has also been an increase in the number of farms that are not the owners' only source of income. The owners have another job off the farm. Hobby farms are run as a means of recreation. This does not mean that these farms are not productive; many are very successful (Fig. 3.14).

18 What are the implications of having an ageing farm workforce?

19 Why has the number of people employed on farms dropped since the 1950s?

20 Why have farming families looked more and more to themselves alone in running their farms?

21 What has caused the slight increase in the number of people moving out of cities and into country areas, particularly along the coast?

iStockphoto.com/Dcwcreations

Figure 3.14 This successful hobby farm runs a small herd of Hereford cattle

Agriculture as a consumer

Agriculture is itself a consumer of resources, goods and services. These include land, labour, fertiliser, water, **capital**, electricity, chemicals and machinery. The people who live and work on the farms are also consumers of the products produced by other sectors of the economy, such as cars, electrical goods, clothes, entertainment, holidays and anything else that they feel they need to bring them satisfaction in life.

22 Explain how the agricultural sector of the economy is a consumer.

Agricultural trade

Trade is an important activity for any country, providing the following benefits.

- Trade gives us the chance to choose from the most competitively priced goods and services from around the world.
- Trade allows Australia to specialise in the production of goods and services in which we have comparative advantage, thereby maximising our economic growth.
- Australia's exports are equivalent to more than 21% of our GDP, building the nation's wealth and prosperity.
- Foreign investment plays an important role in our economic development and provides capital to fund business expansion.

Figure 3.15 on page 50 indicates the percentage each industry sector contributed to Australian exports for 2012 on a balance of payments basis.

ISBN 9780170265560

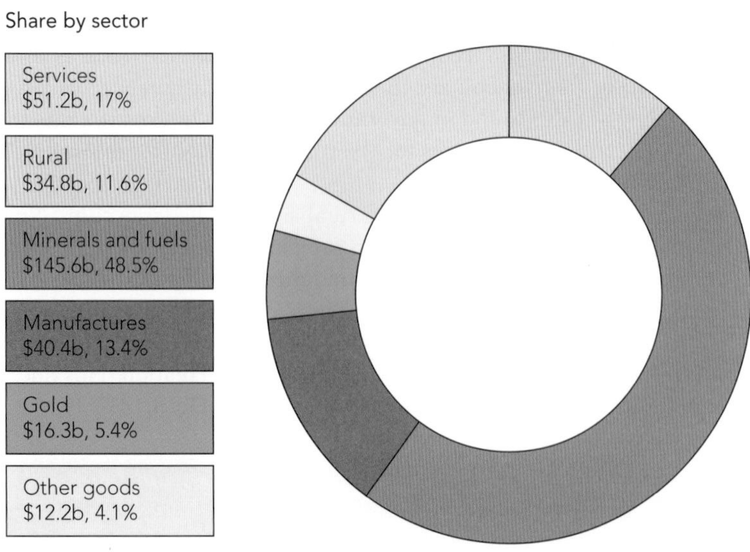

Figure 3.15 Exports of goods and services from Australia, 2012*

*Balance of payments basis. Based on ABS cat. no. 5368.0.

connect

Australia's trade agreements – DFAT

The Australian Government continuously tries to reduce distortions in global agricultural trade and provide better market access for Australian exporters. It achieves this mainly by lowering or eliminating tariffs through the use of multilateral, regional and **bilateral trade** agreements. Refer to Australia's trade agreements on the DFAT website.

Average global tariffs for agricultural goods are more than three times higher than for non-agricultural goods, some agricultural tariffs are as high as 800%, with some countries continuing to provide approximately 50% of farm income through government subsidies. Australia has reduced its own tariff levels and other protection on agricultural and food products since the early 1970s.This has resulted in Australia being one of the world's most efficient agricultural producers.

Australia has concluded seven bilateral and regional trade agreements with:

- Malaysia
- Association of Southeast Asian Nations (ASEAN)–Australia–New Zealand
- Chile
- United States
- Thailand
- Singapore
- New Zealand (the Australia–New Zealand Closer Economic Relations Trade Agreement [known as ANZCERTA or CERTA])

Australian free trade agreements under negotiation, or not yet in force, are:

- Australia–China
- Australia–Gulf Cooperation Council (GCC), with Saudi Arabia, Qatar, Bahrain, Oman, Kuwait and the United Arab Emirates
- Australia–India Comprehensive Economic Cooperation Agreement
- Indonesia–Australia Comprehensive Economic Partnership Agreement
- Pacific Agreement on Closer Economic Relations (PACER) Plus, with Cook Islands, Micronesia, Kiribati, Nauru, New Zealand, Niue, Palau, Papua New Guinea, Republic of Marshall Islands, Samoa, Solomon Islands, Tonga, Tuvalu and Vanuatu
- Regional Comprehensive Economic Partnership, with ASEAN, China, India, Japan, Republic of Korea and New Zealand
- Trade in Services Agreement
- Trans-Pacific Partnership Agreement, with Brunei, Canada, Chile, Japan, Malaysia, Mexico, New Zealand, Peru, Singapore, the United States and Vietnam
- Korea–Australia Free Trade Agreement
- Japan–Australia Economic Partnership Agreement.

 ISBN 9780170265560

Quarantine and food safety

In world trade operations one common problem is ensuring consumers are being supplied with food that is safe to eat. 'Safe' is defined by the standards the importing country considers appropriate and these must be balanced to ensure that that strict health and safety regulations are not being used as an excuse for protecting domestic producers.

All World Trade Organization members (including Australia) set their own level of sanitary and phytosanitary (plant health) protection in relation to quarantine and food safety. The Australian Phytosanitary certificate is used to certify that the Australian plants or plant products have been inspected according to appropriate procedures, and they are considered to be free from quarantine **pests**, practically free from other injurious pests, and conform with the current phytosanitary regulations of the importing country. It also requires quarantine and food safety measures to be based on science and not industry protection. This ensures that members cannot use unfair and unjustified quarantine or food safety restrictions to block trade.

The importance of global value chains in trade today

A 'value chain' is the full range of activities involved in designing, producing and delivering a product or service. The nature of trade is changing, with value chains becoming 'global' and crossing more borders than ever. The Organisation for Economic Co-operation and Development (OECD) estimates that more than half of the value of world exports is attributable to trade within global value chains (GVC).

Value chains have become global for a number of reasons, including:

- increasing information and communication technology capabilities
- changing production costs (for example Asia's manufacturing cost advantage)
- lower trade and transport costs and improved international logistics
- lowering barriers to trade (for example tariff and quota reduction).

The growing importance of GVC has further increased the incentive for countries to cut barriers to trade. Many industries now source inputs to production from every corner of the globe as part of GVC.

23 Assess the likely role of agriculture in the Australian economy in the future.

Resource adjustment

A significant number of farms are faced with reduced incomes, which in part are due to the cost–price squeeze. That is, the costs of farm inputs are rising faster than the prices received for farm outputs. Farmers find themselves in the situation of not being able to make a comfortable living. They are faced with what is known as the 'farm problem', where they find it difficult either to increase production and so generate more income or to move out of farming altogether into some other occupation. This problem is in fact one of the resources of labour and capital not adjusting quickly enough in the economy in response to changes in costs and prices.

The cost–price squeeze

The 'cost–price squeeze' has been part of the farming scene for at least the last 60 years. During this period the cost of farm inputs has risen rapidly, often due to inflation or energy costs as well as raw material costs. At the same time the price of farm outputs has only gradually risen or been reduced, because of the market factors of competition and monopolies, and market forces arising from the fact that only a certain amount of agricultural produce can be consumed at any one time, which tends to stabilise food prices. Consequently, if increased production occurs due to favourable seasons or improved technologies being adopted then supply will increase and market prices will fall. This means that a real decline in farmers' **terms of trade** has occurred. Farmers have stayed in business by becoming more efficient and increasing production. It is a complex problem affecting farmers (who are receiving lower prices) and, in turn, the suppliers of farm materials, processors, distributors, marketing firms and research establishments.

24 What is the 'cost–price squeeze'?

The farm problem

25 What is the 'farm problem'?

The farm problem, then, is a resource adjustment problem. There are difficulties with moving labour out of farming into other areas of employment and with obtaining finance (money) for development to increase production.

Difficulties with relocating labour

Farmers who decide to leave their farms and farm labourers who are displaced, along with their families, need to be employed elsewhere in the economy. This is most likely to be in large urban areas: the cities and the larger country towns.

Difficulties with increasing production and reducing costs

To increase production and/or reduce costs, the farmer has two strategies available:

1 increase the size of the farm by buying more land
2 adopt new technology that generates higher outputs and/or reduces costs.

With a bigger farm, more can be produced. Technology that increases outputs and/or reduces costs includes a broad range of things, such as labour-saving machinery (Fig. 3.16), fertilisers and high-yielding crop varieties. Modern haymaking machinery in the last 15 years has reduced the amount of labour required to make and store hay. One person can cut, rake, bale and stack the hay in the shed and only ever have to get off the tractor to change machines, whereas 15 years ago two or three people were required, especially in the loading and storing of bales. Sheep-handling machines have meant that less labour and time are required to carry out husbandry operations, such as drenching, foot paring and crutching. The introduction of **hybrid** maize varieties increased maize production by some 20%.

26 What options are open to farmers to increase their production?

To adopt either or both these strategies, an injection of capital (investment money) is needed. This is not easy to come by.

Alamy/Terry Mathews

Figure 3.16 This modern rotary dairy requires only three people to milk 600 cows twice a day

Reluctance to move out of farming

People find it difficult to leave the farm and find employment elsewhere. This can be for one or a number of reasons.

- *Skills.* People may lack skills for any other sort of employment. They have skills in animal husbandry and crop production, but these are not transferable to other occupations and industries.

 ISBN 9780170265560

- *Education.* People may have little or no educational qualifications. Only 25% of Australia's farm workforce has school-leaving (School Certificate or Higher School Certificate), trade or tertiary qualifications. This does not compare well with New Zealand (50%) or Europe (90%).
- *Finance.* People may not have the finance to re-establish themselves and their families in a new situation, particularly in the cities, where housing prices are higher than in the country.
- *Lifestyle.* People may not want to change their lifestyle. They perceive the alternatives to be unsuitable for them. The city is seen to be not a very pleasant place to live and bring up a family (Fig. 3.17).

27 Explain why people find it difficult to leave the farm and find employment elsewhere.

Dreamstime.com/Lucidwaters

Figure 3.17 Some aspects of city life are not attractive

Problems with finance

There are problems on two sides: financial institutions are reluctant to lend capital for investment in farm development, and farmers are reluctant to borrow capital for development.

Lending institutions

The lending institutions – banks and other financiers – are reluctant to lend capital for farm expansion and investment in technology because they consider that the farmer has insufficient **collateral security** to offer as backing for the loan. This is partly because of the uncertainty attached to agricultural production. This uncertainty arises from two sources.

- *Agricultural prices.* The prices received for agricultural products are subject to considerable variation. Australia depends heavily on world markets for selling its agricultural produce. Therefore the prices that Australian farmers receive are subject to the forces of supply and demand in the world market and to the policies of other countries that affect world prices (for example, tariffs).
- *Physical environment.* The physical environment includes such aspects as the weather, pests and disease, which can alter the amount of production and therefore the returns dramatically. Floods and droughts adversely affect all kinds of plant and animal production (Fig. 3.18, page 54). Severe cold conditions can kill unprotected livestock, especially the young and **sheep off shears**. Frost at the time a wheat crop is coming into head can reduce its yield dramatically. Hail can destroy a fruit crop in a matter of seconds. Outbreaks of disease or pests (for example black stem rust of wheat, heliothis insects in cotton) can severely reduce yield and returns.

Shutterstock.com/Dan Schreiber

Figure 3.18 Extensive flooding on this farm severely limits production in some years

Farmers

Farmers are reluctant to go into debt for several reasons.

- They may not be able to get reasonable terms for the loans that they want. Interest rates might be too high. The timing of repayments might not coincide with the income generated by the improvements. Fruit trees, for example, do not bear fruit for a number of years after planting. The repayment time might be too short.
- Farmers may also fear increases in interest rates. Many farmers ran into such difficulties in the 1980s, when interest rates spiralled upwards while prices for farm produce remained relatively low.
- A perceived cultural stigma in the rural community to being in debt might also prevent farmers from taking out loans.

Farm Finance program

The Australian Government Farm Finance program aims to build the financial viability of farm businesses, using the following strategies:

1 short-term assistance through concessional loans for productivity improvements or debt restructuring
2 funding of additional counsellors within the Rural Finance Counselling service
3 establishing a nationally consistent approach for debt mediation.

Such methods are designed to:

- foster the development of a more profitable farm sector that is able to operate competitively in a deregulated financial and market environment
- improve the competitiveness of the farm sector in a sustainable manner
- promote better financial, technical and management performance from the farm sector
- provide support to farmers who have prospects of sustainable long-term profitability with a view to improving the productivity of their farm units
- provide support in a way that ensures that the farmers who are supported become financially independent of that support within a reasonable period
- provide that support through:
  - a grants for the purposes of subsidies for interest payable on and the associated costs of loans, whether the loans are provided by a state or by another person
  - b grants for the purposes of farm training, planning, appraisal, support services and rural adjustment research.

28 Why do lending institutions find it difficult to lend capital to farmers to expand their production?

29 Why are farmers reluctant to go into debt to finance the expansion of their farm's production?

30 What is the aim of the Australian Government Farm Finance program and what strategies will be used?

 ISBN 9780170265560

Natural disasters

Farmers affected by floods, fires, cyclones and droughts can get assistance through the Exceptional Circumstance (EC) events program. This applies to rare events that are outside those that a farmer could be normally expected to manage, and that last for more than 12 months.

Assistance is in the form of:

- low-cost loans for carrying on until income can be made again, buying stock to replace those lost and rebuilding damaged fences, buildings and structures
- freight subsidies for transporting fodder and stock and for carting water.

Drought has been the major cause of farmers seeking this kind of assistance. A recent upgrade to the EC program relating to drought is providing farmers with the following support measures:

- income support
- drought concessional loans
- finance to install water related infrastructure
- additional funds to allow pest management in drought affected areas
- increased access to social and mental health services in communities affected by drought.

31 How do Natural Disaster Relief Arrangements assist farmers in times of natural disaster, such as severe drought?

Alamy/Tim Cuff

Figure 3.19 Farmers can gain assistance under the Natural Disaster Relief Arrangements to help them recover from floods, fires, cyclones and droughts.

ISBN 9780170265560

Chapter review

Things to do

1. Draw up a list of the members of your family, and compare each of their roles in life with the roles of the members of the Green family.
2. Make a list of all the possible sources of information that the Greens could use in making decisions on their farm.
3. Choose an agricultural commodity, and trace all the steps that it must go through from the time that it leaves the farm gate until it reaches the consumer.
4. Choose an agricultural commodity that has both a domestic and an export market, and every week for 6 months record the price received by the farmer producing the product and the price paid by the consumer. Each week, note events that could affect the farm price (for example weather, international events).
   a Plot the farm price and the retail price over the 6-month period.
   b Explain the changes and general trends in price that you observed.
   The prices of several agricultural commodities are published in weekly rural newspapers. In some cases it may be possible to obtain weekly wholesale prices as well.
5. Find the property section of the classified advertisements of a weekly rural newspaper. Look at the advertisements for a number of farms for sale, and record their size, location and price.
   a Does location affect the price of farms?
   b What other factors affect the value of farms?
6. Design and administer a survey in your local community to discover the general perception that people have of farmers, city dwellers, country life and city life.

Things to find out

1. What are the trends in farm size and ownership in Australia? Are the number of family farms increasing or decreasing?
2. China is one of Australia's most important trading partners.
   a Describe the products that are exported to China, and what proportion of these products are from agriculture.
   b Describe the products that are imported from China.
3. What is happening to the rural population? Is there still a drift to the cities?
4. What proportion of farmers utilise other forms of employment off the farm to supplement their incomes?
5. Obtain information from the ABS (Agriculture Services) on a selected region of your state. Draw up a profile of agricultural activity and productivity in the region.
6. Review what new technology is available to increase production of an agricultural industry in your region.
7. Investigate what organisations a farmer can go to in order to borrow capital for improving the farm.
8. Determine what educational opportunities are available to a young person at the end of secondary school in each of the following communities:
   a a small country town of up to 10 000 people
   b a large rural centre with a population of 30 000 or greater
   c a capital city.

Extended response questions

1. Compare and contrast the lifestyles of teenagers sitting for the Higher School Certificate in country and city situations.
2. Select an agricultural industry in Australia.
   a Briefly describe its history to the present.
   b What problems now face this industry?
   c What are its prospects in the short term (next 2 years) and the longer term (5–10 years)?

ISBN 9780170265560

UNIT 2
ANIMAL PRODUCTION

CHAPTER 4

Outcomes

Students will learn about the following topics.

1 Regionally significant animals.
2 Factors influencing animal production.
3 General principles of animal welfare.
4 The design and research of a simple animal trial.

Students will be able to demonstrate their learning by carrying out these actions.

1 Identify the two influences on animal production on a farm as animal efficiency and environmental factors.
2 Describe how the productivity of an animal depends on its rate of growth and development, and its rate of reproduction (and lactation if it is a mammal).
3 Outline the main limiting factors that influence animal production (nutrition, disease, genetics, climate and management).
4 List the main methods for measuring animal productivity; for example, kilograms of wool produced per sheep per year.
5 Understand the general principles of animal welfare.
6 Identify a range of regionally significant farm animals.
7 Design and conduct an experiment and analyse the results for an animal trial.

ISBN 9780170265560

FACTORS INFLUENCING ANIMAL PRODUCTION

Words to know

brucellosis a bacterial disease in cattle that causes abortion

coccidiosis a disease caused by an organism that can damage the lining of the small intestine and affect absorption of nutrients

concentrates feeds with high concentrations of major nutrients

development the changes in the proportions of various parts of an animal's body as the animal gets older

embryo the developing organism in early pregnancy; in plants, the small immature plant found in the seed

genotype the genetic make-up of an individual organism, consisting of its chromosomes and genes, half of which were inherited from each parent

growth the increase in size and weight of an animal as it gets older

hereditary disease disease passed on to the offspring by one of the genes

heredity the transfer of genetic traits or factors from parents to offspring

heterosis or **hybrid vigour** the increased vigour of crossbred progeny, or hybrids

immunity the ability of an organism to resist the attacks of pathogens by producing antibodies

metabolic disease a disease that occurs when one section of the body is not working normally

metazoal disease a disease caused by a metazoan, which can be seen with the naked eye (e.g. flatworm or roundworm)

microbial disease a disease that occurs when a pathogen enters an animal

monogastric an animal with only one stomach (without a rumen)

pathogen a disease-causing organism (can be a micro-organism or an invertebrate)

phenotype (in agriculture) the appearance of an animal or a plant including the production brought about by the interaction of the genotype and environment

postnatal the period after birth

prenatal the period before birth, which includes the embryonic and foetal stages

productivity a measure of the efficiency of a production system that is concerned with the rate at which conversions of inputs to outputs occur

roughages bulky feeds that are high in fibre and vary in protein, depending on the source

ruminant a cud-chewing animal with four stomachs, including a rumen that is inhabited by millions of microbes

vaccination an injection of a substance that produces immunity or resistance to a specific disease

connect

NSW Department of Primary Industries – Livestock

connect

Livestock for Western Australia

Introduction

The information contained in this chapter is introductory, with a more in-depth coverage in Chapters 5 to 13. In addition, information can be obtained from the NSW Department of Primary Industries and the Livestock for Western Australia websites.

Regionally significant animals will differ between states and countries. They might include poultry, goats, sheep, pigs, dairy cattle, beef cattle, alpacas or fish.

The quantity of animal products – meat, wool, mohair, eggs or milk – that can be produced on a farm is influenced by two main factors.

1 *The efficiency of the animal.* This depends on the animal's rate of **growth** and **development**, and its rate of reproduction (and lactation if it is a mammal).

   Productivity is affected by stocking rate; that is, the number of animals per area.

   Animal productivity varies between animals of the same breed and between different breeds, depending on genetic factors. It can be measured using the following yardsticks:
   - kilograms of wool produced per sheep per year
   - number of eggs laid per hen per year
   - weight gain in kilograms per animal per year
   - number of lambs born per ewe per year
   - litres of milk produced per cow per lactation.

2 *Limiting factors.* The main limiting factors are nutrition, disease, genetics, climate and management. Management is the action taken by farmers to control the stocking rates, growth and development, and reproduction in their animals. This action can be direct or indirect.

Certain aspects of these two areas of influence on animal production will be discussed in more depth in the following sections.

Efficiency of the animal

The productivity of an animal depends on its rate of growth and development, and its rate of reproduction and lactation.

iStockphoto.com/Rossmcf

Figure 4.1 Angus cow with a calf obtained by artificial insemination of the cow with semen from a Canadian bull

ISBN 9780170265560

Growth and development

The rate at which animals grow and reproduce, and the size, shape and composition of the carcases are all aspects of growth and development that affect the profitability of a farming business.

The following sequence of events covers the normal growth pattern in all mammals – conception, development of the **embryo**, development of the foetus, birth, puberty, reproduction and maturity, old age and finally death.

As a young animal gets older, two changes occur.

1 It increases in size and weight. This is called growth.
2 The proportions of various parts of its body change. This is called development.

The growth that occurs before birth is called **prenatal** growth. There are two main stages during prenatal development – the embryonic and the foetal. The size of an animal at birth is affected by the following factors:

- the number of young born
- the size of the dam (mother)
- the age of the dam
- the sex of the litter members
- the level of nutrition
- the breed (within species)
- the size of the sire.

The growth that occurs after birth is called **postnatal** growth. At birth the young animal appears to have a relatively large head, long legs, a small body and small hindquarters. The nervous tissue and the organs that are of greatest use at this time are relatively well developed. The growth of an animal after birth will be affected by:

- its size at birth
- the sex of the animal
- its **genotype** or breed
- its environment
- the level of nutrition.

If periods of weight loss are prolonged or if the amount of weight loss is great, permanent stunting of growth might occur.

Reproduction

Efficient reproduction is essential if an animal production system is to be profitable. High fertility and survival rates are the results of good animal management, and this is based on a sound knowledge of reproduction.

Fertile animals are those that reproduce efficiently. Fertile females are able to ovulate successfully, become fertilised and produce young. Most of the factors that influence fertility are under the control of the farmer, and therefore knowledge of these factors will help in reducing infertility.

Infertility is the failure of animals to reproduce. The fertility of farm animals (as discussed further in the following chapters) is affected by:

- genetics
- nutrition
- climate
- disease
- management.

A number of techniques are available for increasing the rate of reproduction and for increasing the use of superior genotypes or genes. The techniques, outlined in the following chapters, include the following:

- pregnancy diagnosis
- artificial insemination
- synchronisation of oestrus
- embryo transfer
- increasing the ovulation rate.

connect

Pregnancy testing of beef cattle

ISBN 9780170265560

1 What are two areas of influence on animal production on a farm?
2 Define the terms 'growth' and 'development'.
3 List five factors affecting fertility in farm animals.
4 What does 'lactation' mean?

connect

Factors that affect production in pig enterprises

connect

Selenium and vitamin E deficiencies in sheep

Lactation

'Lactation' means the secretion of milk. It is the final event in the reproduction of mammals. Milk supplies the young mammal with all the nutrients it needs for survival and growth.

Domesticated animals often produce milk that is surplus to the needs of their young. Using management practices, farmers have been able to obtain the surplus milk from cattle, sheep and goats in order to provide milk and milk products for human consumption.

Hormones control the growth and development of the udder tissue, the process of milk formation, and the letdown of milk from the udder.

Limiting factors

The main limiting factors are nutrition, disease, genetics, climate and management.

Nutrition

In Australia nutrition or feed is usually the most limiting factor in extensive animal industries (cattle, sheep and goats). An animal needs food for the following functions:

- as a source of energy so that it can maintain the working of its organs (lungs and heart), maintain its body temperature and be able to move about
- as material for building and maintaining body structures, such as bones, muscle, skin, wool, hair and teeth
- for the regulation of body processes, as nutrients are needed to produce hormones
- for the storage of chemical energy for growth, milk, wool or egg production.

Foods are substances that can be digested and absorbed by the body of an animal. The main nutrients, or kinds of food substances, are carbohydrates, proteins, fats, vitamins, minerals and water.

Although animals must be able to absorb all the above nutrients from their intestines, it is not always essential that the diet contains all these substances. **Ruminants** (sheep, cattle and goats), which carry large populations of bacteria, protozoans, and some fungi, are able to make amino acids in their rumen. **Monogastrics** (pigs and poultry) have no rumen, and so the essential amino acids must be added to the feed, or the animals will not grow properly. Animals that are exposed to sunlight do not have to be supplied with vitamin D in their diet. This nutrient is formed under the skin by the action of sunlight.

The requirements of animals for energy and protein can be considered as two-fold.

- *Maintenance requirements.* The maintenance requirement consists of the amount of food needed to keep the animal alive and healthy without producing any product (e.g. milk, eggs and wool) or making any growth (weight gain) and is affected by many factors, including:
  - the animal's weight
  - the animal's body composition (fat or thin)
  - the animal's physiological status (pregnant, lactating or dry)
  - whether the animal is diseased or healthy
  - the surrounding climate.
- *Production requirements.* The production requirement is largely independent of the size of the animal and is proportional to the quantity and quality of the product that it yields. A cow producing 22 litres of milk per day has a production requirement twice that of a cow producing only 11 litres per day and should be given extra feed.

Feed ingredients are classified as either concentrates or fibres/roughages. **Concentrates** have high concentrations of major nutrients and include cereal grains (maize, wheat, oats, sorghum and barley) and by-products, such as oilseeds, meatmeal and fishmeal. **Roughages** contain substantial amounts of fibre and include pastures, forages, hays, silages and fibrous by-products, such as cottonseed hulls and sunflower seed hulls.

The diets of pigs and poultry are normally made up of concentrates, whereas the diets of grazing animals (sheep, cattle, goats and horses) are normally based on roughage, with some concentrates when the nutrient content of the roughage is inadequate (e.g. during a drought or during winter).

5 List and briefly describe four reasons for animals needing food.

 ISBN 9780170265560

Disease

Animal diseases cause losses in animal production. Australia is fortunate not to have some of the more devastating diseases of some overseas countries (e.g. rabies). Some diseases cause major losses in income; for example, mastitis lowers production in dairy cattle and its treatment is expensive. All diseases are costly to some extent. The animal might be lost or there might be a loss in production. Money may have to be spent on veterinary chemicals in order to control or prevent the disease.

connect

Flystrike management tools

An animal disease is any kind of upset to the normal body functioning that has an adverse effect on the animal. Adverse effects include slower growth, weakening lower production, and even death.

There are four main types of diseases – hereditary, metabolic, microbial and metazoal.

1 **Hereditary diseases** are passed on to the offspring genetically.
2 **Metabolic diseases** occur when one section of the body is not working normally. Milk fever is caused by a decrease in the levels of calcium in the blood of an animal. It is a common condition in dairy cattle just before or after calving has occurred. Disturbances in mineral availability are usually due to area deficiencies, such as iodine deficiency in the Gippsland area of Victoria, diet imbalances, such as a lack of iron, copper or calcium minerals being fed to animals, or toxic effects from ingesting excess amounts of minerals or poisons.
3 **Microbial diseases** occur when a **pathogen**, or disease-causing organism, enters the animal. Some examples of microbial diseases are:
   - viral diseases – Newcastle disease of poultry and swine fever of pigs
   - bacterial diseases – mastitis, enterotoxaemia (pulpy kidney) and footrot in sheep and cattle
   - fungal diseases – lumpy jaw and ringworm in cattle
   - protozoal diseases – **coccidiosis** in poultry.
4 **Metazoal diseases** are caused by metazoans, which can be seen with the naked eye. They include:
   - flatworms – liverflukes and tapeworms
   - roundworms – threadworms, nodule worms and barber's pole worms
   - insects – botflies of horses, sheep blowflies and lice
   - ticks and other arthropods – sheep keds (blood-sucking parasites), sheep itchmites, cattle ticks and paralysis tick.

6 List four types of organisms that cause microbial diseases, and name one disease caused by each.

There are many methods used to control or prevent animal diseases. The following are the most important methods.

Eradication

Eradication involves totally ridding a farming system of a disease by testing all animals and slaughtering those that are infected with the disease. This method of disease control has been carried out on an Australia-wide basis in order to eliminate **brucellosis** in cattle, and has resulted in herds being brucellosis-free for several decades.

connect

Control of brucellosis

connect

How to drench a cow

Vaccination

Vaccination involves inoculation (injection) of the host with part or the whole of the pathogenic organism, with the result that the host develops **immunity** or resistance to further infection.

Chemical control

Chemical control takes the form of drenching for the control of internal parasites (Fig. 4.2), and dipping or the external application of pour-on chemicals for the control of external parasites. Insecticides used to kill lice, keds and to prevent fly strike in sheep are usually applied by plunge dipping or shower spraying. Chemicals can be added to the feed ration of pigs and poultry to control diseases of the gastrointestinal tract; for example, **coccidiosis** is controlled by adding a coccidiostat to the feed.

Archives New Zealand-Te Rua Mahara o te Kāwanatanga AANR 6325/W3302 Drawer 1

Figure 4.2 Drenching cattle for internal parasites

ISBN 9780170265560

Biological control

Biological control is being used to eradicate the screw-worm fly from the southern part of the United States. This parasite lays its eggs in open wounds and the larvae burrow into the host's body, causing severe loss of production and death. The female screw-worm flies only mate once. Therefore, large numbers of male flies are bred artificially, sterilised by irradiation and released by aircraft. European rabbit numbers have been controlled by the use of viruses, namely myxomatosis and the calicivirus, to reduce competition for food resources with sheep and cattle in Australia.

Genetic control

Genetic control relies on the fact that the ability of hosts to develop resistance to disease is often inherited. It is therefore possible to select and breed animals that have a high degree of resistance to parasites. Zebu cattle are more resistant to ticks than European cattle.

Management control

Management control involves the planning of the layout and relative position of facilities on the farm; for example, chickens should be kept separated from old stock. Sheep, cattle and goat parasites can be starved out by resting paddocks and using pasture rotation. Areas that harbour snails (involved in the lifecycle of the liver fluke) can be fenced off. These are usually swamp areas.

Genetics

If animals lack the genes permitting rapid growth and high reproduction rates, their production will be limited. Progress has been made in the poultry industry by the breeding of fast-growing meat strains and prolific egg-laying strains. Improvement has been made in the dairy and beef industries by importing semen from high-producing overseas bulls.

The ability of an animal to produce meat, wool, mohair, eggs or milk is governed by its **heredity**, or ancestry. The genetic make-up of an animal sets an upper limit to its possible production. However, if production in an animal industry is limited by an environmental factor (e.g. nutrition), improvement of this factor will have a much greater effect on increasing productivity than attempts to improve the quality of livestock through genetics.

The genetic make-up of an animal is described as its **genotype**. Each animal is a product of its genes and the environment. The environment includes all non-genetic factors, such as climate, nutrition, disease and stress. The interaction of the genotype and the environment results in the **phenotype**. The phenotype includes the production of the animal as well as its shape or appearance.

Where there are no serious limiting factors, considerable progress can be made in improving the quality of livestock by using proven methods of animal breeding. Robert Bakewell (1725–95) was the first to begin the planned improvement of livestock by breeding. He selected animals according to how much meat or wool they produced and mated 'the best with the best'. He based selection on productivity rather than on the appearance of the animal.

In selective breeding the farmer chooses the animals that are to be mated. There are three main types of selective breeding systems.

1 *Inbreeding*. This involves the mating of close relatives (brothers with sisters, mothers with sons and so on). It produces a uniform line of animals. It can also bring together undesirable genes, with such results as dwarfism in cattle and furlessness in rabbits.
2 *Line breeding*. This is a type of inbreeding based on a single common ancestor (a sire or dam) used over several generations of mating. A high degree of uniformity of type and production are obtained in a herd or flock.
3 *Crossbreeding*. This involves the mating of unrelated animals of different breeds of the same species. In this system new genes are brought into the flock or herd. The crossbred progeny, or **hybrids**, are usually more vigorous than either of the parents. This phenomenon is called **hybrid vigour**, or **heterosis**.

7 List and briefly describe the three main types of breeding systems.

ISBN 9780170265560

Climate

Climate is often a limiting factor in animal production. In northern Australia it is common to find high temperatures causing decreases in the productivity of cattle. This has been overcome to some extent by introducing tropical breeds from overseas, such as Brahman cattle from the USA (these animals having their origin in India), that are better suited to the tropical environment. These imported cattle are now used extensively in crossbreeding with European breeds.

Animals are greatly affected by climate. The following important climatic factors affect animal production: temperature, humidity, solar radiation and day length.

connect

Weather and climate

Temperature

Temperature extremes can have adverse effects on animal production. In Australia heat stress caused by high temperatures is a major problem. Animals suffering from mild heat stress will drink more water and eat less food. Their productivity will be reduced. In dairy cattle, heat stress causes a reduction in milk production, in poultry it causes thin-shelled eggs and in rams it results in poor quality sperm. Heat stress can be minimised by planting shade trees and using breeds that are appropriate to the environment. Rams to be joined during hot summers should be provided with adequate shelter and carry some cover of wool.

Cold temperatures are an important factor in the survival of newly hatched chickens because the chickens have trouble controlling their body temperatures. In large broiler sheds the temperature is controlled through the use of brooders and insulation (in the walls and roof). Severe cold conditions can cause heavy loss of newborn lambs.

Humidity

Humidity levels that are too high reduce the amount of heat lost from an animal by evaporation. The evaporation of sweat helps cool the animal and high humidity reduces the evaporation of sweat and so its cooling effect. Pigs and sheep have poor sweating ability, and poultry have no sweat glands and have to rely on the cooling of the moist air passages of the trachea, lungs and air sacs. Problems of high humidity, which occur in totally enclosed poultry sheds, can be reduced by installing fans or using computerised tunnel ventilated sheds.

8 Briefly describe how animals are affected by temperature and humidity.

Solar radiation

Solar radiation can affect animal growth and reproduction. Ultraviolet rays can cause sunburn, skin cancer and eye cancer. Sunburn occurs in pigs (Fig. 4.3), with the white-skinned breeds most likely to be affected. Eye cancer is almost unknown in cattle breeds such as the Santa Gertrudis, in which a ridge of protective bone shades the eyes.

Shutterstock.com/Robin Williams

Figure 4.3 Pigs suffer from sunburn

ISBN 9780170265560

Day length

Changes in day length affect reproduction. A decrease in daylight hours causes increased sexual activity in rams. The decrease in daylight hours has a greater effect on ewes, which begin to cycle as the days shorten, as they are seasonally oestrous animals. The reproduction of sheep is affected in the tropics, and the animals will breed during all seasons of the year. The shedding of the long winter coat in cattle is controlled by day length.

Management

Management includes the decisions and routine practices performed by farmers to control the stocking, growth, reproduction and survival rates of their flocks or herds. Farmers can change the diets of their animals, they can vaccinate them to prevent disease and they can introduce new genetic material by purchasing new animals and artificial insemination. The ways that farmers manage their animals include all the decisions they make about feeding, breeding, husbandry practices (drenching, mating, vaccinating, weaning and so on) and selling stock.

The behaviour of farm animals can sometimes have marked effects on their production. Farmers need to know something about animal behaviour if they are to manage their animals well and obtain the highest production from them.

Animals respond to the type of handling they receive. Farmers must be consistent in their approach and behaviour when handling animals. They must be quiet but firm so that they can win and maintain the confidence of the animals.

All domestic animals are creatures of habit and quickly become accustomed to repetitive procedures. Sheep, cattle and goats that are always worked in the same way through a set of yards quickly become used to this way, and are easier to handle than animals that are worked in different ways.

There are a number of routine husbandry procedures or operations to which sheep, cattle and goats are subjected, in order to maintain their productivity. These include drenching, jetting, mating, marking of offspring (ear tagging and castrating males), weaning, vaccinating and foot care. Apart from these special procedures, all animals should be given routine inspections and the following questions asked.

- Are all animals grazing normally?
- Are there any lame animals in the mob or herd?
- Are any animals scouring?
- Are there any animals separated from the main mob or herd?
- Are there any signs of discharge from noses or eyes?
- Are any animals rubbing unduly on posts or trees?
- Are any animals moving so as to indicate their sight is impaired?

After the general inspection a closer inspection should be made, with as little disturbance as possible. If any animal is found to have a health problem, steps should be taken to treat the problem.

9 List five questions that should be asked when carrying out a routine inspection of sheep, cattle and goats.

connect

Cattle handling

Animal welfare

Specific advice and assistance on management and disease control in animals can be obtained from qualified advisers, whose services are available through private and government agencies.

Whatever the type of animal production, owners, agents and managers are responsible for the health and wellbeing of the animals in their control.

Sound animal husbandry principles are an important ingredient to meet the welfare requirements of animals. Stockhandlers need to be flexible in their approach to caring for animals.

The most important factors affecting welfare in a herd or flock are the behaviour and the attitude of the manager. The manager needs to anticipate situations in which welfare may be at risk and to recognise early signs of ill-health or distress in animals, so that preventive action can be taken.

connect

Pig housing in Australia

Discuss the factors raised in this video.

ISBN 9780170265560

The basic requirements for the welfare of animals are:

- a level of nutrition adequate to sustain good health and vigour
- access to sufficient water of suitable quality to meet the animals' needs
- social contact with other animals; but with sufficient space to stand, to lie down, stretch their limbs and perform normal behaviour
- protection from predators
- protection from injury and disease, and treatment if they occur
- protection from adverse extremes of weather where possible
- provision of reasonable precautions against the effects of natural disasters; for example, fodder storage to protect against drought
- handling facilities that, with normal use, do not cause injury but do minimise stress to the animal.

connect

Animal welfare

Agricultural careers

There are many career opportunities within the field of agriculture. The journey from farm to household of food, fibre and natural building materials is achieved through the work of many people, from farming, trade, research, economic, engineering, education and other vocational backgrounds. For example, the cattle industry includes people in trades and professions such as agronomist, animal nutritionist, geneticist, climatologist, butcher, farm hand, farm manager, rural banker, veterinarian and people employed in the marketing, wholesale and retail trades. What is very obvious is that there are not enough people entering the world of agriculture in farming technical, economic, retail and marketing or research roles to meet the food demands of a growing world population in a technologically efficient and sustainable way. Throughout this book, opportunities will be taken to introduce the reader to various career pathways in agriculture.

Visit the Dairy Australia – Education and careers website to learn more about this developing concern.

connect

Dairy Australia – Education and careers

connect

Why do scientists work with farmers?

Read the information and then answer the questions in 'Things to find out'.

ISBN 9780170265560

Chapter review

Things to do

1. Weigh a lamb or chicken each week for 2 months. Construct a table showing the weight each week and the weight gained per week. Graph these results.
2. Visit a dairy farm. List the feeds eaten by the cows and classify the feeds as concentrates or roughages.
3. Visit a sheep stud on which the farmer uses line breeding. List the advantages and disadvantages of this breeding system.
4. Visit a broiler farm and describe how the temperature is controlled in the shed.
5. This is a simple trial, with appropriate methodology.

connect

Animals in schools: Animal welfare guidelines for teachers

Experimental design and research – A 10-day chicken trial

Note: A trial involving a variation in diet needs to conform to specific animal welfare guidelines as outlined in 'Animals in schools: Animal welfare guidelines for teachers'.

Aim: To carry out a 10-day broiler chicken (meat bird) nutrition trial comparing two feeds

A variation in diet can be achieved by using commercially prepared foods of varying formulas.

Method

1. Obtain a group of 20 chickens of the same breed and sex, and randomly divide them into two groups of 10 each. This is best done by tossing a coin for each chicken, heads become group A, tails group B. Continue until you have 10 chickens in one group.
2. Weigh each chicken.
3. Place group A in a pen and feed them on a diet containing normal protein feed for broilers.
4. Place group B in another pen and feed them a slightly higher protein feed (e.g. 2% more protein than the feed used for group A chickens) from a different feed company.
5. Record the weights of chickens in groups A and B. Every day, for 10 days, feed and water are topped up.
6. Record the amount of food added for each group daily.

Questions

1. Calculate the average weight gain for group A and group B daily.
2. Calculate the total amount of food consumed for each group in kilograms for the 10 days.
3. Graph the average daily weight gain for group A and group B against time (days).
4. Calculate the growth rate for each group of chickens from day 1 to day 10 in grams per day.

$$\text{Growth rate} = \frac{\text{Day 10 average broiler weight (g)} - \text{Day 1 average broiler weight (g)}}{\text{10 days}}$$

5. Calculate the food conversion ratio (FCR) for each group. This ratio shows the quantity of food consumed compared to live weight gained and is calculated as follows.

$$\text{FCR} = \frac{\text{Total feed consumed by 10 chickens over 10-day period (kg)}}{\text{Total increase in weight of 10 chickens over 10-day period (kg)}}$$

The lower the FCR the better, because it means that the chickens are converting food into body weight more efficiently.

6. Draw as many conclusions as you can from your data, graphs and calculations.
7. How could the design of the experiment be improved?

ISBN 9780170265560

Things to find out

1. Describe how pigs are artificially inseminated.
2. Describe the procedure of jetting and how it is carried out.
3. Outline the husbandry procedures or operations that are carried out at lamb marking.
4. Visit a beef cattle property. Find out how the farmer weighs the animals.
5. In Europe, foot-and-mouth disease is a serious problem. Find out the causes and symptoms of this disease, and how it has been kept out of Australia.

Extended response questions

1. The quantity of animal products – meat, wool, mohair, eggs or milk – that can be produced on a farm is influenced by the efficiency of the animals. This depends on their rates of growth and development, and their rates of reproduction (and lactation if referring to mammals). Discuss this statement.
2. 'Animal production is influenced by nutrition, disease, genetics and climate.'
   a. Describe how nutrition, disease, genetics and climate can limit animal production.
   b. Describe how nutrition, disease, genetics and climate can be managed by a farmer to optimise production.

CHAPTER 5

Outcomes

Students will learn about the following topics.

1. The basic nutritional requirements of animals.
2. The basic anatomy and physiology of monogastric and ruminant digestive systems.
3. The basic anatomy and physiology of digestive systems in a range of farm animals.
4. Beneficial relationships between microbes and animals in the process of digestion.
5. The fate of energy in animal nutrition.
6. Managing the nutritional requirements of monogastric and ruminant animals in terms of their digestive physiology.

Students will be able to demonstrate their learning by carrying out these actions.

1. Define the term 'nutrition'.
2. Explain why water, carbohydrates, proteins, fats, vitamins and minerals are needed by animals.
3. Define the term 'essential amino acid'.
4. Explain the role of fats in increasing the energy concentration of diets.
5. Outline the role of vitamins A, D, E and K in an animal's diet.
6. Outline the role of calcium, phosphorus, sodium, iron and copper in an animal's diet.
7. Describe the cause of and treatment for a metabolic disease, such as milk fever.
8. Describe the significant structural and functional differences between the alimentary tracts of monogastrics and ruminants.
9. Describe the relationship between the ruminant and rumen microbes.
10. Label a flow chart that shows the energy losses from the body during digestion and metabolism of food.
11. Explain the terms 'gross energy', 'digestible energy', 'metabolisable energy' and 'net energy'.
12. Describe the energy requirements for maintenance and production.
13. Explain the term 'feed conversion ratio' (FCR).
14. Design and explain a ration to meet animal nutritional requirements for a particular stage of production.
15. Explain the terms 'absorption', 'digestion', 'digestibility', 'intake', 'fermentation' and 'fibre'.

ISBN 9780170265560

NUTRITION AND DIGESTION

Words to know

absorption the passage of digested nutrients through the membrane of the alimentary canal into the bloodstream

amino acids organic compounds that combine to make proteins

anaerobic without oxygen

balanced ration a ration containing a balance of nutrients. Small quantities of mineral salts and vitamins are often added to give the correct balance of nutrients

diet a general description of the types of feeds eaten by an animal

digestibility the proportion of the food that is not excreted in faeces, and is assumed to be absorbed by the animal

digestible energy energy absorbed by an animal after the digestion of food (allowing for energy loss in faeces)

digestion the breakdown of large insoluble particles into simpler, soluble substances within the digestive (or alimentary) tract so that they can be absorbed

essential amino acid an amino acid that the animal is not able to make (or synthesise) in quantities sufficient for growth and development

fermentation the breakdown of starches and cellulose to sugars under anaerobic conditions; that is, in the absence of oxygen

fibre a feed material consisting mainly of cellulose (from plant cell walls); feeds high in fibre are called roughages

gross energy energy content of a food before digestion

intake the type and amount of food eaten or water drunk in a period of time

intravenous injection the injection of fluid into a vein

major minerals minerals present in an animal's body in large amounts (e.g. calcium, phosphorus, potassium)

maintenance energy amount of energy derived from food that is needed to keep an animal alive and healthy, but does not allow for growth or production

metabolisable energy energy available to an animal for use by the body after energy loss in urine and methane production in ruminants

net energy energy used by an animal for maintenance and production

nutrition the study of foods and the food needs of animals

production energy additional energy needed by an animal for forms of production, such as growth, pregnancy and lactation

ration a quantitative measure of the feeds being eaten

trace minerals minerals present in an animal's body in very small amounts (e.g. iron, zinc, copper)

Introduction

Nutrition is the study of foods and the food needs of animals. It is concerned with the chemical composition of food, and the processes of digestion and absorption. It also involves looking at the quantities of nutrients required for animals, and how inadequate feeding affects an animal's production. The energy, proteins, fibre, minerals, vitamins and water obtained from foods are used to maintain life function, allow energy for movement, production of products such as milk, wool and meat, and reproduction.

Food

Food can be defined as anything, either solid or liquid, that when swallowed, digested and absorbed will supply the body with energy and materials needed for growth and maintenance of good health. All animals require certain nutrients in their food to meet the needs of the body. These nutrients are classified into the following groups: water, carbohydrates, proteins, fats, vitamins and minerals.

Water

Water is vital to life. Animals constantly lose water in perspiration, so water intake must balance this loss. An animal will die more quickly from lack of water than from the lack of any other nutrient. Water has many important functions in the body. It helps in:

- maintenance of body fluids
- transport of nutrients and in the elimination of waste products
- regulation of body temperature
- digestion by hydrolysis.

Animals obtain water mainly through drinking, but also in solid food and from metabolic activity in the body.

Carbohydrates

Carbohydrates include sugars, starch and cellulose. They are composed of carbon, hydrogen and oxygen. An animal uses carbohydrates to provide it with energy. This energy is used to maintain the working of organs, regulate body temperature and enable the animal to move.

All animals can use sugars and starch as food, but only ruminants are able to break down cellulose to obtain energy. Ruminants are animals that have four stomachs, the first of which is called the rumen. Microbes inside the rumen break down the cellulose, which is found in all plant cell walls. Cattle and sheep are ruminants.

Grains such as maize, sorghum, oats, wheat and barley contain high levels of starch, and are a concentrated source of carbohydrates (Fig. 5.1).

Figure 5.1 Grains are a concentrated source of carbohydrates

ISBN 9780170265560

Proteins

Proteins contain carbon, hydrogen, oxygen, nitrogen and sometimes sulfur. Proteins are formed from smaller molecules called amino acids. There are more than twenty different amino acids present in food proteins. Of these, more than half are classified as essential amino acids, as they cannot be synthesised by non-ruminant animals. Complete protein foods contain all the essential amino acids and therefore have a high biological value. They include foods derived from animal proteins, such as meatmeal. Plant-based proteins are of low biological value and include legume seed and linseed meal.

A non-ruminant animal cannot synthesise (or make) essential amino acids within its body, and therefore they must be present in the animal's food. Protein needs vary from animal to animal; for example, pigs require lysine and threonine in their diet, chickens need glycine. However, a ruminant animal is able to synthesise essential amino acids in the rumen.

Proteins are used to make muscle, tissue, wool and hair. Proteins are the main constituent of enzymes, hormones and cell protoplasm in an animal's body.

They are especially important for growing animals, pregnant animals and for milk production.

Protein can be obtained from plant or animal sources. The protein content in young pasture grasses and clovers is 20–24%, but it decreases to 4% in old dry pasture, and to 1% in oaten straw. Protein concentrates, such as meatmeal (60%) and fishmeal (63%), can be fed as a supplement to animals whose diets are low in protein.

Protein foods are expensive to feed and deteriorate in storage. Excess protein is not stored in the animal's body but broken down in the liver to form urea and fatty acids.

Non-protein nitrogen foods are sources of nitrogen that ruminants can use, such as urea, usually supplied as biuret. It is usually mixed with molasses when fed to ruminants. Urea releases ammonia when ingested in the rumen and this is released into the blood system and incorporated into amino acids. Because of the high nitrogen levels produced such foods are too toxic for monogastrics.

Fats

Fats contain carbon, hydrogen and oxygen, and are used as a source of energy. They contain more energy than carbohydrates. One gram of fat or oil produces 2.25 times the energy value of 1 g of carbohydrate. Beef tallow (fat) is a cheap nutrient and may be included in rations to increase the energy level.

Plants and seeds contain fats. Oilseeds, such as linseed oil and peanut oil, have a high fat content, and are often fed to animals.

There are three essential fatty acids – linoleic, linolenic and arachidonic. Fatty acids regulate body metabolism. If fatty acids are deficient in an animal's diet, the animal may grow slowly or develop skin conditions.

Fats can be classified as saturated or unsaturated. The difference between saturated and unsaturated fats relates to the types of chemical bonds between the carbon atoms found in fats. Saturated fats have no double bonds in their structure. They are found in beef, beef and lamb fat, and in butter. Unsaturated fats have one or more double bonds between carbon atoms. They are found in most vegetable oils, fish oils, and are used to make polyunsaturated margarine.

Some fat is desirable in the diet of all animals. Too much fat in the diet causes an excess of fat on the animal. It also causes the fat of the animal to be soft and oily rather than firm.

Fats are needed in the diet of animals for:

- improving food palatability
- enabling the body to more efficiently absorb calcium and vitamin A from foods
- some fatty acids are essential to the normal functioning of an animal's body; for example, linolenic acid, linoleic acid.

Fats often retain smells or taints, which make foods greasy and rancid.

Vitamins

Vitamins can be defined as organic compounds that are required in small amounts for normal growth and maintenance of animal life. Vitamins perform a wide range of metabolic functions. They can be classified as fat soluble or water soluble. Vitamins A, D, E and K are fat soluble, whereas the B vitamins and vitamin C are water soluble.

Adequate amounts of certain vitamins should be present in the diet. Vitamin deficiencies are rare in grazing animals, but can occur in housed animals if the food ration is deficient.

Vitamin A

Vitamin A is made by animals from a substance called carotene. It can be stored in the liver in large amounts. It is essential for health of the eyes and for protection of the mucous membranes. If animals eat enough green plants they will obtain adequate carotene from which to make vitamin A (Fig. 5.2).

If animals are deficient in vitamin A, they develop night blindness and a poor resistance to infection. Intensively housed pigs and poultry may have these problems if their ration is not formulated correctly. Vitamin A deficiency can be corrected by feeding animals carrots, fish oil or green feed. Vitamin A injections can also be used.

Vitamin D

Vitamin D is made by the animal's body when sunlight acts on a substance in the skin called ergasterol. Vitamin D is needed by the body to allow proper formation of bone tissue.

Without vitamin D the bones of animals become deformed, and young animals can develop rickets. Only animals deprived of direct sunlight for months would be affected. In Australia, only housed animals (pigs and poultry) would be likely to be deficient. Vitamin D deficiency can be corrected by feeding the animal fish oil and exposing it to natural sunlight. Synthetic vitamin D is also available.

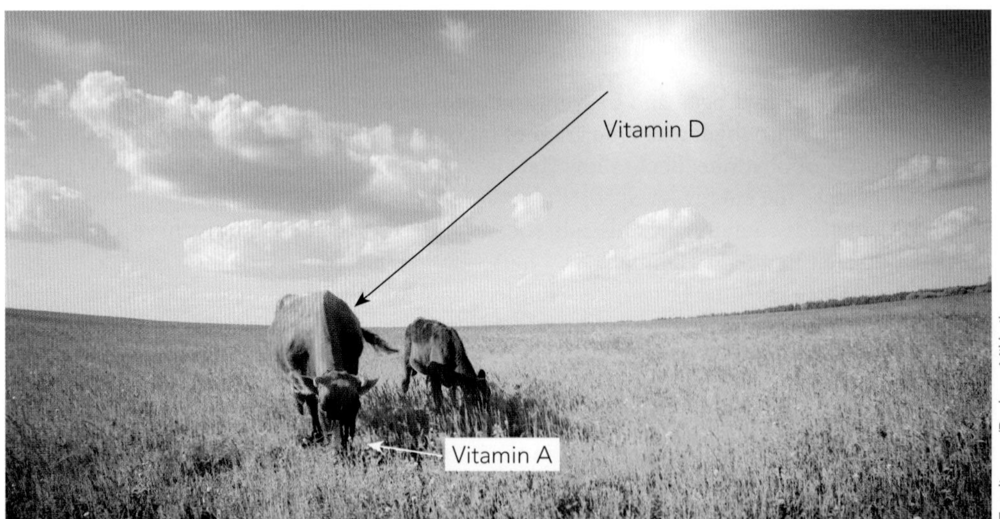

Figure 5.2 Vitamins A and D are supplied in abundance if animals have access to green feed and sunlight

Vitamin E

Vitamin E is an antioxidant, and it prevents damage occurring to cell membranes. It is not stored in the body.

Animals deficient in vitamin E have reduced fertility. In chickens a deficiency also results in muscle wasting and poor coordination, causing the chickens to stagger around. This deficiency can be corrected by feeding the chickens cereal grain and synthetic vitamin E.

Vitamin K

Vitamin K is involved in blood clotting and it occurs in green leafy plants, fishmeal and egg yolk. Vitamin K deficiencies are rare.

 ISBN 9780170265560

B vitamins

Ruminant animals are able to synthesise vitamins in the B complex by the action of micro-organisms in the rumen. There are several different vitamins within the B group.

Thiamine (vitamin B1) is involved in carbohydrate metabolism. Animals deficient in thiamine show loss of appetite and reduced brain and muscle function. Thiamine deficiency can be corrected by feeding cereal grains.

Riboflavin (vitamin B2) is involved in carbohydrate metabolism and hydrogen transport. Riboflavin-deficient animals have poor appetite and skin problems. Riboflavin deficiency in poultry causes lower egg hatchability, curled-toe paralysis and clubbed down (feathers). Riboflavin deficiency can be corrected by feeding yeast and green leafy crops.

Pyridoxine (vitamin B6) is involved in protein metabolism and antibody production. Pyridoxine deficiency symptoms in pigs and chickens include retarded growth rate and skin problems. Pyridoxine deficiency can be corrected by feeding cereal grains.

Cyanocobalamin (vitamin B12) is made up of complex molecules that contain cobalt. It is involved in protein metabolism and the metabolism of propionic acid. Deficiency of cyanocobalamin causes reduced growth rate. Ruminant animals are able to synthesise cyanocobalamin if adequate levels of cobalt are present in their bodies. Ruminant animals can be fed cobalt pellets to assist the production of cyanocobalamin.

Minerals

Minerals are chemical substances needed by all animals for satisfactory growth and development. Sixteen mineral elements are known to be essential. Minerals can be divided into two groups. The **major minerals**, such as calcium, phosphorus, potassium, sodium, chlorine, sulfur and magnesium, are present in the body in relatively large amounts, and are required as a relatively high proportion of the diet compared to other minerals. The **trace minerals** are present in very small amounts in the body. These include iron, zinc, copper, manganese, iodine, cobalt, molybdenum, selenium and chromium.

Calcium

Calcium is the most abundant mineral in an animal body. It occurs in the bones and teeth. It is involved in the transmission of nerve impulses and the contraction of muscle fibres.

If calcium is deficient in the diet of young animals then adequate bone formation does not occur, and the condition known as rickets may develop. The symptoms of rickets are misshapen bones, enlargement of the joints, lameness and stiffness. Milk fever is a condition that may affect high-producing dairy cows soon after calving. It occurs if the cow is unable to get sufficient calcium from the bones to produce large quantities of milk. Symptoms of milk fever are low blood-calcium levels, muscular spasms and, in extreme cases, paralysis followed by death. Normal levels of blood calcium can be restored by **intravenous injections** of calcium borogluconate.

Green leafy crops, especially legumes, are a good source of calcium. A shortage of calcium in the body may occur during a drought if animals are handfed on grain alone, as grain is low in calcium. This can be corrected by feeding a calcium supplement of 1% ground limestone. Animal by-products containing bone, such as fishmeal, meatmeal and bonemeal, are excellent sources of calcium.

Phosphorus

Phosphorus is closely associated with calcium in the animal body. Phosphorus plays an important part in carbohydrate metabolism. Like calcium, phosphorus is required for bone formation. Phosphorus deficiency can cause rickets, as can a calcium deficiency. Also, pica (or depraved appetite) has been observed in cattle that have had a deficiency of phosphorus in their diet. Affected animals will chew on wood, bones and other objects. Meat products containing bone, milk, cereal grains and fishmeal are good sources of phosphorus. The ratio of calcium to phosphorus is important as these two elements make up 75% of mineral matter in the body and 90% of this is in bone. Most animals need the ratio to be close to 1 : 1, but poultry require a ratio of 3 : 1 calcium to phosphorus.

ISBN 9780170265560

Sodium

Most of the sodium in an animal body is present in the soft tissues and body fluids. Sodium is concerned with the acid–base balance and osmotic regulation of the body fluids. A deficiency of sodium in the diet reduces the use of digested proteins and energy, and retards the growth of animals. In hens, egg production is also reduced. Meatmeal and foods of marine origin are rich sources of sodium. Rock-salt licks can be used or common salt can be added to the ration.

Iron

Iron is essential to the structure of haemoglobin (the pigment of red blood cells) and many enzymes. Iron deficiency results in anaemia. Anaemia occurs in piglets unless they have access to soil or receive an injection of iron shortly after birth.

iStockphoto.com/DaydreamsGirl

Figure 5.3 Piglets need a source of iron

1. List three functions of water in the animal body.
2. Name three grains that are sources of carbohydrates.
3. What are proteins used for in the animal body?
4. Name three essential fatty acids needed by animals to regulate body metabolism.
5. Where is vitamin A stored in the body?
6. Describe the effects of vitamin E deficiency on chickens.
7. How is a cow with milk fever treated?
8. Why is iron an important mineral in an animal's diet?

Copper

Copper is an important component of many enzymes and it occurs in the blood plasma. A copper deficiency results in loss of pigmentation of feathers, wool and hair. It can be corrected by the injection of copper sulfate or the use of licks containing copper.

Digestion

Digestion takes place in the mouth and digestive tract (or alimentary tract) of the animal. This process is both physical and chemical in nature. Feeds are broken into smaller particles by the physical action of chewing. The chemical breakdown of feeds involves the action of acids and enzymes.

In monogastric animals (such as pigs, poultry and horses) there is only one stomach, and the animals depend on the activity of enzymes produced by their own digestive glands. In ruminant animals (such as sheep, goats and cattle) there are four stomachs, including a large rumen. The rumen contains a large population of bacteria, protozoa and anaerobic fungi, which are responsible for much of the digestion of the feed.

Monogastric digestion

In pigs, poultry and horses the oesophagus leads to a single stomach (Fig. 5.4). The duodenum, into which the bile and pancreatic ducts empty, connects the stomach to the small intestine, where most of the absorption of nutrients occurs. The small intestine leads to the caecum and large intestine. The caecum varies from a double structure in a bird to a large single organ in a horse.

ISBN 9780170265560

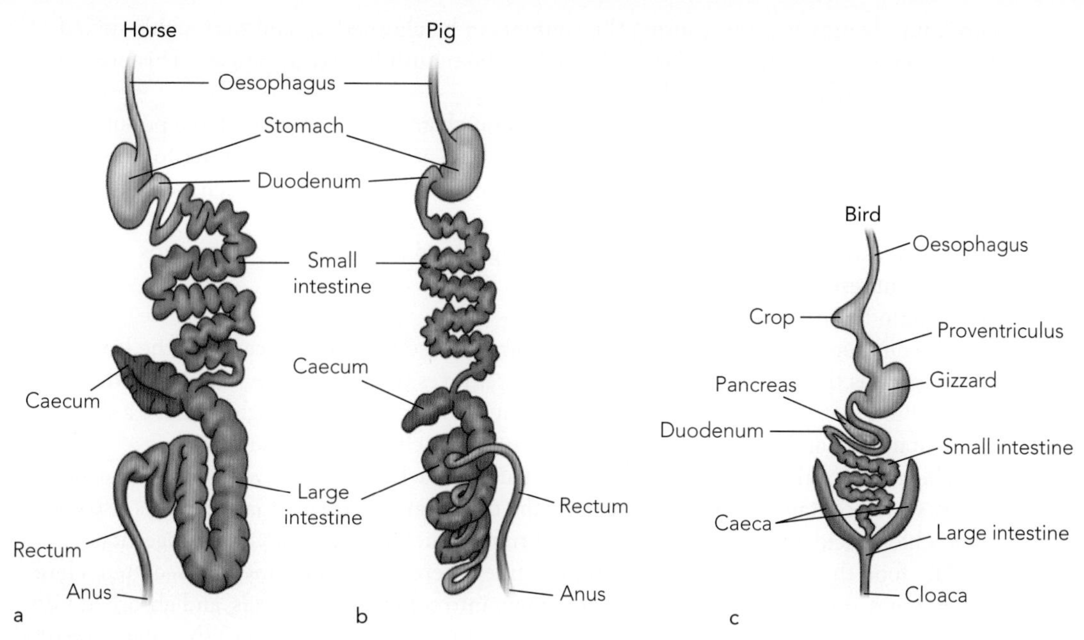

Figure 5.4 Digestive tracts in **a** horse, **b** pig and **c** bird

The digestive tract in birds has some unusual features. The bird has no teeth and must swallow its food whole. Food may be stored temporarily in the crop. The crop secretes mucus, which moistens and softens the food before it passes to the proventriculus.

The proventriculus is the only true stomach. It secretes gastric juice. Gastric juice consists of hydrochloric acid and the enzyme pepsin (for protein digestion). The food, mixed with gastric juices, then passes into the gizzard. The gizzard is unique to the bird. The walls of the gizzard are muscular and strong, and contract to break up the food mechanically. The diet of poultry should include pieces of grit. The grit passes through to the gizzard where it assists in the breaking up of food particles.

After the food leaves the gizzard it enters the duodenum, the first part of the small intestine. Pancreatic juice and bile enter at the lower end of the duodenum and mix with the partly digested food. The enzymes in the pancreatic juice act to break down the carbohydrates, fats and proteins in the food. The food is broken down into simple components, including monosaccharides, amino acids, glycerol and fatty acids, which are absorbed into the bloodstream as they pass through the small intestine.

Unabsorbed food particles pass into the large intestine, where some of the water is removed. The paired caeca, at the beginning of the large intestine, appear to be unimportant in the digestive process in the bird.

Waste products of the digestive process (faeces) and urine from the urinary tract are released via a common opening called the cloaca. The cloaca is the external opening where the digestive, urinary and reproductive tracts end.

9 List three examples of monogastric animals.

10 List three examples of ruminant animals.

11 What is the function of the crop in a bird's digestive tract?

12 Why are pieces of grit included in the diet of birds?

Ruminant digestion

Ruminant animals include sheep, goats and cattle. The ruminant animal feeds on bulky, fibrous foods that would be indigestible to the monogastric animal. The ruminant is able to break down the cellulose in the fibre with the help of the bacteria, protozoans and anaerobic fungi that live in the rumen, or first stomach. (A certain amount of fibre is also required in the diet of the monogastric animal to ensure movement of material through the digestive system.)

As the food is swallowed it is mixed with saliva. This lubricates the food and keeps the stomach contents liquid. The saliva is alkaline, which helps to prevent the stomach contents from becoming too acid.

There are four stomachs – the rumen/paunch, the reticulum/honeycomb, the omasum and the abomasum (Fig. 5.5, page 78).

The rumen and the reticulum are closely joined together and have the same function. They are often referred to as the paunch. The paunch has strong muscular walls, which are

ISBN 9780170265560

continuously moving, thus causing the contents to be churned up and thoroughly mixed. Food from the paunch is regularly returned to the mouth for extra chewing. This process is called rumination or chewing the cud.

The rumen-reticulum acts like a **fermentation** chamber. It contains a large population of bacteria, protozoans and anaerobic fungi. Bacteria are the most numerous, followed by protozoa then anaerobic fungi. These organisms secrete enzymes, which assist in three important processes:

1 the breakdown of carbohydrates (starch, sugars and cellulose) into fatty acids
2 the synthesis of proteins
3 vitamin B synthesis.

The omasum (or third stomach) is sometimes called the bible, because of the leaf-like partitions that line its walls. The omasum removes 60–70% of the water from the rumen fluid that enters it. In fact, the ruminant animal could be thought to have a divided fermentation chamber of three parts and a true stomach, the abomasum.

The abomasum or fourth stomach is the true stomach. Cells in its walls secrete gastric juices, which contain an enzyme that starts the digestion of proteins. Gastric juices also contain hydrochloric acid, which kill the majority of rumen microbes and starts their digestion.

The food then passes into the small intestine where several enzymes are secreted. Here the starches, proteins and fats are broken down into soluble compounds, and absorbed into the bloodstream through the villi. The undigested food residues pass into the large intestine. In the large intestine other bacteria digest the food and more water is removed from the food residues. These residues then pass into the rectum and out of the body through the anus as dung, or faeces.

13 What are the names of the three groups of organisms that live in the rumen?

14 What is rumination?

15 What is the function of the omasum?

16 What happens to food in the small intestine?

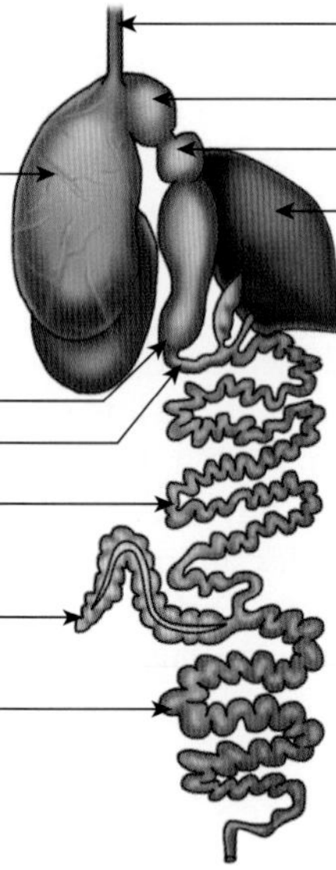

Figure 5.5 Digestive tract of a sheep

Carbohydrate digestion

The diet of the ruminant contains considerable quantities of cellulose, starch and sugars. Mature pastures contain mainly cellulose, whereas high concentrate feeds (such as cereal grains) contain mainly starches.

The breakdown of carbohydrates in the rumen may be divided into two stages (Fig 5.6). In the first stage, complex carbohydrates are digested (or broken down) into simple sugars. This process is carried out by enzymes secreted by microbes in the rumen. In the second stage, these simple sugars are taken up (or 'eaten') and metabolised by the microbes in the rumen. The main end products of the metabolism of carbohydrates by rumen microbes are acetic, propionic and butyric acids, and carbon dioxide and methane. The acids are absorbed into the blood through the rumen wall, and carried to the liver. They are used as the main energy source by the ruminant.

 ISBN 9780170265560

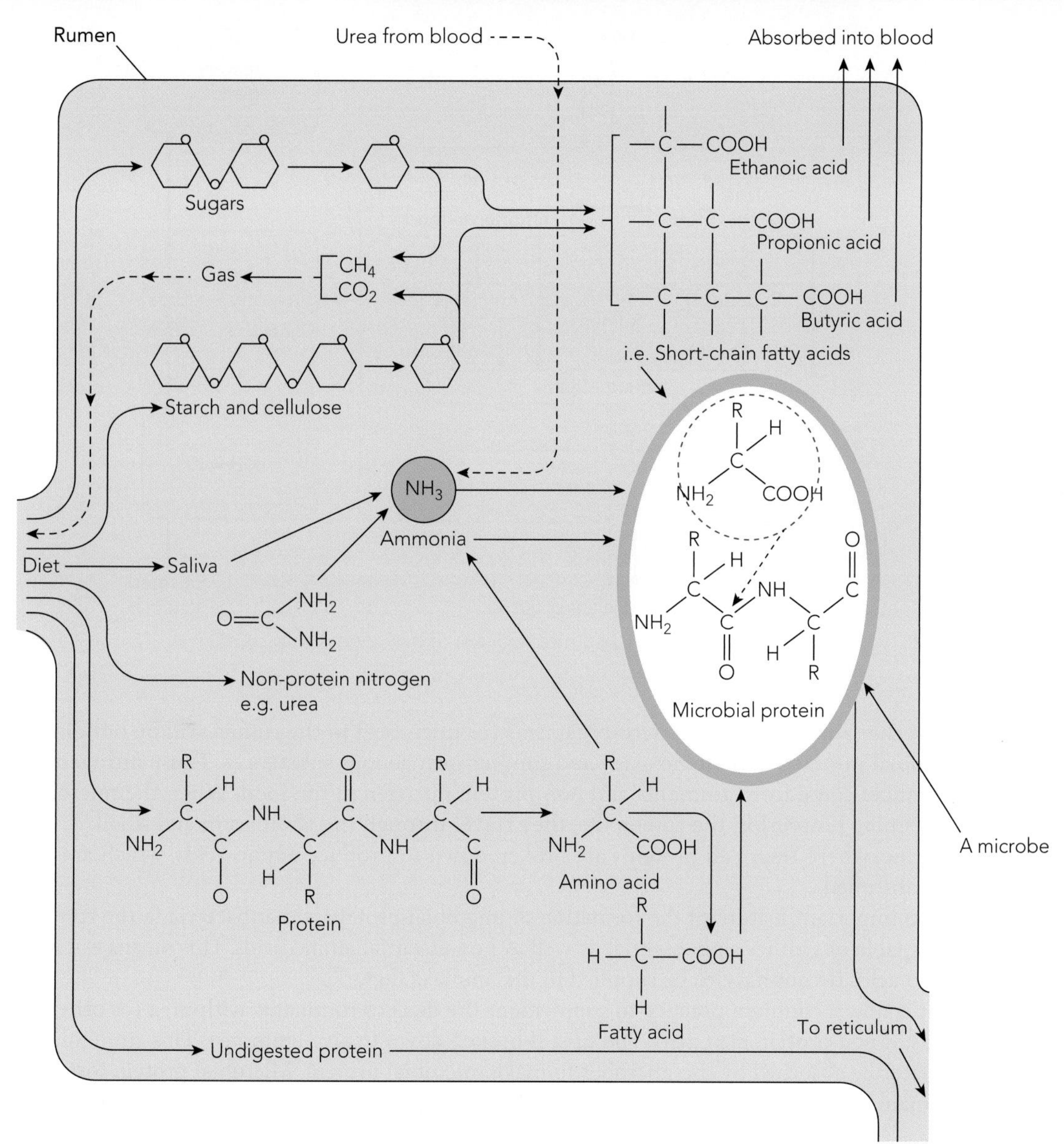

Figure 5.6 Some of the changes occurring in the rumen

If methane gas builds up in the rumen, it causes a condition known as bloat. This occurs when cattle graze pastures rich in clover or lucerne.

Protein digestion

Food proteins are broken down into peptides and amino acids by rumen micro-organisms (Fig. 5.7, page 80). Some of the amino acids are further broken down into organic acids, ammonia and carbon dioxide. Ammonia produced from the breakdown of amino acids may be absorbed from the rumen into the blood, carried to the liver and then converted into urea. Some urea is returned to the rumen in saliva and also directly through the rumen wall, but most of it is secreted in the urine.

17 List the three acids produced in the second stage of carbohydrate digestion.

18 What causes bloat?

19 What is produced when food proteins are broken down?

ISBN 9780170265560

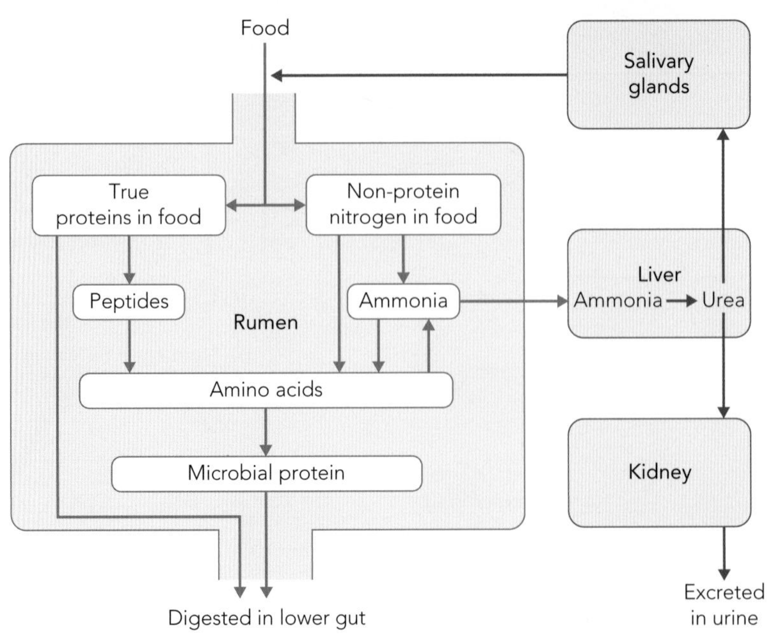

Figure 5.7 Digestion and metabolism of nitrogen compounds in the rumen

At the same time, other micro-organisms (or microbes) in the rumen will be building microbial protein from amino acids and simpler nitrogenous substances. These nitrogenous substances come from ammonia and non-protein nitrogen in the food. When the microbes (containing protein) in the rumen die, they travel through the abomasum and small intestine, where their cell proteins are broken down to produce amino acids, which are then absorbed.

An important feature of the formation of microbial protein is that bacteria in the rumen are capable of synthesising essential as well as non-essential amino acids. This means essential amino acids do not have to be supplied in the animal's diet.

It is now a common practice to supplement the diets of ruminants with urea (or other forms of non-protein nitrogen). The urea is broken down to give ammonia. This ammonia can later be absorbed by the microbes to make microbial protein. Microbial protein forms the main source of protein for the ruminant and has a biological value of 80 (the top of the scale is egg white with 100). This means ruminants can be fed low-quality protein foods and these will be upgraded in quality by the action of microbes. For a ruminant animal to obtain the full benefit from high-quality protein food sources, these foods need to be protected from microbial attack by a coating so the food reaches the small intestine to be digested. These foods are called 'by-pass' proteins.

Theoretically, this means the ruminant could exist without true protein in the diet. If too much urea is included in the diet of sheep or cattle, they may die of urea poisoning.

Undigested protein from the rumen, together with microbial protein, passes from the rumen, through the omasum and abomasum to the small intestine. The protein is broken down into amino acids, which are absorbed through the small intestine wall and taken to the liver. The liver uses some of the amino acids for protein synthesis and releases some back into the circulation system for protein synthesis in other tissues.

The functions of the rumen micro-organisms (bacteria, protozoa, anaerobic fungi), and the benefits gained by their host, include:

- rumen microbes synthesise vitamins B and K
- rumen microbes improve the quality of protein, ensuring that the host eventually receives all the essential amino acids
- rumen microbes can synthesise microbial protein from non-protein nitrogen sources, such as urea.

connect

Ruminant digestive system

connect

A cow's digestive system

20 What is microbial protein?

21 Explain how ruminants are able to use non-protein sources of nitrogen, such as urea.

ISBN 9780170265560

Nutritional requirements of animals

When food is eaten, the products of digestion are absorbed and metabolised in body tissues to provide energy for all life processes, which include:

- muscular work (walking, breathing and so on)
- keeping the body warm
- storing chemical energy in the body during growth
- supplying chemical energy contained in milk or eggs.

The amount of feed given to an animal over a 24-hour period (or a set time) is known as a **ration**. A **diet** refers to any mixture of foods an animal normally eats. A **balanced ration** contains foods in proportions that will properly nourish the animal.

There are two main types of ration:

- *maintenance ration*, which supplies enough nutrients to nourish a resting animal with no gain or loss of weight
- *production ration*, where sufficient nutrients, energy and proteins are provided for additional energy production and growth, resulting in outputs such as eggs, wool and milk.

22 List three uses of the energy obtained from digestion and absorption.

23 Name the gas produced by ruminants during digestion.

Feeding standards

Feeding standards are tables showing the amounts of food and nutrients that should be provided in the rations of different species for different purposes. These purposes may be growth, fattening or lactation (milk production).

Energy requirements

The energy content of a food is measured by combustion in a bomb calorimeter. This gives a measure of the total, or **gross energy**, of a food. Only a part of the total energy in food is available for use by the body, as losses of energy occur for a number of reasons (Fig. 5.8).

Normally, a certain proportion of food (e.g. some of the fibre) is not digestible and appears in the faeces. **Digestible energy** is the 'gross' energy minus the energy contained in faeces.

Of the digestible energy available to the animal, some energy is lost as urine. In ruminants some energy is also lost as methane gas. When these energy losses are subtracted from the digestible energy, the remainder is metabolisable energy. **Metabolisable energy** is the energy available to the animal for use by the body.

Some energy is also lost as heat. This heat loss is subtracted from the metabolisable energy to give the **net energy**. The net energy might be used for maintenance of body functions, or it might be stored as new tissue in the body, or converted into products such as wool, milk and eggs.

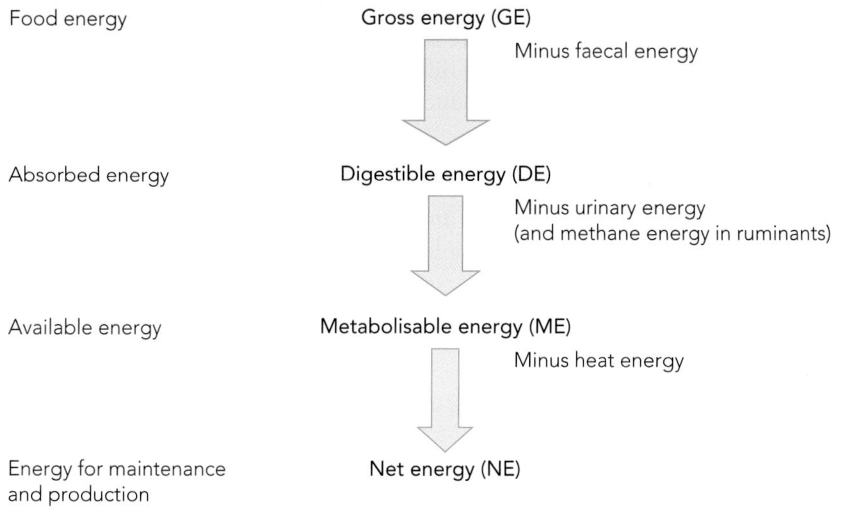

Figure 5.8 The energy losses from the body during digestion and metabolism of food energy

ISBN 9780170265560

24 What is the function of production energy?

The energy requirements of animals are usually considered as two-fold.

- The **maintenance energy** requirement consists of the amount of energy needed to keep the animal alive and healthy, but does not allow for growth or production. This requirement depends on the live weight of the animal.
- The **production energy** requirement is the additional energy needed for forms of production, such as growth, pregnancy and lactation. This requirement depends on the size of the animal, and the quantity and quality of the product being produced.

The system used in Australia for feeding standards for the energy requirements of ruminants is the metabolisable energy (ME) system. The daily requirements of an animal for energy are expressed as the number of megajoules of metabolisable energy needed per day for maintenance and production. Some examples of the metabolisable energy content of Australian feedstuffs are given in Table 5.1.

Table 5.1 Metabolisable energy content of some Australian feedstuffs

Feed	(MJ/kg of dry matter)
Wheat grain	13.0
Oat grain	12.5
Lucerne hay	8.5
Oaten hay	9.3
Peanut meal	11.0

Table 5.2 Daily metabolisable energy requirements of sheep and cattle

Animal and its production state	(MJ)
Sheep (40 kg) Maintenance	7.4
Pregnant ewe (40 kg) 2 weeks before lambing	8.2
Lactating ewe (40 kg) With single lamb	16.3
Beef steer (200 kg) Maintenance	27.0
Beef steer (450 kg) Maintenance	49.0

From Table 5.2 it can be seen that a pregnant ewe, of 40 kg body weight, 2 weeks before lambing needs 8.2 MJ of metabolisable energy per day. This can be satisfied by feeding her 1 kg of lucerne hay (see Table 5.1).

The energy requirements of an animal will depend on these factors:

1 the body weight of the animal
2 the amount the animal is producing
3 the environment of the animal (e.g. if the air temperature falls, the energy needs will increase)
4 the degree of stress the animal is experiencing (e.g. if the animal is being bullied by other animals, its energy needs will increase).

Energy requirements for poultry are also based on the metabolisable energy system. However, the ME values for poultry feed are slightly different from the ME values of the same feed used for ruminant nutrition. The difference stems from an allowance made for the fact that poultry (monogastrics) do not produce methane.

Protein, mineral and vitamin requirements

Animal requirements for protein, minerals and vitamins for both maintenance and production vary according to species, physiological status, environment and the production rate required.

The requirements for protein, minerals and vitamins are usually expressed as the percentage concentration in the diet. They can also be expressed as the weight required by the animal per day.

Some examples of the protein content of Australian feedstuffs are shown in Table 5.3.

ISBN 9780170265560

Table 5.3 Protein content of feeds

Food	Protein content (%)
Linseed meal	31
Prime lucerne hay	13
Poor lucerne hay	8
Wheat grain	10
Oaten hay	5
Grass pasture	5

Table 5.4 Protein requirements of animals

Animal	Protein requirement in feed (%)
Very young chicken	22
Laying hen	15
Hen not laying	7
Store (lean) sheep	6
Fattening sheep	10
Cow in full milk (lactating)	15
Dry cow in calf	10

25 Name three factors that affect the amount of nutrients required by an animal.

26 Explain why a lactating ewe with a single lamb has a higher daily ME requirement than a pregnant ewe (see Table 5.2).

27 Young chickens that are actively growing have a high protein requirement in their feed. Look at Table 5.3 and suggest a feedstuff with a high protein content that could be included in chicken starter crumble.

Some examples of protein requirements of animals are shown in Table 5.4.

From Table 5.4 it can be seen that a dry cow in calf, that is, a cow not lactating, but pregnant, needs approximately 10% protein in her feed. If the cow was kept on grass pasture, which contains 5% protein (as seen in Table 5.3), the cow would lose weight, because she would not be getting sufficient protein.

Protein levels in pastures vary according the age of the pasture. When grass is young, protein levels in pastures can be as high as 30% but by flowering time this has reduced to 15%. As the plants dry out after flowering, protein levels fall to 12% and they continue to fall as the plant dies. Should it rain when the plants are dead, the protein levels of pasture grasses rapidly decline to 3–4%. Dairy cattle require 15% to 18% protein in the food when producing, so farmers maintain high protein levels in pastures by slashing them regularly.

Feeding of animals

The diet of an animal is a general description of the types of feeds eaten by the animal, whereas the ration includes quantitative measures of the feeds being eaten. Thus, the diet of a dairy cow might contain hay, barley and coconut meal, while the ration of the cow would consist of 4 kg of hay, 1 kg of barley and 0.5 kg of coconut meal.

Feed ingredients are classified as either concentrates or roughage. The diets of pigs and poultry are normally made up of concentrates. The diets of cattle and sheep are based on roughages, mainly pasture, with some concentrates added when required.

A knowledge of the nutritional make-up of various feeds helps in maintaining the health and growth rates required for both production and performance. Consider the common horse feeds shown in Table 5.5 on page 84.

The following factors should be considered when formulating a ration:

- the age of the animal – young animals that are growing quickly will have greater nutritional requirements
- the climate and weather – cold conditions increase an animal's food requirements because more heat must be produced to keep the animal warm
- disease or stress – disease can reduce an animal's food intake, and stress, such as bullying from other animals, will increase an animal's energy needs
- the level of production – an animal that is producing meat or milk will need more feed than an animal that is maintaining its own body weight
- the cost and nutritional level of the food – cost per tonne of feed and the metabolisable energy content of the feed must be considered
- the palatability of the food – this relates to the variety of food stuffs available
- succulence or moisture content – high moisture content dilutes the energy value of the food; this is particularly important when grazing animals on young, actively growing pasture or pasture regrowth after rain
- freedom from poisons or taints – for example, sorghum will affect the nervous system, capeweed will induce hair balls and horehound taints products such as milk
- suitability of the food to the animal – cracked wheat is more readily digestible than whole wheat, which tends to pass through cattle.

Ration proportions can be calculated easily using a computer spreadsheet.

ISBN 9780170265560

Table 5.5 Common horse feeds

Food	Energy MJ/kg dry matter	Carbohydrate	Protein	Calcium (Ca)	Phosphorus (P)	Ca:P ratio	Uses
Ryegrass/clover pasture	4.2	40%	12%, supplies 9 of 10 essential amino acids	4.3 g/kg dry matter	1.0–4.0 g/kg dry matter	2:1	Roughage
Lucerne	9.2	45%	16%, has all essential amino acids	8.0 g/kg dry matter	2.0 g/kg dry matter	4:1	Protein source and roughage
Oats	12.5	60%	10%, has only 3 essential amino acids	0.8 g/kg dry matter	3.3 g/kg dry matter	1:4	Energy source
Barley	13.8	70%	10%, has only 2 essential amino acids	0.5 g/kg dry matter	2.8 g/kg dry matter	1:5	Energy source
Soybean meal	15.6	42%	46%, has all essential amino acids	30 g/kg dry matter	7.2 g/kg dry matter	1:2.4	Protein source but unpalatable in quantity
Bran	10.5	40%	12%, has 6 essential amino acids	1.8 g/kg dry matter	10.5 g/kg dry matter	1:6	Source of P but must be mixed with lucerne chaff for correct Ca:P ratio

28 From Table 5.5, which feeds supply horses with adequate amounts of energy?

29 From Table 5.5, which feeds supply all the required essential amino acids, and why is this important?

connect

Feeding and nutrition of livestock

connect

Australian feedlot industry

connect

Feeding and nutrition

connect

Nutrition

All rations should be palatable, or attractive, to the animal. Palatability is judged by the senses of taste, smell and vision. Molasses is often used as a feed ingredient to increase palatability and stimulate intake.

The ration should contain a balance of nutrients. Small quantities of mineral salts and vitamins are often added to give the correct balance of nutrients. The mixture is often pelleted to prevent the animal selecting particular foods within the mix, and to overcome the problem of dust. Pellets are also easy to store and transport.

The feed conversion ratio (FCR), or feed conversion efficiency (FCE), is used to calculate the animal's efficiency in gaining weight from food eaten:

$$\text{FCR or FCE} = \frac{\text{weight of food eaten by animal}}{\text{weight gained by animal}}$$

Animals that have a low FCR are considered efficient users of feed; for example, poultry can convert 2–3 kg of feed into 1 kg of live weight. Sheep and cattle are less efficient and have higher FCRs; they can consume greater than 8 kg of feed and convert it into 1 kg of body weight.

Rations for selected animals at particular stages of production

Typical feeds to be included in rations can be divided into:

- roughage feeds, which are high in fibre but low in **digestibility**, such as hay and straw
- succulents, which are high in moisture but low in energy, such as silage and green crops
- concentrates, which are high in particular levels of soluble carbohydrates, fats and proteins, such as cereal grain, fishmeal or oil cakes.

Table 5.6 illustrates the percentage composition of rations developed for particular levels of production in poultry.

ISBN 9780170265560

Table 5.6 The percentage composition of production rations for poultry

Composition of ration (%)	Broiler (meat bird)	Layer	Breeder
Energy foods; e.g. ground wheat, maize meal	70%	84%	76%
Protein; e.g. fishmeal, soybean meal	24%	13%	19%
Vitamins and minerals; e.g. salt, calcium, potassium	6%	3%	5%

Table 5.7 Stage of production related to protein and energy requirements of dairy cattle and broiler chickens

Stage of production	Feed requirements	Protein requirements	Energy requirements	Bodily requirements
Heifer 6 months	Kikuyu grass plus white clover pasture mix. Calf grower pellets as a supplement.	The pellets supply crude protein at 18%, urea 1%, crude fat 2% and crude fibre at 12%. All high levels.	Pasture supplies medium levels of protein and medium metabolisable energy levels.	Rapid muscle and bone growth and development.
Dairy cow mid-pregnancy and lactating	Kikuyu, white clover pasture mix. Rumen by-pass protein pellets. Dairy meal concentrate pellets.	The by-pass protein pellets supply very high levels of crude protein (31%), crude fat and crude fibre. The concentrate pellets supply high levels of protein (19%), urea 1.5%, crude fat and crude fibre.	The pasture supplies medium levels of protein and medium metabolisable energy levels.	Mature maintenance needs plus energy for production of milk and for growth of calf.
Broiler chicken 1 week old (meat bird)	Broiler starter crumbles.	Crude protein is 22%, which is very high. Crude fat is 3.8%, which is high.		Rapid bone and muscle growth and development.
Broiler chicken 4 weeks old	Broiler finisher pellets.	Crude protein at 19% and crude fat at 3.5%.		Rapid bone and muscle development. Limited fat deposits.

The amount of food needed by a dairy cow is largely determined by her energy requirements and, consequently, by the energy content of the feed she eats. Measurements of an animal's energy requirements and the energy content of various feeds are available and expressed in terms of megajoules (MJ). The metabolisable energy (ME) value has been calculated for many feeds and for various animals at different stages of growth and production. Table 5.8 indicates the energy content of some common feeds and Table 5.9 the daily energy needs of a mature dairy cow (page 86).

ISBN 9780170265560

30 List three examples of concentrates.

31 What do roughages contain?

32 List five factors to be considered when formulating a ration.

33 What substance can be added to a ration to make it more palatable?

34 Why are feed mixtures often pelleted?

Table 5.8 Nutritive value of feed materials

Food	ME (MJ/kg of dry matter)
Green pasture grass	10.6
Dry pasture grass	9.9
Forage oats	8.7
Legume hay	9.4

Table 5.9 The daily energy needs for maintenance of mature dairy cows

Body weight (kg)	ME required (MJ)
400	44.8
500	52.3
600	59.8
700	66.5
800	74.1

Table 5.10 Energy requirements expressed in MJ/L milk at SNF and various butter fat levels

Solids non-fat %	Fat content % 3.5	Fat content % 3.6	Fat content % 3.8	Fat content % 4.0	Fat content % 4.2
8.6	4.77	4.9	5.04	5.17	5.30
8.7	4.84	4.98	5.10	5.24	5.37
8.9	4.91	5.04	5.17	5.31	5.44

The following example demonstrates how this information can be applied to feeding dairy cattle.

From a herd of milkers, the data in Table 5.11 is obtained, showing energy requirements in MJ per litre of milk produced.

Table 5.11 Dairy cattle energy production requirements

Tag number	Body weight (kg)	Maintenance energy needs (MJ)	Pregnancy stage	Milk production (litres/day)	Solids non-fat %	Butter fat %
240	500	52.3	5	15	8.6	3.5
350	590	55.75	7	11	8.7	3.8
210	500	52.3	Not pregnant	14	8.7	4.0
420	550	56	4	16	8.5	3.3

35 In Table 5.7, why would a cow be fed protein pellets that bypass the rumen?

36 Calculate the total energy requirement for cow 240 in Table 5.11, using information from Table 5.9 and the Solids column of Table 5.10.

The total energy requirement for cow 210 is 52.3 MJ for maintenance (see Table 5.9) plus a production energy requirement of 14 litres × 5.24 = 73.36 MJ (see Table 5.10).

Total metabolisable energy needs for cow 210 are 125.66 MJ per day, based on production levels only to provide energy for basic bodily functions and milk production.

The cow is fed pellets during milking that supply 53 MJ toward this requirement, leaving 72.66 MJ of metabolisable energy to be obtained from pasture grazed during the day.

A cow eats approximately 3–4% of its body weight in pasture per day. Consequently, cow 210 will eat approximately 17–20 kilograms of pasture per day.

This is fresh grass with a reasonable moisture content of approximately 70%. One kilogram of fresh pasture contains 300 g (0.30 kg) of dry matter. The energy supplied from this is 0.30 × 10.6 MJ = 3.18 MJ (see Table 5.8) of energy/kg of pasture eaten.

Consequently, over the day, cow 210 will obtain 20 kg × 3.18 MJ = 63.6 MJ/day.

The cow is eating below her daily energy requirements, so must consume more pellets or be given a high-energy supplement during the milking period as the pasture can only contribute 63.6 MJ of the 72.66 MJ of metabolisable energy required. Failure to do this would stress the body of the cow due to lactation.

ISBN 9780170265560

Chapter review

Things to do

1. Examine and learn to recognise the following feed samples:
   a. cereal grains – maize, sorghum, oats, wheat, barley
   b. protein concentrates – fishmeal, meatmeal
   c. oilseeds – linseed
   d. roughage – oaten straw, hay, silage.
2. Describe a sample of the feed used for poultry on the school farm. Note whether the birds are fed pellets or crumbles. Suggest why the feed might change as the birds get older.
3. Observe rumination of cattle on the school farm.
4. Observe the grazing habits of sheep on the school farm.
5. Examine a dissected hen. Note:
   a. the main features of the digestive tract
   b. the consistency of contents in the different parts of the digestive tract.
6. Examine the digestive tracts of a ruminant, such as a sheep, and a monogastric animal, such as a pig.
7. To determine the amount of protein required in a ration, use this simple method, called the Pearson Square method.

   A farmer requires a ration of 16% protein. Feeds available are wheat (10%) protein and meatmeal (50%) protein.

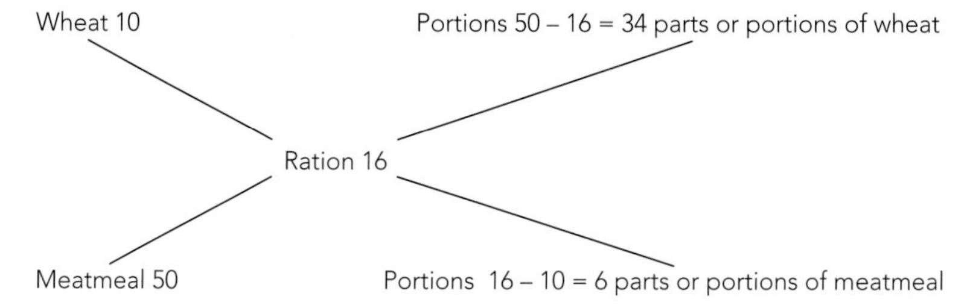

   Now determine the parts required to make a ration with 18% protein from barley (10%) protein and soybean meal (42%).
8. Look at the feed labels below and answer the questions.

Broiler starter crumbles	
For broiler chickens: 1 day old to 4 weeks old	
Crude protein min.	22.0%
Crude fat	3.80%
Urea max.	0
Crude fibre max.	3.5%
Salt	0.50%
Fluorine	0.02%

Compounded from wheat, sorghum, fishmeal, meatmeal and bonemeal, soybean meal, cottonseed meal, sunflower seed meal, tallow, limestone, salt, oxides of zinc and manganese, sulphates of iron, copper, potassium iodide, sodium selenite, vitamins A, D3, E & B12, menandionine, sodium bisulphite, riboflavin, calcium pantothenate, niacin, pyridoxine, folic acid, biotin, choline chloride, ethoxyquin, methionine and either Avotec 90 ppm or Elancoban 90 ppm or Coxistac 60 ppm as a coccidiostat.

ISBN 9780170265560

Broiler finisher feed	
Crude protein min.	19.0%
Crude fat	3.50%
Urea max.	0
Crude fibre max.	3.50%
Salt	0.50%
Fluorine	0.02%

Compounded from wheat, sorghum, fishmeal, meatmeal and bonemeal, soybean meal, cottonseed meal, sunflower seed meal, salt, lime, vitamins A, D3, E, K & B12, niacin, calcium pantothenate, folic acid, pyridoxine, choline chloride, riboflavin, oxides of zinc and manganese, sulphates of iron and copper, potassium iodide, ethoxyquin, methionine and either Coxistac 60 ppm or Elancoban 90 ppm or Avotec 90 ppm as a coccidiostat.

Calf grower pellets	
Crude protein min.	18.0%
Crude fat min.	2.0%
Crude fibre max.	12.0%
Salt	0.4%
Urea max.	1.0%
Fluorine max.	0.035%

Ingredients include: wheat, sorghum, triticale, barley, oats, corn, millrun, rice pollard, field peas, faba beans, lupins, meatmeal, bonemeal, soybean meal, sunflower meal, canola meal, cotton seed meal, vegetable oils, tallow, molasses, limestone, rock phosphate, dicalcium phosphate, vitamins A, D & E, iodate cobalt sulphate, ferrous sulphate, manganous oxide, sodium, urea.

Dairy pellets	
Crude protein min.	19.0%
Crude fat	2.5%
Urea	1.5%
Crude fibre max.	10.0%
Salt	1.15%
Fluorine	0.02%

Ingredients include wheat, sorghum, triticale, barley, oats, corn, pollard, bran, hominy, lucerne, soybean meal, sunflower meal, safflower meal, cottonseed meal, canola meal, meatmeal and bonemeal, fishmeal, limestone, molasses, tallow, salt, bentonite, cobalt carbonate, vitamins A & D3, sodium sulphate, copper oxide, urea

a Compare the energy and protein content of calf grower pellets with that of dairy pellets.
b Relate each feed's energy and protein content to the needs of a ruminant animal.
c Compare the energy and protein content of broiler starter pellets with that of broiler finisher crumbles.
d Relate each feed's energy and protein content to the needs of monogastric animals.
e From any of the four feeds identify sources of concentrated energy and protein.

Things to find out

1 Obtain a label from a bag of poultry or pig feed. Find out the following information:
a the net weight of the bag
b the product name
c the name and address of the manufacturer
d the guaranteed analysis in relation to crude protein (minimum percentage) and crude fat (minimum percentage)
e the ingredients in the feed.

ISBN 9780170265560

2. Describe the symptoms of deficiency of the following nutrients in pigs or poultry:
   a. vitamin A
   b. riboflavin (vitamin B12)
   c. calcium
   d. iron.
3. Describe the cause of and treatment for a metabolic disease, such as milk fever.
4. By-pass protein feed supplements are being included in the ration of sheep, beef and dairy cattle.
   a. Describe how these products work.
   b. What is by-pass protein?
5. Explain why ruminant animals rarely suffer from vitamin B deficiency.
6. Design a ration for an important class of farm animal that you have studied.

Extended response questions

1. There are significant structural (physiological) differences between the alimentary tracts of monogastrics and ruminants.
   a. Describe the structural differences between the tracts.
   b. Describe the functional (physiological) differences between the tracts.
   c. Discuss how these differences affect dietary requirements and food-conversion rates.
2. Only a part of the total energy in food is available for use by the body.
   a. Draw a flow diagram to show the losses of energy from the body during digestion and metabolism of food.
   b. Explain the terms 'gross energy', 'digestible energy', 'metabolisable energy' and 'net energy'. What is the function of net energy?
3. Evaluate the relationship between the ruminant animal and the rumen microbes.

CHAPTER 6

Outcomes

Students will learn about the following topics.

1. The basic anatomy and physiology of reproductive systems in mammals and poultry.
2. The role of hormones in the regulation of animal reproductive processes and behaviour.
3. Factors limiting the fertility of farm animals.
4. Management techniques used by farmers to manipulate reproductive processes in farm animals.

Students will be able to demonstrate their learning by carrying out these actions.

1. Describe the anatomy of reproduction in mammals and poultry.
2. Describe the physiology of reproduction in mammals and poultry.
3. Label a diagram of the reproductive tract of a mammal and a bird.
4. Describe reproduction in birds.
5. Label a diagram of the reproductive tract of a bird.
6. Describe the role of hormones in reproduction.
7. Explain the interaction of hormones and an animal's oestrous cycle.
8. List the major factors that affect fertility in animals.
9. Explain how a farmer can manipulate the following factors to increase reproductive efficiency: genetics, nutrition, climate, disease and management.
10. Describe the techniques used by farmers to manipulate reproductive performance in animals.
11. Evaluate the various techniques used to manipulate reproductive performance in animals.

ISBN 9780170265560

REPRODUCTION

Words to know

anatomy the study by dissection of the structure of the body of an organism

dissection the act of cutting an organism into parts to show its structure

dystochia a difficult birth

embryonic mortality the death of an embryo

fertilisation the union of male and female sex cells; in plants, the union of the male and female sex cells in the ovule

gestation period the period of pregnancy

hormone a chemical substance secreted by the ductless, or endocrine, glands directly into the bloodstream to control body actions or processes

mammal the class of animal that nourishes its young with milk from the mammary glands

maturity the state of being fully developed

neonatal mortality the death of a young animal soon after it has been born

oestrogen a hormone produced by the ovary and responsible for the development of female sexual characteristics; also responsible for the signs of heat

parturition the act of giving birth

physiology the way in which organisms, or parts of organisms, function

polyoestrous in some non-pregnant animals oestrus recurs again and again throughout the year (e.g. pigs, cattle)

puberty the age at which a young animal's reproductive organs are functional (sexual maturity)

reproduction the formation of new individuals by the fusion of two sex cells to form a zygote

seasonally polyoestrous describes the situation where animals usually breed during particular months of the year (e.g. goats, sheep, horses)

ISBN 9780170265560

Introduction

The rate at which animals reproduce will affect the profitability of a farming system. To control the breeding of animals, the farmer must understand the anatomy and physiology of reproduction, and the factors that affect fertility in farm animals.

A farmer can increase reproductive efficiency by manipulating the following factors: genetics, nutrition, environment, pests and disease and management.

There are a number of ways of improving reproductive performance in animals. These include pregnancy diagnosis, artificial insemination, synchronisation of oestrus, embryo transfer and techniques for increasing the ovulation rate.

Anatomy and mammalian reproduction

The female

The female reproductive tract of a mammal consists of the ovaries, fallopian tubes (or oviducts), uterus, cervix, vagina and vulva (Fig. 6.1).

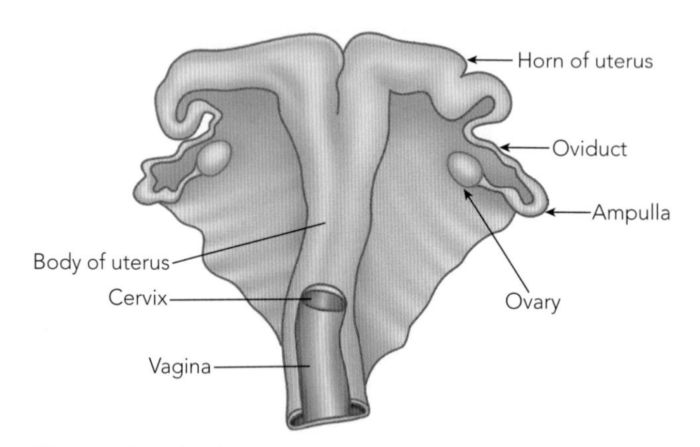

Figure 6.1 The female reproductive organs (ewe)

There are two ovaries, located one on each side of the abdominal cavity. Each ovary has two main functions – the production and release of ova, or eggs, and the secretion of hormones (oestrogen, progesterone and relaxin) required for conception and pregnancy.

The fallopian tubes, or oviducts, are the fine tubes that carry the ova from the ovary to the uterus, where development of the embryo takes place. In farm animals the anterior end of the oviduct is expanded into a funnel, or fimbria, which guides eggs shed from the adjacent ovary.

connect

Anatomy of bovine female reproductive tract

The uterus, or womb, consists of a body and two horns. Eggs enter the uterus 3 or 4 days after the time of egg release, or ovulation. If fertilised, the embryo attaches to the wall of the uterus, where it stays for the duration of pregnancy.

The cervix is a muscular and fibrous tube connecting the body of the uterus to the vagina. During pregnancy it is sealed to protect the foetus from infection. At around the time of mating and ovulation the cervix opens to allow sperm to pass through its narrow passage way. The cervix also dilates at birthing time.

The vagina is the connection between the cervix and the vulva. The urethra, which carries urine from the bladder, opens into the floor of the posterior part of the vagina.

The vulva is the entrance to the reproductive system. There are two muscular folds, which are normally closely opposed, keeping the inner surface of the vagina clean. Swelling and/or colour change, especially in sows, indicate the animal is on heat or ready for mating.

1 What are the two main functions of the ovary?
2 Name the structure that prevents the uterus from becoming infected.

ISBN 9780170265560

The male

The male reproductive tract consists of the testes, epididymis, vas deferens (seminal ducts), accessory glands, urethra and penis (Fig. 6.2).

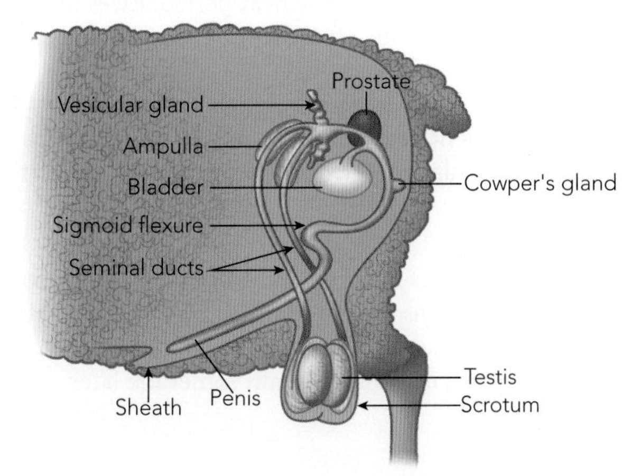

Figure 6.2 The male reproductive organs (ram)

The testes develop inside the abdomen and then descend into the scrotum. Each testis contains seminiferous, or sperm-producing, lobules. The testes have two functions – the production of sperm and the secretion of sex hormones. The scrotum is the skin-covered pouch that contains and supports the testes. Its main function is to keep the temperature of the testes at several degrees below body temperature.

The epididymis consists of three parts – a head, body and tail. The sperm produced by the testes are stored in the epididymis until mating. Water is also reabsorbed through the walls of this structure, thus concentrating the sperm.

The vas deferens (seminal ducts) are thin tubes connecting the epididymis to the penis. It is through the vas deferens that the sperm pass.

The accessory sex glands are situated behind the neck of the urinary bladder. They consist of the seminal vesicle, the prostate gland, Cowper's gland and the glands of the ampulla. At ejaculation, seminal fluid from these accessory glands is released into the urethra and mixed with sperm from the testes. The penis is the organ of copulation. It has two functions – depositing semen into the female reproductive tract and emptying the bladder during urination.

3 List the two main functions of the testes.
4 Where are the sperm stored?
5 Name the male accessory glands.

Physiology of mammalian reproduction

The female

Reproduction involves three cycles in the female. These are the lifecycle, the annual breeding cycle and the oestrous cycle.

The lifecycle follows the pattern of prenatal life, birth, infancy, **puberty**, the prime of life, senility and death. The period of reproductive activity is confined to the prime of life.

The annual breeding cycle is the yearly rhythm that animals go through during the prime of life. The breeding cycle is controlled by environmental influences. The main one is the seasonal rhythm of the increase or decrease in the hours of daylight. The decreasing daylight hours in autumn are a trigger for the start of the breeding season in sheep. Some animals, such as dogs and cats, only have one period of sexual activity per cycle; for example, female dogs come on heat once every 6 months. These animals are called monoestrous animals. Other animals such as pigs or rabbits are called **polyoestrous** and can breed the whole year round. **Seasonally polyoestrous** animals include sheep, cattle and goats and for them, breeding is confined to one season during the year, although there are many heat periods during this season.

ISBN 9780170265560

The oestrous cycle occurs in the breeding cycle of females; each oestrous cycle consists of a period of sexual activity followed by a period of sexual inactivity. If the female does not become pregnant, these cycles continue throughout the breeding season. Details of the oestrous cycle change with different species. For mares the length of the oestrous cycle is 21 days with egg release 1 day before the end of the heat period. Ewes have a 16–17 day cycle, sows 21 days with egg release 36–48 hours after the onset of heat and cattle 21 days with egg release 13–15 hours after the end of heat.

The female reproductive system produces ova. Further, it provides a suitable environment for **fertilisation** and the development of the embryo and foetus. The stages of reproduction are puberty, the oestrous cycle, ovulation, fertilisation, pregnancy and birth.

Puberty

The first step in this series of processes is the reaching of puberty. Puberty is the age at which the young animal's reproductive organs are functional. The age at which puberty occurs is affected by the weight of the animal. Although the female may become pregnant after reaching puberty, sexual **maturity** is not reached until sometime later.

The oestrous cycle

The female reproductive organs work in repetitive cycles called oestrous cycles, which are different for each species. The female will accept the male for only a limited period of each oestrous cycle. At this time the female is said to be in oestrus, on heat or in season. Table 6.1 shows reproductive characteristics of some animals.

Table 6.1 Reproductive characteristics of some farm animals

Animal	Onset of puberty (months)	Gestation period (days)	Length of oestrous cycle (days)	Length of heat or oestrus
Doe	4–8	150	20	36 hours
Ewe	6–12	147	17	30 hours
Sow	4–9	114	21	48 hours
Cow	6–18	280	21	18 hours
Mare	10–24	336	21	6 days

Ovulation

Ovulation is the shedding or release of an egg, or eggs, from the ovary. As ovulation approaches, one (or more) of the Graafian follicles (mature follicles) enlarges rapidly and inside it, the egg develops to maturity. During or shortly after oestrus the large follicle ruptures to release an egg, which enters the fimbria or funnel of the fallopian tube. The cow releases only one egg; the ewe and doe, one, two or three; and the sow, up to 20. For this reason, twins are scarce in cattle but common in sheep, and pigs usually produce litters with a large number of piglets.

Fertilisation

After mating, or joining, the sperm rapidly swim through the uterus to the fallopian tubes. Fertilisation occurs when a single sperm penetrates one egg. The external membrane of the egg changes and becomes impenetrable to other sperm. Fertilisation usually occurs in the upper one-third of the fallopian tube. The fertilised egg, or zygote, undergoes cell division and passes into the uterus. After further cell division it develops into an embryo. At first the embryo lies free within the uterus and can move about. The developing membranes then become attached to the wall of the uterus. This is called implantation.

Following implantation the placental membranes develop. The main purposes of these membranes are to carry nutrients and oxygen to the foetus and to take away waste products. The length of pregnancy, or **gestation period**, varies with the species, as shown in Table 6.1.

ISBN 9780170265560

Birth

Birth, or **parturition**, is under the control of hormones. Towards the end of pregnancy, changes in the hormone content of the blood cause the passage through the cervix and vagina to enlarge. Continuous contractions of the uterine wall then force the foetus through the cervix into the vagina and finally through the vulva. Parts of the placenta, or 'afterbirth', are then expelled.

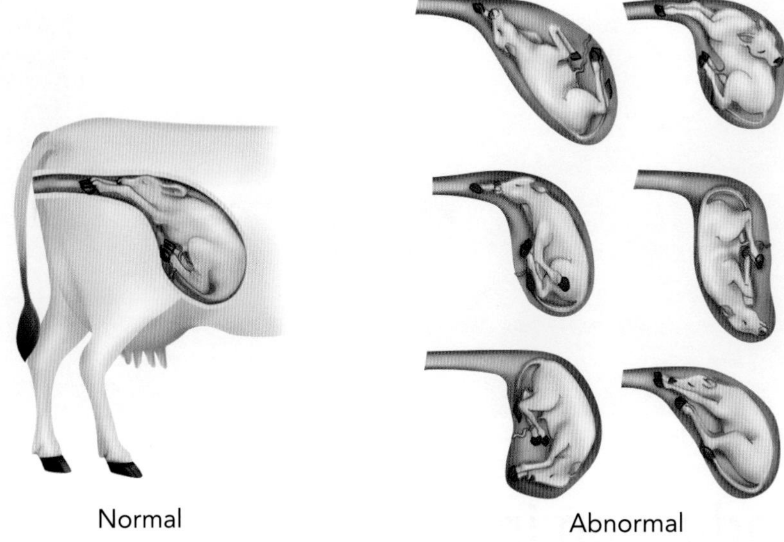

Figure 6.3 Normal and abnormal birth positions in cattle

6 Define the term 'puberty' in relation to female mammals.
7 Consider Table 6.1.
   a What is the length of the oestrous cycle in cattle?
   b What is the gestation period in sheep?
8 What is implantation?
9 What is the afterbirth?

The male

The male reproductive system produces semen. This fluid has two components – the sperm and seminal plasma. Sperm are produced in the testes during the reproductive life of the male. Seminal plasma is the secretions from the accessory glands, and it is added to the sperm at ejaculation.

10 Why should a 9-month-old bull not be mated regularly?
11 What is the function of the vesicular gland, or seminal vesicle?

Puberty

The age of puberty and sexual maturity varies between species. Bulls are capable of producing semen at 9 months of age, but the quantity and quality are often poor. They should not be mated on a regular basis until several months after puberty. A young ram is capable of producing fertile semen at 6 months of age, and a young boar at 7 months.

Spermatozoa production

Sperm are formed within the long coiled seminiferous tubules, which lie in segments of the testes (Fig. 6.4). The sperm are formed from groups of cells that line the tubules. Sperm made in the seminiferous tubules then enter the epididymis for storage.

Seminal plasma or fluid

Seminal plasma varies in volume and composition between species and contains several substances. Nutrients such as fructose are secreted from the vesicular glands, or seminal vesicles, and provide the sperm with energy. Secretions from the prostate gland excite the sperm to swim and secretions from Cowper's gland help lubricate the end of the penis.

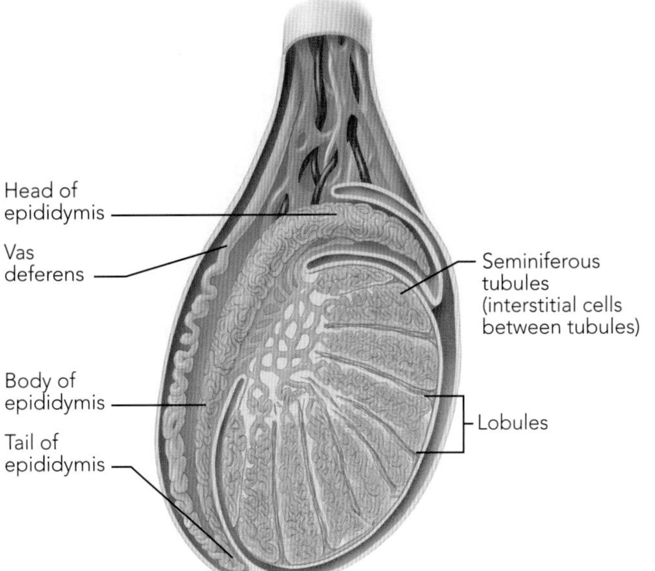

Figure 6.4 The structure of the testis

ISBN 9780170265560

Libido

12 Name four factors affecting libido.

Male sexual drive, or libido, is the desire and ability to mate with a female. Good semen production and strong libido are both essential if females are to be effectively mated. Libido is affected by age, body weight, temperature and recent sexual activity.

Shutterstock.com/Chrislofotos

Figure 6.5 A breeding bull

Reproduction in birds (avian reproduction)

The male

In the male bird the testes are paired and attached to the kidneys. They do not descend into a scrotum as in other farm animals. In proportion to body size, the testes of male birds are larger than other farm animals. Thin tubes take sperm from the testes to the cloaca, where sperm may be stored in small pouches. There is no penis, and semen is deposited by the extension of the cloaca.

Shutterstock.com/PCHT

Figure 6.6 A rooster

The female

In the hen the reproductive organs are also paired, but only the organs of the left side develop properly. The left ovary of a hen is different from a mammalian ovary. It consists of two lobes, and each follicle is attached to the ovary by a long stalk. The follicles are also different because nearly all the space inside them is taken up with the relatively large ova. Each ovum takes approximately 10 days to mature under the control of the follicle-stimulating hormone.

Ovulation occurs when the fully developed ovum breaks out from the follicle. The ovum then begins its passage down the oviduct. The oviduct is a long tube with elastic walls, consisting of five distinct regions (Fig. 6.7 on page 97). As the ovum passes through the oviduct, the egg white, or albumen, shell membranes and shell are added.

ISBN 9780170265560

Following is an outline of what happens in each of the regions of the oviduct.

1 *Infundibulum*. The ovum first passes through the short infundibulum, or funnel, in which it may be fertilised by sperm and in which the anchoring cords, or chalazae, are attached.
2 *Magnum*. The ovum then passes into the magnum, which has thick glandular walls. Here it receives the albumen coating (a white substance).
3 *Isthmus*. As the ovum passes through the isthmus, the two shell membranes are added around the albumen.
4 *Shell gland*. The ovum then enters the shell gland where the shell is added.
5 *Vagina*. Finally, the egg passes through the vagina where it receives the 'bloom' that seals the pores of the egg shell.

The whole process of egg formation in the average hen takes approximately 26 hours, most of which is spent in the shell gland (approximately 20 hours).

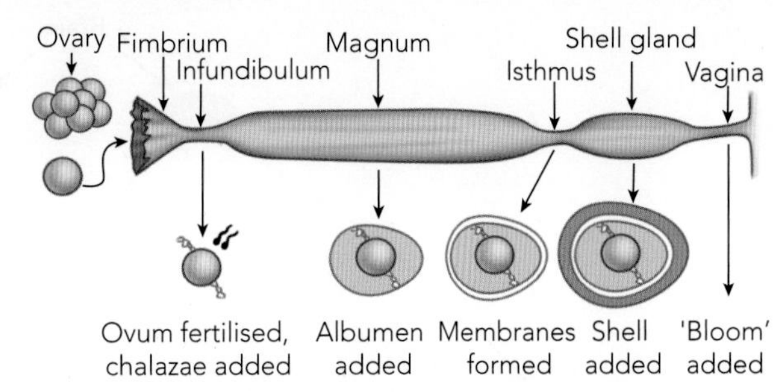

Figure 6.7 The reproductive organs of a hen

13 Where are the testes found in birds?

14 Describe how the ovary of a hen is different from a mammalian ovary.

15 What is the function of the isthmus?

The role of hormones

Hormones in mammals are chemical substances secreted by the ductless, or endocrine, glands directly into the bloodstream to control body actions or processes.

The main endocrine glands controlling reproduction are the pituitary gland (sometimes called the master endocrine gland), the gonads and, in pregnant females, the placenta. The pituitary gland is located in a bony depression at the base of the skull. It has two distinct areas – the anterior and posterior pituitary. The gonads include the ovaries and testes.

The pituitary gland

The anterior pituitary secretes several hormones, including three that are concerned with reproduction (Fig. 6.8).

1 Follicle-stimulating hormone (FSH) stimulates the growth and development of follicles in the ovary. In the male it stimulates production of spermatozoa in the testis.
2 Luteinising hormone (LH) in the female acts upon the mature follicle (called a Graafian follicle), causing it to rupture, releasing an ovum, and to develop into a corpus luteum. In males LH stimulates the production of testosterone (the male sex hormone) in the testis.
3 Lactogenic hormone (prolactin) is concerned with the maintenance of lactation in the females of some animals.

The posterior pituitary gland produces one hormone that affects reproduction. This is oxytocin and it has two roles. It stimulates the letdown of milk in the lactating animal, and it stimulates muscular contraction of the uterus at mating and at birth to help expel the foetus.

The gonads

The gonads are the sex glands (the ovaries in the female and the testes in the male) that produce the sex hormones.

Ovaries

The ovaries produce three main hormones of reproduction.

1 Oestrogen is produced by the maturing follicle and has three main functions:
   i the stimulation of oestrus, or heat, so that the female will accept the male
   ii the promotion of growth of the uterus, enlargement of the vulva and secretion of mucus from the glands in the cervix
   iii the promotion of growth and development of ducts and tissue in the mammary gland in conjunction with progesterone.

ISBN 9780170265560

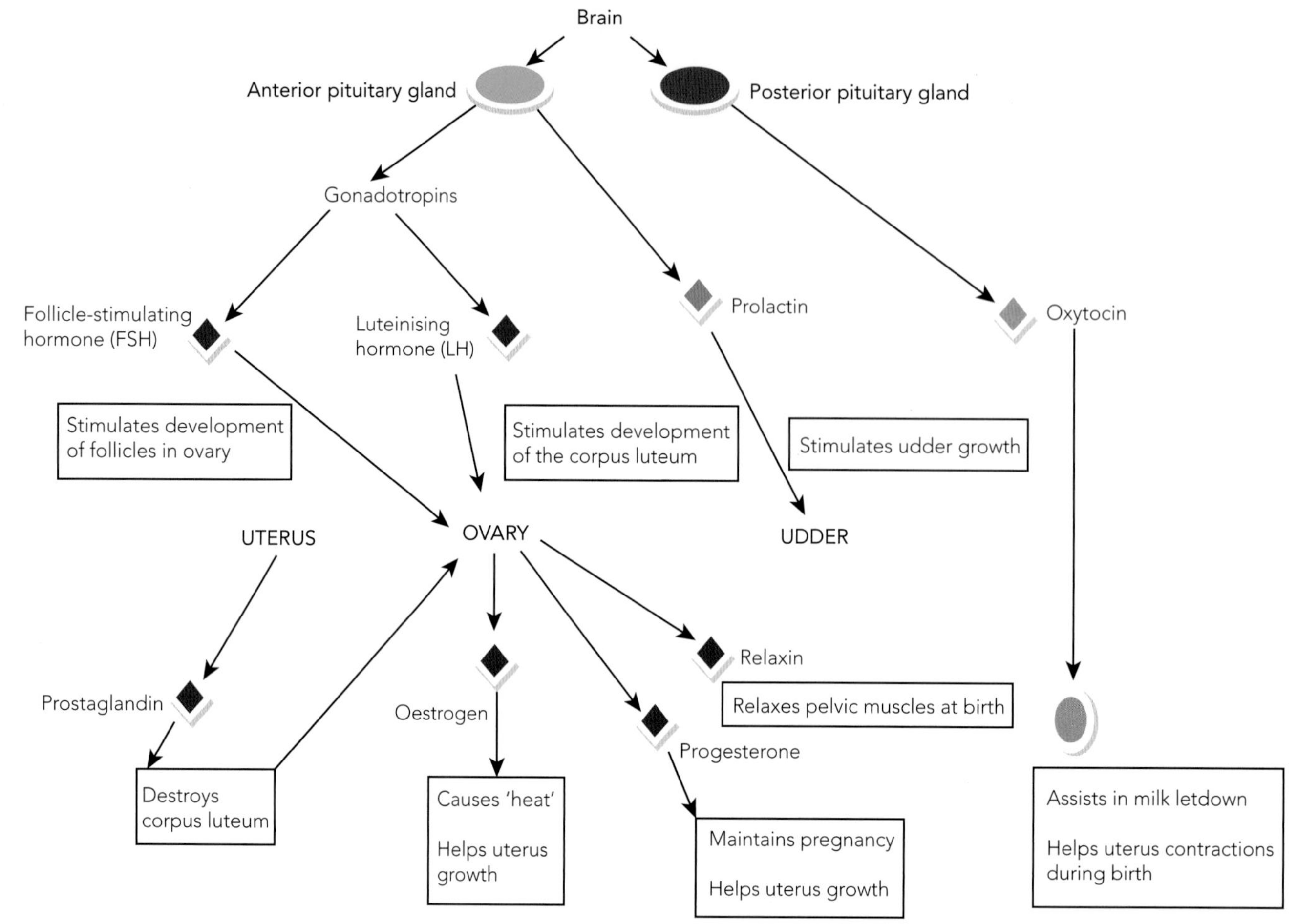

Figure 6.8 The endocrine system – the anterior pituitary acts as a master gland, controlling the activities of other endocrine glands.

2 Progesterone is produced by the corpus luteum, which develops within the ruptured follicle after an egg has been released (called ovulation). Progesterone has a number of functions:
  i the maintenance of pregnancy
  ii the stimulation of development of the wall of the uterus
  iii the prevention of oestrus and ovulation by the inhibition of FSH production.
3 Relaxin is produced mainly by the ovaries. It causes the pelvic ligaments to relax at birth.

Testes

The testes produce one main hormone of reproduction – testosterone. The production of testosterone is constant after puberty as males do not exhibit cyclic breeding activity, and most show no seasonal breeding activity. Testosterone is responsible for:

- the development and maintenance of the sexual organs and accessory glands (prostate, seminal vesicle, Cowper's gland and ampulla)
- the later stages of sperm production
- regulating libido (sex drive).

ISBN 9780170265560

Hormones and the oestrous cycle

The anterior pituitary gland produces follicle-stimulating hormone (FSH), which stimulates a follicle containing an ovum (egg) to mature in the ovary. This mature follicle is called a Graafian follicle. The Graafian follicle produces the hormone oestrogen, which causes the female animal to stand for mating by the male (oestrus). Oestrogen triggers the release of luteinising hormone (LH) from the anterior pituitary gland. Luteinising hormone causes the Graafian follicle to rupture and release the mature ovum (egg). This is called ovulation. In the ruptured follicle the corpus luteum develops and produces progesterone. Progesterone blocks the production of FSH by the anterior pituitary gland so that no further follicles mature while the female animal is pregnant. If the female does not become pregnant the uterus produces prostaglandin, which causes the reduction in size of the corpus luteum and thus the amount of progesterone production. The reduced level of progesterone in the blood stimulates the anterior pituitary gland to start producing FSH again and so the cycle begins again. Prostaglandin can be injected into cows to synchronise heat.

Reproductive behaviour in male and female animals is influenced by the action of hormones. In females oestrogen will cause the female on heat to stand for mating by the male. Testosterone stimulates mating behaviour in males by the regulation of libido, or sex drive.

16 Define the term 'hormone'.

17 What is the function of FSH?

18 State two functions of oxytocin.

19 Where is progesterone produced?

20 What is the function of testosterone?

21 Explain the interaction between hormones in the animal's oestrous cycle.

Factors affecting fertility

Fertile animals are those that reproduce efficiently. They are able to ovulate successfully, become fertilised and produce young. Most of the factors that influence fertility are under the control of the farmer, and therefore knowledge of these factors will help in reducing infertility.

Infertility is the failure of animals to reproduce. There are many factors affecting the fertility of farm animals. They can be conveniently considered under the following headings – genetics, nutrition, environment, pests and disease, and management. More information on dairy cattle infertility can be found at the following websites.

connect

Dairy cattle infertility

connect

Factors that affect animal fertility

Genetics

There are genetic differences in fertility between species of animals. For example, a sow produces two litters per year, with approximately 11 piglets in each litter, producing a total of 22 offspring per year. A ewe produces one or two lambs per year. A cow usually produces one calf per year.

There are also genetic differences between breeds of the same species. Some breeds of sheep, such as the Border Leicester, have a higher percentage of twins than other breeds, such as the Merino. Some animals are infertile because they have a genetic abnormality in the anatomy of the reproductive or endocrine glands.

Nutrition

The rate of development of the reproductive organs and the onset of puberty are determined more by body weight than age.

Mature female animals on a low plane of nutrition and below normal body weight (Fig. 6.9) will have irregular oestrous cycles, lower ovulation rates and decreased fertility. If such animals are pregnant their offspring will tend to be smaller and weaker.

Shutterstock.com/KreativKolors

Figure 6.9 A cow on a low plane of nutrition

Ewes are often given more feed shortly before and during the joining period. This is known as flushing and will increase the ovulation rate and fertility if it causes an increase in body weight and condition. The effect of flushing is only noticeable if ewes are in poor condition before the joining period.

Feeding ewes a high plane of nutrition in the later stages of pregnancy increases foetal growth and may result in higher birth weights. However, overfeeding may lead to oversized lambs, which leads to **dystochia** (difficult birth) and the loss of young and sometimes the mother.

The effects on reproduction of poor nutrition in males are less evident than in females. A deficiency of vitamin A prevents normal sperm formation in bulls and rams. Overfeeding and excessive fatness may reduce libido.

22 Name five factors affecting the fertility of farm animals.

23 Explain the term 'infertility'.

24 What is flushing?

ISBN 9780170265560

Environmental factors – climate

Two main aspects of the climate affect reproduction: temperature and day length.

Temperature

High temperatures increase **embryonic mortality** (death of the embryo) and reduce birth weights in piglets and lambs. Lambs with a low birth weight are weak and have a greater chance of death. This is one cause of low survival of lambs until marking (ear tagging, castration of males, tail docking) in Australia.

High temperatures also have a harmful effect on sperm production. The number of sperm is reduced and many sperm are abnormal. Rams that are to be joined during a hot summer should be provided with adequate shelter and water, and they should carry some cover of wool.

A combination of low temperatures and wet and windy weather will lead to high **neonatal mortality** in newborn lambs. This combination occurs in the high country of New South Wales and Victoria.

Day length

The breeding season, which depends on day length, is less marked in rams than ewes. In British breeds there is a marked decline in semen volume and quality out of season. In several wild animals there is a distinct mating season called the rut, or rutting period.

In some non-pregnant animals oestrus recurs again and again throughout the year. These species are called polyoestrous, and include pigs and cattle. Others, including goats, sheep and horses, usually only breed during particular months of the year and are seasonally polyoestrous.

The stimulus that controls breeding activity in seasonal breeders is the change in the ratio of hours of daylight to hours of darkness. The length of the breeding season in sheep differs between breeds and can be related to the latitude of their country of origin.

Pests and disease

Any disease that affects the health and vitality of an animal may reduce its reproductive capacity. Certain diseases may affect the reproductive organs themselves. Some types of infection may stop the production of either sperm or ova, or prevent the passage of the sperm or the ova along the reproductive tract. Other diseases may result in lack of implantation of the embryo or abortion of the foetus during pregnancy.

Venereal diseases often reduce the fertility of flocks and herds. They include:

1 *vibriosis*, which is caused by the bacterium *Vibrio foetus*, and results in infertility and abortion in ewes and cows.
2 *leptospirosis*, which is caused by organisms in the *Leptospira* genus, and produces high fever and abortion in cows.

Pests, such as sheep blowfly (which causes fly strike), and internal parasites, such as worms, can reduce the vigour of the animal and therefore reduce its fertility.

A major disease affecting the fertility of sheep is oestrogenic infertility. It occurs if ewes eat oestrogenic pasture, usually some types of subterranean clovers. The ewes fail to conceive.

Management

An animal is fertile if it produces an offspring, but it does not show fecundity until it produces a number of offspring. Farmers can improve the fertility and fecundity of their animals, and thereby increase the efficiency of reproduction, by using the following management practices.

1 Accurately detect whether they are on heat when trying to inseminate cows, ewes or sows.
2 Selectively cull to remove infertile cows, ewes or sows and infertile bulls, rams or boars.

ISBN 9780170265560

3 Determine the optimum mating time so that the rams are joined to ewes at the time of the year when their breeding activity is at its highest (March, April and May, and August, September and October). This practice works well providing the market can support it. It is best to adopt a whole systems analysis with any farm activity and to consider all of the factors that can influence a management decision. Climatic influences, market conditions and social demands for the product often affect the timing of a particular farm-management decision, such as when to mate animals.
4 Prevent and control pests, disease and parasites through vaccinating, drenching and dipping.
5 Provide animals with adequate nutrition – the relevant types and quantities of feed.
6 Provide sufficient male animals so that all females are mated – for sheep the ratio is three rams for every 100 ewes, and for beef cattle a similar ratio applies.

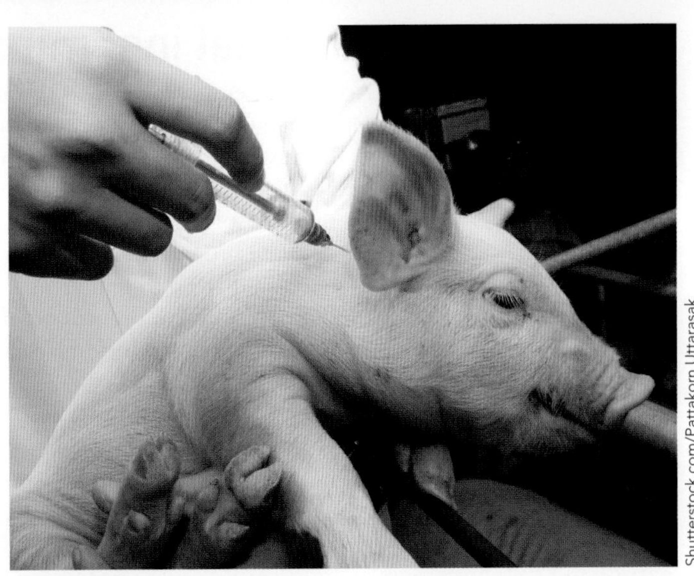

Shutterstock.com/Pattakorn Uttarasak

Figure 6.10 Vaccinate to prevent disease

25 How does temperature affect embryonic mortality in pigs and sheep?

26 List the management practices used by farmers to increase the efficiency of reproduction.

27 Name two venereal diseases that often reduce the fertility of flocks and herds.

Techniques for improving reproductive performance

A number of techniques are available for increasing the rates of reproduction and the rate of use of superior genotypes or genes. The techniques include pregnancy diagnosis, artificial insemination, multiple ovulation, flushing, synchronisation of oestrus, embryo transfer and increasing the ovulation rate.

Pregnancy diagnosis

Diagnosing or testing early after mating, or joining, enables more efficient stock management. Pregnant animals can be given better pastures or supplementary feeding, which will facilitate optimum foetal growth. Non-pregnant animals can be checked and mated again or culled.

Pregnancy diagnosis can be carried out by a number of methods. In cattle the procedure is rectal palpation. This involves placing a gloved hand in the rectum (45 to 60 days after mating) and feeling through the rectal wall for the foetus and placental attachments. Another method is to use a blood test. The blood sample (Fig. 6.11) is sent to a laboratory for analysis. In both cattle and sheep an ultrasound device is used to detect the presence of a foetal skeleton.

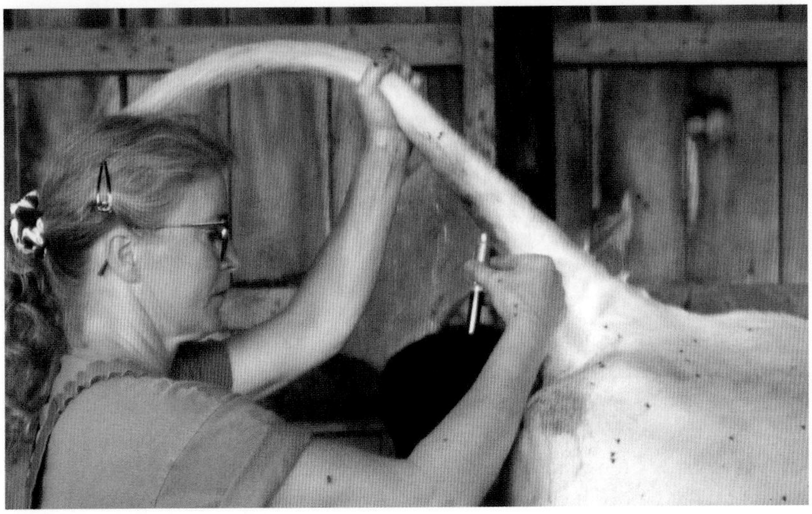

Getty Images/Portland Press Herald

Figure 6.11 Pregnancy testing

ISBN 9780170265560

Artificial insemination

connect

Artificial insemination

Artificial insemination (AI) is the act of using instruments to deposit semen in the female reproductive tract with the aim of achieving pregnancy.

In Australia, AI is used extensively with dairy cattle (Fig. 6.12) and to a lesser extent with beef cattle, sheep, pigs and goats. The benefits of using AI include:

- the widespread use of superior sires
- accurate sire selection for AI enables the widespread testing of unproven bulls over many herds in many different areas
- the eradication or prevention of venereal diseases
- the introduction of new bloodlines from other countries through the use of 'overseas' bulls
- the allowance for simple crossbreeding with one or several breeds, without the expense of keeping a number of bulls of different breeds.

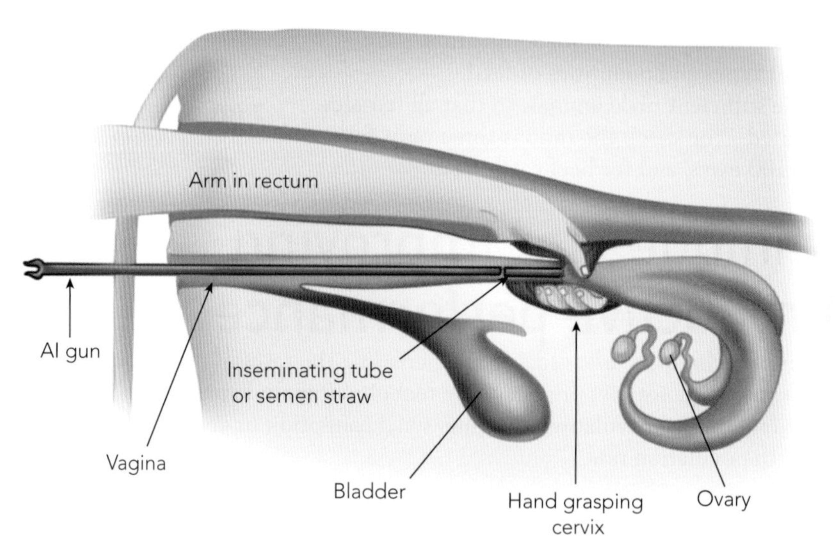

Figure 6.12 The technique for artificial insemination of a cow

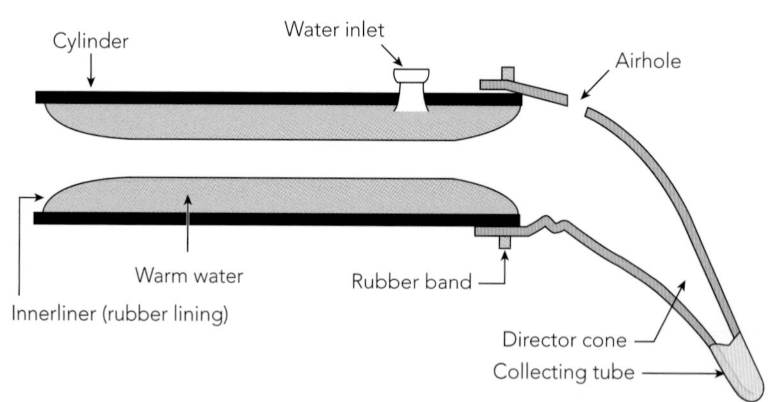

Figure 6.13 Cross-section through an artificial vagina

The disadvantages of using AI include:

- heat detection and yarding cows for insemination takes extra time
- damaged semen or poor inseminator technique will reduce fertility
- inseminators careless with hygiene can easily spread disease from one cow to another.

When AI is to be used, semen is collected from bulls using a container called an artificial vagina (Fig. 6.13). The semen is then examined for its quality and fertilising ability. Fertility is determined by the semen's colour, volume, sperm concentration and motility.

A special diluent is added to the semen. It consists of skim milk, glycerol, fructose and antibiotics. The semen is then placed in plastic straws. Each straw is identified with the bull's name and date of collection. The straws are slowly frozen in liquid nitrogen and transferred to large storage vats. The semen is stored at –196°C.

ISBN 9780170265560

Synchronisation of oestrus

Synchronisation of oestrus involves bringing all the animals in a herd or flock into oestrus at the same time. In Australia most commercial sheep and beef cattle properties are extensive enterprises and heat detection is difficult. Without using oestrus synchronisation it would be necessary to inseminate every day for approximately 20 days with sheep or 30 days with cattle, to make sure each animal was fertilised. In dairy cattle it is relatively simple to observe oestrus as cows are handled twice each day for milking. There are signs, or symptoms, of oestrus in dairy cattle.

- The vulva is moist and red.
- There is a clear mucus discharge from the vulva.
- The cow stands to be ridden.

The benefits of oestrus synchronisation are several.

- If all animals in a flock or herd are brought into oestrus on a chosen day, they can be yarded and artificially inseminated together.
- Lambing and calving can be condensed into a shorter period.
- Lambs or calves can be sold as an even pen or lot.

Oestrus synchronisation involves treating the females to be inseminated with hormones so that they will all come into oestrus and ovulate at about the same time. There are two common methods.

1 *Progesterone treatment*. A progesterone-releasing intravaginal device (PRID) attached to a plastic coil is inserted into each cow's vagina on the same day and removed 10–12 days later. Insemination occurs 56 hours after withdrawal of the coil. In sheep a pessary (sponge) containing progesterone is placed in the ewe's vagina and removed after 14–16 days. The ewe is then inseminated 48–60 hours later.
2 *Prostaglandin injection*. With this method oestrus and ovulation occur after giving two doses of prostaglandin. The injections are given 10–12 days apart. The cow is then inseminated 48 hours later.

The disadvantages of oestrus synchronisation include:

- synchronisation is expensive (for injections or implants)
- the farmer needs to have high level organisational skills, especially when carrying out an embryo transfer program.

connect

Heat detection in beef cattle

Embryo transfer

The embryo transfer (ET) technique enables embryos to be transferred from one female to another. It enables greater use of superior females in breeding programs.

Normally a cow will produce six or seven calves in her lifetime. She usually produces a single egg, and thus a single calf, each time she ovulates. By the use of hormones she can be superovulated, or made to produce several eggs (multiple ovulations). These can be collected and implanted into several foster mothers, or recipient cows. Using this method a single cow might produce 50 calves in her lifetime. Figure 6.14 on page 104 shows the steps in embryo transfer. Further information can be found on the Steps in embryo transfer website.

There are various benefits associated with embryo transfer.

- It is a method of obtaining many more calves, far more quickly, from particularly valuable or prized cows than is possible by the normal reproductive processes.
- The embryos can be frozen, stored and used later.
- The frozen embryos can be transported from place to place or country to country.
- The embryos can be split to give four or more new embryos.
- The embryos can be sexed so that, for example, in the dairy industry only female embryos are implanted.
- Herd genetic improvement is rapid compared with AI or paddock mating where a bull runs in a paddock with a herd of cows.

connect

Embryo transfer

connect

Cow embryo flush

connect

Steps in embryo transfer

ISBN 9780170265560

28 How is pregnancy diagnosis performed in sheep?

29 List five benefits of using AI.

30 At what temperature is semen stored?

31 Explain the term 'synchronisation of oestrus'.

32 Briefly describe the two common methods of oestrus synchronisation.

33 What is the main advantage of embryo transfer?

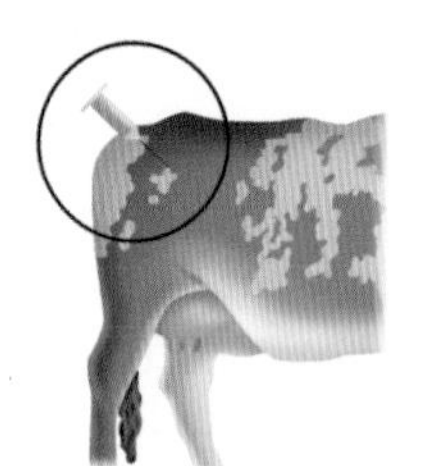
Superovulation of donor with hormones

Artificial insemination (5 days after initiating superovulation)

Non-surgical recovery of embryos (6-8 days after mating) using a Foley catheter

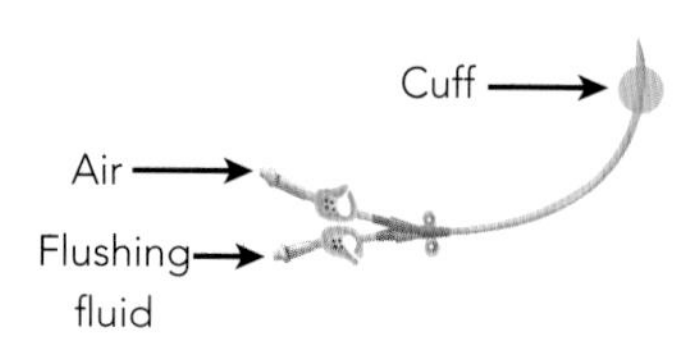

Foley catheter for recovery of embryos

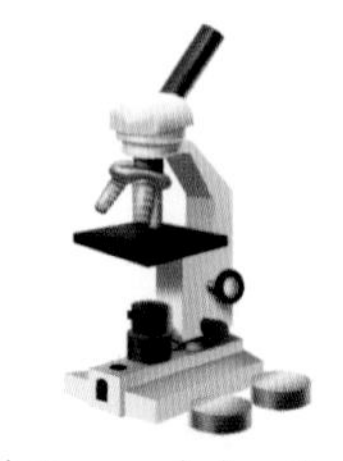
Isolation and classification of embryos

Storage of embryos indefinitely in liquid nitrogen or at room temperature for a few hours

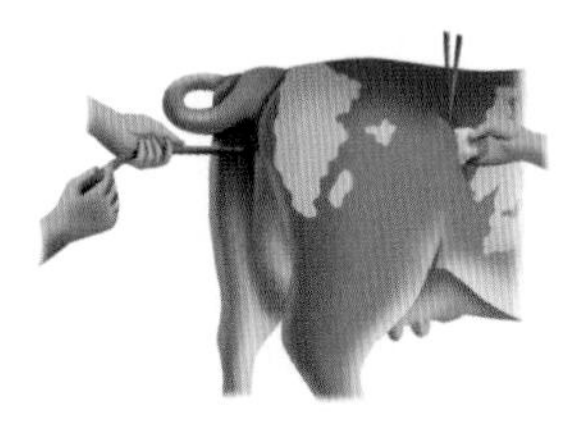
Transfer of embryos to recipients surgically or non-surgically

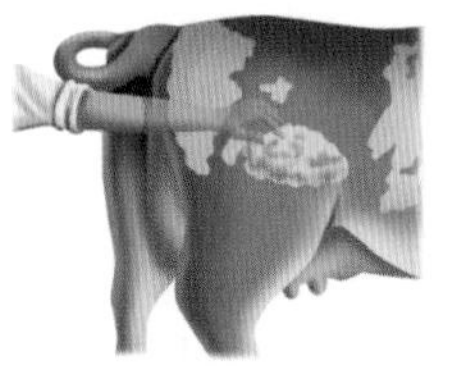
Pregnancy diagnosis by palpitation through the wall 1–3 months after embryo transfer

Birth (9 months after embryo transfer)

Figure 6.14 The steps in embryo transfer

Disadvantages include the following points.

- Specialised and expensive labour is needed to carry out the procedure or practice (usually a veterinarian specialising in ET).
- Embryo implantation rates and calving percentages can be low.
- Recipient cows need their oestrous cycles to be synchronised so that they are ready to receive an embryo.

Increasing the ovulation rate

In some intensive areas of sheep production it is often desirable to increase the number of lambs born per 100 ewes joined. There are a number of ways of increasing the number of multiple births in sheep. Each method acts to increase the ovulation rate.

The ovulation rate can be increased permanently in future generations by:

- crossing with other breeds that have a naturally higher incidence of multiple births (twins, triplets)
- selectively breeding within one flock, by breeding animals that are born as multiples or have given birth to multiples.

The ovulation rate can be increased temporarily by:

- injecting ewes (or cows) with small doses of pregnant mare serum (PMS), which is a source of FSH
- treating ewes with a vaccine that immunises them against certain of their own gonadal hormones, such as progesterone
- placing the female animal on a high plane of nutrition prior to mating or joining (this procedure is known as flushing).

34 List two ways of increasing the ovulation rate permanently in future generations.

ISBN 9780170265560

Chapter review

Things to do

1. Examine the reproductive tract from a ram, boar or bull. Draw and label the main organs.
2. Examine the reproductive tract from a ewe, sow or cow. Draw and label the main organs.
3. Observe the **dissection** of a hen. Draw and label the reproductive organs.
4. Observe semen samples using a monocular microscope (magnification x400).
5. Observe the calving, lambing, kidding or farrowing process.
6. Describe the devices used for heat detection in sheep and cattle; for example, the raddle for rams and the chin-ball headstall for vasectomised bulls.
7. Discuss the use of the chemicals and equipment used for heat synchronisation in cattle; for example, the PMS injection and the progesterone coil.
8. Observe the AI of a cow.
9. Draw and label the equipment used for AI – the AI gun, straw and sheath.
10. Pregnancy test a cow using rectal palpation (feel).
11. Describe the equipment used for castration of lambs and calves.
12. Use records from the school farm or neighbouring property to calculate the age of the cows at first mating, their age at first calving and calving intervals.

Things to find out

1. Outline the function of the cervix in a female reproductive tract.
2. Name three structures through which sperm pass from the time they are formed in the testes until they are released from the penis.
3. Several million lambs die each year on Australian sheep properties. Find out how these lambs die and suggest how the losses could be reduced.
4. Why does embryo transfer have only a limited application on commercial cattle properties, but an extensive application on stud properties?
5. Determine which synthetic hormones are used for manipulating animal reproduction in your district.
6. Research how semen and embryo sexing is carried out.

Extended response questions

1. Compare and contrast the structural and functional features of mammalian and avian reproductive tracts.
2. Describe the role of reproductive hormones in the regulation of the physiology and behaviour of farm animals.
3. Describe how the following factors affect the fertility of farm animals: genetics, nutrition, environment and climate, disease and management. Illustrate each factor with an example.
4. Describe and explain the interaction between hormones in an animal's oestrous cycle.
5. The following techniques can be used to manipulate the reproductive performance of a herd: oestrus synchronisation, artificial insemination and embryo transfer.
   - a Describe why a cattle producer would use each of the techniques.
   - b Discuss any problems or disadvantages with these techniques.
6. 'Increased productivity in the livestock industries is dependent on improvements in reproductive efficiency.' Discuss this statement.
7. Evaluate two management techniques that are used by farmers to manipulate reproduction in farm animals; for example, AI or ET. For each technique:
   - a briefly describe the technique
   - b list the advantages of the technique
   - c list the disadvantages of the technique
   - d outline your judgement about these techniques.

CHAPTER 7

Outcomes

Students will learn about the following topics.

1. The anatomy and physiology of lactation in dairy cattle.
2. The role of hormones in the regulation of lactation.
3. The composition of colostrum and milk.
4. The factors affecting milk composition and yield.
5. The factors affecting milk quality.
6. Maintenance of lactation by the farmer.

Students will be able to demonstrate their learning by carrying out these actions.

1. Define the term 'lactation'.
2. Describe the anatomy of a mammary gland.
3. Label the main structures on diagrams of an udder, teat and alveolus.
4. Discuss the role of hormones in lactation.
5. Describe how milk is secreted.
6. Describe milk letdown.
7. Outline the composition of colostrum and milk.
8. List the factors affecting milk composition and yield.
9. Outline the factors that affect milk quality.
10. Label a typical lactation curve.
11. Outline herd-recording procedures.

ISBN 9780170265560

LACTATION

Words to know

adrenalin a hormone synthesised by the adrenal glands that produces a fight-or-flight response in animals

alveoli the small glands where the milk is made

antibody a chemical substance (a protein) made by animals in response to bodily invasion by a pathogen, which combines with the pathogen and renders it harmless

colostrum a protective substance containing antibodies, which is passed on in the first milk

lactation the secretion of milk

oxytocin a hormone produced by the posterior pituitary, which is responsible for milk letdown

prolactin a hormone that stimulates the alveolar cells to secrete milk

ISBN 9780170265560

Introduction

Lactation means secretion of milk. It is the final event in reproduction in mammals. The purpose of milk is to supply the young mammal with all the nutrients it needs for survival and growth.

Domesticated animals often produce milk that is surplus to the needs of their young because of the farmer's manipulation of the production system. Using management practices, farmers have been able to use cattle, sheep and goats to provide milk and milk products for human consumption.

Discussions in this chapter will be focused largely on the cow, in view of its economic importance.

The mammary gland

The mammary gland has a similar structure in all highly evolved mammals, although the number and shape of glands might vary between species; for example, sheep and goats have two mammary glands, while pigs might have up to 18 separate mammary glands.

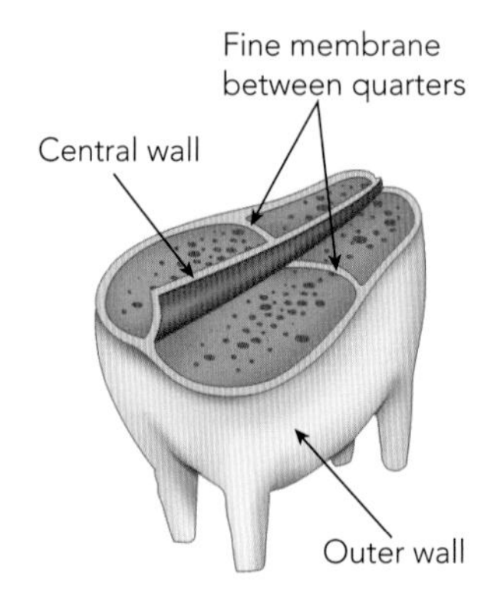

Figure 7.1 A transverse cross-section of a cow's udder

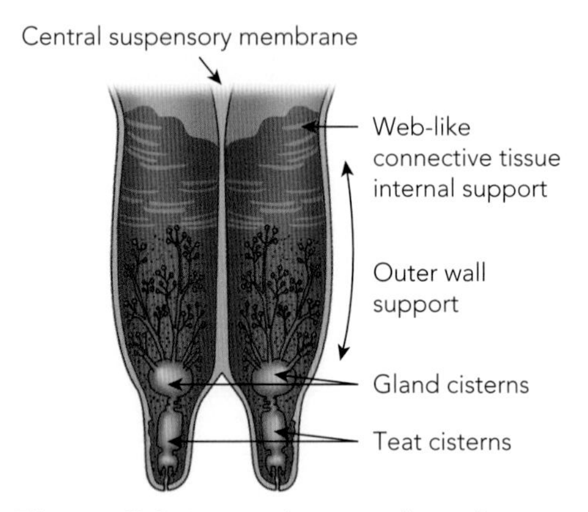

Figure 7.2 A vertical section through a cow's udder

1. Why is milk important to young animals?
2. How many mammary glands do sheep, goats and pigs have?
3. What is the name of the small glands where milk is made?

The udder of the cow has four separate mammary glands, or quarters, and each has its own teat and duct systems (Figs 7.1 and 7.2). The duct systems are independent; milk formed in one gland cannot pass into adjacent glands. The rear glands are usually larger than the front glands.

In high-yielding dairy cows a large volume of milk is made and stored in the udder between milkings. Ligaments hold and attach the udder to the body. If all these ligaments are strong, the udder is held up against the abdomen. If the attachments are weak, the udder sags down, which results in damage to the udder and a higher likelihood of disease.

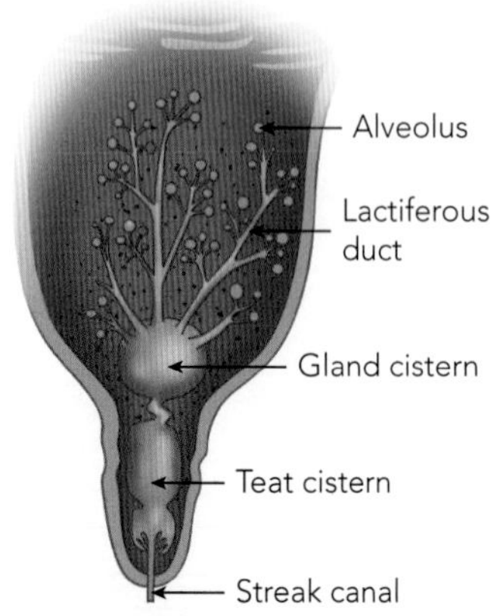

Figure 7.3 Main structures of a cow's udder

Milk is made in the small glands called **alveoli** (Fig. 7.3). These glands are scattered throughout the udder tissue in each of the four quarters. Milk from each alveolus empties into very fine ducts, or tubes, and then into 8–12 larger ducts called lactiferous ducts. The milk is then emptied into a large cavity called the gland cistern (or udder cistern). The gland cistern holds approximately 500 mL of milk.

The connection between the gland cistern and the teat canal (or teat cistern) is very narrow and the tissue folds grow inwards. Milk flow is reduced if these folds are pushed together and close off the passage. This can happen at the end of milking if the teat cups are left on the cow too long, resulting in the teat cups creeping up onto the udder (Fig. 7.4).

ISBN 9780170265560

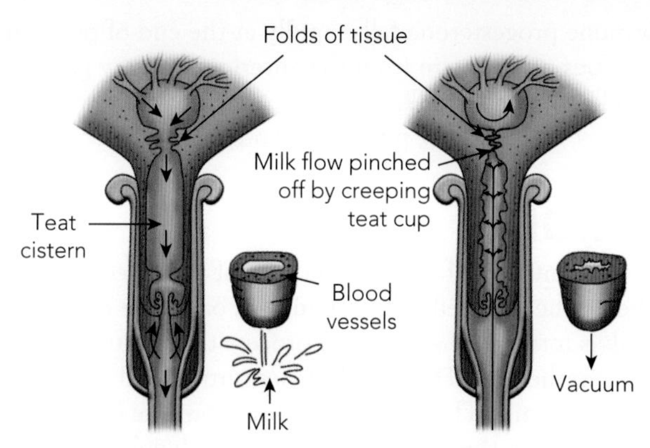

Figure 7.4 If teat cups are left on the cow too long (right diagram), they can cause serious damage

Milk travels down through the teat canal (Fig. 7.5). This cavity forms the inside part of the cow's teat. Loss of milk between milkings is prevented by a number of folds of skin at the lower end of the teat canal, which form the Furstenberg's rosette. Just below these folds of skin, the teat canal is closed by a sphincter muscle. This muscle prevents dirt and foreign bodies entering the teat. The teat walls contain arteries and veins that supply blood to the teat tissue.

The milk secreting glands (or alveoli) are very small. Each consists of a hollow sphere of cells surrounded by a network of muscle fibres called the myoepithelium. The myoepithelium contracts during milking to force the milk into the lactiferous ducts. Figure 7.6a shows two alveoli surrounded by their network of blood capillaries and myoepithelium. In Figure 7.6b the myoepithelium is shown contracting, thus reducing the size of the alveolus. The main blood supply to the udder is the mammary artery, which branches off from the aorta. Blood leaves the udder in the mammary veins.

4 What can happen if the milk cups are left on a cow for too long?
5 What are the muscle fibres surrounding each alveolus called?

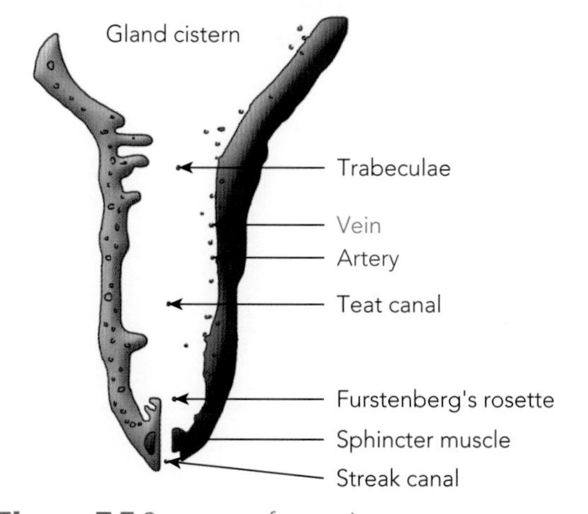

Figure 7.5 Structure of a cow's teat

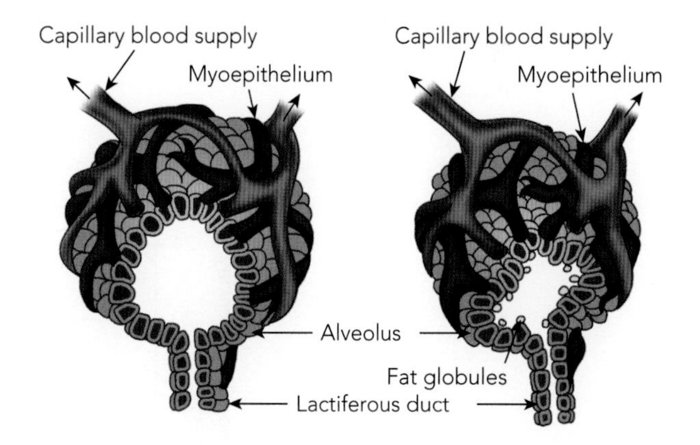

Figure 7.6 Structure of an alveolus

Hormones and milk production

Hormones control the growth and development of the udder tissue, the process of milk formation and the letdown of milk from the udder. The pituitary gland produces hormones that influence other endocrine glands and organs. These in turn produce hormones that affect lactation.

Follicle-stimulating hormone, secreted by the anterior pituitary gland, stimulates the growth of follicles in the ovary. The follicles then produce the hormone **oestrogen**, which stimulates udder growth. The placenta also produces oestrogen.

6 Name one hormone produced by the anterior pituitary gland.
7 What is the function of oxytocin?

The level of the hormone progesterone falls rapidly at the end of pregnancy, permitting the release of large quantities of **prolactin** from the anterior pituitary gland. Prolactin stimulates the alveolar cells to secrete milk.

Oxytocin is produced by the posterior pituitary gland and is responsible for milk letdown.

Milk secretion

Milk secretion involves the synthesis of milk by the cells of the alveolar epithelium, and the passage of milk from the epithelium cells into the lumen, or centre of the alveolus (Fig. 7.7).

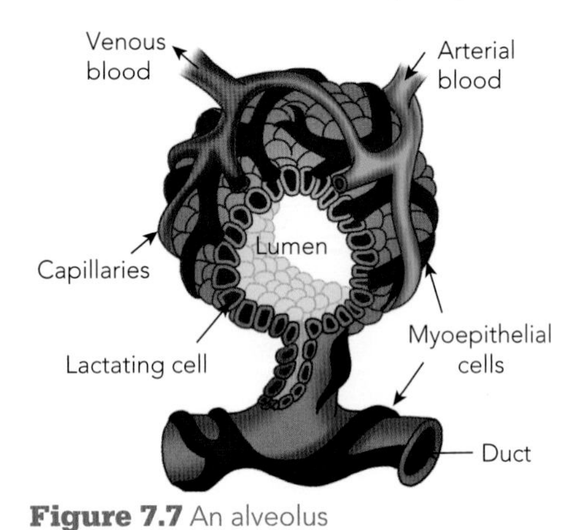

Figure 7.7 An alveolus

Milk is formed by a chemical process that occurs within the cells of the alveolar epithelium. The blood carries nutrients that are necessary for the formation of milk. These nutrients pass from the blood capillaries outside the alveolus into the epithelial cells of the alveolar wall.

Some milk constituents are made by the alveolar cells from nutrient raw materials in the blood. Most of the protein, all of the lactose and some of the fats are made in this way. Other milk constituents that are carried in the blood (e.g. the minerals calcium and phosphorous) pass through to the alveolar cells and appear unchanged in milk.

At the beginning of lactation many of the cells of an alveolus are functioning. As lactation proceeds, more and more of the cells begin to secrete milk until the maximum production is reached. In cattle the maximum production is reached 3–6 weeks after calving (Fig. 7.8). Then, as lactation proceeds, more and more of the cells in each alveolus cease to function.

Lactation in cattle lasts for up to 11 months. Lactation in sheep usually lasts for only 12–16 weeks. In pigs, lactation can last for 9 weeks, at which time the litter is weaned. Normally, piglets are weaned at 4 weeks.

8 Name the small structures where milk is formed.
9 Name two milk constituents in the blood that appear unchanged in milk.

Cows are normally milked twice per day. Some large commercial dairies prefer to milk their cows three times per day (at 8-hourly intervals) to increase production. It is thought that the increased stimulation from milking three times per day causes the secretion of more oxytocin. Milking more frequently also decreases udder pressure, which promotes milk secretion.

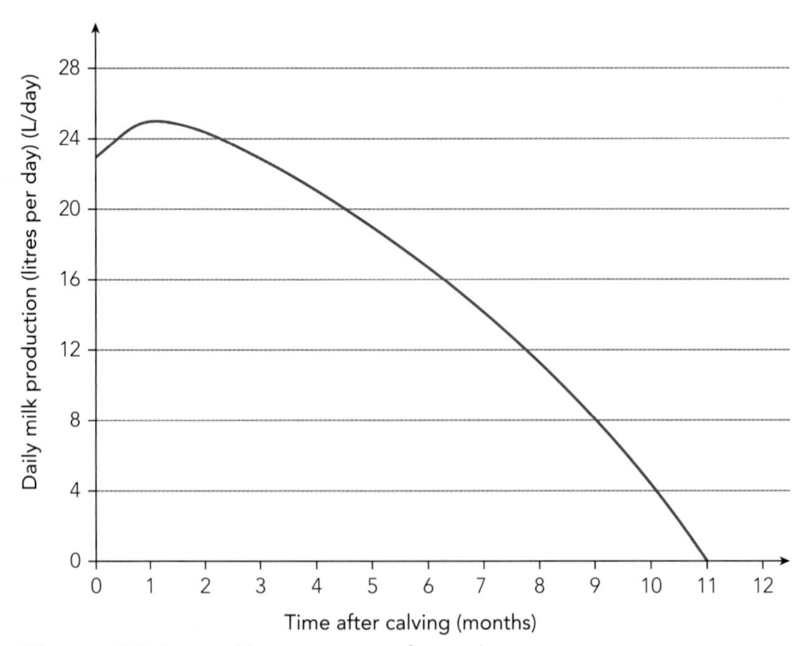

Figure 7.8 A typical lactation curve for cattle

ISBN 9780170265560

Milk letdown

Milk letdown involves all the processes whereby milk contained in the mammary gland at the time of sucking or milking is made available for withdrawal. It is a complex reflex, involving the nervous system, the circulatory system, glands and hormones. It is often called a neuro-hormonal reflex.

The skin of the udder, particularly the teat, is sensitive to physical stimulation. Physical stimulation can be provided by a sucking animal, hand massage or teat cups. When such stimulation takes place, the nervous system carries an impulse to the brain, and the hormone oxytocin is released into the bloodstream by the posterior lobe of the pituitary gland (Fig. 7.9).

The oxytocin causes the muscle fibres surrounding each alveolus, that is, the myoepithelium, to contract. Milk is then forced out of the alveoli into the duct system, and is readily available for withdrawal. This is called milk letdown, or milk ejection. The interval between simulation and milk letdown is approximately 40 seconds. If a cow is given a fright or suffers pain, the hormone **adrenalin** is released into the bloodstream. This causes a restriction of the blood supply to the skin and the udder, thereby blocking the effect of oxytocin on the alveoli. Hence, the animal is prepared for 'fight or flight' but not for milk letdown.

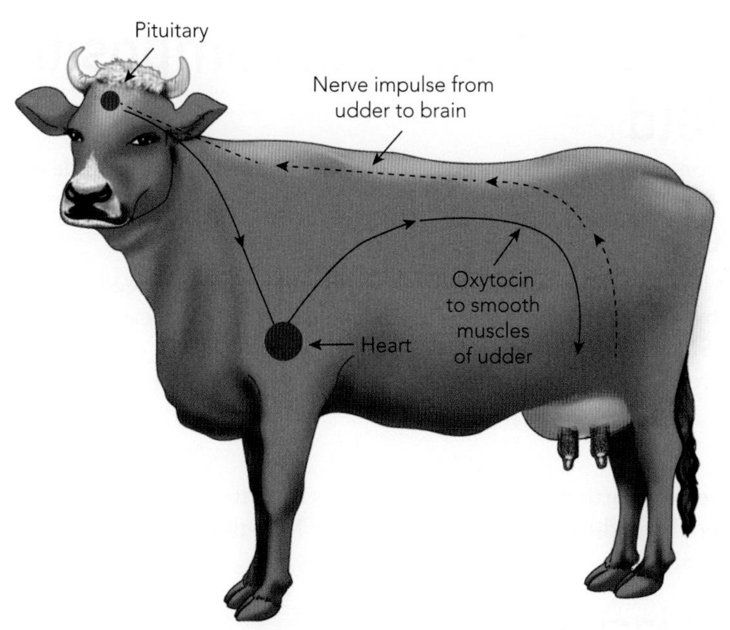

Figure 7.9 The milk letdown reflex

10 How is oxytocin involved in milk letdown?

11 Briefly explain how a barking dog could affect milk letdown.

Composition of colostrum and milk

Colostrum

In late pregnancy and for the first few days after birth the mammary gland secretes a substance called **colostrum**. The components of colostrum are important for the survival of the newborn animal for the first few days of life.

Colostrum contains higher concentrations of certain immunoglobulins, vitamin A and vitamin D than normal milk. The immunoglobulins, which are proteins, contain **antibodies**. These antibodies are absorbed through the small intestine during the first one or two days of life. They give the newborn animal passive immunity to bacterial diseases. This is particularly important for calves, lambs, kids, foals and piglets in which immunity is not acquired in the uterus by placental transfer. Colostrum also contains less lactose than normal milk, and this helps to prevent scouring.

A few days after birth, the composition of colostrum gradually changes to that of normal milk.

Milk

The analysis of an average cow's milk is shown in Table 7.1.

Table 7.1 Milk composition

Constituent		Average (%)	Range (%)
Water		87.25	84.0–89.5
Butterfat		3.75	2.0–7.0
Protein	Solids-not-fat	3.20	2.8–4.0
Lactose	Solids-not-fat	4.80	4.5–5.2
Ash and minerals	Solids-not-fat	0.75	0.6–0.8
Other constituents	Solids-not-fat	0.25	–

12 Why is it important that a newborn calf, lamb, kid, foal or piglet gets colostrum from its mother?

13 What does SNF stand for?

Solids-not-fat (SNF) refers to the solid components in milk that are not fats. Thus, SNF does not include water or butterfat.

The secretion of milk in a cow will continue until her production has fallen to a level where milking is no longer profitable. This is usually 9–11 months after calving.

Factors affecting milk composition and yield

The composition and yield (or quantity) of milk are affected by a multitude of factors.

Nutrition

The butterfat level is mainly influenced by the fibre content of the feed, while the SNF level is influenced by the energy content of the feed. Underfeeding will cause a drop in milk yield, butterfat and SNF content. Grazing young, succulent pastures will cause an increase in milk yield but a lowering of butterfat content, because of a reduced fibre intake.

14 How does underfeeding affect milk yield, butterfat and SNF content?

15 Which breeds produce the highest protein percentages in their milk?

Breed

Jersey and Guernsey cattle have higher percentages of butterfat, SNF and protein in their milk, but generally produce less milk than Friesians. Milk that is high in butterfat is also usually high in SNF.

Stage of lactation

Butterfat and SNF percentages steadily decline over the first 2–3 months of lactation. The lowest point corresponds with the cow's maximum milk yield point. Butterfat then slowly rises until the cow is dried off. SNF levels remain low until near the end of the lactation.

Age

Both butterfat and SNF percentages decrease after the first lactation.

Season and weather

Seasonal changes alter milk composition by their effect on the availability of feed. In high temperatures cows tend to eat less. This causes a lowering of total milk volume, SNF and protein levels. However, butterfat percentages are increased. In low temperatures the reverse occurs.

ISBN 9780170265560

Disease

Diseases such as mastitis reduce milk yield. The lactose content also decreases, causing a decrease of SNF. Butterfat is also reduced. Teats are dipped after the milk cups are removed, to prevent infection. For more information on mastitis visit the Dairy Australia website.

connect

Dairy Australia

What is mastitis?

Stress

Any factor that frightens or upsets a cow during or immediately prior to milking will affect milk yield and composition. Cows might also suffer stress due to oestrous, resulting in lower milk production during this time.

Milking technique

The first milk to be let down is usually lower in butterfat, while the last milk to be milked out of the udder is higher in butterfat. Therefore, if the cups are removed before the cow is completely milked out, the butterfat percentage of the milk will be lower. It is important for milking machines to be well maintained. Automatic milking systems have been critically trialled and are slowly being introduced to improve farm efficiency.

connect

Automatic milking systems

Factors affecting milk quality

Milk quality refers to more than just the chemical composition of milk. Milk can be contaminated while in the udder, during milking or during storage. High-quality milk should not be tainted, or contain bacteria, sediments or traces of disinfectants. There are several ways in which milk quality can be affected.

Contamination in the udder

Milk might become tainted by the cow grazing certain types of feed, silage or weeds. Diseases, particularly mastitis, can cause bacterial contamination of milk in the udder.

Contamination during milking

If they are dirty, udders should be washed and dried properly prior to milking. Milk can also be contaminated if the milking cups and milk lines are not cleaned properly at the end of each milking session.

Contamination during storage

To prevent contamination, the bulk storage tank should be cleaned carefully after it is emptied, using a disinfectant such as iodine, and then thoroughly rinsed.

16 What disease can cause bacterial contamination of milk?

17 How can contamination during milking be prevented?

Maintenance of lactation

A 'lactation curve' is a graph of milk yield against time (Fig. 7.10, page 114). In most animals, milk yield rises for a period following birth of offspring, reaches a peak and then steadily declines.

In the dairy cow, milk yield reaches a maximum usually 6–8 weeks after calving. The length of the lactation period is approximately 10–11 months. Milk yield is usually low towards the end of the lactation.

To ensure maximum milk yield, the dairy farmer aims for each cow to produce a calf every 12 months. To achieve this, a cow should be mated 3 months after the birth of her last calf. The gestation (or pregnancy) period for cattle is 9 months. For the 1–2 months prior to the birth of the next calf, the cow is usually 'dried off' and not milked. This allows time for the cow's body reserves to build up so that she can produce a healthy calf and a high milk yield in her next lactation.

ISBN 9780170265560

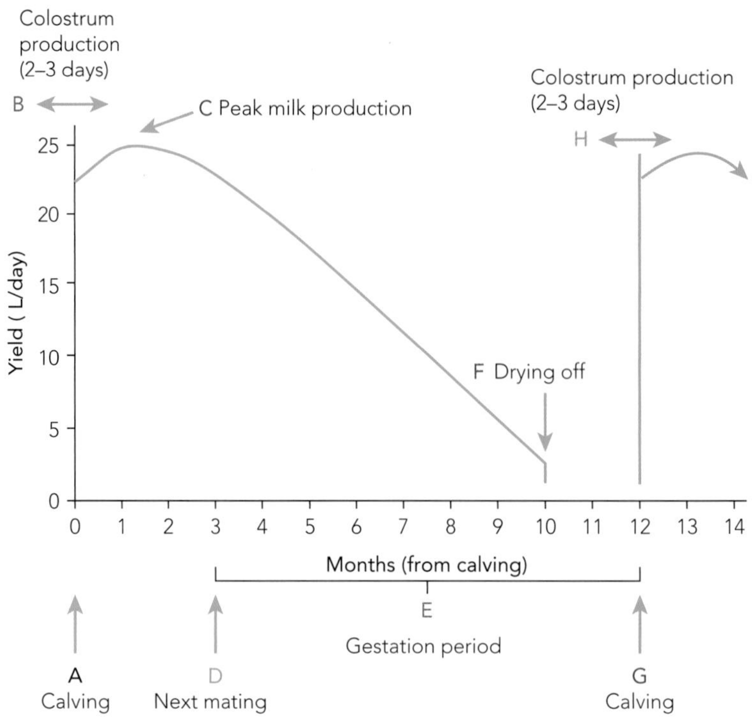

Figure 7.10 A typical lactation curve, indicating ideal times for mating and drying off

connect

Dairy Express

18 List three pieces of production information recorded for each cow on a herd-recording report.

19 List three uses for the information contained in a herd-recording report.

Herd recording

Some farmers use a program where each cow in a herd is tested once per month. The quantity she produces is measured in litres by a milk-flow meter. A sample of her milk is also tested for percentages of butterfat and protein. This can be done electronically.

A report is compiled of all the cows tested in a herd and their production. This information is valuable for culling or for selecting cows for breeding. Other aspects of management, for example feeding, can be altered to improve herd productivity. Further information can be found at Dairy Express. Here information from individual farmers is placed on a database that can be accessed via a password. Farmers can use this information to improve their management processes.

ISBN 9780170265560

Chapter review

Things to do

1. Look at the udder of a dry cow and a milking cow. Describe the main differences.
2. Hand milk a cow. Is it easier if you wash the cow's teats and udder first?
3. Use a Rapid Mastitis Test Kit to test fresh cow's milk for mastitis. Describe the consistency of the milk after adding the test chemical.
4. Visit a local dairy. Draw a floor plan of the shed. Label the main features that make milking more efficient.
5. Place a small quantity of milk on an agar plate. Place it in a warm position for a couple of days. Describe the type of micro-organisms that develop.
6. Study the information in Figure 7.11. Which cow should be culled because of low production?

DEPARTMENT OF AGRICULTURE NSW

HERD No	157 2157	Dairy Herd Improvement Program							
SAMPLING DATE	17/6/15	PRODUCTION REPORT				DHIP-I0			
VISIT NUMBER	147	MANAGEMENT ACTION LIST OPTION				NO		PAGE 0003C00869	
SHED NAME	COW RECORDING No.	LAST SAMPLING					SOMATIC CELL COUNT	ACTION	
		MILK L	% FAT	FAT KG	% PROTEIN	PROTEIN KG			DATE
PONTIAC EMERALD	266	18.5	2.8	0.53	2.7	0.51			
CIT STARERIGHT	267	12.0	3.8	0.47	3.3	0.41			
ACE ROSEBLOOM	268	25.0	3.4	0.88	3.0	0.77		CALVED	17/6/15
SHOWGIRL	269	22.0	3.2	0.73	3.0	0.68			
CITATION LADY 2	270	18.0	3.7	0.69	3.1	0.57		CALVED	15/6/15

Figure 7.11 Part of a herd-recording report

Things to find out

1. Why do farmers often have a few Jersey cows in a Friesian herd?
2. What happens to milk when it is homogenised?
3. Why do farmers dry off their cows at about the 10th month of a cow's lactation? How do they do this?
4. Describe how mastitis can be controlled in a dairy herd.
5. Why do farmers teat dip their cows after removing the milk cups?

Extended response questions

1. Lactation is a significant process that has to be managed.
   a. Discuss briefly the role of hormones in lactation.
   b. Describe the factors that affect milk composition and yield.
   c. Describe how the management of a milking shed or dairy can affect the quantity and quality of milk produced.
2. Mastitis is a disease that causes inflammation of the udder. Discuss:
   a. what causes it
   b. how it affects milk quality and quantity
   c. how it is spread to other cows
   d. how it is treated.
3. Some farmers use herd recording systems.
   a. Briefly describe how these are used and what information is collected.
   b. How can the herd recording report be used to promote better farm management?

CHAPTER 8

Outcomes

Students will learn about the following topics.

1. Factors affecting growth and development.
2. Changes in the proportion of muscle, fat and bone during the life of an animal.
3. Management practices to optimise growth and development.

Students will be able to demonstrate their learning by carrying out these actions.

1. Define the term 'growth'.
2. Define the term 'development'.
3. Explain that animal growth can be monitored by measuring changes in live weight.
4. Show with the help of a graph that the growth of the foetus speeds up rapidly as the time of birth approaches.
5. Show with the help of a graph that the growth of an animal after birth follows an S-shaped curve.
6. Name the two main stages during prenatal development.
7. Explain that when young animals are born their size will be affected by:
   a. the number of young born
   b. the size of the dam
   c. the age of the dam
   d. the sex of the litter members
   e. the level of nutrition
   f. the breed within species.
8. Explain that the growth of an animal after birth will be affected by:
   a. its size at birth
   b. the sex of the animal
   c. the genotype of the animal
   d. the environment
   e. the level of nutrition.
9. Manage and monitor the growth and development of a farm animal.
10. Explain the terms 'muscle', 'bone', 'fat' and 'carcase'.
11. Show with a diagram how the various tissues of an animal's body compete for the supply of nutrients.
12. Explain how the major carcase components (muscle, bone and fat) change as an animal develops and relate these changes to consumer needs.
13. Describe the term 'compensatory growth'.
14. Describe how hormones control animal growth.
15. Evaluate management techniques available to farmers to manipulate growth and development, including use of hormone growth promotants (HGPs), feed additives and genetics.

ISBN 9780170265560

GROWTH AND DEVELOPMENT

Words to know

bone hard connective tissue that makes up most of the skeleton of vertebrates

carcase what remains of the animal body after the head, feet, hide, tail and internal organs of the abdomen have been removed

development the changes in the proportions of the various parts of an animal's body as the animal gets older

embryo the developing organism in early pregnancy

fat a chemical substance made from glycerol and fatty acids

foetus the developing organism in later pregnancy (the embryo becomes a foetus)

growth the increase in size and weight of an animal as it gets older

hormone growth promotants compounds similar to sex hormones, which increase growth in farm animals; also called anabolic steroids

muscle tissue consisting of elongated cells (muscle fibres)

somatotrophin growth hormone

ISBN 9780170265560

Introduction

The rate at which animals grow and reproduce, and the size, shape and composition of their carcases, are all aspects of growth and development that affect the profitability of a farming business. The following sequence of events covers the normal growth pattern in all mammals: conception, development of the embryo, development of the foetus, birth, puberty, reproduction, old age and finally death. Farm animals can be slaughtered at different stages depending on the markets available.

As a young animal gets older, two changes occur.

1 It increases in size and weight. This is called growth.
2 The proportions of various parts of its body change. This is called development.

Growth curves

Although the growth of the animal body is complex, the time relationships of growth can be shown by growth curves.

The actual growth curve

The simplest and most common means of analysing growth is to plot live weight against age. The growth curve produced is sigmoid in shape and is much the same for all animal species (Fig. 8.1).

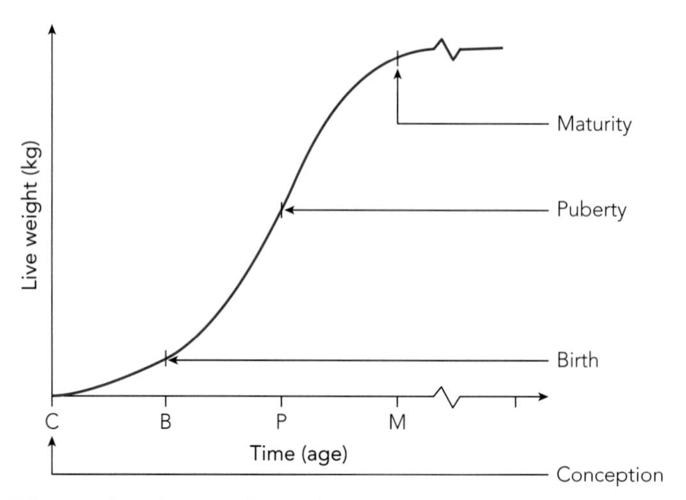

Figure 8.1 The actual growth curve

Setbacks in growth may occur because of poor nutrition, disease or stress (bullying).

During early pregnancy the growth of the foetus is slow, but during the last one-third of pregnancy growth becomes rapid. After birth the growth rate continues to increase until it reaches a maximum around the time of puberty. This occurs at the point of inflection on the actual growth curve. From then on growth rate decreases as the animal reaches its mature size.

For most farm animals puberty (when the animal becomes sexually active) occurs when they reach approximately 30% of their mature live weight. Under conditions of adequate feeding, the graph of live weight against age is S-shaped, as shown in Figure 8.1. The steeper slopes indicate higher growth rates. The period during which maximum growth rate is obtained is the age at which the animal is most efficient at converting feed into live-weight gain. If animals are slaughtered before they reach this stage, the chance for exploiting maximum biological efficiency is lost. If animals are kept after they have reached mature size, live-weight gain is slow and the conversion of feed to meat is less efficient.

1 Define the term 'growth'.
2 What is the shape of the actual growth curve?
3 Why is it poor management to continue keeping animals after they have reached mature size and before they are sold for slaughter?

The growth rate curve

The growth rate curve is obtained by plotting growth rate (increase in live weight per unit of time) against time (Fig. 8.2). Maximum growth rate occurs at a point corresponding to the point of inflection of the simple growth curve. This is also the time of puberty.

 ISBN 9780170265560

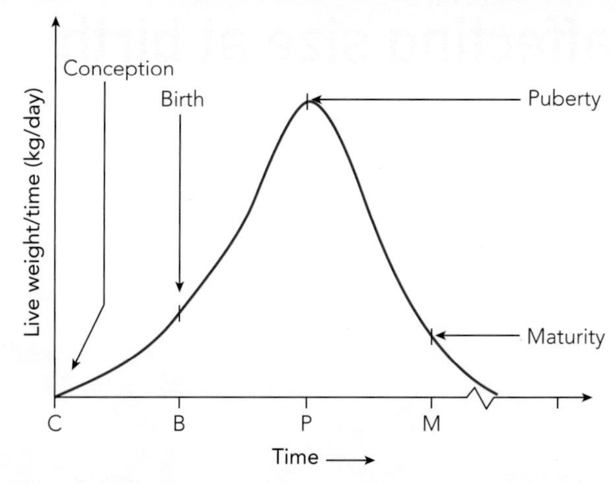

Figure 8.2 The growth rate curve

Prenatal growth

The growth that occurs before birth is called **prenatal** (see page 59) growth. There are two main stages during prenatal development – the embryonic and the foetal stages.

The embryonic stage

During the embryonic stage, cell division occurs without a noticeable increase in volume or weight. The cells form three distinct 'germ-cell layers', which develop into the various organs and tissues of the animal.

By the end of the embryonic stage, organs and tissues have developed from the germ-cell layers, and the organism is called a foetus.

The foetal stage

The various organs and tissues show marked changes in size and shape during the foetal stage. A rapid increase in size and weight of the foetus occurs during the later stages of pregnancy. Figure 8.3 shows the prenatal growth curve in sheep. An important feature is that the unborn lamb puts on approximately three-quarters of its weight in the last month of pregnancy.

The growth of the organs within the foetus follows the needs of the foetus. Organs that are needed during foetal life, or soon afterwards, develop earlier than the organs that are needed in postnatal life. The heart, brain, kidney, liver and spinal cord all have their fastest growth rates in prenatal life. The sex organs develop later.

4 Name the two stages of prenatal development.

5 When does a rapid increase in the size and weight of the foetus occur?

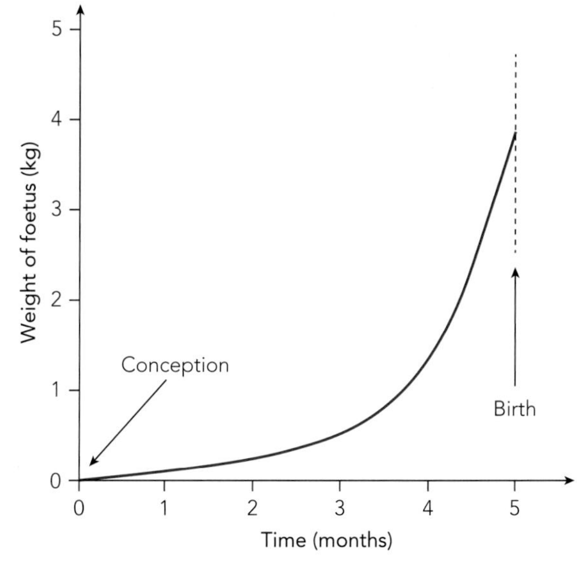

Figure 8.3 Prenatal growth curve in sheep

ISBN 9780170265560

Factors affecting size at birth

Number of young born

Generally, as the number of young increases, the tendency is for the birth weight to decrease. Usually individuals of large litters are lighter than those of small litters. In sheep, twins are normally lighter than singles, and triplets are lighter than twins.

Source text to come

Figure 8.4 Litter size affects the weight of individuals

Size of the dam

Large dams usually have offspring that are heavier at birth than those of smaller dams. Mature Charolais cows (630 kg live weight) produce calves weighing 48 kg at birth, while smaller Angus cows (450 kg live weight) produce calves weighing 30 kg.

In crosses made between dams and sires of different sizes, the offspring are proportional to the size of the dam and not midway between the two parents.

Age of the dam

The age of the dam influences birth weight. Young dams generally produce lighter offspring. As the dam matures she produces heavier offspring, and at the end of her reproductive life she again produces lighter offspring. In cattle, calves from heifers are lighter than those from mature cows. In pigs, young sows usually produce smaller litters with smaller average birth weights than older sows. In sheep, lambs from 2-year-old ewes are lighter than lambs from mature ewes.

Sex of the litter members

For single births, the males are usually larger at birth than the females. In the case of multiple births, males and females are a similar size.

Level of nutrition

The level of nutrition is important at mating and in late pregnancy.

Nutrition at the time of mating may affect the number of ova, or eggs, released and therefore the number of young. In ewes a low plane of nutrition prior to mating, or joining, may reduce the number of twins conceived. If the level of nutrition is very poor it may affect the number of ewes conceiving.

6 What general rule explains how birth weight is affected by the number of young born?
7 Give an example to show how the size of the dam affects the birth weight of an animal.
8 How is the birth weight of an animal affected by the age of the dam?
9 How does poor nutrition during late pregnancy affect the birth weight of a lamb or calf?

ISBN 9780170265560

Low levels of nutrition in late pregnancy may lead to smaller birth weights of lambs or calves, and this will affect their survival rate. Overfeeding in late pregnancy may lead to oversized lambs or calves, which result in a difficult birth. This is called dystochia, and can result in the death of the young and sometimes the dam.

The birth weight of an animal influences its chances of survival. Figure 8.5 shows that the newborn lambs most likely to die in the first few weeks of life are either too light or too heavy. The lightweight lambs cannot generate sufficient body heat to stay alive on cold, windy nights. Heavyweight lambs are more likely to have birth difficulties.

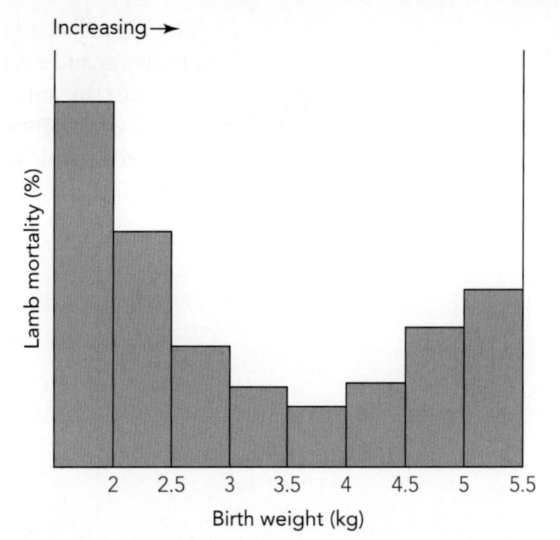

Figure 8.5 Birth weight and mortality in lambs

Breed within species

There is a large variation within a species in the birth weight of offspring. Generally, the larger the adults of the breed, the larger the offspring. Charolais cattle are a larger breed than the Angus. At birth the Charolais calf is larger than the Angus calf. Corriedale ewes usually have bigger lambs than Merino ewes.

Postnatal growth and development

The growth that occurs after birth is called **postnatal** (see page 59) growth. At birth the young animal appears to have a relatively large head, long legs, small body and small hindquarters. The nervous tissue and the organs that are of greatest use at this time are relatively well developed. It is important that the young animal has good nervous control and mobility at an early age.

Organs of the body grow at different rates. Essential organs needed for maintenance of life, such as the heart, kidney, oesophagus, lungs and intestines, develop early and are well developed at birth. These organs have a slower growth rate in post-natal life. Other organs, such as the udder and sex organs, are late developing. The carcases of young animals will contain a lot of bone, intestines and other viscera, which are waste products.

The various body regions also change in their relative proportions during postnatal life (Fig. 8.6). The body increases in depth relative to length, and the head and legs become shorter relative to the body as a whole. The head and legs grow first (they are early-developing parts); while the body, loins and hindquarters grow later (they are late-developing parts). The last part of the body to develop to full size is the loin region.

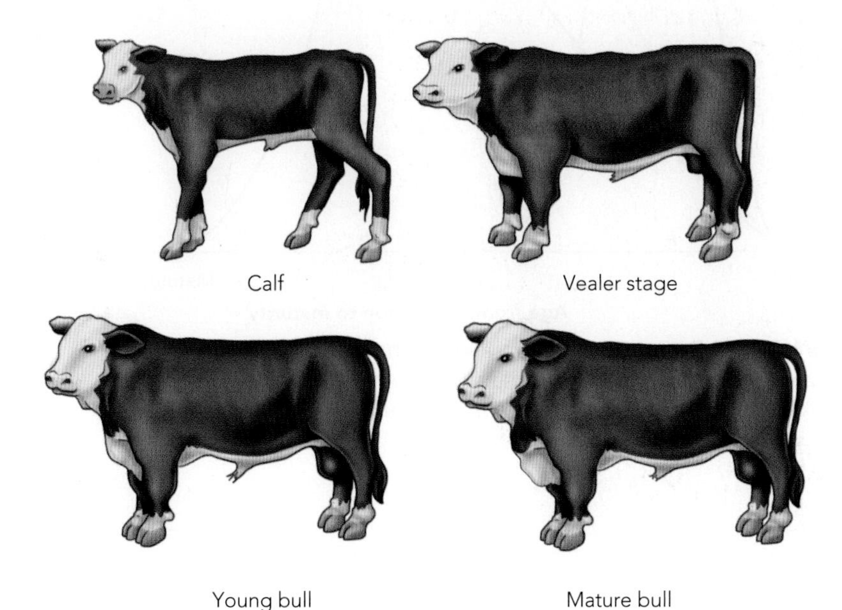

Figure 8.6 Changes in body shape as an animal ages

ISBN 9780170265560

The animal's body is made up from various tissues. These tissues develop in a certain order – brain and nerves, **bone**, **muscle** and **fat** – and compete for the supply of nutrients coming into the animal's body (Fig. 8.7). In a young animal the brain and bone cells take up most of the nutrients.

The age at which body tissues develop is affected by the plane of nutrition and the maturity of the animal (Figs 8.8 and 8.9).

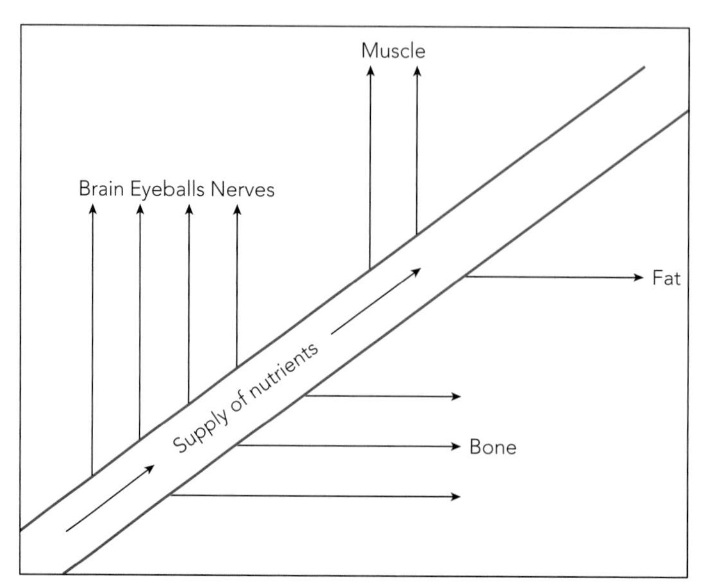

Figure 8.7 How the tissues of an animal's body compete for the supply of nutrients. In a very young animal, the developing brain, eyeballs and nerves demand more nutrients than the laying down of fat.

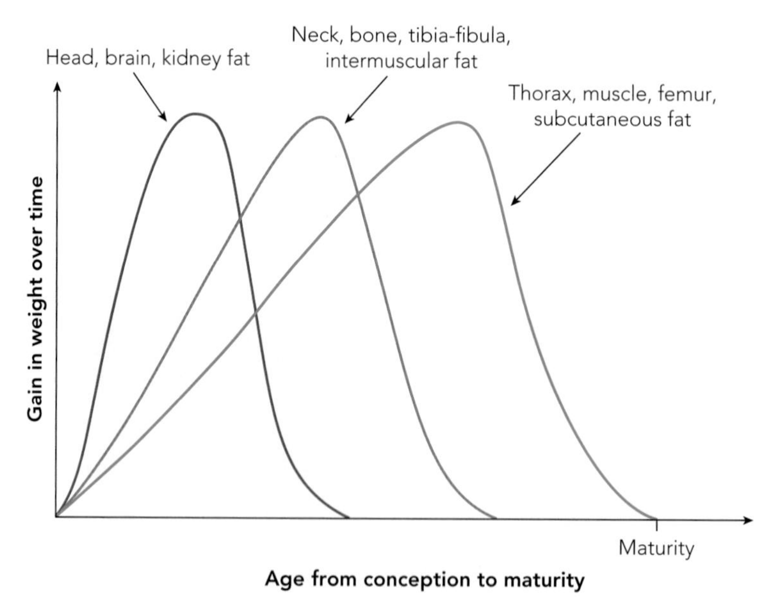

Figure 8.8 The rate of increase in weight, showing the order of development of different parts and tissues of the body, in an early-maturing animal or one under a high plane of nutrition

ISBN 9780170265560

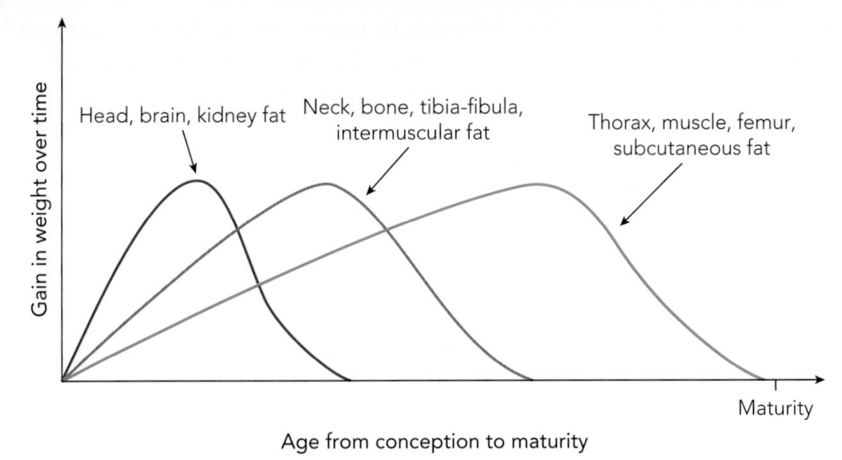

Figure 8.9 The rate of increase in weight, showing the order of development of different parts and tissues of the body, in a late-maturing animal or one under a low plane of nutrition

Fat deposits are formed in the body in a particular order and in several locations.

1 Abdominal fat forms near the kidneys and intestines.
2 Subcutaneous fat forms over the outside of the body muscles just under the skin. This fat fills out the shape of the body and covers any bony projections of the hip, backbone and shoulder.
3 Intermuscular fat occurs in the connective tissues between the muscles or groups of muscles.
4 Intramuscular fat begins to be formed within the muscle or meat fibres. This causes white or creamy patches of fat to appear scattered through the red meat. This appearance is called marbling in beef and gives better flavour to meat.

10 Describe the appearance of the head, legs, body and hindquarters of a young animal at birth.
11 Name three early-developing organs of the animal body.
12 What is the last part of the animal body to develop?
13 Draw a sketch (similar to Fig. 8.7) to show how the tissues of an animal's body compete for the supply of nutrients.
14 What is marbling in beef?

Factors affecting growth after birth

Size of the animal at birth

Generally, the larger the newborn, the faster it will grow and the larger it will be at maturity.

Sex of the animal

Male animals grow faster and reach greater mature sizes than females. Castrated male animals grow at rates midway between those of male and female animals. Males castrated at a young age do not normally develop secondary sex characteristics, and the bones do not develop the same thickness, although the growth in length is not affected. They do not have the broad head and thick neck typical of an entire (uncastrated) male.

Genotype, or breed

Genotype (or genetic makeup) (see page 59) sets an upper limit to an animal's growth and development. Some breeds and strains of breeds will reach mature shape and tissue development at lighter weights and earlier ages. These animals are early maturing, and they can be sold earlier and the cost of feeding them will be less than late-maturing breeds and strains.

Environment

The genotype of an animal sets the upper limit to its possible growth rate and development, whereas the environmental factors will determine its actual rate of growth and development. If the environment is non-limiting, the animal will reach its genetic potential for growth and development.

ISBN 9780170265560

15 List the four factors that affect the growth of an animal after birth.

16 What is the main reason for some farmers producing early-maturing animals?

17 How does climate affect the growth of an animal?

18 How can animals be fed to produce large but lean carcases?

The environment is made up of the following factors: nutrition, climate, disease and stress.

- *Nutrition*. This refers to the amount and quality of feed available to the animal. Supplements can be given when the nutrition is inadequate.
- *Climate*. Extremes in climate may affect growth. In very hot weather the animal may stop grazing. Shade should be provided. In very cold, windy weather the animal uses energy to keep warm. Shelter should be provided.
- *Disease*. Diseases may stunt growth. They can be prevented or controlled by vaccination or drenching.
- *Stress*. This can be reduced by adjusting the stocking rate and using appropriate husbandry practices.

Nutrition has a major effect on postnatal growth. Animals raised in a controlled environment and provided with optimum nutrition will closely approach their genetic potentials for growth and development. The nutrition of an animal affects the development of its parts, organs and tissues. Late-developing parts, organs and tissues are stimulated by a high level of nutrition and stunted by underfeeding.

By controlling or changing the level of nutrition of an animal, the farmer can produce a carcase that has the desired proportions of bone, meat and fat. Feeding an animal on a high level of nutrition in its early stages will produce good growth of bone and muscle (Fig. 8.10). If the quantity of feed is then reduced in the later stages, a carcase will be produced that is large but lacks excessive fat (lean).

Alamy/David Page

Figure 8.10 An animal needs a high level of nutrition in its early stages

Permanent stunting of growth

If periods of weight loss are prolonged or if the amount of weight loss is great, permanent stunting of growth might occur. If starving occurs early in an animal's life, it may be stunted permanently. During a drought young animals should be given feed first, as older animals may be able to recover from a period of starvation.

If an animal is growing without adequate nutrition, bone growth may continue at the expense of muscle and fat, leading to the development of an animal with long legs, a large head and a small body.

ISBN 9780170265560

Compensatory growth

The term 'compensatory growth' is used to explain the accelerated growth shown by animals following periods of mild nutritional restriction (e.g. a short drought). When the feed restriction is over and the animal starts eating again, it is more efficient at using its feed for growth. This new efficiency appears to be partly due to an increased appetite.

During a drought, adult stock may be allowed to lose weight and are then fed a maintenance ration. When the drought breaks and feed improves the animals will show accelerated or compensatory growth.

19 What conditions can cause permanent stunting?

20 During a drought, which animals should be fed first?

21 What is compensatory growth?

22 What type of ration are animals fed during a drought?

Body composition

Farm animals are economically important for supplying meat and other products.

- Intestines are used for making casings for smallgoods, surgical sutures, and strings for musical instruments and tennis racquets.
- Hides are used in the production of leather goods.
- Horns and hoofs are used as sources of protein and neat's-foot oil.
- Pancreas, adrenal and pituitary glands are used in the manufacture of drugs and hormones.

Carcase components

The most valuable part of the animal's body is its carcase. The carcase is what remains of the animal's body after the head, feet, hide (or skin), tail, heart, lungs and all the viscera (internal organs of the abdomen) except the kidneys and the kidney fat are removed. The carcase consists of muscle, bone and fat. An ideal carcase should contain the maximum amount of muscle, an optimum amount of fat and the minimum amount of bone.

The relationship between live weight, carcase weight and dressing percentage is shown by the following formula:

$$\text{Dressing \%} = \frac{\text{carcase weight}}{\text{live weight}} \times \frac{100}{1}$$

The higher the dressing percentage figure, the less waste. Mature animals have relatively small heads and legs, and are able to lay down fat instead of other tissues. This enables them to produce carcases with a higher dressing percentage and higher fat content than young animals.

The proportions of the major carcase components (bone, muscle and fat) change as the animal develops. The proportion of bone and muscle decreases and the proportion of fat increases as the animal becomes heavier.

The ideal carcase

An ideal carcase has a maximum muscle development with some fat, depending on consumer preferences but minimum bone as these last two components make up the less valuable part of a carcase.

Beef producers can use this knowledge to manipulate growth to best fit market specifications. Consumer preferences or tastes are increasingly seeking products that are:

- hormone free
- lean meat cuts with minimum fat
- consistent quality.

Further information on these aspects can be found on the Beef cattle appraisal website.

connect

Beef cattle appraisal

Factors affecting carcase composition

Carcase composition is affected by:

- *Breed*. The carcases of dairy breeds of cattle are usually leaner than those of beef breeds.
- *Sex*. At a given body or carcase weight, females are usually fatter than castrated males, which are fatter than entire males.

ISBN 9780170265560

- *Nutrition*. If growth is restricted by providing inadequate feed, fat is the tissue most affected, followed by muscle, while bone is relatively unaffected.
- *Age*. Earlier maturing breeds (such as Angus and Hereford) also have a larger amount of fat than later maturing animals (such as Limousin and Charolais) at the same age. As animals age, their carcase contains a higher proportion of muscle and fat to bone, as shown in Figure 8.11.

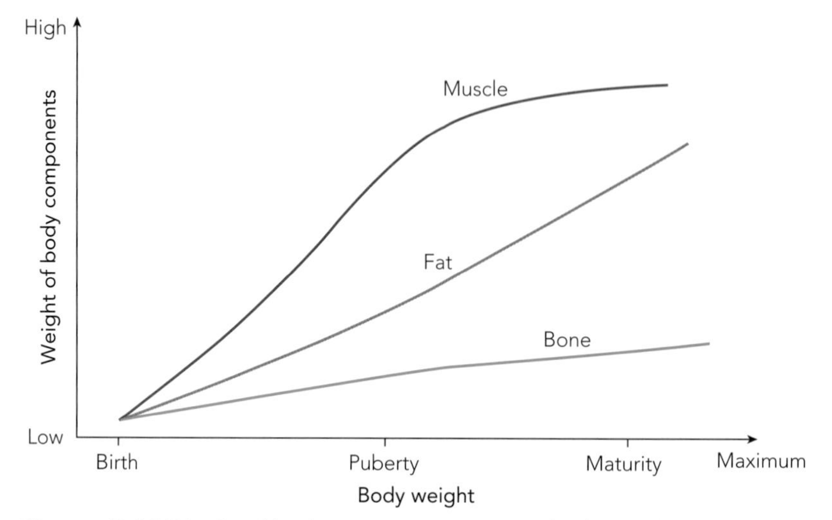

Figure 8.11 Weight of body components against body weight

connect

Market specifications for beef cattle

23 Name four products that come from an animal.

24 Describe the components of an ideal carcase.

25 If a steer had a live weight of 500 kg and a carcase weight of 270 kg, calculate the dressing percentage.

26 List four factors that affect carcase composition.

From newborn to puberty the animal rapidly puts on muscle. From puberty onwards the amount of muscle development decreases and fat increases. Fat is the latest of all tissues to mature but in later life fat increases more rapidly than any other tissue. Cattle are sold at different times to meet the different demands of consumers; for example, vealer, yearling and breaking ox. This means that there should be rapid growth, giving a heavy carcase at a young age. Ideally, the carcase will be blocky with maximum muscle and minimum amount of fat and bone.

Hormones and growth

Several hormones affect the growth of an animal.

- *Growth hormone* (**somatotrophin**) promotes the growth of bone and muscle, and the breakdown of fat. Excessive production of growth hormone leads to gigantism, whereas inadequate production results in dwarfism.
- *Insulin* increases the growth of bone and muscle tissue and is involved in the formation of fatty tissue.
- *Thyroid* hormones are concerned with the regulation of energy metabolism and the metabolic rate of an animal.
- *Sex hormones* include testosterone and oestrogen. Testosterone, a male sex hormone, increases the formation of muscle protein and probably bone growth. Oestrogen, a female sex hormone, inhibits bone growth (particularly the long bones). This helps to explain why female animals are smaller than male animals.

27 What condition results from excessive production of growth hormone?

28 What affect does testosterone have on growth?

ISBN 9780170265560

Management practices that optimise growth and development

The following practices or techniques are used by beef cattle farmers to manipulate animal growth and development.

- *Growth promotants: **hormone growth promotant** (HGP)*. Growth promotants have been used in recent years to increase growth in farm animals. These compounds are similar to sex hormones, and are called anabolic steroids.

  Growth promotants will only be profitable when feed supplies are plentiful.

  In steers the commercially available pellets are implanted into the ear of the animal. Small amounts of the steroid are released gradually over a 2-month period. To reduce the chance of there being drug residues in the meat, the treated animal should be withheld from slaughter for sufficient time to enable the carcase level to decrease. Many countries ban the import of meat that has been treated with anabolic steroids.

  Some examples of HGP substances include:
  - oestrogen compounds that increase protein synthesis to influence growth rates and muscle development hence weight gain
  - progesterone plus oestrogen, which improves weight gain in steers
  - testosterone, which increases growth in heifers and is used to finish off steers.
- *Feed additives*. Urea blocks are used as a source of non-protein nitrogen for ruminants where pastures are protein deficient. Minerals such as sodium and sulphur are necessary mineral supplements for all cattle grazing forage sorghums. This combats the effect of hydrogen cyanide in the stems of sorghum plants which would otherwise be fatal to the animal.
- *Selective breeding (genetics)*. Selective breeding is used to improve the genetic quality of the animal, and therefore increase the upper limit to their growth. Cross breeding (mating of two different breeds of the same or related species) can produce hybrid vigour, where the offspring or progeny grow at a faster rate than either of the parent breeds. Outbreeding involves the use of semen from other countries to widen the genetic pool of the current stock. This increases genetic variability and increases favourable characteristics.

29 Where are hormone growth promotant pellets implanted in steers?

30 Why do hormonally treated steers have to be withheld from slaughter for a short period after being implanted?

31 Describe three techniques used by beef farmers to manipulate animal growth and development.

ISBN 9780170265560

Chapter review

Things to do

1. Weigh a chicken, piglet, lamb or calf each week for 10 weeks. Record weight gain against time.
2. Measure the girth, shoulder height, body length and head length of a lamb or calf each week for 10 weeks. Which measurement shows the greatest increase?
3. The following table shows how the body weight changes with the age of the pig.

Age (weeks)	Body weight (kg)
Birth	1
1	3
5	11
10	25
15	38
20	55

   Graph the body weight against age. (Age is on the x-axis.)

connect

Poultry Hub: stages in growth and development of a chicken

4. The stages in the growth and development of a chicken for the first 21 days of life are shown on the Poultry Hub website.

   After viewing the video, undertake the following procedure to study the development of chicken embryos.

   a Obtain fertilised eggs from a hatchery.
   b Examine the eggs after 2, 9 and 12 days.
   c Place the eggs in large Petri dishes.
   d Open part of the shell and membrane with forceps and scissors.
   e Draw and describe the main features of the embryo.
5. Perform a 'carcase analysis' on a broiler chicken at market weight. Determine the percentage of feathers, bone, muscle, fat and wastes.

Things to find out

1. Which countries import marbled beef from Australia?
2. List the farm management practices that may increase the proportion of muscle and decrease the proportion of fat on an animal's carcase.
3. Discuss why many cattle breeders join their heifers to small beef bulls like the Angus and Murray Grey.
4. How does the birth weight of animals affect their survival? (In your answer, mention both light and heavy offspring.)

connect

Prices and markets

5. Research the ideal time to sell steers to meet consumer demand for vealer, yearling and breaking ox markets. The Prices and markets website may assist you.
6. During recent years some farmers have used growth promotants (anabolic agents) to increase the growth of their farm animals.
   a On which animals have these growth promotants been used?
   b Does this practice hold any potential dangers for human health?

ISBN 9780170265560

Extended response questions

1. Efficient animal growth and optimum development are essential aims for all profitable animal enterprises. For a farm animal you have studied:
   a. describe the normal process of growth and development
   b. explain how a farmer can use an understanding of the growth and development process to improve the efficiency of this farm animal.
2. Many livestock producers use growth hormones in their animal production systems.
   a. Discuss how these hormones work.
   b. Explain how a knowledge of natural hormone functions can be used in animal management.
   c. Explain why some producers give synthetic hormones to their animals.
3. The growth of an animal after birth is affected by a number of factors.
   a. Discuss these factors.
   b. Explain how farmers can use a knowledge of these factors to achieve optimum growth rates in their animals.
4. Evaluate management techniques available to farmers to manipulate growth and development, including use of hormone growth promotants (HGPs), feed additives and genetics.
5. Write an essay describing the factors that affect the size of an animal at birth.

CHAPTER 9

Outcomes

Students will learn about the following topics.

1. The nature of chromosomes and genes.
2. Gene action.
3. Genotype and phenotype.
4. Heritability.
5. Selection.
6. Progeny testing.
7. Performance testing.
8. Breeding systems and the genetic basis to improve quality and production of animals.
9. Genetic engineering.

Students will be able to demonstrate their learning by carrying out these actions.

1. Define the term 'genetics'.
2. Describe the work of Robert Bakewell.
3. Describe the role of genes and chromosomes in genetics.
4. Explain the terms 'homozygous' and 'heterozygous'.
5. Show how the genetic control of characters can be determined by only one gene pair or many gene pairs.
6. Outline examples of genetic abnormalities.
7. Explain the terms 'genotype' and 'phenotype'.
8. Discuss the concept of heritability.
9. Discuss how selection is used to improve production.
10. Describe progeny testing and performance testing.
11. Outline the main mating or breeding systems that are used with animals.
12. Discuss the use of breeding systems in animal production systems including crossbreeding and line breeding.
13. Outline the role of objective measurement and heritability on breeding programs for a named industry.
14. Explain the term 'genetic engineering'.

ISBN 9780170265560

BREEDING AND GENETICS

Words to know

allele different form of one gene (e.g. the two forms of the gene for plant height in peas, 'T' tall and 't' dwarf)

carrier an animal that can transmit a hereditary characteristic or disease to its offspring, without itself showing the characteristics or any symptoms of the disease

chromosome a thread-like structure within the nucleus of a cell, which carries genes

crossbreeding the mating of unrelated animals of different breeds or the crossing of unrelated plants (i.e. plants with different genotypes)

deoxyribonucleic acid (DNA) the biological material that stores genetic information for an organism; one chromosome is one DNA molecule, a double stranded molecule in the shape of a helix; along each strand is a sequence of bases comprising many genes

first-cross first generation hybrid offspring produced from mating of parents with different characteristics

gamete sex cell

gene a part of a chromosome that determines what a particular characteristic will be; it is a sequence of bases on a DNA strand; one gene produces one polypeptide or protein

genetic engineering the manipulation of an organism's genetic material by removing, adding or reversing a gene or genes to produce a new type of organism

genetics the study of inheritance

genotype the genetic makeup of an individual organism, consisting of its chromosomes and genes, half of which were inherited from each parent

heterozygous having two different alleles for any one gene; that is, having one dominant and one recessive gene for a particular character or trait

homozygous having two identical alleles for any one gene; that is, having either two dominant or two recessive genes for a particular character or trait (the genes within each pair are the same)

line breeding a type of inbreeding based on a single common ancestor (a sire or dam) used over several generations of mating

selection differential a measure of the superiority of the selected group over the group before selection was made

transgenic animals or plants that have been genetically modified by the addition of a gene or genes from another species

ISBN 9780170265560

Introduction

The ability of an animal to produce meat, wool, mohair, eggs or milk is governed by its heredity, or ancestry. The genetic makeup of an animal sets an upper limit to its possible production. However, if production in an animal industry is limited by an environmental factor (e.g. nutrition), improvement of this factor will do much more to increase productivity than attempts to improve the quality of livestock through **genetics**.

Where there are no serious limiting factors, considerable progress can be made in improving the quality of livestock by using proven methods of animal breeding. Animal genetics has achieved excellent results in the poultry industry in meat and egg production.

Robert Bakewell (1725–95) began the planned improvement of livestock by breeding. He based selection on productivity rather than the appearance of the animal. Bakewell's methods were used by other breeders in his era. Unfortunately, a period followed where breeders concentrated on 'pure breeds'. In order to maintain the quality of the new breeds, they set up breed standards. These standards acted as a barrier to genetic progress, because breeders were concerned with selecting animals of good appearance rather than high productivity.

Rapid progress in raising production per animal can be made where the breeder has one main aim – to improve production. To achieve this aim the production of animals must be measured or assessed as accurately as possible. In addition, a flock or herd must have a high level of reproduction in order to maintain itself, and to enable selection of replacement animals and sale of surplus animals.

connect

Bull selection

1 How did Robert Bakewell select his animals?
2 Name and briefly describe the two main types of observable characteristics that are used when selecting animals.
3 On how many characteristics does a farmer select breeding animals in order to make rapid genetic progress?

Every animal has a large number of observable characteristics (traits), which can be used when selecting animals to be mated. The main characteristics can be classified as follows.

- Qualitative characteristics can be given a name or description, but not numerically expressed (e.g. polled condition of sheep and cattle, and the colour of coat or feathers).
- Quantitative characteristics can be expressed numerically or measured. They include:
  - discrete characteristics, which can be counted (e.g. the number of pigs in a litter, teats on a cow, or eggs laid by chickens)
  - continuous characteristics, which can be measured but not counted (e.g. litres of milk per cow per day).

Progress in animal breeding will be more rapid if the farmer selects breeding animals on only a few characteristics.

Chromosomes and genes

Genetic material is located in the nucleus of every cell in the body. Each cell (with the exception of sperm and ova) contains the same amount of genetic material. The genetic makeup of one individual is different from that of all other individuals in the species, unless the individual has an identical twin.

The basic unit of genetic materials is the **gene**. The gene is a unit of hereditary material located on a chromosome that, by itself or with other genes, determines a characteristic of an organism. The genetic makeup of an animal is called its **genotype**.

In the cell nucleus, genes are strung together along long strands called **chromosomes**. Except for the sex chromosomes (X and Y chromosomes), chromosomes occur in matching pairs. The chromosomes in a pair are the same length. When an animal is conceived, each parent provides half the chromosomes of the individual through its sperm or egg.

The same number of chromosomes appears in all body cells of animals of the same species. There are differences between species in the number of chromosomes.

4 What is a gene?
5 What is a chromosome?

Table 9.1 The number of chromosomes and number of chromosome pairs in farm animals

Species	Number of chromosomes	Number of chromosome pairs
Sheep	60	30
Cattle	60	30
Pigs	38	19
Poultry	78	39

ISBN 9780170265560

Gene action

Genetic control of characteristics can be determined by only one gene pair or many gene pairs working together.

The action of single genes

Some characteristics, such as coat colour or the presence of horns in sheep and cattle, are controlled by single gene pairs. The relationship between the genes in a pair can be either dominant–recessive or co-dominant.

Complete dominance

Consider one gene pair. Each gene of that pair can be of two possible forms. The two possible forms are slightly different in chemical structure. One form is dominant and the other is recessive. The presence of the dominant form masks the effect of the recessive form.

The occurrence of horns on cattle is an example of complete dominance. A dominant gene is represented by an upper case letter and a recessive gene by the same letter, but in the lower case. Thus, in cattle, horns are caused by the recessive gene p, but the polled gene P is dominant to the horned gene p. A horned animal therefore has the genes pp, whereas one that is polled may have genotypes PP or Pp. Figure 9.1 shows what happens when a pure-breeding, or **homozygous**, polled animal (PP) is crossed with a pure-breeding, or homozygous, horned animal (pp). Animals that are not purebred for a characteristic are called **heterozygous** for that character.

Inheritance of horns in cattle can also be shown using a grid system called a Punnett square (designed by R.C. Punnett, a British geneticist), as shown in Figures 9.2 and 9.3 (page 134).

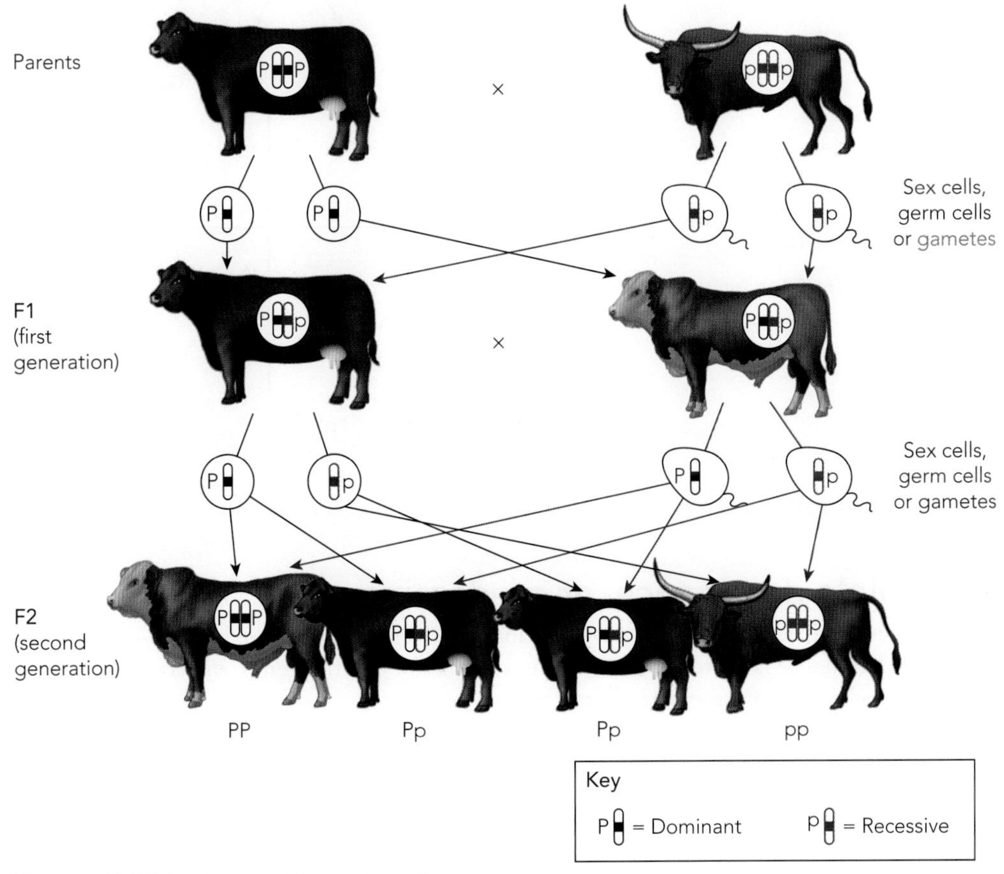

Figure 9.1 Inheritance of horns in cattle

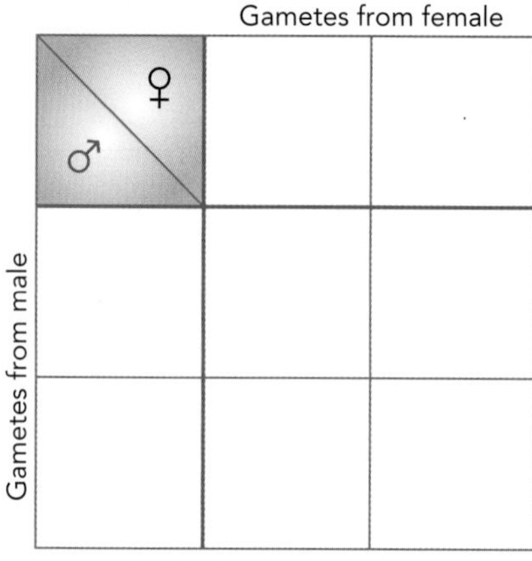

Figure 9.2 A Punnett square

Parents (2n, or full complement of chromosomes in cells)

Polled PP x Horned pp

Gametes (n, or half the chromosome number as found in sperm or ova)

P P p p

Gametes

F1 PP / pp	P	P
p	Pp Polled	Pp Polled
p	Pp Polled	Pp Polled

All the F1 will look like the dominant form (i.e. phenotype ratio: all polled)

All the F1 have both genes (i.e. genotype ratio: all heterozygous (Pp))

If two of these FI, heterozygous cattle (Pp) are crossed, the F2 phenotype and genotype would be:

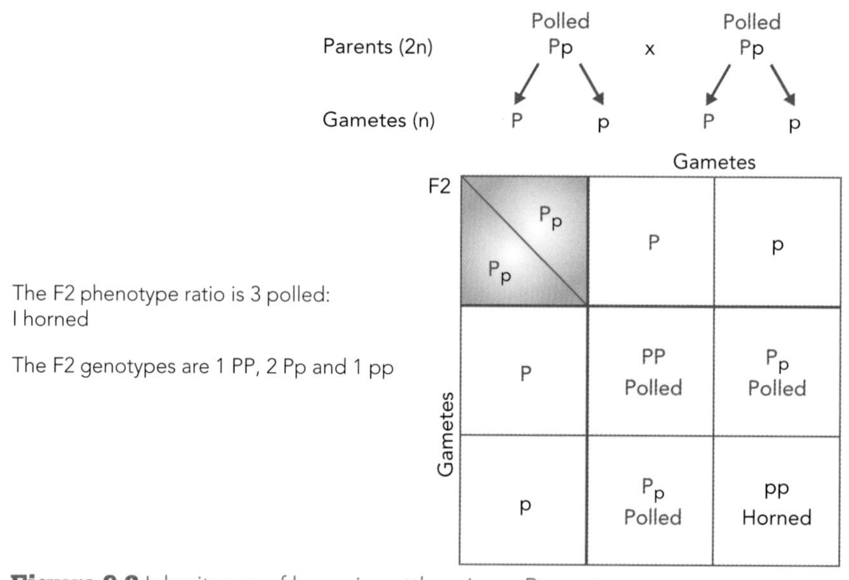

Figure 9.3 Inheritance of horns in cattle using a Punnett square

Incomplete dominance, or co-dominance

Not all genes have a dominant–recessive relationship. Some gene pairs exhibit co-dominance, where the dominance is shared. Neither gene is dominant and the genes' *effects* sometimes 'blend' to produce an intermediate characteristic.

Coat colour in Shorthorn cattle is an example of incomplete dominance; when a red Shorthorn bull is mated with a white Shorthorn cow, the progeny are roan in colour.

Red colour is controlled by a single pair of genes, which can be called RR. White colour is controlled by a single pair of genes, which can be called WW. If a red bull, RR, is mated with a white cow, WW, at fertilisation R and W come together forming an RW pair of genes in the calf. A mixture of red and white colour is obtained, producing a roan calf.

6 Name two characters that are controlled by single gene pairs.

7 Name one example of complete dominance.

8 What is incomplete dominance, or co-dominance?

ISBN 9780170265560

If a roan bull (RW) is mated with a roan cow (RW), the colour of the calves can be red, white or roan. Each male gamete (germ cell, or sex cell) from the roan bull contains either an R or a W gene. Similarly, each female gamete from the roan cow contains either an R or a W gene. The possible combinations of these genes are shown in Figure 9.4. If sufficient roan bulls and cows are mated the proportion of red : roan : white will be 1 : 2 : 1.

9 What happens if a roan Shorthorn bull is mated with a white Shorthorn cow? (Show all working.)

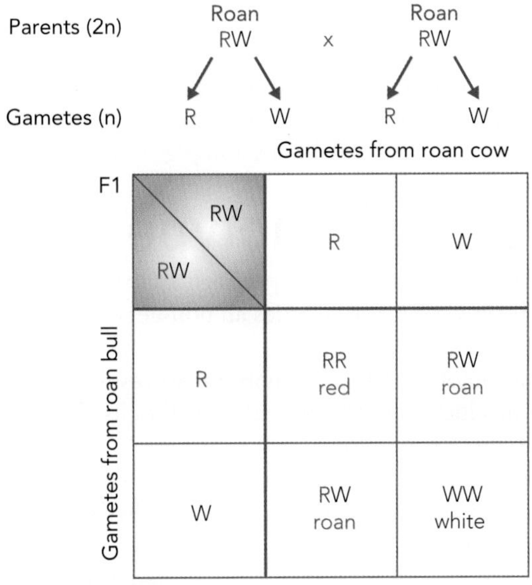

Figure 9.4 Inheritance of roan coat colour

Multigene inheritance

Most of the economically important characters, such as productive efficiency and production of wool, meat and milk, are controlled by many genes acting together. This is called multigene, or polygenic, inheritance.

When a character is controlled by one gene, there are three possible genotypes (AA, Aa and aa). If a character is controlled by two genes, there are nine possible genotypes. It is thought that milk production in dairy cattle is controlled by at least 100 genes, and therefore there are a large number of possible genotypes.

When the possible number of genotypes becomes very large, the effect on any single gene is too small to detect. Milk yield is a continuous variable, which ranges from very low quantities per lactation to very high quantities. If the milk yield of cows in a herd is plotted, the distribution will be as shown in Figure 9.5. Some cows will produce a very small or very high yield; most will produce an average yield or slightly higher or lower than average.

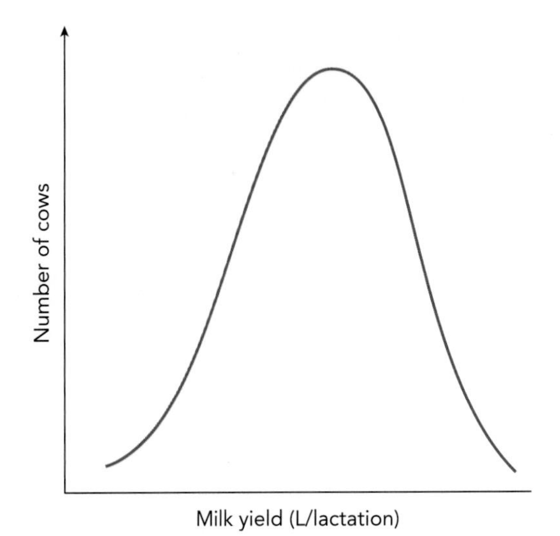

Figure 9.5 A curve of normal distribution

The **phenotype** (see page 59) of an animal (e.g. the measured milk yield) is affected not only by the genotype, but also by the environment (feeding, disease, climate and management). Therefore, the variation in milk yield of a group of cows is due partly to differences in genotype and partly to differences in phenotype.

Abnormalities

A large number of mutations have occurred and are occurring in domestic animals. They give rise to defective characters, or fancy points. A mutation arises when nuclear material called **deoxyribonucleic acid (DNA)** replicates and the new DNA is not a direct copy of the original. Such changes involve the change of a gene from one **allele**, or gene form, to another. These new genes are most often deleterious (harmful) genes. The causes of mutations vary. Some mutations are spontaneous and their origins are unknown. Others are caused by ionising radiation (including gamma rays and X-rays) and chemical mutagens. (A mutagen is a chemical capable of causing mutation.)

10 Name three characteristics that are controlled by multigene inheritance.

Gene mutations usually give rise to abnormal development, which adversely affects production of meat, milk, wool and so on. Many gene mutations give rise to abnormal development in the embryo, and so the young often fail to develop properly and foetal death results. These gene mutations are called lethal factors. They are frequently found in poultry and cause death in the shell before incubation.

White colour is often associated with abnormalities. White Shorthorn heifers often suffer from an underdevelopment of the uterus and vagina, which renders them infertile. This is known as 'white heifer disease'.

Mutations that cause embryonic death lead to decreased fertility in the breed or strain. Other mutations cause defects or abnormalities that make the animal less vigorous and less efficient commercially; for example, scrotal hernia in pigs retards growth.

11 What is a mutation?

12 Name two causes of mutations.

13 What abnormalities are associated with white colour in animals?

Genotype and phenotype

The genetic makeup of an animal is its genotype. Each animal is a product of its genes and the environment. The environment includes all non-genetic factors, such as climate, nutrition, disease and stress.

The interaction of the genotype and the environment results in the phenotype. The phenotype includes the production of the animal as well as its shape or appearance. This interaction can be shown by the equation:

Genotype + environment = phenotype

The genotype sets the upper limit of production. Whether this will be reached depends on the environment; for example, a dairy cow with a genetic potential to produce 30 litres of milk per day might only produce 22 litres if she is underfed or diseased.

Animal production can be improved by altering the environment of the animal or by altering the genotype of the animal (through selective breeding) to better suit the environment.

The Australian Milking Zebu (AMZ) has been selectively bred to meet the need for acceptable milk production in areas of high temperature and humidity. The AMZ is tick resistant, heat tolerant, fertile and can produce milk from poor quality pasture. The breed was developed by the CSIRO, by crossing Sahiwal and Red Sindhi cattle (both breeds from Pakistan) with Jersey cattle.

14 What factors are covered by the term 'environment'?

15 What is the genotype of an animal?

16 What is the equation linking genotype, environment and phenotype?

Heritability

When an animal strongly resembles its parents for a character that is little affected by the environment, the character is said to be highly heritable. When the environment easily affects a character of an offspring, the character is said to be of low heritability. The heritability of a character is a measure of how much the genotype influences the phenotype. If the influence is strong, a high-producing animal will have high-producing genes. Therefore, heritability estimates are a measure of how much the superiority of a parent is passed on to the offspring.

The basis of increasing productivity through selective breeding is the selection of animals of superior genotype as parents of the next generation. Heritability estimates make it possible to predict the possible rate of progress in breeding.

The symbol for heritability is h^2. Heritabilities are described as either a decimal or a percentage. If a characteristic is said to have a heritability of 0.16, or 16%, this means that only 16% of the variation found in that character in an animal population is due to differences in genotype. The rest of the variation is due to the environment.

The highest possible heritability would be 100%. Any heritability over 30% is considered to be high, and selecting for characteristics with high heritability percentages will result in rapid progress. If a cow has a measured milk yield 500 litres above average and she is mated to a bull with a history of siring daughters who have similarly superior milk production, the offspring of this mating would be expected on average to inherit 80 litres of that superiority (that is, 500 L × 0.16), assuming a 16% heritability.

In general, characters such as growth rate and milk yield have heritabilities ranging from 0.15 to 0.50, while characters relating to fertility have lower heritabilities.

ISBN 9780170265560

Identical twins are useful for experimental purposes, because they can be used to measure the effects of nutrition and management independent of genetic influences. For example, by placing one twin cow in a high-yielding dairy herd and the other in a low-yielding herd, it can be shown that any differences in yield between the two cows are due to feeding and management rather than genetics.

17 What is heritability?

18 What characters have high heritabilities?

Selection

Selection is choosing animals to be mated. In selecting individuals the aim is to choose superior animals, which will not only maintain their superiority for their lifetime, but will also pass the superiority on to their offspring. Selection aims to raise the average value of phenotypic characters by changing the genotype; that is, choosing animals that are genetically superior.

The **selection differential** is a measure of the superiority of the selected group over the group before selection was made. If the distribution of milk production in a large herd is plotted, a curve would be obtained as shown in Figure 9.6. If it is decided not to keep any cow for breeding that produced less than 18 litres per day, the mean yield of the selected group would be higher than the mean before selection was made. The difference between the mean of the selected group and the mean of the unselected group is called the selection differential. The equation is:

Selection differential(s) = mean of selected group – mean of unselected group

If the mean milk yield before selection was 21 litres per day and the mean after selection was 23 litres per day, the selection differential would be 2 litres.

19 What is selection?

20 If the pressure of selection is strong and only the best animals are kept, what happens to the rest?

21 Write the equation for the selection differential.

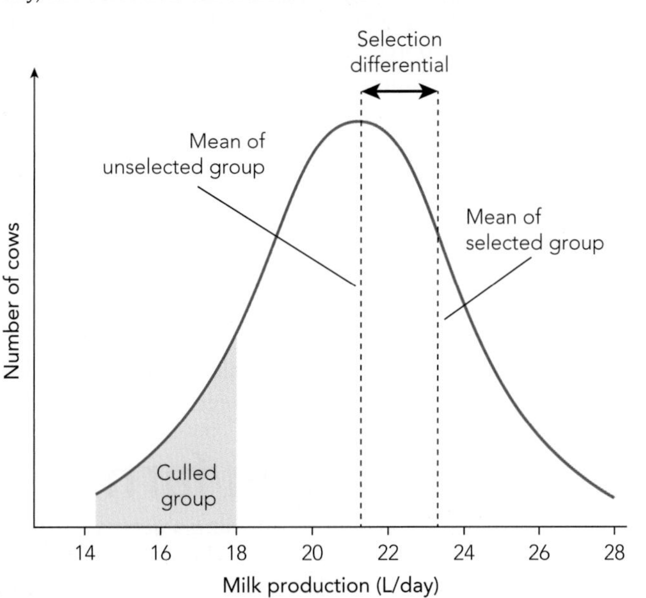

Figure 9.6 The selection differential

Progeny testing

Progeny testing measures the genetic value of the animal by the production records of its offspring or progeny. Since a male produces many more offspring during his lifetime than a female, the progeny test is applied more widely to males than females.

The progeny test is important for characteristics that are not applicable to the male himself, but can be transmitted via his genes to his female offspring, such as milk and egg production. If, for example, the character is milk production, which cannot be measured in the male, estimates can be made of the male's genetic value by looking at the production records of his daughters. The average production of his daughters must be compared with the average production of other cows. If the bull's daughters are better producers than the other cows in the same herd, he is a good bull; if they are worse producers, he is below standard.

connect

Genomics

Progeny testing is expensive and takes a number of years to carry out. Outstanding sires are discovered this way and farmers can buy semen from these superior animals. By using artificial insemination the semen can be introduced into the farmer's animals, and rapid genetic progress can be made.

22 What is progeny testing?

23 For which products is progeny testing important?

24 What is performance testing?

25 What characters is performance testing used for?

Performance testing

Performance testing measures the characters in the sire himself, as contrasted to progeny testing, which measures the characters in the offspring. The aim of performance testing is to measure the genetic value of an animal in relation to a particular character.

Its greatest use is in selecting males for characters such as growth rate and feed conversion efficiency, which can be measured on a sire without slaughtering him.

When a character is fairly strongly inherited and where the animals are maintained under identical environmental conditions and management, performance testing will give quicker results than progeny testing

Breeding systems

Mating is quite distinct from the selection of breeding stock. The farmer still has to decide which animals are paired together. Normally, mating follows selection.

Mating or breeding systems can be classified as:

- *Random*. Selected animals are mated at random without regard to their relationship.
- *Inbreeding*. This involves the mating of close relatives – brothers with sisters, mothers with sons and fathers with daughters. It produces a uniform line of animals. Inbreeding can also bring together undesirable genes; for example, dwarfism in cattle.
- ***Line breeding***. This is a type of inbreeding based on a single common ancestor (a sire or dam) used over several generations of mating. Its value depends on the degree of superiority of the outstanding individual used. A high degree of uniformity of type and production is obtained. Valuable traits of an animal are passed on to several generations.
- ***Crossbreeding***. This involves the mating of unrelated animals of different breeds to produce **first-cross** offspring with different characteristics from their parents. In this system new genes are brought into the flock or herd. The crossbred progeny, or hybrids, are usually more vigorous than either of the parents. This phenomenon is called hybrid vigour, or heterosis.

  Heterosis is shown, for example, by poultry (White Leghorn crossed with Australorp birds produce more eggs than purebred Leghorn or Australorp birds) and cattle (Brahman crossed with Hereford calves grow at a faster rate than purebred Brahman or Hereford cattle).

connect

Beef cattle breeding systems

26 Describe one disadvantage or problem that can result from inbreeding.

27 What is heterosis, or hybrid vigour?

Breeding programs on farms

Good breeding programs are based upon defined breeding objectives. A breeding objective describes the 'ideal' animal a producer aims to breed and is influenced by market requirements. Market specifications for animals being sold form the basis of market requirements that reflect consumer demand. Female animals kept for breeding would be selected on the basis of production characteristics, for example quantity and quality of milk, and conformation, for example udder structure and ease of calving.

The main decision then becomes which animal is retained within the herd for breeding and to which sire or dam it will be mated. Historically, selection was based on visual assessment but more recently breeding programs based on the measurement of production and confirmation characteristics have become more important.

Subjective assessment, objective measurement and heritability

Selection should consider both subjectively measured traits, such as those measured through visual assessment, and objectively measured traits, such as those identified through measurement or genetic assessment.

ISBN 9780170265560

The difficult task of selecting breeding stock based on genetic assessment is made easier and more precise through Estimated Breeding Values (EBV) for cattle and goats, Australian Sheep Breeding Values (ASBV) for sheep and Australian Breeding Values (ABV) for dairy cattle.

- *Dairy cattle*. Objective measurement involves giving a measurement of productivity that quantifies the characteristic; for example, litres of milk or butterfat percentage in milk. Within the dairy industry there are two types of objective measurement.
  1. *Herd recording*. This form of objective measurement is used to improve milk production (quantity and quality). The farm manager can:
     - cull low milk producing cows
     - cull cows that have low milk fat and protein percentages in their milk
     - use information to select cows for embryo transfer.
  2. *ABVs*. These values can be used when selecting straws of semen from bulls to improve milk yield, protein percentage in milk, fat percentage in milk, ease of calving, temperament and conformation.
- *Beef cattle*. Performance recording of beef cattle started in 1985, within individual herds and later expanded to a cross herd scheme. Breedplan® calculates estimated breeding values (EBV) for a range of traits, such as:
  - weight – birth weight, mature cow weight
  - fertility – scrotal size, days to calving, calving ease
  - carcase – carcase weight, eye muscle area, marbling

  Included in the calculation of EBVs are the animal's own performance, the performance of known relatives, the heritability of each trait and the relationships between the traits. All breeds of beef in Australia use Breedplan®.
- *Sheep*. Breeding programs within the sheep industry focus on the production of fibre or meat. Sheep Genetics Australia is a database that brings together objective genetic measurements for both meat and wool.
  - Meat – LAMBPLAN measures live weight and fat percentages of carcases.
  - Wool – various state departments of agriculture have operated performance recording services for wool traits.
  - Sheep Genetics Australia (SGA) calculates ASBVs via programs called MERINOSELECT for the wool and LAMBPLAN for the prime lamb sectors of the market. A wide range of ASBVs will be reported to breeders including wool, growth rates, carcase quality, reproduction, temperament and internal parasite resistance.

connect

Genetics and breeding

connect

Australian Dairy Herd Improvement Scheme

connect

Breedplan®

connect

Sheep genetics

Genetic engineering

Genetic engineering involves the manipulation of an organism's genetic material. Scientists have developed techniques to isolate, manipulate and reproduce identified segments of DNA from viral, bacterial and animal sources. Genetic engineering has two distinct purposes.

1. *Gene transfer*. An unlimited number of copies of a particular segment of DNA are taken from any species. This 'cloned' DNA may be transferred to other members of the same species from which it was derived, or to members of another species in order to replace a defective gene. The cloned DNA may enable identification of individuals whose corresponding segments of DNA are defective as a result of mutation.
2. *Gene products*. Unlimited quantities of gene products; for example, antigenic proteins of foot-and-mouth virus (for use in vaccines) or rare proteins like growth hormone, have been mass-produced in engineered forms of the bacterium *Escherichia coli*.

Genetically engineered pigs

Pigs can be given extra genes to make them grow faster, eat less feed and produce leaner meat than their unmodified or 'normal' cousins. These so-called **transgenic** pigs are made using a process that injects extra genes for pig growth into pig's eggs that have recently been fertilised. The added pig growth hormone gene is designed to be switched 'on' when zinc is added to the pigs' diet. These animals provide greater profits for farmers.

28 What is genetic engineering?

29 Briefly explain how gene transfer and gene products are used in genetic engineering.

Chapter review

Things to do

1. Visit a piggery. How does the owner select:
   - a boars for mating?
   - b sows for mating?
2. Thirty bulls were housed under identical conditions at a testing station and fed identical rations. The bulls were tested for average daily weight gain and feed conversion ratios.

Bulls	Average daily weight gain (kg/day)	Feed conversion ratio (FCR/FCE)
Group average	1.14	7.04
Midnite	1.18	6.15

   Would Midnite be a good bull to buy? Why or why not?
3. Mannosidosis is a lethal inherited disease occurring in Angus and Murray Grey cattle. The disease results in the death of some calves at birth. Surviving calves show an awkward gait, tremors of the head and aggressive behaviour.

   The mannosidosis enzyme is controlled by a single pair of genes, one from each parent. The pair of genes of normal pure-breeding animals is described as MM, while the defective gene pair in diseased animals is mm. Heterozygous animals with one normal and one defective gene (described as Mm) appear normal, but are **carriers**.
   - a Work out the genotype and phenotype of:
     - i a carrier bull and a normal cow
     - ii a carrier bull and a carrier cow.
   - b If blood tests to find out the genotype were taken of all bulls before they were used for breeding, what should happen to carrier animals?

Things to find out

1. Give three examples of genes that are lethal in animals.
2. What does the term 'genetic engineering' mean? Give two examples of how this can be used to increase animal production.
3. Some mutations are not particularly abnormal or defective, and have been used by breeders as the basis of new 'fancy' breeds. List some examples of fancy breeds (e.g. porcupine pigeon).
4. Breedplan® has been producing estimated breeding values (EBVs) for beef cattle for several years.
   - a What is Breedplan®?
   - b What are Breedplan® EBVs?
   - c Describe two ways that a farmer could use EBVs in herd management.
   - d Explain the heritability/accuracy figures given to EBVs on beef bulls.
5. Why have certain breeds of cattle, for example Brangus, become very popular in recent years?
6. Explain the genetic basis for the A2 milk product available in stores and which group of consumers this product is being marketed to.

connect

Beef cattle breeding and selection

ISBN 9780170265560

Extended response questions

1. There are a number of breeding systems, including: inbreeding, line breeding and crossbreeding.
   - a Briefly describe each breeding system.
   - b Discuss the advantages and disadvantages of each breeding system.
   - c Discuss the genetic basis for each breeding system.
2. Genetic engineering offers prospects for manipulation and modification of livestock that are not possible with conventional breeding procedures.
   - a Outline some of these genetic engineering methods.
   - b Discuss some of the problems that may arise.
3. Outline the role of objective measurement and heritability on the breeding programs of farms, using one specific industry program as an example.
4. Discuss why genetic engineering is an ethical issue relevant to a named animal production system.
5. 'Animal production can be improved by altering the environment of the animal or by altering the genotype of the animal (through selective breeding) to better suit the environment.' Discuss this statement, using examples you have studied.

ISBN 9780170265560

CHAPTER 10

Outcomes

Students will learn about the following topics.

1. The nature and impact on animal production systems of microbes, invertebrates and pests.
2. Animal diseases.
3. The effects of disease on animals.
4. Types of diseases.
5. Causes of disease.
6. Infection and disease.
7. Resistance and immunity in animals.
8. Important animal diseases in Australia.
9. Disease control.
10. Integrated pest management (IPM).
11. An IPM program for an animal production system.
12. The evaluation of an IPM program.

Students will be able to demonstrate their learning by carrying out these actions.

1. Define the terms 'disease' and 'pest'.
2. Describe the effects disease has on animals.
3. Explain the difference between infectious and non-infectious diseases.
4. Explain the difference between endoparasites and ectoparasites.
5. Describe the main types of diseases – hereditary, metabolic, microbial and metazoal.
6. Outline the primary and secondary causes of disease.
7. Explain the terms 'vaccine', 'antibody' and 'antitoxin'.
8. Describe the complex interrelationship, or interaction, between the host, the pathogen and the environment for an animal disease.
9. Explain how infective agents enter the body.
10. Discuss resistance and immunity in animals.
11. Describe an important animal disease and an animal pest.
12. Evaluate methods that can be used to control and prevent animal pests and disease.
13. Research an IPM program for an animal production system and evaluate its effectiveness.

ISBN 9780170265560

ANIMAL DISEASES AND PESTS

Words to know

antibody a chemical substance (a protein) made by animals in response to bodily invasion by a pathogen, which combines with the pathogen and renders it harmless

antiserum or **antitoxin** the preserved serum of an animal that has previously had a specific disease or has been injected with a vaccine; the serum contains a high concentration of antibodies

disease triangle a conceptual model that shows the interactions between the environment, the host and an infectious agent

ectoparasite an organism that lives permanently or temporarily on the surface of the host's body

endoparasite an organism that lives in the internal organs of the host

infectious disease a disease caused by a pathogen transmitted from a diseased individual to a healthy one

integrated pest management (IPM) the use of two or more methods to control pests or diseases

perennial a plant that continues to grow year after year

pest any organism that injures, irritates or damages livestock, livestock products or plant products, and can adversely affect productivity

phagocyte a specialist cell that engulfs bacteria and foreign particles and digests them

vaccine a biological preparation that improves immunity to a particular disease; a vaccine typically contains an agent that resembles a disease-causing micro-organism, and is often made from weakened or killed forms of the microbe, its toxins or one of its surface proteins

ISBN 9780170265560

Introduction

An animal disease is any kind of upset in the normal body functioning that has an adverse effect on the animal. This upset or deviation from normal leads to signs and symptoms of disease. In animals, symptoms can take the form of an increase in body temperature (or fever), an increase in the number of white blood cells, loss of appetite, mucus discharge, skin rashes, or loss of body weight and production.

Diseases are caused by micro-organisms, animal parasites or plant poisons. Diseases in animals can be infectious or non-infectious. In **infectious diseases**, the disease-causing agent, or **pathogen** (see page 59), is transmitted from a diseased individual to a healthy one. Non-infectious diseases cannot be transmitted from one individual to another. These diseases are not caused by pathogens, but have a number of causes. Some are the result of genetic disorders and are called hereditary diseases. Others are caused by one section of the body not working normally and are called **metabolic diseases**. A distinction is often made between endoparasites and ectoparasites. **Endoparasites** live in the internal organs of the host, whereas **ectoparasites** live permanently or temporarily on the surface of the host's body.

1 Explain the difference between infectious and non-infectious diseases.
2 What is a pathogen?
3 Explain the difference between endoparasites and ectoparasites.
4 What is the difference between a disease and a pest?

The effects of disease on animals

Animals used in agricultural production systems are susceptible to infection by a wide range of pathogens, including bacteria, protozoa, viruses, fungi, insects, arachnids, platyhelminths and nematodes.

Disease and **pests** in animals may have the following effects:

1 death of a small or large number of the affected animals (e.g. caused by tetanus, blackleg or pulpy kidney)
2 weakening and weight loss (e.g. caused by ephemeral fever or three-day sickness)
3 stunted growth (caused by internal parasites, such as worms, in calves, lambs and pigs)
4 lower production (e.g. cows with mastitis produce less milk)
5 infertility and lower calving or lambing percentages (e.g. caused by vibriosis or leptospirosis)
6 reduced sale price of animal (e.g. if the animal has warts or lice).

All diseases are costly to the farmer and result in lost production or increased costs of production. When an animal dies, the farmer loses the capital value of the animal and it has to be replaced. An animal may also have to be culled if it has a serious disease. Some animals that are diseased will never reach their maximum production potential. When an animal has a disease it may need treatment. Costs include the medicine and the labour required to treat the animal.

connect

Animal diseases

Make a summary of the information.

5 Name six effects that diseases have on animals.
6 Why are diseases costly to the farmer?

Types of diseases

There are four main types of diseases – hereditary, metabolic, microbial and metazoal.

Hereditary diseases are passed on to the offspring by one of the parent's genes. A featherless condition in poultry is passed on to chickens in this way and so is dwarfism in cattle.

Metabolic diseases occur when one section of the body is not working normally. Milk fever in cattle occurs when calcium in the blood drops to a low level, usually just after calving. This disease can be treated by an intravenous injection of calcium borogluconate.

Microbial diseases occur when a pathogen enters an animal. A pathogen can be a virus, bacterium, fungus or protozoan. Some examples of microbial diseases are:

- viral diseases – Newcastle disease of poultry, swine fever of pigs and ephemeral fever in cattle
- bacterial diseases – mastitis, enterotoxaemia (pulpy kidney) and footrot in sheep and cattle
- fungal diseases – lumpy jaw and ringworm in cattle
- protozoal diseases – coccidiosis in poultry.

Metazoal diseases are caused by metazoans (pests) that can be seen with the naked eye. These organisms include:

- flatworms – liver flukes and tapeworms
- roundworms – threadworms, nodule worms and barber's pole worms
- insects – botflies of horses and sheep blowflies
- ticks and other arthropods – sheep keds, sheep itchmites and cattle ticks.

7 Name the four main types of diseases.
8 When do metabolic diseases occur?
9 Name two flatworms and two roundworms.

ISBN 9780170265560

Primary and secondary causes of disease

The primary cause of a disease is what actually causes the disease, whereas the secondary causes assist the disease to occur. The most common secondary causes are poor nutrition, poor hygiene and overcrowding.

1 *Poor nutrition* can result from underfeeding, either during a drought or when paddocks are overstocked. When animals are competing for feed, those at the bottom of the pecking order receive less feed.
2 *Poor hygiene* can occur when animals are intensively housed and waste removal is inefficient. Animals are often exposed to poor hygiene at vaccination, castration and drenching.
3 *Overcrowding* is most common in housed animals. The weaker animals are not able to compete for feed and water and become stunted. They are also more susceptible to disease. Overcrowding also leads to a build-up of waste materials.

A primary cause will cause the disease, but may need certain conditions; whereas a secondary cause is sometimes necessary for the disease to occur, but cannot cause the disease itself. For example, the primary cause of mastitis in dairy cattle is an infection by bacteria, while secondary causes would include poor dairy hygiene and poorly adjusted milking machines (Fig.10.1). Lack of feed, cold weather and rough handling would lower the cows' resistance to disease and would also be considered secondary causes. Another example of primary and secondary causes relates to the disease tetanus. The primary cause is the bacteria, which produce the toxin, or poison. The secondary cause is a cut or wound that allows the bacterial spores to enter into the body tissues where they can multiply and produce the toxin.

Primary causes of disease can be viral, fungal, parasitic, toxic or metabolic.

10 What are the two basic types of cause of disease?

11 What is the primary cause of a disease?

12 Name three common secondary causes of disease.

13 What are the primary and secondary causes of mastitis?

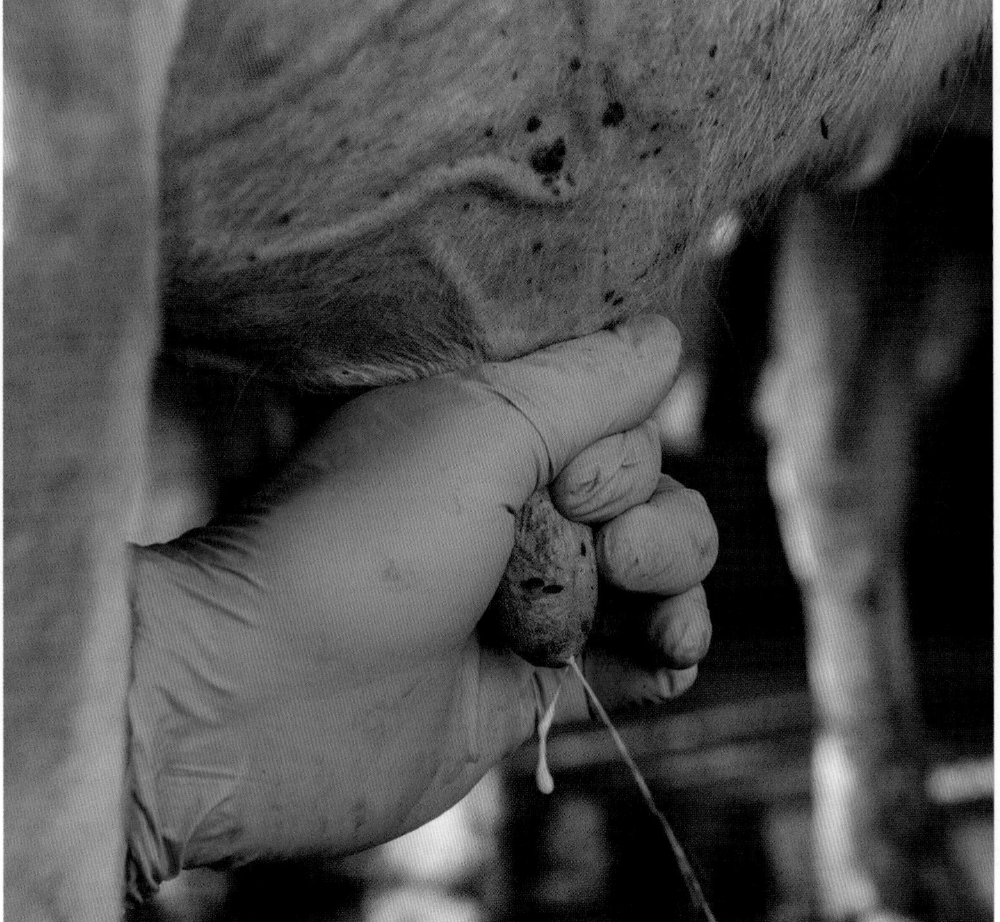

Dairy Australia

Figure 10.1 Mastitis is indicated by the thickened discharge from the teat

Infection and disease

Micro-organisms cause disease by entering a host, multiplying and leaving the host. For a particular host, infection by an organism depends on:

- the existence of a source of the pathogen
- the transferral of the pathogen
- the invasion of the host by the pathogen as it overcomes the host's barriers to infection
- the establishment of the pathogen within the host long enough for it to cause the disease.

Once inside the animal host, the extent of damage is determined by the growth of the organism, or by the production of toxins or enzymes that are able to decompose host tissue. The severity of the disease can also be influenced by the environment. If the environment is unfavourable, the disease will not occur. This interrelationship, or interaction, is shown in Figure 10.2, and is known as the **disease triangle**. Animals affected by the activities of micro-organisms show various symptoms according to the particular interaction of host, pathogen/parasite and environment.

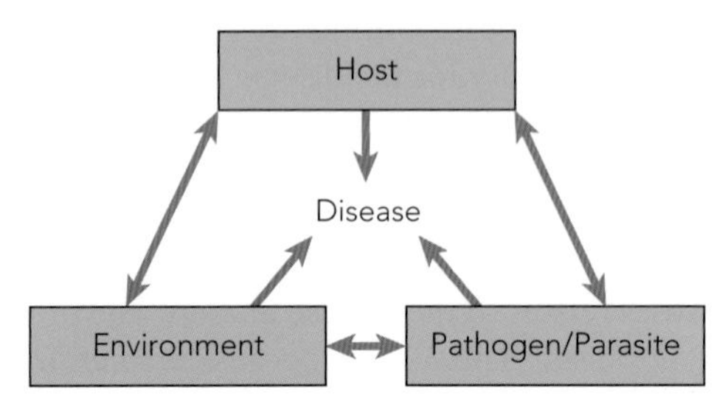

Figure 10.2 The interrelationships of host, environment and pathogen/parasite in disease

The symptoms show the extent to which the normal functions of the host are disrupted. Sometimes the symptoms are not distinctive. Lameness has a number of causes in sheep and cattle. The extent of the symptoms depends on the activity of the pathogen/parasite and the resistance of the host. Infective agents may enter the body in a number of ways. The most common method of entry for pathogens is via the mouth. Pathogens are swallowed with food or drinking water. All the internal parasites, such as roundworms and tapeworms of the stomach and intestines, enter the body through the mouth when the animal eats the eggs or larvae on food or pasture plants, or drinks contaminated water.

Some pathogens may enter the body through a wound or break in the skin. These wounds or breaks can be caused by injury, needles or bites. The bacteria that cause tetanus can enter through cuts made during lamb marking.

Some pathogens may enter the body through the respiratory tract. Infectious pneumonia is caused by organisms that are inhaled.

Mastitis is one animal disease that can be investigated by looking at the complex interaction between the problem organisms (bacteria), a susceptible host (cow) and a suitable environment for the disease to occur (Fig. 10.3).

14 Draw a diagram to show disease as an expression of host, environment and pathogen/parasite interactions.

15 List the ways in which infective agents can enter the body.

Host
Older lactating cow

Environment
Poor milking shed procedures, such as not washing and drying the udder before applying cups, using the same cloth to wipe all udders, inefficient milking machines causing teat canal erosion, muddy laneways

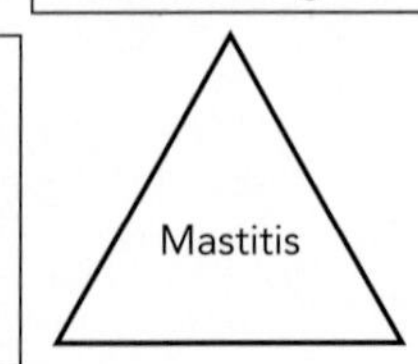

Pathogen
Bacteria (mostly *Staphylococcus* and *Streptococcus*), from other infected quarters, other infected cows, cup liners or wiping cloths, mud, urine and faeces splashes

Figure 10.3 Disease triangle for mastitis

ISBN 9780170265560

Spread of pathogens to new hosts

To spread disease, pathogens must be transmitted to new and susceptible hosts. Methods by which this is achieved in animals include the following.

1. *Aerial contamination.* Respiratory infections, such as colds and influenza, are spread by sneezing. Droplets are discharged when an animal coughs, sneezes or breathes. If the droplets are breathed in by other animals, the pathogen may enter a new host.
2. *Contaminated food.* The ingestion of food contaminated with *Clostridium botulinum* produces a toxin, which causes botulism. A small amount of the toxin is enough to cause paralysis or death.
3. *Wounds or breaks in the skin.* The bacteria that cause tetanus can be spread to other lambs during lamb marking if the tail-docking knife is contaminated. Vaccinating needles that have penetrated the skin of an infected animal can spread the disease to others.
4. *Direct contact.* Infectious diseases may be spread by sexual intercourse. In cattle, a bacterium that causes infectious infertility is spread by sexual intercourse between diseased and healthy animals.
5. *Faeces.* Parasitic worms are spread in faeces. The adult lives in the host animal's body and produces eggs and larvae, which pass to the outside. The eggs then hatch and the larvae develop into young worms ready to return to another host.

Knowledge of the ways that pathogens can be transmitted from one animal to another is important when trying to develop methods to control disease.

16 List the methods by which animal pathogens are spread to new hosts.

17 Why is it important to know how pathogens are transmitted from one animal to another?

Resistance and immunity in animals

The animal's body is constantly under threat of attack by pathogens. To overcome this, the body has a number of defence mechanisms that function at different stages of the attack.

Outer defences

The first defence mechanism by which animals limit the activity of pathogens is to have intact skin. This prevents pathogens from entering the body. Healthy membranes of the mouth, nose and windpipe help stop invading pathogens, while stomach acids reduce the number of bacteria entering with food.

Inner defences

If pathogens penetrate the first line of defence, they are met by the second line. **Phagocytes** are specialist cells in the body fluids, liver and spleen that are able to engulf bacteria and foreign particles and digest them.

Immune system

The third line of defence is the body's immune system. **Immunity** (see page 59) is the body's ability to resist the attacks of pathogens by producing antibodies. **Antibodies** are produced in response to antigens, such as viruses and bacterial toxins, and neutralise their damaging effects. Different pathogens have different antigens, and one type of antibody generally combines with only one particular antigen.

Individual immunity can be active, where the body makes its own antibodies, or passive, where ready-made antibodies are introduced to the body. Both active and passive immunity can be naturally or artificially brought about.

When an animal has been attacked by a certain disease, it normally builds up a supply of antibodies and will stay resistant for some time, or immune to a second attack. This active immunity can be stimulated in an animal either by an attack of the disease or by injection with a **vaccine**. A vaccine consists of a killed suspension of the pathogen, or a living suspension of a weakened, or attenuated, strain of the pathogen that is no longer capable of causing the disease.

When antibodies pass from the body of the mother into her unborn young, the unborn young receive a natural passive immunity. An **antiserum**, or **antitoxin**, is an example of artificial passive immunity. If an animal is sick, it can be treated with antiserum. This would

18 What is the body's first defence mechanism against the activity of pathogens?

19 What are phagocytes?

20 What does a vaccine contain?

ISBN 9780170265560

give it a greater chance of recovery. Antiserum is the preserved serum of an animal that has previously had the disease or has been injected with a vaccine. The serum contains a high concentration of antibodies. Antiserum is expensive compared with vaccine.

Some important animal diseases in Australia

connect

Cattle disease

Select three diseases of cattle and outline causes, symptoms and control measures.

Animal diseases in Australia are caused by a range of conditions and a number of organisms. Many diseases are classified according to the nature of their causes. Following are examples of some important diseases in Australia.

Hereditary disease – dwarfism in cattle

Dwarfism is transmitted by a recessive gene carried by both the sire and the dam. Therefore, to eliminate it, both bulls and cows producing dwarfs must be culled. Culling will result in the gradual dilution of the gene and, providing dwarfs are not used for breeding, the gene will eventually disappear.

Metabolic disease – milk fever

connect

Milk fever

Summarise the cause, symptoms and control measures.

Milk fever (hypocalcaemia) occurs in cattle (cows) and, much less commonly, in pigs and sheep. It is caused by a sudden drop in the level of calcium in the blood. It occurs just before or after calving. Cows are not usually affected until they have had two calves.

Affected cows first show unsteadiness on their hind feet, and finally go down on the ground. The 'classic' signs of this disease are the cow lying with her body in the normal resting position but with her neck bent sideways, and a complete lack of response to stimuli.

The fever is treated with an injection of a calcium solution into the neck vein.

Toxic disease – phalaris staggers

Phalaris staggers occurs in sheep, and is caused by a toxic substance that accumulates in the young shoots of the **perennial** grass phalaris. When phalaris is grazed by hungry sheep or after rain, the sheep may develop a staggering gait and have convulsions.

Phalaris staggers can be prevented by giving the sheep cobalt 'bullets' for some weeks before and after the period of new growth in the phalaris grass, which follows seasonal rain.

Viral disease – ephemeral fever

Ephemeral fever (three-day sickness) occurs in cattle. It is caused by a virus spread by insects. The level of infection varies from year to year. It can spread very quickly and affect a large number of herds.

Symptoms start with a fever, running eyes and nose, accompanied by stiffness and lameness. After three days the animal usually recovers, which is why the disease is called three-day sickness.

connect

Ephemeral fever

Outline the symptoms, cause and treatment.

It affects some animals more severely than others, particularly bulls and pregnant heifers. They will go down and remain down for several days. The fever can cause infertility in bulls.

Animals are treated with large doses of common aspirin, and by providing them with shade and plenty of drinking water. Cattle can now be vaccinated to prevent this disease.

Bacterial disease – mastitis

Mastitis occurs in cattle. It is caused by bacteria entering the udder. Mastitis is an important cause of loss in milk production.

Mastitis can affect one or more quarters of the udder. An infected quarter becomes swollen and sore, and the milk contains pus. Most simple cases of mastitis appear as a few clots of pus in the milk.

ISBN 9780170265560

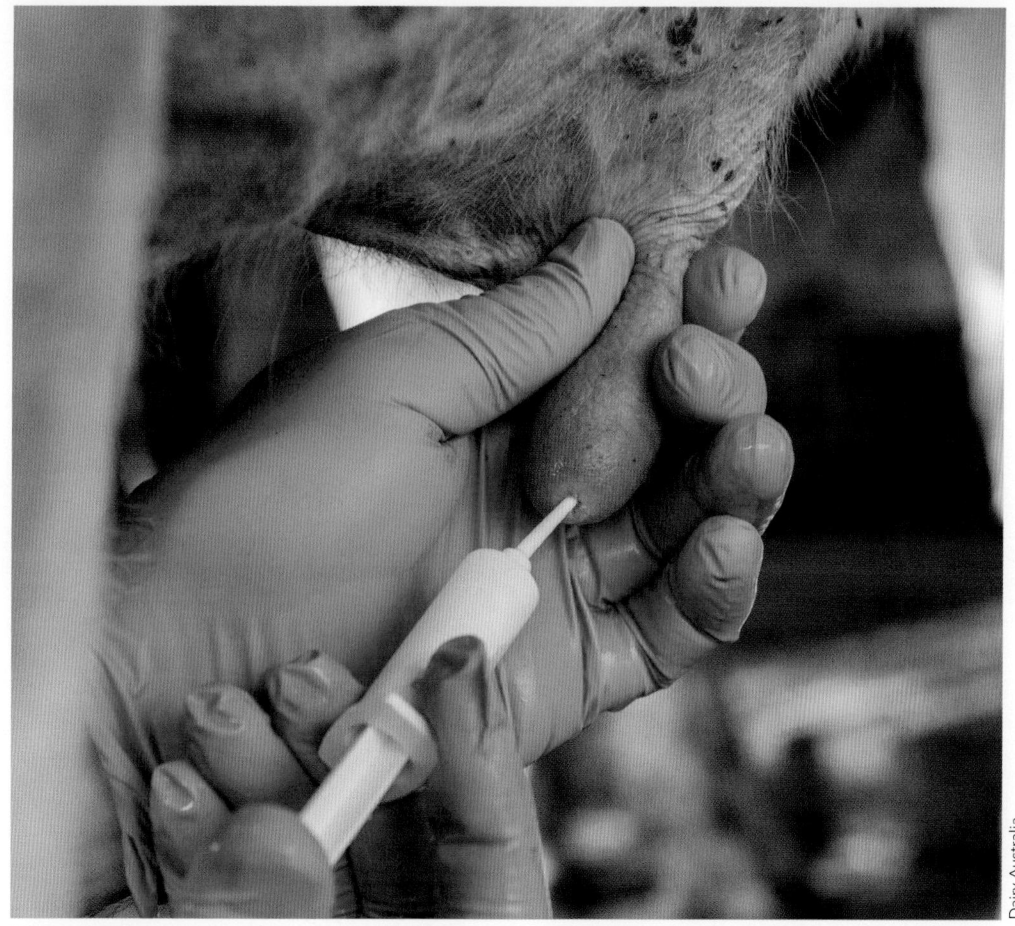

Dairy Australia

Figure 10.4 A cow with mastitis being treated with penicillin

Mastitis is spread through a herd by lack of cleanliness in the dairy, dirty milking cups, muddy yards and poor milking techniques.

Infected cows can be treated with penicillin, which is inserted into the teat canal. Cows that do not respond to penicillin are culled.

connect

More information on mastitis

Fungal disease – lumpy jaw

Lumpy jaw occurs in cattle. It is caused when a fungus, which is present in the mouth and tonsils of normal cattle, invades the bony tissue, possibly when milk teeth are shed or lost.

The bones of the jaw are affected by a hard, immovable swelling, which may secrete pus from one or more openings. Lumpy jaw can be treated with penicillin.

connect

Lumpy jaw

Protozoal disease – coccidiosis

Coccidiosis occurs in poultry, and is caused by a protozoan parasite that enters the bird during feeding. The parasite multiplies in the intestine and burrows into the wall of the intestine and caeca.

Infected birds close their eyes, do not eat and have dirty, ruffled feathers.

The disease can be controlled by including a drug, called a coccidiostat, in the feed.

connect

Coccidiosis

Create a summary of the information.

Flatworm – liver fluke

Liver fluke is a flatworm that occurs in cattle, sheep and goats. The flat, leaf-like worms infest the bile ducts of these animals. The parasite is only found in areas where permanent streams allow the lifecycle to be completed.

Large numbers of liver flukes cause weight loss, anaemia and eventually death.

ISBN 9780170265560

connect

Liver fluke

Use this information to develop a report on the control of liver fluke in sheep.

A liver fluke has a complicated lifecycle, involving a stage in a particular type of snail that lives in swampy areas (Fig.10.5). Sheep, cattle or goats concentrated in or around wet areas during dry periods may pick up heavy fluke burdens.

Control is achieved by draining or fencing off swampy areas. Treatment involves drenching affected animals.

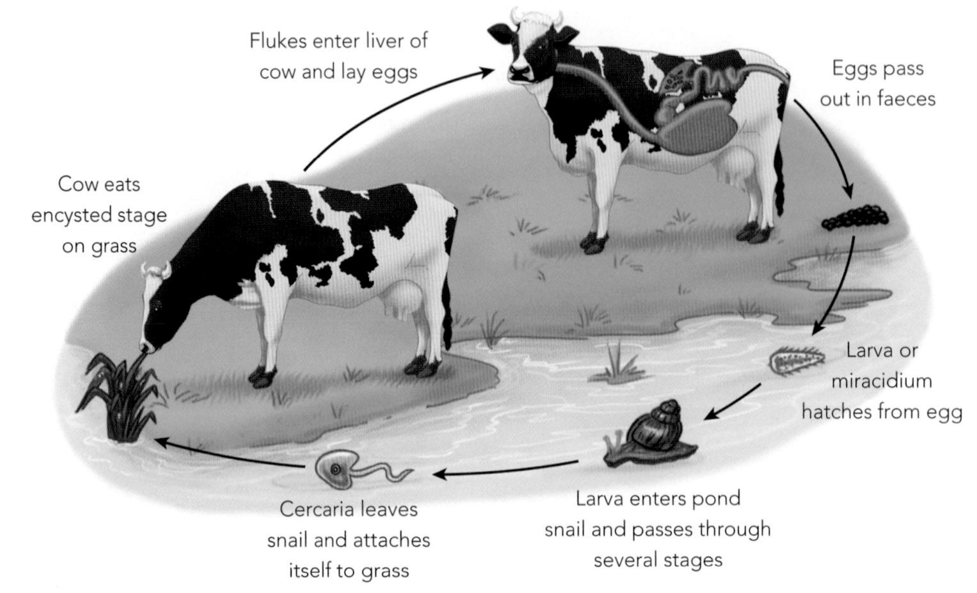

Figure 10.5 The lifecycle of the liver fluke

connect

Internal parasites in sheep

Summarise the information presented.

Roundworm – barber's pole worm

connect

Roundworm

Barber's pole or stomach worm is a disease that occurs in sheep, goats and sometimes cattle. The small nematode lives in the abomasum, or fourth stomach.

The adult worm sucks blood from the lining of the abomasum. This loss of blood results in anaemia, which is shown by paleness of skin and eye membranes. The other symptom is lack of stamina.

Warm, moist conditions favour the rapid development of the eggs into infective larvae. Control can be achieved by rotational grazing. Paddocks should be spelled to allow worm larvae to die off.

Treatment involves drenching affected animals to kill the adult worms. Information on worm infestations can be found on the following website.

Insect – sheep blowfly

connect

Blowflies

Breech flystrike prevention program

Blowfly strike is a common disease affecting Australian sheep. The green blowfly is the type that most often causes strike in sheep. This blowfly and several other species are called 'primary' flies because they can start a strike on any sheep that provides favourable conditions. 'Secondary' flies usually only attack a sheep on an area already struck by primary flies.

Warmth and moisture are necessary for blowfly strike, therefore it is seen mostly in humid weather.

Strike occurs in the breech region, on the body or head. It occurs if the area becomes an attractive place in which to deposit eggs. Female flies are attracted to sheep by the smell that comes from contaminated and decomposing wool. The larvae soon hatch and attack the skin.

Blowfly strike can be controlled by breeding plainer breeched sheep, mulesing, crutching, jetting and dipping. Refer to the Blowflies and Breech flystrike prevention program websites.

21 When does milk fever occur and how is it treated?

22 What are the symptoms of ephemeral fever and how is it treated?

23 How is coccidiosis in poultry controlled?

24 What are the symptoms of barber's pole worm and how can the disease be treated?

25 What environmental conditions favour the spread of blowfly strike?

Disease control

Many methods are used to control or prevent animal diseases. **Integrated pest management (IPM)** involves using several methods to control a pest or disease. Two or more methods are usually used together or integrated. The most important of these methods are described below.

ISBN 9780170265560

Eradication

Eradication involves the complete removal of a disease from a farming system by testing all animals and slaughtering those that are infected with the disease. This method has been used to eliminate brucellosis in cattle in Australia.

Vaccination

Vaccination is a way of protecting animals from a number of bacterial and viral diseases. The host is inoculated (or injected) with a part or the whole of the pathogenic organism. Immunity to further infection results, for the vaccine stimulates the body to produce antibodies against the disease-causing organism.

In dead vaccines the organisms are killed. In live vaccines, although the organisms are living, they are in a weakened or attenuated form.

Vaccination is usually a cheap and effective method of preventing disease.

Chemical control

Internal parasites can be controlled by chemicals administered as drenches, and external parasites can be controlled by chemicals applied as dips. A number of drenches are now available that are given by injection under the skin or into the rumen. Others can be poured along the animal's back and are absorbed through the skin.

Insecticides to kill lice and keds and to prevent fly strike in sheep are usually applied by plunge dipping or shower spraying. External parasites can be controlled by applying insecticides to certain areas of the body, such as the head, back and breech. This is called jetting, and is used with sheep.

Chemicals can also be added to the feed ration of poultry and pigs to control diseases of the gastrointestinal tract (e.g. coccidiosis is controlled by adding a coccidiostat to the feed).

Chemical control would be more effective if it were used in combination with other disease-control methods. However, it is the preferred method for many farmers. The main advantages of chemical control are:

- its convenience and effectiveness
- its quick results
- its comparative cheapness
- the lack of availability of other effective methods.

The main disadvantages of chemical control are:

- the developing resistance of parasites to some chemicals
- its toxicity to the farmer/operator
- risk to consumer health due to residue and failure to comply with withholding periods
- its increasing costs (often related to the cost of research)
- broader environmental effects on desirable organisms.

Biological control

Biological control is the use of one living organism to control another organism.

This approach has been used to eradicate the screw-worm fly from the southern part of the United States. This parasite lays its eggs in open wounds and the larvae burrow into the body, causing severe loss of production and death. The female mates only once. Large numbers of male flies are bred that are artificially sterilised by irradiation and then released by aircraft. When the female screw-worms mate with the sterile males, no offspring are produced and the species dies out.

Rabbit numbers in Australia have been controlled through the release of calicivirus.

The main advantages of biological control are:

- its selectiveness (only the target pest is affected)
- the unlikelihood of organisms developing resistance
- it can be self-perpetuating in nature (organisms will locate the pest and increase in number)
- its friendliness to the environment.

connect

Parasite control

Create a summary of the information.

Goat health

Select three diseases and discuss the treatment method for these diseases.

Poultry health

Select three diseases and discuss the treatment method for these diseases.

Vaccination program when raising pigs

Summarise the information.

The main disadvantages of biological control are:

- the length of time taken to research different methods
- its slowness in acting
- the difficulty and expense of developing and applying methods.

Genetic control

The ability of animals to develop resistance to disease is often inherited. It is possible to select and breed animals that have a high degree of resistance to parasites.

Zebu cattle (breeds originally from India) are more resistant to ticks than European cattle; therefore, Zebu cattle are used in crossbreeding programs in northern Australia to breed animals that have a higher degree of resistance to ticks. This will reduce the need for dipping these cattle.

The main advantages of genetic control are:

- selective targeting of disease or pest
- possible implanting of genes that code for antiviral proteins specific to the disease causing agent
- reasonably permanent change is achieved via the genetic mechanism avoiding future treatment costs where animals become either resistant to or tolerant of particular diseases

The main disadvantages are:

- the length of time required to research and test the impact of control strategies on the wider environment
- slow process to breed animals with a resistance to the disease
- costly to breed animal with a genetic resistance to disease
- animals may become tolerant to levels of infection and continue to spread disease in the natural community as carriers.

connect

Worm resistant sheep

Summarise the information.

Management control

Good management can reduce the incidence or level of disease on a farm.

An important method of disease control is the planning of the layout and relative position of buildings and silos on the farm. On poultry farms chickens should be kept separate from older birds so that the young stock do not pick up disease. Access roads and feed silos should be located so that trucks entering the farm do not have to pass poultry sheds. Diseased birds should be isolated in separate sheds.

On grazing properties parasites can be starved out of grazing areas by resting paddocks and using pasture rotation. Areas that harbour snails can be fenced off (this will break the lifecycle of the liver fluke).

26 How has brucellosis been eliminated in Australia?

27 Explain how vaccination protects an animal from disease.

28 List the main advantages of chemical control.

29 What is biological control? Give one example of this method.

30 Why are Zebu cattle used in crossbreeding programs in northern Australia?

Quarantine

Quarantine measures have probably prevented the introduction of many animal diseases to Australia from other areas of the world. Animals imported from overseas must be held in quarantine for a period of time long enough to allow the symptoms of any suspected disease to develop.

The most important exotic diseases are caused by viruses. Viral diseases include foot-and-mouth disease of sheep, cattle, goats and pigs, and swine fever in pigs.

Integrated pest management

Integrated pest management (IPM) refers to the use of two or more methods to control pests or diseases. The outcome may be the complete elimination of the pest or disease or a reduction in the number of disease-causing organisms to tolerable levels for the health of the animal and the economic viability of the industry. Because the main focus of IPM is to combine methods, strategies could include elements of cultural, chemical, genetic and biological controls and regulatory requirements; for example, quarantine.

The main advantages of this method of control are:

- the acknowledgement that disease-causing organisms and pests may exist but at levels below that which have significant effects on animal health and economic viability; this promotes the development of an environment where the development of resistance to control measures is greatly diminished

 ISBN 9780170265560

- minimal resistance problems
- the establishment of environmentally sustainable management practices.

The main disadvantages are that:

- disease is present within the population at a minimal level but some animals could be carriers
- the farmer has to be aware of the interactions between the pest or disease and the animal to know when best to apply treatments
- results are often slow and complete elimination of the pest or disease is often not the intended outcome.

Basic steps involved in the development of any IPM program include:

- correctly identifying the disease-causing organism or pest
- identifying the level of economic loss to the animal's productivity should no control methods be applied and determining when the animal is most vulnerable to disease
- knowing what control methods are currently available and in what combination these control strategies should be used; for example, to control worms, infected paddock areas need to be identified and fenced off and animals strategically drenched; breeding more resistant animals could also be incorporated as a long-term control strategy
- monitoring the level of disease in the herd via, for example, somatic cell counts for mastitis, worm egg counts from faecal samples for sheep to assess the best time to commence control actions for worm infestation. This decision is mainly based on when the damage caused by the pest or disease is more costly than the control strategies:
- selecting control strategies depending on their cost, the farmer's skill and their knowledge (of the disease organism, the host animal and the environment interactions) and the environmental impact of the control measures
- carrying out control measures at a time that will optimise their effectiveness and according to safety and government regulations to prevent injury and environmental damage.

Examples of IPM programs

Table 10.1 outlines an example of an IPM program for mastitis in dairy cattle.

Table 10.1 IPM program for mastitis in dairy cattle

Prevention
Wiping, washing and drying udders before applying milking cups.
Spraying all teats with Iodophor (or alternative disinfectants) after milking.
Applying softeners (e.g. glycerine) to the udder to prevent surface cracking.
Clipping tails, singeing hairs off udder.
Ensuring milking machine pressure, etc. is correct to minimise damage to udders.
Replacing rubber milking cup liners when required or at recommended time intervals.
Control
Culling animals that have clinical mastitis more than three times in one lactation.
Antibiotic treatment of cows with clinical cases during lactation.
Dry cow therapy – all cows are given a course of antibiotics at drying off time to treat subclinical cases and prevent new infections.
Monitoring
Using the rapid mastitis test to identify new cases.
Checking the in-line milk mastitis indicator for clotted or discoloured milk.
Analysing bulk milk cell counts and individual cow somatic cell counts to identify potential cows.
Watching each animal at milking time for fever, listlessness, or hot, swollen and hard udder.

connect

IPM steps in controlling sheep blowfly

31 Define the term 'integrated pest management'.

Chapter review

Things to do

1. Visit a piggery. Draw a plan of the layout of the sections of the shed. Describe how disease is controlled in the piggery.
2. Visit a dairy farm. Find out the common diseases of dairy cattle. Perform a rapid mastitis test on a sample of milk.
3. Perform the following husbandry operations designed to reduce the incidence of disease in animals – foot paring, drenching, crutching and vaccination.
4. Examine a number of animals and assess their health using the 'vet check' list below:
   - a *Behaviour.* Does the animal demonstrate normal quiet behaviour?
   - b *Posture.* Does it have an arched back?
     Is its head too high or low?
   - c *Gait.* Does it walk normally?
     Can it stand up?
   - d *Condition.* Is its skin supple and clean?
     Does it have a good colour?
     Is its hair shiny and clean?
     Is the animal the same size as others of its age?
     Does it have well-covered bones?
     Are its eyes bright and clear?
     Are the number and condition of its teeth normal?
   - e *Discharges.* Are the urine and faeces normal?
     Are there any signs of diarrhoea, mucus or blood?
     Is there mucus coming from the vagina?
     Is there a nasal mucosal discharge?
   - f *Abnormal body features.* Are there any swellings, colouration of skin, cuts or bruises?
     Is there any worn skin?
     Are the knees, flanks, tail and all body appendages normal?
   - g *Appetite.* Does the animal eat normally?
   - h *Clinical signs.* Does the animal have a temperature?
     Are its pulse rate and respiration normal?
5. Perform the following husbandry operations designed to reduce the incidence of disease in animals: foot paring, drenching, crutching and dipping.
6. Obtain 10 lambs of similar ages and carry out the following trial.
   - a Randomly divide the lambs into two groups. Each group will have five lambs.
   - b Weigh each of the lambs.
   - c Drench one group every 4 weeks with a broad-spectrum drench.
   - d Place the other group in a similar-sized paddock, with similar pasture, but do not drench them.
   - e Record the weights of the lambs in both groups each week.
   - f Calculate the average weight gain for each group per week.
   - g Graph the average weekly weight gain for both groups.
7. Research one animal pest and disease and prepare a report using the following headings: causative agent, method of transfer, symptoms, effects on production and control/prevention strategies.
8. Research using secondary sources an IPM program for a named animal production system.

ISBN 9780170265560

Things to find out

1. Phagocytes and antibodies are both found in normal animal blood. What are they and why are they important?
2. What is the difference between active and passive immunity?
3. Four types of pathogens are viruses, bacteria, fungi and protozoans. Name one animal disease caused by each type and answer the following questions for each disease.
   a How is the disease spread?
   b What are the symptoms of the disease?
   c How can the disease be prevented or controlled?
4. Investigate what conditions might prevent an animal contracting a disease a second time?
5. A large number of chemicals capable of killing most parasites are now available. However, not all parasites will be completely killed by the use of these chemicals. Some parasites will develop resistance to the chemicals. Evaluate what alternative methods of control can be used by farmers.

Extended response questions

1. Describe the mechanisms of resistance and immunity in animals.
2. Describe the methods used to control diseases in animals. Mention quarantine, chemical control, biological control, genetic control and management.
3. For one named farm animal:
   a describe one important disease and pest, indicating causative agent, method of transfer, symptoms and effects on production
   b evaluate methods that can be used to control and prevent animal pests and diseases.
4. Disease results from the interaction, under desirable conditions, of the pathogen or parasite with the host.
   a Name an animal disease that you have studied.
   b For this disease, describe the interaction between the pathogen or parasite and the host.
   c Describe how a knowledge of this interaction would enable farmers to prevent or control the disease.
5. Evaluate an IPM program naming the target organism and the animal host. Include at least two methods of control. For each method develop a brief description, advantages, disadvantages and a judgement of this process.

CHAPTER 11

Outcomes

Students will learn about the following topics.

1 The behavioural patterns of farm animals in relation to management.
2 The legal requirements of managing animal production systems.
3 The ethics of managing animal production systems.
4 The important principles of animal welfare.

Students will be able to demonstrate their learning by carrying out these actions.

1 Understand an animal's physical and behavioural characteristics.
2 Apply knowledge of behaviour to animal management.
3 Monitor the physical aspects of the environment of a selected farm animal.
4 Perform safe handling and management techniques for the care and welfare of animals.
5 Evaluate factors that need to be considered when carrying out a particular animal husbandry practice.
6 Describe and assess the ethical issues relevant to animal production systems.
7 Investigate animal welfare legislation for a specific farm animal.
8 Discuss the implications of legislation for animal production systems.
9 Assess what practices a sensible farmer might adopt.

ISBN 9780170265560

ANIMAL PRODUCTION, ETHICS AND WELFARE

Words to know

animal ethics the study of human–nonhuman relations; includes animal rights, animal welfare and animal conservation

animal welfare the physical and psychological wellbeing of animals

innate behaviour behaviour that exists in the animal from birth – it is not learned

receptor that part of the nervous system that is responsible for detecting changes in an animal's environment

stimuli (singular: stimulus) a change in the external or internal environment of an organism that excites a receptor

ISBN 9780170265560

Introduction

Management of animals covers all the decisions made about feeding, breeding, husbandry practices (e.g. drenching, branding and vaccination) and selling livestock. There are different ways to handle the many problems that arise on a farm. The farmer must choose the way that will give the best results for the animals. For this to occur, farmers need to know about animal behaviour if they are to manage their animals well and achieve production levels that are economically viable. There are a number of behaviour patterns in farm animals.

Animal management and behaviour

Behaviour can be defined as anything that an animal does; in most cases, some form of movement is involved. Animals respond to **stimuli** in their environment. By constantly adjusting to changes in their habitat, animals improve their chances of survival. Many animals have special sensors, or **receptors**, such as sight (eyes), sound (ears), smell (nose) touch (skin, hair) and taste (tongue). Through these receptors, animals receive stimuli so that they can feed, find shelter, mate, reproduce and escape predators.

Types and patterns of behaviour

Behaviour can be innate (inborn) or learned. **Innate behaviour** includes a chicken pecking its way out of its shell, a newborn calf sucking from its mother and a mother licking the coat of a newborn animal to dry and mark it (Fig. 11.1). Innate behaviour occurs even if the animal is isolated from others of its own kind. Learned behaviour is defined as a relatively permanent change in behaviour as a result of a previous experience. It allows an animal to choose the best response in a particular situation. Most animal behaviour is learned. A very young animal begins to learn how to recognise its mother; fears are learned as well as reactions to other animals and people.

1 List the two main categories of behaviour and provide an example of each.
2 Outline examples of receptors and how they relate to animal behaviours.
3 There are 12 videos linked to this chapter, commencing with the video at the website below. For three videos of interest to you, take your time and make notes on the behaviours that need to be recognised and husbandry practices that have developed to achieve acceptable animal welfare standards.

connect

Preparation for mustering

Shutterstock.com/Lakeview Images

Figure 11.1 A newborn calf is cared for and protected by its mother

The videos for this chapter introduce you to ways of safely handling animals based on knowledge of behaviour and **animal welfare** best practice techniques.

Some animals prefer each other's company and tend to move about in; for example, sheep, which graze in a herd or mob.

If any group of farm animals is brought together, the animals will quickly determine a social order among themselves. In doing so they will peck, bunt or bully each other. One of them will become the leader of the group. With dairy cattle and sheep when the leader moves in a particular direction or to another paddock the remainder of the herd will copy or mimic this behaviour and follow.

At certain times all animals show a degree of aggressiveness toward each other. Females with young at foot often show this behaviour. Animals may also become frightened by quick movements and noise or by pests, such as foxes, or poorly trained dogs.

ISBN 9780170265560

There are distinct patterns of behaviour relating to the functions of feeding, elimination of waste, shelter seeking, fighting, mimicking, reproduction and investigative responses. Knowledge of animal behaviours enables safe and predictable stock movement and handling. Animal movements are aided by effective farm design in relation to walkways, fencing, watering points and gate placement. Safe handling practices are enhanced by suitable race, pen and crush designs.

Cattle and sheep tend to graze early in the morning and late in the afternoon. During the day, time is spent resting and ruminating (chewing the cud), so shady shelter spots in paddocks that are accessible in the middle of the day reduce stress. If there is no shade, sheep will gather in small groups and put their heads in the shade created by other animals.

Grazing animals usually eat plant parts that are high in protein first and leave other parts, preferring soft plants, such as clover and lucerne, instead of grasses.

Apart from grazing in mobs, sheep always graze into the wind; consequently, the prevailing wind in a district will determine the best place for gates and yards. Ewes with long wool will not be affected by a sudden cold change and usually will not seek shelter. If they have a newborn lamb at foot it might die unless the ewe seeks shelter. For this reason many farmers shear their ewes before lambing so that the ewes are more sensitive to adverse weather conditions and will seek shelter.

Both sheep and cattle select and keep to certain paths through pastures following fence lines on high spots, which can lead to soil erosion problems. These animals also deposit large quantities of dung and urine in the camps or spots where they sleep at night. These areas are usually on a hilltop or in the corner of a paddock. The long-term effect of such behaviour is to increase fertility in these areas at the expense of the rest of the paddock. As a result, pasture production in paddocks with set stocking for long periods becomes very uneven.

Pigs tend to defecate away from their sleeping area. This allows modern piggeries to be cleaned quickly because the manure can be hosed away easily if pen design incorporates allowance for this pattern of behaviour.

Aggressive behaviour

Aggressive behaviour is more confined to males and usually during breeding seasons, although a frightened animal can become aggressive. Sheep will lower their heads and move away from their opponent then charge, fowls fight with their spurs and claws (Fig. 11.2) and can easily run up a farmer's leg to attack the upper area of the body if alarmed. Bulls will bellow and paw the ground prior to charging.

When animals are introduced into a new paddock or pen they explore it by sight and smell and by moving around it. If a new object is placed in the environment, the animals nearby will approach the object cautiously until their curiosity is satisfied; cattle will always walk the boundary fence when put in a new paddock

Sight is very important. Cattle tend to move in straight lines between water troughs and camp sites whereas sheep move parallel to fence lines using the fence for orientation.

Figure 11.2 Aggressive behaviour: Fowls fight with their spurs and claws

Care-giving behaviour

Care-giving behaviour among domestic farm animals is the sole responsibility of the female. It is stimulated by female sex hormones. Before giving birth, ewes, cows and mares seek isolation from the flock or herd. During periods before and after birth the female might seek shelter from bad weather or from predator attack. On seeing this behaviour, farmers will often bring female animals that are close to giving birth to paddocks closer to the homestead or to areas that allow easy and frequent observation. After the birth of their offspring, cows and ewes, but not sows, will lick and clean their young. Smell, taste and sound are important for the development of the mother–offspring bond. By placing some distance from the flock or herd the mother attempts to limit interference from other animals in the bonding process. The bond is very strong at first but later breaks down, so by weaning time it no longer exists.

Social behaviour

Some animals live naturally in groups, and others are kept in groups by farmers. Social behaviour is the behaviour of animals living in groups and includes how they organise their groups and how they behave toward each other. Poultry live in flocks. The animals quickly work out a social order among themselves. The animals will peck one another, soon one will become group leader and they will order themselves down the line. This hierarchy of dominance and submission is known as 'peck order'. Chickens usually form a peck order by the time they are 7 weeks old. This can also be used to an advantage when mustering and moving stock. Sheep will always walk in a circle until the sheep who is the leader is at the head of the mob and makes the change.

Piglets form a dominance hierarchy or 'teat order' within 2 days of birth (Fig. 11.3). Each piglet sucks from their teat, with the more dominant piglets sucking from the most productive teats.

4 Describe five different behaviour patterns that can be seen in farm animals.

5 List four management practices that can reduce stress in animals.

iStockphoto.com/mikedabell

Figure 11.3 Within 2 days of birth, piglets have formed a dominance hierarchy

Many animals are also territorial and will attack other animals or people that enter their territory. Some dairy cows prefer to be milked in a particular bail and farmers should respect this preference along with the herd hierarchy for entrance to the milking shed if they wish to get maximum production from their cows.

The farmer can reduce stress in animals by:

- handling the animals frequently when they are young
- handling the animal gently and quietly; e.g. cattle, which have a range of vision of 300 degrees, should be approached calmly from the side, talking to them while moving
- providing adequate trough space for animals at feeding time
- using effective placement of shelters, fence lines, watering points and animal walkways
- using well designed yards, races and crushes for animal husbandry activities, such as a curved handling race leading to a crush for cattle so they can see the animal ahead of them and not feel isolated as they move to the crush

In addition, for their own safety it is important that farmers treat all animals with caution; e.g. being aware of possible injury situations, such as working with Dorset horn sheep in pens and races when legs can be caught up in the animal's horns.

 ISBN 9780170265560

Ethics and welfare considerations of animal production systems

Animal ethics is a term used to encompass a range of factors relating to how animals should be treated by humans. When animals are looked after by people, basic questions need to be considered in relation to why such a relationship should occur, the purpose for interacting with an animal, how the animal is treated and what management decisions must occur to meet the basic needs of the animal. Most people when caring for an animal accept the concept that they must look after the welfare and needs of the animal, but this is often based on a flexible set of principles. Such moral codes focus on the welfare of the animal primarily, the purpose for which the animal is kept and that it is treated in a humane manner.

The following websites offer insight into the concept of and legislative guidelines for animal ethics.

Animal welfare is concerned with the wellbeing of animals. The principles that define animal welfare are best reviewed by viewing the following websites, which summarise the approach to animal welfare as developed by Professor John Webster. These are known as the 'five freedoms' and allow an animal to live as naturally as possible, be stress free and maintain health and fitness.

The five freedoms are:

1. Freedom from thirst and hunger – by ready access to fresh water and a diet to maintain full health and vigour
2. Freedom from discomfort – by providing an appropriate environment, including shelter and a comfortable resting area
3. Freedom from pain, injury, and disease – by prevention or rapid diagnosis and treatment
4. Freedom to express most normal behaviour – by providing sufficient space, proper facilities, and company of the animal's own kind
5. Freedom from fear and distress – by ensuring conditions and treatment that avoid mental suffering.

The websites given contain information relating to Commonwealth, state and industry requirements for animal welfare. View them and make a short summary of the animal welfare information contained in each.

6 What is the role of ethics committees in looking after animals in research and teaching institutions? Refer to the websites below.

connect
Code of practice

connect
Animal ethics committees

7 View the 'five freedoms' websites below, which describe the five principles or freedoms that form the cornerstones of animal welfare. Summarise the husbandry practices that promote the implementation of these freedoms.

connect
Free from hunger and thirst

connect
Free from discomfort

connect
Free from pain, injury and disease

connect
Free to express normal behaviour

connect
Free from fear and distress

Animal welfare considerations and farm husbandry operations

Any husbandry procedure must be necessary for the animal's welfare for it to be carried out. How the procedure is performed and by whom is an ethical consideration and directly relates to the welfare of the animal. Any farm procedure dealing with animals must be carried out at an appropriate time because the animal's health, fitness and degree of stress it is subjected to through the husbandry procedure all have consequences for the management procedures before, during and after the event. For example, consider the following husbandry operations in relation to both ethics and animal welfare requirements.

Procedure	'7 in 1' vaccination of dairy cows
Ethical reason	To protect the cow from bacterial diseases that can easily be acquired from pastures and soil.
Procedure	The farmer should be able to effectively and quickly give the vaccine, ensuring that it is subcutaneous (hence effective) and that the needle has not punctured muscle or hit bone that might produce an abscess.
Welfare	Correct procedure should be followed, and needles should be sterilised, sharp and not burred. The vaccine should be within its use-by date. Adult animals are given a yearly booster 4 weeks before calving to boost their immune system to provide the calf with antibodies through the mother's milk. The cow should be moved back into high-quality pasture as soon as possible.

connect
Welfare of livestock

connect
Dairy Australia – Animal welfare

connect
Standards and guidelines

8 Summarise the information by writing four main points provided on the NSW Department of Primary Industries website in relation to animal welfare.

9 Using information from the Dairy Australia website, list two welfare practices that need to be considered when managing either cows or calves.

10 What is the role of the Commonwealth Government in relation to animal welfare?

Procedure	Dehorning and castration
Ethical reason	To prevent bruising of meat, operator and animal injury, unwanted breeding and reduce aggression due to herd behaviours.
Procedure	For animals less than 6 months old, farmers are able to dehorn or castrate an animal. Over this age a veterinarian must perform the procedure. Correct techniques to minimise pain must be used (analgesics and anaesthetics). Infection to be minimised by use of disinfectants. The animal must be correctly restrained to prevent injury to itself and the operator.
Welfare	Legal regulations must be adhered to. Follow-up observations and care are necessary. The animal must be supplied with fresh food and water regularly and stress is to be minimised.

Animal ethics, welfare, legal issues and requirements

There are many examples of ethics issues being debated within animal production systems. These include the practice of mulesing on sheep, live exports of livestock, battery cage egg production, use of hormone treatments in beef cattle production and the use of farrowing crates in piggeries to mention a few. All these practices have some foundation in what was regarded as acceptable management practice, but ethical and welfare concerns soon arose about each practice. Market requirements now limit or eliminate the use of some practices. Consider the example of battery cage egg production, which is a form of intensive farming.

The hens are well fed individually in a controlled climate that:

- eliminates heat and water stress
- provides clean hygienic conditions and reduced disease risk (because there is no contact with wild birds that can spread avian flu)
- increases production levels (because energy is used for egg production and not wasted in movement)
- prevents natural behaviour patterns (because hens are prone to cannibalism)
- offers easy access to food and water
- allows the use of antibiotics in animal production to control disease via food or water.

This is also an efficient way of producing eggs because labour costs are lower due to the use of technologies in environmental control, feeding, watering and egg collection.

The disadvantages include:

- inability of the hens to stretch their wings because of the small cages
- inability to take dust baths, scratch or peck in soil
- the development of foot problems because the birds stand on mesh wire
- no privacy to lay eggs
- prolonged daylight hours are often used by lighting cycles to maximise egg production
- inability of the birds to form a natural peck order.

Farmers are responsible for the health and wellbeing of animals under their care. The farmer as a manager learns to recognise the behaviour patterns of animals, to anticipate situations and recognise signs that may lead to the welfare of the animal, the farmer or their staff being at risk. There are many regulations that apply to animal welfare. In addition, there are model codes of practice developed for many industries.

11 View the website below and list the main Acts and regulations relating to animal industries.

connect

Animals in schools

Regulations protecting the welfare of sheep include the following.

- The *Prevention of Cruelty to Animals Act 1979*. Lamb marking: lambs must be younger than 6 months of age when castrated. Tail docking and mulesing can only be carried out on lambs.
- The *Veterinary Practice Act 2003*. Mulesing lambs: it is an offence to mules sheep that are over 12 months of age, unless you are a vet.
- The CSIRO publication *Model Code of Practice for the Welfare of Animals: Sheep*.

ISBN 9780170265560

Chapter review

Things to do

1. Describe how temperature, humidity levels and lighting conditions are regulated in an intensive animal production shed for either poultry or pigs.
2. Using data loggers and various probes to measure environmental factors, assess the physical aspects of the environment of a selected animal; for example, for chickens, slightly vary the temperature of the brooder and record behavioural changes, then restore brooder to optimal operation.
3. Debate the usefulness of the five principles that define good animal welfare practice.
4. Evaluate the need for animal ethics committees.

Things to find out

1. For a named animal husbandry practice (e.g. lamb marking, drenching, castration, dehorning, hoof trimming or pairing, or crutching), list how the following factors can be managed to reduce stress and minimise the risks to farm animal welfare:
   - use of appropriate equipment
   - skill of the operator
   - timing of the animal practice
   - management of the animals after completion of the practice.
2. For a named animal species, identify two physical and two behavioural characteristics for which a farmer's knowledge can assist with the management of the animal.
3. Outline two examples showing how an understanding of an animal's physical and behavioural characteristics can assist in the management of the animal. Some examples include panoramic vision of animals, flight zones, and herd or mob instinct.
4. Using examples, discuss how knowledge of animal welfare practices has led to better outcomes for animals.
5. Describe and assess the ethical issues relevant to animal production systems.
6. Discuss why a sensible farmer would follow the legislation and guidelines for the benefit of the animals and the farming business.

Extended response questions

1. Investigate animal welfare legislation for a specific farm animal and discuss the implications of the legislation for the relevant production system.
2. Discuss the ethical or moral issues associated with ONE of the following: mulesing of lambs; live sheep, beef cattle or goat export; battery cage egg production; or the use of farrowing crates in intensive piggeries.
3. Evaluate the role of legislation in the development of animal management systems.

CHAPTER 12

Outcomes

Students will learn about the following topics.

1. Consumer and market requirements for commercial animal products.
2. The use of technology in the production of animal products.
3. The use of technology in the marketing of animal products.
4. Recent research findings that contribute to animal production systems.
5. The role of research in animal production systems.

Students will be able to demonstrate their learning by carrying out these actions.

1. Recognise the features of animal products that are important to consumers.
2. Use a range of sources to gather information about a specific agricultural problem or situation in animal production systems.
3. Outline the impact of research on agricultural production systems.

ISBN 9780170265560

ANIMAL PRODUCTION: MARKET REQUIREMENTS, TECHNOLOGY AND RESEARCH

Words to know

bibliography a list of source materials used or consulted in the preparation of a work

biotechnology any technique that changes living organisms at the molecular level (genes and chromosomes) to produce useful products, such as medicines (e.g. insulin), an insect-resistant plant or a calf through embryo transfer

consumer person who buys and uses a product to satisfy their needs or wants

exotic not native

gene patents the ability to register and charge for the use of newly created genes

innovation a new concept which has not existed previously, or a new application of an idea or form of technology

innovators those who are early adopters of new technologies

markets places where farm produce is sold

producer a person who produces goods or services

sexed semen semen that has been processed in a laboratory into two groups; one group is likely to produce male offspring and the other group only female offspring

sexed embryos embryos that have been separated in a laboratory according to their sex

survey the collection of sample opinions, facts, etc., made in order to estimate the overall situation

sustainable able to maintain production levels over the long term

technology the practical application of knowledge, such as the use of machinery, computers and/or techniques for undertaking agricultural practices (e.g. the machinery and the method used to grow a wheat crop)

variable the particular feature being investigated in an experiment; all other factors are kept constant

Introduction

It is an easy task for many non-farming members of the community to state simply that farmers should diversify into better income-earning production systems and not remain shackled to historically established enterprises. Unfortunately, uncertainty in markets, the impact of drought and many well-written entrepreneurial prospectuses over the past few years have encouraged people to enter into production systems that had little hope of success.

Diversification of farm enterprises is a sound method of combating uncertainty in agricultural production. The development of new enterprises is also a rewarding activity and, for those farmers who are successful, a very profitable activity. However, very sound guidelines should be followed when diversifying into new farm activities.

Farmers are now realising that product oversupply and the market protection policies of other nations will bite more deeply into the profitability of Australia's traditional agricultural industries. This will stimulate the development of other farming activities and consequently lead to the development of new markets for Australia.

Consumer and market requirements

Consumer preferences or tastes are an increasingly important marketing factor as **producers** and retailers seek to diversify **markets** for their products. A visit to any supermarket will allow you to identify various marketing strategies. In the marketing of beef, examples of the following requirements can be seen:

- meat identified as being hormone free
- lean meat cuts with minimal fat
- no bruising, punctures or evidence of carcase damage while in transport
- disease free
- consistent quality of meat cuts as highlighted by use of words such as 'prime meat'
- attractive packaging including moisture absorbing pads on the bottom of meat trays
- meeting quality assurance standards indicating best practice
- consumer interest in where their meat has come from; that is, 'paddock to plate' concept.

Technology

When agricultural farming systems were first established in Australia, they were mere copies of European systems, incapable of providing a **sustainable** base on which a diverse industry could grow. The **technology** (practical applications of theory) imported by the first colonial farmers was not appropriate to the semi-arid conditions of many parts of this country. Livestock could not adjust to the heat and humidity; crops grew poorly in the fragile, low-fertility soils. These problems were compounded by land-use policies based on the complete removal of tree and brush cover and overstocking with animals, which caused soils to compact and erode. The consequences are still clearly visible in agricultural areas today.

Technological developments that have occurred since these early farming days have eventually led to the establishment of a technology and system of land-use management that is more appropriate to Australia's delicate environmental balance of soils, climate and living organisms. This now underpins much of Australia's economic development.

1 What factors prompt farmers to diversify farm production possibilities?

Adoption of technology

Most technological advances in agriculture are financed by the government. Due to the competitive nature of Australian agriculture, adoption of technological advances by farmers is crucial to its long-term productivity.

Pattern of adoption

Technological changes in the more biological aspects of farm systems are rapidly adopted by farmers, especially where such changes can be easily integrated into existing farm management structures. Examples include using improved varieties of crops, pasture plants or animal species, or more effective drenches or fertilisers. In these cases, only those who adopt

 ISBN 9780170265560

these newer technologies (the **innovators**) receive any benefit. Biological **innovations** are less disruptive than mechanical innovations, as they require little or no reorganisation of the farm.

The initial people to adopt the new mechanical-type technologies are risk takers. They are innovative people, who often contract with dealers to be the first in a region to use a new style of tractor, harvester or plough. They undertake high initial establishment and tooling-up costs in the hope that the increased productivity gained through using the new equipment will not only cover costs but also provide an increase in profit. Extra profit would be achieved through increased efficiency, gained by farming more land, by doing tasks less often or more quickly (in the case of equipment such as combine harvesters), or by doing farm tasks more effectively (e.g. ploughing, injection of fertiliser into the soil and pest control).

Figure 12.1 indicates the pattern of adoption of new techniques on farms.

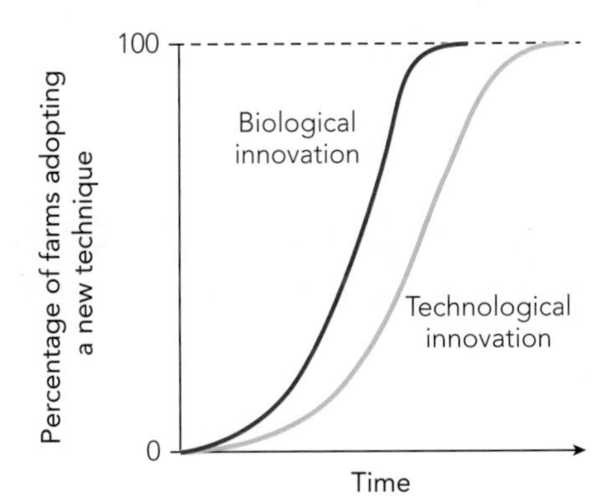

Figure 12.1 The rate of adoption of new techniques on farms

Innovators usually enable companies or research organisations to demonstrate their ideas or products at field days. Due to competitive pressure, other farmers will adopt proven technology because of gains in either time use or profit once the benefits of adopting this new technology can be assessed. Eventually, even the most conservative of farmers are forced to adopt new proven technology – or else sell up due to poor profit margins in comparison with surrounding farms.

Advantages and disadvantages

New technology increases the level of product supply onto the market. Consequently, market price falls unless demand also can be stimulated, as indicated in Figure 12.2.

Producers may get some benefit from adopting new technology in the following ways.

- Increased efficiency gained by the adoption of new technology usually results in a lowering of operational cost.
- Lower selling prices for farm goods in some instances may assist where natural products compete with synthetic products.

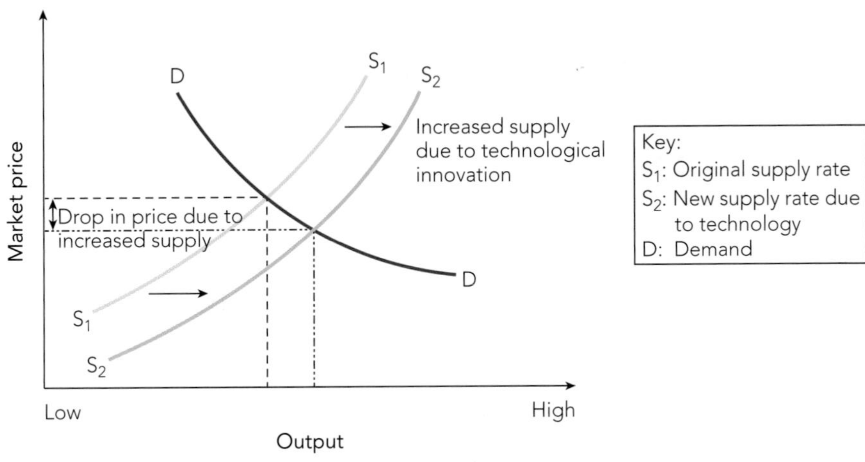

Figure 12.2 The impact of technology on market price

ISBN 9780170265560

2 Why were early colonial agricultural systems subject to problems in their long-term development?
3 Explain the difference in the rate of adoption of biological and mechanical technological innovation by farmers.
4 Outline the benefits farmers might gain by the early adoption of new technologies.

- Where a product is exported overseas, increased production levels may increase potential returns from overseas.

Farmers might in some instances be worse off from advances in technology. Production levels might increase to such an extent that product prices fall, and the money saved from acquiring and using the machinery might be less than the fall in market prices.

New technologies and innovations in industry

The work of individuals and agricultural scientists is not the only reason for Australia's increased capacity to produce food. New technologies or innovations in industry also have been important. Cost, training opportunities, willingness to change and the level of understanding and skill required to successfully adopt a new technology limit the rate of adoption by many farmers of new technologies and innovations.

Use of technology in producing animal products

Many technologies are currently being applied to animal production systems.

Robotic milking systems

These systems are innovative forms of dairy farming in which milking becomes a background operation administered by the AMS (automated milking system) and the cows themselves (complete automation of the milking process).

- Cows voluntarily come to the milk harvesting area.
- Milking is distributed over a 24-hour period.
- One cow is milked at a time.
- No human intervention as teat cups are attached and removed from each quarter independently.
- The milking unit comprises a milking machine, a teat position sensor, a robotic arm for automatic teat-cup application and removal, and a gate system for controlling cow traffic.

Effects on production

Robotic milking systems offer some major advantages to farmers. They can:

- make dairy farming a more socially attractive occupation so it becomes easier to attract and retain staff in industry
- minimise labour input, because the farmer is freed from milking (with its rigid schedule)
- increase milking frequency (e.g. to three times daily), resulting in less stress on udder and increased comfort for cow because, on average, less milk is stored
- make detection of clinical mastitis easier and the system can separate unmarketable milk
- improve herd management through use of computer; the farmer can change feed in response to milk production, and individual cow histories can be examined for illness or injury.

connect

Future dairy

View the video clip narrated by Max Roberts and write a summary.

Robotic shearing of sheep

View the video on the Robotic shearing website and summarise the main advantages of this technology.

connect

Robotic shearing

Sheep – fat scans

View the video on the Fat scans website. How does this innovation contribute to animal production?

connect

Fat scans

Dairy cattle – ultrasound pregnancy scanning

Visit the Pregnancy testing products website. How does this innovation contribute to animal production?

connect

Pregnancy testing products

ISBN 9780170265560

Breedplan®

Visit the Breedplan® website. How does this innovation contribute to animal production?

Sexed embryos and semen

Choosing superior sires and cows based on performance characteristics as sources for genetic material (**sexed semen** or **sexed embryos**), heifer calves can be produced that are genetically superior. Using reproductive processes, such as multiple ovulation, many heifer calves can be produced in a short period of time, which enables selection based on performance and conformation. This results in rapid genetic improvement.

Rumen protein by-pass pellets

When a dairy cow is fed rumen protein by-pass pellets, the protein is not broken down by the rumen microbes, so it is not lost by being converted to microbial protein but travels to the small intestine to be digested. This allows more protein of a higher quality to be available to the animal, boosting both milk production levels and milk quality.

Technology in marketing an animal product

Technology is a vital component in many marketing systems for agricultural products. With the aid of technology many agricultural products can be graded, according to purity, size, colour, protein quality, fibre diameter or other types of objective measurement. Because of this objective assessment, certain commodities can be sold by description or on the basis of the sample (e.g. wool). When goods are sold by description buyers simply need to have the information in front of them; they do not have to travel to the marketplace to buy. Objectively assessed product can be sold by telephone or internet; Auctionsplus is an example of this method, used to sell cattle. Computer systems can also be used to help promote and sell agricultural products. Some examples of the application of technology to marketing animal products are listed below.

- *Wool sale by description*.
- *Embryo video sales*. Farmers can buy cattle embryos to transplant or transfer into a cow that they own. Search for cattle embryo sales on the internet. For a particular farm or property that you find, summarise the advertisement.
- *Marketing beef cattle*. View the website link and read the information. How would this feedback enable the farmer or producer to improve animal production?
- *Modified milks*. For example, 'Lite White' and 'Omega 3'. View the Modified milks website. How does technology improve the marketing of modified milks?
- *Internet marketing of animal products*. The internet allows a direct link to be made between consumers and producers. Farmers are now developing their own markets based on particular consumer preferences, such as for organic beef. This has resulted in the formation of a number of niche markets.

Future directions and the role of research

Future directions in innovation and new technology in agriculture include:

- vaccine development
- **genetic engineering** (see page 139)
- more sustainable agricultural systems
- environmental considerations.

Use these points as starting issues along with the activities at the end of this chapter to research one agricultural enterprise and appraise the technological or alternative innovations that have occurred.

connect

Breedplan®

connect

Wool classing

Read and summarise the information.

connect

Modified milks

connect

Beef grading

5 List three examples of changes in technology that benefit farmers.

6 What factors limit the adoption of new technology by farmers?

Vaccines

New technologies are responsible for the development of vaccines to combat diseases that currently cause widespread economic damage to livestock. The basis of developing these new types of vaccines lies in advances in **biotechnology**. Vaccines are being researched for cattle tick *(Boophilus microplus)*, blowfly and buffalo fly, as illustrated in Figure 12.3 (page 170).

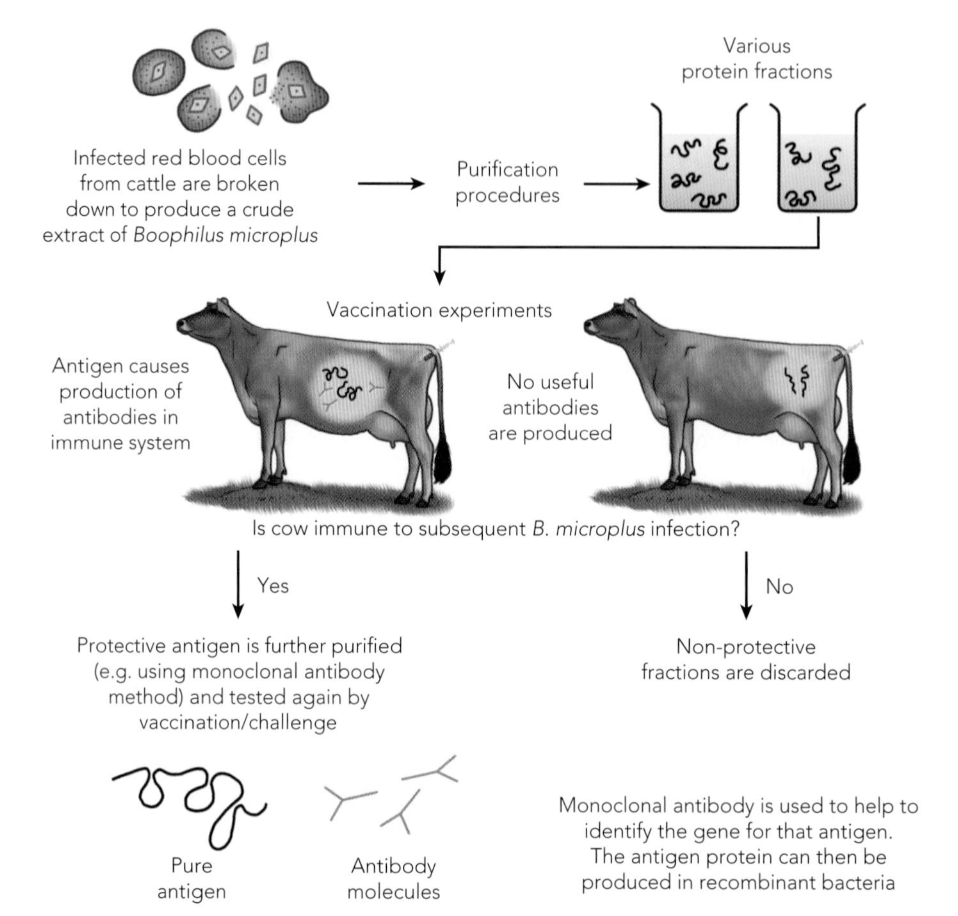

Figure 12.3 Isolating and producing tick fever antigens

connect

Biotechnology

Summarise the information.

connect

Modern biotechnology

Genetic engineering

The manipulation and synthesis of new genes by artificial means (genetic engineering) are aspects of a rapidly developing technology in agriculture. The advent of **gene patents** creates the option to register and charge for the use of newly created genes. The technology is developing in both plant and animal industries.

For example, artificial splitting of an embryo can be used to produce identical twins. Although this process manipulates genetic material it is different from genetic engineering. Refer to the FDA website on page 171 for further information on genetic engineering.

The new era of tailor-made livestock involves modifying the gene responsible for the production of growth hormone. These **transgenic** (see page 139) animals attain accelerated growth rates. The key is the ability of genetic engineers to replace the DNA codes that give a gene its function with other codes that alter this function. **DNA** – deoxyribonucleic acid – is the biological material that stores genetic information for an organism. The modified gene containing altered DNA is then inserted into a fertilised ovum and replaced in the oviduct of an animal to develop naturally.

ISBN 9780170265560

Visit the Genetic engineering website and answer the following questions.

- What is genetic engineering (GE)?
- What types of GE animals are in development?
- How do genetically engineered animals differ from conventional animals?
- What is the difference between animal clones and genetically engineered animals?

connect

FDA Genetic engineering

Sustainability

Innovations in agriculture are also concerned with making agricultural production systems more **sustainable** (able to maintain production levels over time). This involves both the study of problems such as soil acidity and land salination, and the development of land-use techniques that restore land suffering from various forms of degradation to stable and productive levels over a period. These forms of land degradation need to be managed to increase animal production.

View the videos on the Target 100 website and identify farming practices that are making the farms or properties more sustainable.

connect

Target 100

The environment

The maintenance of natural areas on farms and the ability to develop wildlife corridors on properties will eventually allow a much better coexistence of manipulated and natural systems, as shown in Figure 12.4.

a Farm with few natural areas

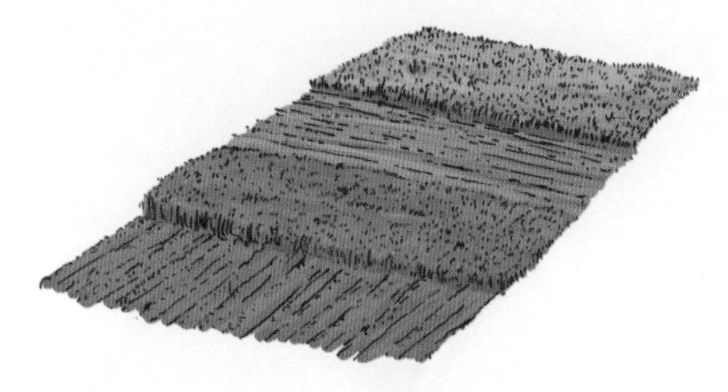

b Corridor development on farms

Figure 12.4 Farms and natural wildlife corridors

Successful alternatives to conventional farm management systems can come about only through education and manipulation of the market by pricing policies, marketing arrangements and taxation incentives.

ISBN 9780170265560

7 Discuss one example of biotechnology and its implications for agricultural production systems.
8 What does the term 'gene patent' mean?
9 Outline how agricultural systems are seeking to develop a more integrated land management system.

Future farm management systems should concentrate on technologies and innovations that reduce reliance on environmentally disruptive inputs (e.g. pesticides and excessive fertiliser use) and increase the use of crop rotation and integrated pest-management strategies and the more efficient use of resources (e.g. water and energy). Genetically engineered plants and animals will also have a place, provided that other scientists can maintain the historically diverse genetic base from which the present plant and livestock industries have evolved.

New approaches

Australian agriculture is composed of many forms of traditional farming systems; however, new or alternative agricultural industries or methods of production are emerging.

Before adopting a new enterprise, it is prudent to determine whether a long-term sustainable market exists for the product, or whether it is necessary to advertise and negotiate a new market possibility. The consumer must want the product, and it takes time and money to develop consumer preferences. How long might it be before people accept kangaroo meat as readily as they accept beef and lamb, for instance (Fig. 12.5)?

iStockphoto.com/Dirk Freder

Figure 12.5 Kangaroo meat is great for the environment, but not yet widely accepted by the public

 ISBN 9780170265560

Products require processing as well as a market outlet. The ability to process a product cheaply, and possibly produce a value-added output, is very important. Many new ventures in plant and animal industries have suffered from lack of a processing factory for the product.

New farm enterprises also require the farmer to adopt new management techniques. This, in turn, might require some retraining or an acceptance of different standards and ways of doing things; for example, growing **exotic** (foreign) animals such as ostriches.

Research

Innovations in technology used in producing animal products are necessary for the ongoing improvement of agricultural production. Important areas of research include applications allowing scientists and farmers to:

- trial new developments in agriculture (for example, new animal varieties)
- identify problem areas in agriculture
- improve product quality and improve yield
- increase the efficiency of production
- produce sustainably.

Research is the basis for progress and meeting the challenges that modern agriculture faces as it adapts to the changes in the economic, environmental and social environments. It is very important that the results and findings of research are accurate and reliable so that decisions based on them are sound.

During your study of agriculture you will be required to understand, implement and evaluate experimental data, current research projects and outcomes in relation to agricultural systems, problems and situations. Reasons why research is important and how you undertake and analyse research projects are outlined below.

There are many ways of going about research. Putting it another way, there are a number of research methodologies. The methodology that is applied to conduct research depends on the nature of the problem or issue under investigation. To find out if a new breed of sheep is going to be better than the existing varieties would involve using a scientific methodology using experiments. How to go about these kinds of experiments has been discussed in Chapter 2. Finding out what consumers want in wheat products involves using surveys, questionnaires, interviews and discussion groups. These are social or economic methodologies.

10 Explain why research is important to agriculture.

Finding a research question

To start research there has to be a problem or issue that needs an answer or explanation. Research questions might arise from a number of sources. Some of these could be television programs on rural affairs, articles in newspapers (rural and metropolitan), problems on the school farm, something that arose out of a field day (such as the claims of the promoter of a new product), journals (popular and scientific), or some issue that is concerning the public in general that is related to agriculture.

Finding information on the question

Once an area of research or a question has been decided upon it is time to find out more. To do this there are many places to search for information. The library is a good place to start. Look for books and journal articles that deal with the area you are researching. You will be able to build up an understanding of the background and context of the problem. The internet will also be a source of information. If you can specify your searches you will save time. Care should be taken in evaluating the authenticity of information obtained from the internet. Information published on the internet is not always reliable and correct.

Talking to an expert or experts in the field is very valuable and email is a good way to communicate with them. They can be found where research is being carried out, such as at universities, the CSIRO, national and state departments of agriculture, the Australian Bureau of Agriculture and Resource Economics, the Australian Bureau of Statistics and the Australian Quarantine and Inspection Service.

ISBN 9780170265560

Talk to your teachers about the issue; not just the agriculture teachers but teachers in other disciplines, such as science and economics, who might have insights into it and be able to direct you to relevant information.

As you go through this process your research will become clearer. In fact, your research might be completely revised or replaced with an altogether new line of research. Clues to which research methodology could be best applied to your research will come to light as you go through this information-gathering process. Ideas about the best techniques to use in the research and to measure results might also become obvious as you investigate.

11 List four sources that you could use to help define a research question.

Designing the research

Scientific approach

If your question suits a scientific approach, make sure the experiment you design includes all the principles of good experimental design – randomisation, replication, standardisation and a control. Careful consideration should also be given to the statistical techniques that you will apply to your results to determine whether or not the differences you observe are due to the experimental treatment or merely chance. For most students an experimental design that examines the effects of just one **variable** is recommended so that the analysis of the data remains manageable. A random design or a randomised complete block design lend themselves to moderately easy analysis and ensure that good experimental design principles are used.

Conducting a survey

Where the research question necessitates the use of a **survey**, designing the research involves developing the questionnaire. Draw up survey questions aiming to gather information that will help answer your research question. Discuss the survey questions with your teacher and a few reliable friends who will give you honest opinions about them. Do these people think your survey questions will give information that will answer the research question? Are the questions clear or are they ambiguous? Use the feedback from these people to modify and refine your survey.

As you develop your questions give some thought as to how you are going to manage the information they generate. It might be a good idea to use questions that give the respondent a series of alternative answers that they can tick – much like a multiple choice examination question. Questions might also give respondents a range of choices for a particular statement ranging from 'strongly agree' to 'strongly disagree'. Both these types of question allow you to calculate the percentage of respondents who answer in a particular way. Questions that require written answers are more difficult to collate and analyse.

You will also have to decide what group or groups in the community you are going to give the questionnaire to. Who does your research question apply to? It might be a particular group (e.g. the owners of beef cattle studs) or it might be the consumers in a suburb of a major city. This will affect how you administer your survey. To contact the beef cattle studs it would be necessary to mail or email the questionnaires to them, while consumers can be approached in person in a shopping centre.

When your survey is ready, administer it to a small number of people to try it out. If necessary make some final adjustments to it.

Research and ethics

It is important to examine the planned research in the light of ethics that should be applied. Some questions to consider about the planned research are as follows:

- Are any animals involved, and if so will they be treated in a humane way so that their welfare is not compromised? Will any animal welfare regulations be breached?
- Will the research be carried out in such a way that previous research by others is not claimed as being the result of this research?
- Are the rights and confidentiality of people involved (e.g. respondents to surveys) at risk in any way?

12 Describe the kinds of research that would be appropriate for:

a a scientific experiment

b conducting a survey.

 ISBN 9780170265560

If the answers to these questions or others indicate that the research, if carried out, will result in unethical behaviour or that laws or regulations will be breached, then the design should be reconsidered.

Doing the research

Conducting the experiment or administering the survey is the next step. It is a good idea to make a calendar of events with dates by which each part of the research will be complete. This gives you an idea of the commitment you will need to complete the work and helps you manage your time.

For scientific experiments all the equipment required must be assembled and the experiment carried out. Keep careful records in your process diary/journal. Record anything and everything that you do or that happens, even if you think it is only remotely related to the research question (e.g. observations, results, weather details and unusual events). These records are invaluable when you are writing your report.

Administering your survey means you will have contact with members of the public. It is important that you approach people with courtesy and tact, making them feel at ease and willing to assist you. When they have completed the survey thank them. If you are conducting the survey by mail or email include a covering letter explaining why you are carrying out the research. A survey might also be done on the internet. One student interested in the demand for beef in different international markets used this method and had thousands of responses from over 100 countries.

For the results of your survey to be meaningful you will need 100 people to complete it. This might not always be possible but the more you get, the more accurate your conclusions will be.

13 Compare carrying out an experiment with conducting a survey.

Analysing the results

Scientific experiments

For the scientific experiment the analysis of the results will be determined by the experimental design. A simple but effective method of analysis is given in Chapter 2.

Surveys

The first step is to collate the material that is in the **survey**. This means counting the number of respondents who answered each question in each particular way. It will mean grouping together similar responses to questions where there has been the opportunity to write answers rather than just tick alternatives. Once the numbers have been counted the percentage of respondents who answered in particular ways can be calculated. Questions should now be grouped together so that they give information on certain aspects of the research question. The responses can also be grouped according to criteria such as age, sex, where the respondent lives, occupation or whatever, and this might throw further light on the research question. For example, an older group might have a completely different opinion about an issue such as genetic engineering from that of a younger group. The possibilities are enormous but the researcher must keep to analysis that is relevant to the research question. It is a good idea to discuss your results with your teacher or some other people to help you focus them on your research question.

14 Describe how you would proceed with the analysis of the following survey question for which you have collected 100 responses: 'How often do you eat beef in a meal?'

Every day

Once every two days

Once a week

Once a month

Never

What have you found out?

At this stage for both experiments and surveys it should be clear what has been discovered about your research question. It might be that your results are exactly as you expected, the opposite of what you expected or show something completely unexpected. Organise your results so that they can be used to back up your findings. Graphs and tables are very useful for summarising and showing results but remember they must be relevant to the research question. Many students fall into the trap of making graphs from as much of their data as possible, especially as they can be easily generated with the aid of a computer and a spreadsheet program. More often than not most of these graphs are irrelevant to the research question being investigated and would be best left out.

15 What part do graphs and tables have in helping to describe what a piece of research has found out?

Reporting your findings

The findings of your research need to be communicated to an audience. That audience might be those who are interested in the area of study, farmers, agricultural researchers and consultants. The usual way is to present a written report. This report will clearly set out:

1 what the research question is
2 the background and what is already known about the question (a literature review)
3 how your research was conducted (method)
4 the results or findings of your research
5 a discussion of the results or findings and the methods used
6 a conclusion
7 recommendations of how the findings can be put into practice and what needs to be further researched in the future.

The report will also acknowledge the sources of all information it has drawn upon. This might be in the form of a **bibliography**. It is important that anybody reading the report can find out where the information came from and be able to go and look it up for themselves.

The list above shows the order of the final report. In actually writing the report it is best to use a different order. Write out the research question again. It would have been established some time ago when you started the research. Writing it again focuses your thinking in constructing your report. Now start with the method and results. Next, tackle the literature review and the discussion. The last thing to write is the conclusion and recommendations. As you write, draw up a list of the sources (such as journals, books and emails) that have been used for your report. This list can be converted easily into a bibliography.

16 Describe the process of producing a report on a piece of research.

ISBN 9780170265560

Chapter review

Things to do

1. Find a journal in the library that reports on research and write a description of the way the reports are presented. What are the sections of each report?
2. List the features of animal products that are important to consumers.
3. Carry out your own research project using one of the methodologies described in this chapter.

Things to find out

1. Contact a research institution and find out what is currently being researched and what methodologies are being used. The internet is a good way to do this.
2. Skim through a current rural newspaper or magazine and find out what are the issues under discussion in the rural community. What research could be done on these issues?
3. Describe what the term 'paddock to plate' means. Research how consumers are able to follow the progress of a beef animal from a farm to a dinner plate on their table.

Extended response questions

1. Research and discuss results and impacts of agricultural experiments and research in a named animal production system; for example, sexed semen, cloning of sheep.
2. Use a range of sources to gather information about a specific agricultural problem or situation in animal production systems, such as mastitis, and explain how technological advances have assisted in identifying or managing the problem, such as the National Livestock Identification System (NLIS).
3. Outline the impacts (both positive and negative) of new technology arising from research on agricultural production systems; for example, robotic shearing.

CHAPTER 13

Outcomes

Students will learn about the following topics.

1. The farm as a production unit and the enterprises on the farm.
2. The subsystems operating on a farm.
3. The goals of the farmer and why these have been established.
4. The routine management procedures for production and why each exists.
5. The ways products from the farm are marketed.
6. The relationship between inputs and outputs in farm production.
7. Technology used in management and production on the farm.
8. Safe work practices employed in agricultural work places.
9. The economic performance of this production unit (farm enterprise).

Students will be able to demonstrate their learning by carrying out these actions.

1. Define the terms 'input', 'output', 'processes' and 'boundary'.
2. Observe, collect and record information on the physical and biological resources of the farm.
3. Construct a calendar of operations for an enterprise production cycle.
4. Describe methods of agricultural recording.
5. Identify various measures of performance, including gross margins.
6. Identify problems associated with production on the farm.
7. Identify factors affecting farm management decisions.
8. Describe the effect of demand and the role of consumer trends on farm production.
9. Identify management practices used to address environmental sustainability.
10. Identify marketing strategies.
11. Identify technologies used on the farm.
12. Gather data using appropriate instruments to measure resources, including weather and soils.
13. Explain ways in which technology is used in farm management and production.
14. Recognise and use safe work practices.
15. Identify potential safety hazards in agricultural work places.
16. Outline work, health and safety (WHS) legislative requirements that affect a farm.

ISBN 9780170265560

CASE STUDY: A DAIRY FARM

Words to know

black box model a model showing only inputs and outputs in a system, farm or enterprise

boundary a limitation of a (farming) system (e.g. a fence, money, managerial skills and so on)

dry cow a cow that has completed her lactation and is not producing milk

lactating cow a cow that is producing milk

milk contract a quantity of milk that the dairy farmer has agreed to produce

Introduction

Mr Mark Oliver's dairy farm 'Mosleigh' at Jamberoo is typical of the well-managed family farms in the Jamberoo Valley.

The farm consists of 100 hectares of land owned by Mark Oliver and his family. He also leases an additional 40 hectares from neighbours. He has a dry run that is 110 hectares in size. This land is used by cows that are pregnant, dry and not lactating. This gives him a total of 250 hectares.

The warm, temperate climate provides an ideal dairying environment. Low temperatures reduce the growth of kikuyu in winter.

The main sources of water are a creek and a number of bores. The creek, which runs through the property, has water flowing permanently. Town water is also available. There are water troughs in most paddocks.

1 What are the three main sources of water on the farm?

The soils on the farm are very productive. They include red earth, or red-brown soils, which are formed on the flood plain of the valley from volcanic parent material, and alluvial soils in the lower valley (which is sometimes flooded) that are very fertile.

Description of farm operations

Pastures

The main pastures are a mixture of kikuyu, perennial ryegrass, white clover and red clover. The paddocks used by the 'milkers' are strip grazed with an electric fence. The animals are given a strip, approximately 3 hectares per day, of highly nutritious feed. The electric fence stops the cattle from trampling feed in the rest of the paddock.

These pastures are maintained by harrowing, slashing and mulching after the cattle have grazed each paddock. The harrows spread out the manure, and the slasher/mulcher cuts any tall unpalatable stems or weeds. The paddock is then closed off from the cattle until it regrows.

2 Name the main pasture species.

3 Explain how the paddocks used by the milkers are grazed.

Paddocks are top dressed each year with superphosphate (250 kg/ha) and urea (125 kg/ha). Potash is also used as required.

In spring and summer excess pasture is cut and made into hay or round bale silage, and fed to cattle in winter or in times of drought. At certain times of the year hay is also fed to the milkers at night after milking, to keep up the level of roughage in their diet.

Animals

The stock consists of Friesian and Jersey cattle. Each day, 280 **lactating cows** are milked. At any time during the year, approximately 60 cows will be dry. These **dry cows** produce a calf and then go back into the dairy herd.

There are also approximately 130 heifers and 50 calves of different ages. The heifers are reared as replacements when the milkers are culled. The milkers are culled because of low production, disease or old age. There are also two bulls.

The dairy

Cows are fed grain or dairy pellet concentrates (Fig. 13.1).

The dairy is a herringbone design. There are two rows of 10 cows with 20 sets of cups. The cows face outwards. Washing of the cows' udders and subsequent attachment of the milk cups is done by the dairyman standing in a pit between the two rows of cows. The dairyman does not have to bend down to wash the cows' udders or put on the milk cups.

The herringbone design allows 20 cows to be prepared and milked at the same time. Time spent in the dairy is reduced by having automatic cup removers and automatic feeders (Fig. 13.2 and the video clip of the dairy at Forbes below).

ISBN 9780170265560

Milking is carried out twice each day. The morning milking starts at 5.00 a.m. and the afternoon milking starts at 3.00 p.m. Milking takes about 3 hours. This includes the actual milking, cleaning the dairy and milk lines, and feeding the calves.

Friesian and Jersey cattle are preferred on this farm because Friesians produce a large quantity of milk, while the Jersey cattle produce a smaller volume of milk but with higher butterfat content. The average yield from the two milking periods is 26 litres per cow per day.

The dairy also contains a refrigeration unit and a bulk tank for storing the milk, which is kept at less than 4°C.

connect

Working dairy farm

4 What are the advantages of a herringbone dairy design?

5 Why are Friesian and Jersey cows preferred to other breeds?

Shutterstock.com/branislavpudar

Figure 13.1 Cattle nutrition

Alamy/© Chris Robbins

Figure 13.2 A herringbone dairy

ISBN 9780170265560

connect

Calf rearing

6 Why are the calves on their mothers for approximately 1 week?

Calf rearing

When the calves are born they are left on their mothers for approximately 1 week. This enables them to obtain colostrum in the milk. Colostrum contains antibodies that protect the calf from disease. The calves are then fed milk twice a day (Fig. 13.3). They are fed milk for the next 8 or so weeks and then weaned. Before being weaned, pellets and hay are fed to the calves for 6 months to develop their rumens. The calves have to be ear tagged, drenched and vaccinated.

Getty images/Cath Shannon (CS Photography), New Zealand

Figure 13.3 Calves drinking milk

Disease

The herd is susceptible to several diseases.

connect

Mastitis

- *Mastitis* is caused by bacteria entering the teat and going into the udder. The milk from cows with mastitis cannot be used. The disease is controlled by carefully washing the udder before milking, and teat dipping after milking. Mastitis can be treated by injecting cows with antibiotic in the teat canal. Antibiotic (long acting) and teat seal are used when the cow is dried off.
- *Milk fever* is caused by a lack of calcium in the blood, and often occurs just after calving. The affected cows are injected with calcium borogluconate.
- *Three-day sickness* is caused by a virus. It causes reduced milk production. The symptoms usually last for approximately 3 days. Cows can now be vaccinated to prevent this disease.

connect

Pinkeye

- *Pinkeye*. This condition is a bacterial infection of the eye which produces a toxin that attaches to the surface of the eye and surrounding membranes. Inflammation then occurs. In severe cases blindness may result. Flies help to spread the organism between animals. Pinkeye can be treated with antibiotic eye ointments and antibiotic injections for more severe cases. Prevention is achieved by attempting to control flies with insecticides, separating infected animals from the uninfected, and by vaccination.

connect

Liver fluke

- *Liver fluke*. This is a flatworm that occurs in cattle as well as other animals and infests the bile ducts of these animals. The parasite is only found in areas where permanent streams allow the lifecycle to be completed. Large numbers of fluke cause weight loss, anaemia, poor milk production and eventually death. Control is achieved by draining or fencing off swampy areas. Treatment involves drenching infected animals.
- *Roundworms*. There are many species of roundworms. Most cattle, especially those in high rainfall areas, have a mixed infection. The symptoms include anaemia, rough coat, stunted growth and weight loss. Control can be achieved by spelling or resting paddocks (to allow the worm larvae to die) and by drenching cattle.

7 Name three diseases that can affect the herd.

8 What organism causes three-day sickness?

9 How are cows with pinkeye treated?

ISBN 9780170265560

Management

The farm is operated by Mark Oliver and his two sons. The minimal labour requirement is made possible by efficient farm management and the use of modern equipment and labour-saving technology. For example, permanent and temporary electric fencing is used on much of the farm. This means that the farmer spends less time rounding up cattle, and allows the farmer the flexibility of being able rapidly to alter paddock size and hence grazing areas.

Calendar of operations

Routine management consists of the daily routine and the seasonal routine. The daily routine includes milking the herd twice a day, feeding calves, culling milkers, mating or artificially inseminating cows, and shifting the temporary day and night electric fences. Mating or artificially inseminating cows is a year-round operation so that cows will calve each week and milk can be produced each day to meet the **milk contract**. The cows are artificially inseminated and the heifers are mated with a bull.

Culling of milkers is a continual operation during the year. Cows are sold if they have chronic mastitis, feet problems, poor temperament or when they are too old. Animal health operations, such as drenching and vaccinating, are also continual activities throughout the year. Table 13.1 illustrates a yearly calendar of routine operations on a dairy farm.

Table 13.1 Calendar of routine activities (daily)

Activity	Jan	Feb	Mar	Apr	May	Jun	Jul	Aug	Sep	Oct	Nov	Dec
Calving	X	X	X	X	X	X	X	X	X	X	X	X
Milking	X	X	X	X	X	X	X	X	X	X	X	X
Heat detection	X	X	X	X	X	X	X	X	X	X	X	X
Pregnancy diagnosis	X	X	X	X	X	X	X	X	X	X	X	X
Vet check	X	X	X	X	X	X	X	X	X	X	X	X
NLIS, ear tagging, tattooing	X	X	X	X	X	X	X	X	X	X	X	X
Dehorning	X	X	X	X	X	X	X	X	X	X	X	X
Calf rearing	X	X	X	X	X	X	X	X	X	X	X	X
Calf weaning	X	X	X	X	X	X	X	X	X	X	X	X
Parasite control	X	X	X	X	X	X	X	X	X	X	X	X
Feeding	X	X	X	X	X	X	X	X	X	X	X	X
Vaccination	X	X	X	X	X	X	X	X	X	X	X	X
Herd recording	X	X	X	X	X	X	X	X	X	X	X	X
Rapid mastitis testing	X	X	X	X	X	X	X	X	X	X	X	X
Artificial insemination	X	X	X	X	X	X	X	X	X	X	X	X

All the activities listed in Table 13.1 need to occur every month of the year because cows are calving each month and the farmer has to produce milk each day to meet the milk contract. Some of these activities relate to getting the cows back into calf or pregnant while other activities involve feeding and maintaining animal health of both calves and mothers. A cow's milk production is at its peak 6–8 weeks after calving. Ear tagging identifies an individual animal within the herd on a particular property. The National Livestock Identification Scheme (NLIS) helps trace the animal back to its property if it is sold. This helps with monitoring cattle movement.

Apart from the activities mentioned above, some seasonal operations are necessary. These include slashing pastures, mowing pastures, spraying weeds, silage and haymaking during the spring and summer. Fertilising (topdressing) pastures, harrowing pastures, fencing and supplementary feeding (hay to cattle) are carried out during autumn and winter. Table 13.2 (page 184) illustrates a yearly calendar of seasonal operations on a dairy farm.

ISBN 9780170265560

Getty/Science Photo Library

Figure 13.4 Artificial insemination

Table 13.2 Calendar of seasonal operations

Activity	Jan	Feb	Mar	Apr	May	Jun	Jul	Aug	Sep	Oct	Nov	Dec
Irrigation	X	X	X	X				X	X	X	X	X
Cultivation	X	X	X	X	X	X		X		X	X	X
Sowing	X	X	X	X	X	X		X	X	X	X	X
Haymaking	X	X	X							X	X	X
Spraying		X	X						X	X		
Silage making			X	X								
Topdressing			X	X	X	X	X	X	X			X
General work e.g. fencing					X	X	X					
Machine maintenance						X	X	X				

10 List the daily routine management operations or activities.

11 List the seasonal routine management operations or activities.

12 Name five operations or activities carried out during spring and summer.

Farm records, such as stock records (including artificial insemination (AI) records and returns to service), paddock records, and Dairy Quality Assurance Program records are maintained on a continuous basis. Accounts and other financial records are also kept.

Machinery

The most valuable machine used on the farm is the tractor. Other machinery and equipment include a fertiliser spreader, a slasher, a mulcher and pasture harrows. Haymaking equipment includes a mower, rakes and a round hay baler with silage attachment. A truck is used for taking cows and calves to the sale yards, and for transporting hay.

Marketing

The farm has a milk contract of 4000 litres per day. Milk is collected once a day in a bulk milk tanker, and taken to the Parmalat factory in Sydney.

The milk can be rejected by the factory if it contains dirt, blood, faecal or antibiotic contamination.

Mark receives payment based on his milk contract. He receives one price for all milk that is produced.

Mark receives a monthly payment from the milk factory and therefore has a regular income. Penalties (a lower price) are paid for milk with mastitis, low protein, low butterfat or a high bacteria count. (Incentive or bonus payments are paid for the opposite.)

13 What are four reasons for the factory rejecting milk from the farm?

 ISBN 9780170265560

Gross margin for a dairy farm

A gross margin provides an indication of the profitability of an activity on the farm. The gross margin includes only variable costs (costs that change during production), not fixed or overhead costs. Gross margins are worked out on a per cow or per hectare basis. The formula for a gross margin is:

Gross margin = total income – variable costs

Table 13.3 Gross margin for 'Mosleigh' dairy farm

Gross income	$
Milk	730 000
Culls (cows)	35 000
Calves	7 500
Total gross income (GI)	772 500
Variable costs	
Seed, fertiliser, fodder, pellets	400 000
Electricity	12 000
Stock health (veterinary)	6 000
Artificial insemination	4 000
Herd recording	7 000
Herd registration (stud cattle)	2 500
Freight, cartage	1 500
Dairy requisites	3 100
Bulk milk collection	20 000
Repairs and maintenance	8 000
Total variable costs (VC)	464 100
Gross margin (GM)	
GM = GI – VC = 772 500 – 464 100	= $308 400
GM/cow = GM/340 (i.e. 280 milkers + 60 dry cows)	= $908/cow
GM/ha = GM/250ha	= $1234/ha

14 If the price of milk was to rise by 5% (and there are no other changes to income or cost), recalculate the gross margin per cow and gross margin per hectare.

Consumers and marketing

Consumer preferences have driven how milk and milk products are marketed, including as:

- value-added products, for example cheese and yoghurt
- health-based milk products, for example low-fat milk, iron/calcium or fortified milks.

Regular consumption of milk by consumers has led to the implementation of contracts to farmers to ensure year round supply of this staple product.

Environmental sustainability

Dairy cows create a large amount of effluent. This can be a useful resource if managed effectively. Most dairy farms see effective effluent management as a way of returning nutrients to the soil.

Poorly managed effluent has the potential to pollute creeks and rivers. This will have detrimental effects on the environment.

Methane gas is a waste product of ruminant digestion and is produced in large quantities on dairy farms. This gas contributes to the greenhouse effect, or warming of Earth's atmosphere.

connect

Greenhouse gas emissions

List five ways to reduce greenhouse gas emissions on dairy farms.

ISBN 9780170265560

Farm technology

Technology is used in management and production on the farm. Examples of technologies being used in the management of modern dairy farms include:

- estimated breeding values (EBVs) in breeding programs for selecting suitable male (semen) and female animals based on their production records, for example temperament, milk production and butterfat
- computers for record keeping, gross margins, weather data and herd management programs
- pregnancy scanners
- animal identification implants
- improved health products, for example rumen implants and feed additives
- biotechnologies, for example embryo transfer and oestrus synchronisation
- robotic milking systems
- improved machinery and equipment
- improved irrigation systems.

connect

Innovation

List the types of innovations discussed.

15 Describe two ways in which technology is used in the management and production of dairy farms.

16 Identify three potential safety hazards on a dairy farm and describe safe working practices that can be used to overcome the problem.

Farm safety

On a dairy farm there are many potential hazards and safety is very important. Table 13.4 summarises the main hazards encountered and safe work practices that should be used.

Table 13.4 Safe management of common dairy farm hazards

Hazard	Example	Safe work practice
Operating machinery	Operating tractors, working near belt or power take-off (PTO) driven machinery	PTO covers on all machines. Tractor roll-over protection structure (ROPS) in place. Disengage all equipment when cleaning or inspecting. Operate to manufacturer's advice. Repair and maintain machinery
Animal handling	Working with bulls, AI, carrying out animal treatments, e.g drenching, milking	Awareness of animal behaviour and avoid frightening the animal
Zoonoses (animal diseases that can be transmitted to humans)	Working with cattle infected with Q fever or leptospirosis	Vaccinate cattle for leptospirosis
Manual handling	Lifting fertiliser bags, animals, feed	Use safe lifting procedures
Chemicals	Use and storage of sprays, pour-on insecticides, fuel, detergents	Read labels, use according to directions, wear personal protective clothing (e.g. respirator), store safely and dispose of according to local government requirements
Electricity	Overhead wires, machine movement and long irrigation pipes being moved near overhead power lines	Be aware of locations of overhead wires, signage indicating location of electrical outlets and cables
Workplace	Sun exposure, general dairy environment	Sun aware behaviour, e.g. hat, sunscreen. Eliminate damaged flooring, rails. Tidy shed environments
Movement awareness	Movement of vehicles, animals around sheds and farmhouse, especially in relation to children	Know where young children are at all times. Clear indications when driving machinery or operating powered equipment
Slips and falls	Wet floors in dairy and uneven surfaces in general	Non-slip flooring in dairy
Noise	Tractors, machines, dairy	Use ear muffs
Effluent ponds	Cleaning effluent ponds and solids traps	Work carefully in the presence of another person
Confined spaces	Working in silos, milk vats	As above
Dust	Milling and mixing feed	Use a respirator

 ISBN 9780170265560

A systems approach to a dairy farm

It is important that the farmer identifies goals. Goals are what the farmer is trying to achieve on the farm. The farmer's goal is usually to make a profit. On a dairy farm this would be achieved by producing good quality milk (and obtaining financial incentives for this), meeting the milk contract and minimising expenses. To minimise expenses the farmer tries to reduce the cost of feed, labour, veterinary costs and so on.

Mark Oliver's dairy farm can be examined in terms of a **black box model**. A black box model shows the inputs and outputs of a farm or enterprise (see Figure 13.5).

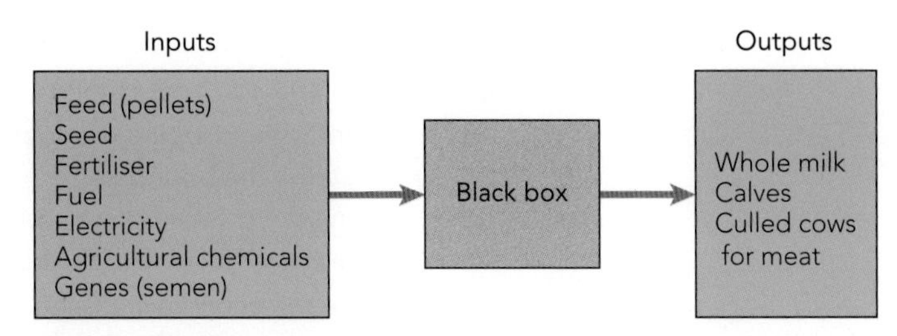

Figure 13.5 Black box model of 'Mosleigh' dairy farm

The parts of the black box model are:

- inputs – the materials that go into the farming system (e.g. fertiliser, seed, fuel, electricity, chemicals, semen, concentrates)
- outputs – the materials produced by a farming system and removed from it (e.g. whole milk, calves, culled cows)
- processes – the activities that change inputs into the desired outputs (e.g. milking, mating, calving, culling). Processes occur in the black box part of Figure 13.5 but are not identified or explained.

A static display model gives more information about the subsystems of the farm and their interaction with one another, and the boundaries (see Fig. 13.6). **Boundaries** are the limitations of the system, and include fences (e.g. the boundary fence), money and managerial skill.

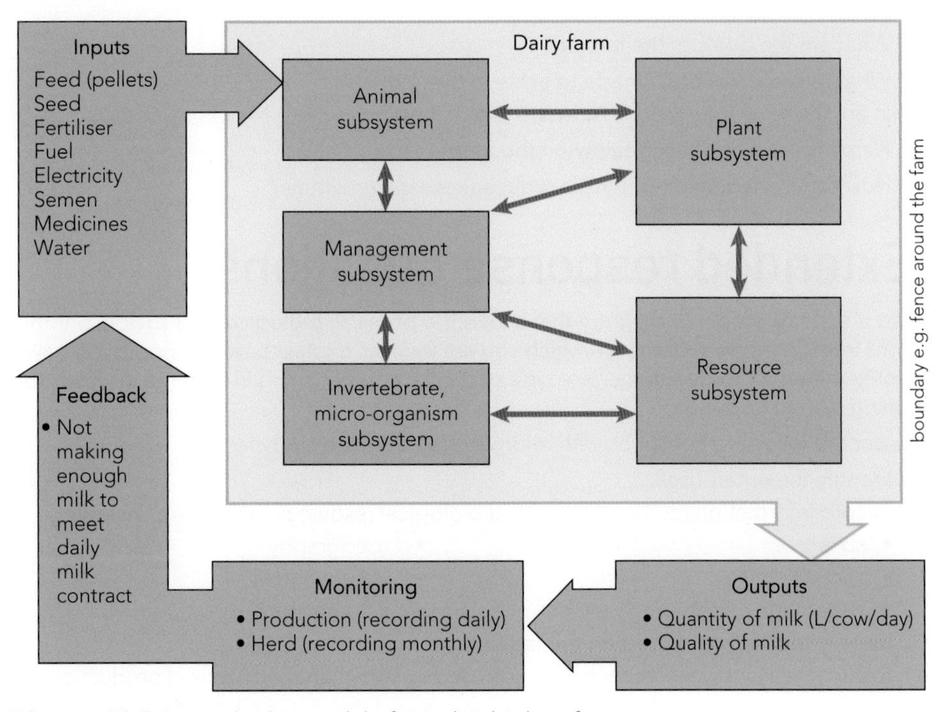

Figure 13.6 Static display model of 'Mosleigh' dairy farm

17 What does a black box model show?

18 What does a static display model show?

19 What is a boundary?

20 Identify the subsystems operating within the farm unit.

connect
Rural studies
Careers in the dairy industry

connect
Careers

connect
Cows create careers
More about agricultural career pathways

connect
Employer of choice

connect
Skilled Ag workers

Careers

There are many career pathways in agriculture relating to working with animals. General advice in relation to careers in livestock industries can be found by visiting the websites at right.

ISBN 9780170265560

Chapter review

Things to do

1. a Describe the annual routine management procedures for a dairy farm.
   b Explain why each exists.
2. Describe what factors a farmer might consider when making farm management decisions.
3. For the farm you have studied:
   a identify the subsystems operating within the farm unit.
   b describe the boundary.
   c name the main inputs, outputs and processes.
4. Analyse the physical relationship between inputs and outputs in farm production.
5. Evaluate the economic efficiency of the farm (by finding out the gross margin per hectare and comparing it with other farms).

Things to find out

Consider the farm you have studied to answer the following questions.

1. Identify instruments used to measure resources such as weather and soils.
2. Identify problems associated with production on the farm that you have studied. Use the following headings:
   Management
   Disease
   Genetics
   Environment
   Nutrition
3. a What are the goals of the farmer?
   b What attempt has been made to achieve these goals?
4. What are the main interactions between subsystems?
5. a Name four areas of uncertainty on the farm.
   b How can management practices reduce these uncertainties?

Extended response questions

1. Draw a systems model or diagram that shows the physical, biological and socioeconomic inputs into a farming system with which you are familiar. Explain how the inputs you have identified interact to determine one type of production occurring in the area in which your farming system is found.
2. Answer the following questions with reference to your case study farm.
   a Identify the enterprises.
   b Record information on the physical and biological resources of the farm, including:
      - soils
      - climate
      - vegetation
      - topography
      - water sources
      - infrastructure.
   c Identify technologies used on the farm.
   d Explain the ways in which technology is used in farm management and production.
3. Draw a systems model or diagram of your case study farm. Explain how analysis of the model could enable you to identify:
   a important interactions between subsystems
   b problems with the operations of the farm
   c problems with the efficiency of the farm
   d possible changes to farm organisation that would reduce problems and increase the efficiency of operation.
4. a For a farm that you have studied, identify potential farm safety hazards, assess the risk and suggest strategies to reduce or overcome these problems.
   b Outline Work Health and Safety legislative requirements that affect a farm.

ISBN 9780170265560

UNIT 3
RESOURCE SUBSYSTEMS AND AGRICULTURE

CHAPTER 14

Outcomes

Students will learn about the following topics.

1. The climate patterns that influence the distribution of agricultural enterprises in Australia.
2. The effect of climate on animal and plant production systems.

Students will be able to demonstrate their learning by carrying out these actions.

1. Distinguish between the terms 'climate', 'weather' and 'microclimate'.
2. Define the terms 'solar radiation', 'solar constant', 'ozone', 'albedo', 'net radiation', 'surface energy balance', 'phototropism' and 'vernalisation'.
3. Describe the importance of solar radiation for plants.
4. Outline the effect of temperature on production systems.
5. Discuss light requirements for plants and animals.
6. Describe how the processes of precipitation, evaporation, run-off and drainage determine the amount of water stored in the soil.
7. Assess the importance of wind and humidity in production systems.
8. Describe various management practices that control the effects of climate on agricultural production.

ISBN 9780170265560

CLIMATE AND ITS EFFECT

Words to know

albedo the amount of solar radiation reflected by a surface

climate the average weather conditions over a long period of time

greenhouse effect the overall increase in temperature caused when solar radiation is unable to escape from Earth's atmosphere because of the presence of atmospheric gases, particularly carbon dioxide

growing season occurs where precipitation exceeds evaporation for a period of time

lodge a term applied to crop plants that fall or are blown over by wind

microclimate the atmospheric conditions near the surface of the ground or adjacent to the crop canopy

net radiation the sum of all radiation energies gained or lost by Earth's surface

ozone molecule produced by the breakdown of ordinary molecular oxygen by sunlight, allowing oxygen atoms to combine with other oxygen molecules to form O_3

photoperiodism the response of organisms to changes in day length, as a result of changes in the ratio of daylight to darkness

photosynthesis the process where carbon dioxide and water are combined using light energy to form high-energy sugar compounds and oxygen

solar constant the average amount of radiation received by a surface at right angles to incoming radiation from the Sun

solar radiation all forms of radiation received by Earth

surface energy balance the balance between radiation gained and lost from the surface of Earth. In daylight hours most of the net radiation energy gained at Earth's surface is used or released as heat

thermoperiod the period of exposure of a plant to a particular temperature

transpiration loss of water from the leaves of a plant by evaporation through the stomata

vernalisation the requirement of a plant for a period of exposure to a particular temperature or range of temperatures (usually low), which stimulates the plant to flower

weather daily changes in the atmosphere in precipitation, temperature, wind, pressure, cloud cover and other factors

Introduction

An area's average weather conditions over a long period of time are described as its **climate**. Because of this long timeframe, the climate of an area remains fairly constant, especially during the lifetime of a farmer. Day-to-day changes in the atmosphere are described as **weather**. These short-term changes have an immediate effect on the level of output from a farm and on the efficiency of any particular farming activity. The main aspects of weather studied by scientists are solar radiation, temperature, rainfall (or precipitation), evaporation rates, humidity and wind effects.

The composition of Earth's atmosphere remains fairly constant up to altitudes of 60–80 km. As altitude increases, the proportion of **ozone** (produced by the breakdown of ordinary molecular oxygen by sunlight, allowing oxygen atoms to combine with other oxygen molecules to form O_3) increases, while carbon dioxide, oxygen and water vapour levels decrease. Earth's lower atmosphere, extending from the land surface to an altitude of 11 km, is the area of clouds, storms, winds and currents.

With increasing height, or altitude, in the lower atmosphere, the temperature drops. Above the cloud zone lies an area virtually free of cloud and storms. High concentrations of ozone can be found in this area. The ozone absorbs radiation, especially the life-damaging ultraviolet radiation, and so provides a protective blanket for life to develop.

The part of Earth's atmosphere consisting of ionised layers is located at altitudes beyond 80 km. Two spectacular effects originating from this section of the atmosphere are the northern lights (aurora borealis) and the southern lights (aurora australis). The outer realm of Earth's atmosphere extends beyond a height of 800 km. Over 50% of Earth's atmosphere is found in the first 5 km above the land surface.

The climate within the first few metres of the ground has an immediate effect on plant and animal production systems. This **microclimate** is very important in agriculture because it directly influences both soil temperature and moisture levels. Changing temperature patterns of the soil and air affect the germination, growth and development rates of plants. Microclimate changes also affect animal production levels because they affect the availability of feed and water, and have general effects on body metabolism and behaviour.

Farming activities modify the microclimate to improve production levels, extend production areas and sustain productivity levels over time.

connect

Energy from solar radiation

View the video and make notes.

connect

What is the greenhouse effect?

Summarise the information in this video clip.

connect

Greenhouse gas emissions

Read and summarise the information.

Solar radiation

Solar radiation refers to the radiation (light and heat) coming to Earth from the Sun. Earth receives heat not only from the direct rays of the Sun but also from short-wave radiation scattered through the atmosphere by the air and other components of the atmosphere (Fig. 14.1). This is an important source of heat for Earth's surface, especially in higher latitudes where the intensity of solar radiation is low during the winter months. Other minor sources of energy exist in Earth itself.

The rate at which solar energy reaches the outside of Earth's atmosphere is almost constant. The average amount of radiation that reaches a surface, as measured at right angles to the incoming solar radiation, is called the **solar constant**.

Scattered incoming radiation is also reflected back into space and lost to production systems on Earth.

The proportion of incoming radiation reflected by clouds or land surfaces is known as **albedo**. Table 14.1 shows albedo values for various surfaces. The amount of radiation actually used by plants to produce sugars for growth is very small (less than 3% of the original incoming solar radiation).

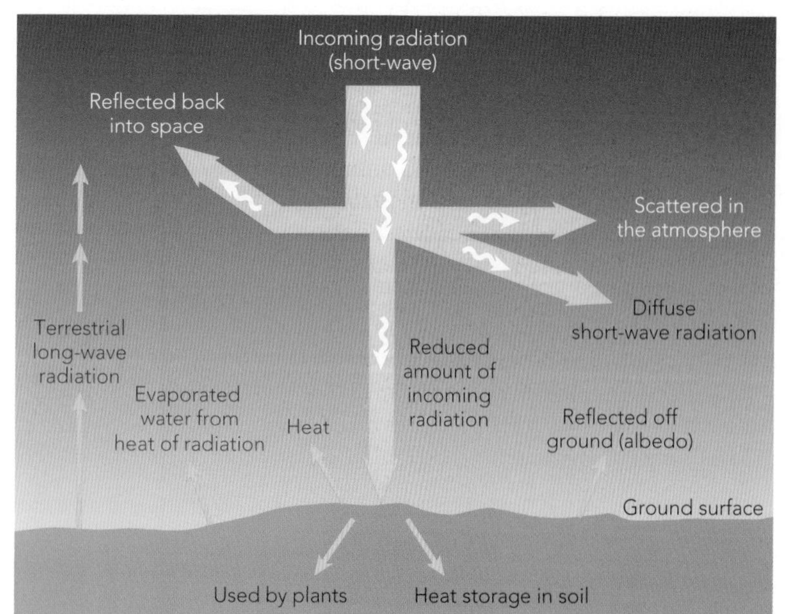

Figure 14.1 The fate of incoming radiation

ISBN 9780170265560

Solar radiation consists of many wavelengths – 98% of the radiation has wavelengths in the ultraviolet, visible and infrared areas of the spectrum. The atmosphere is selective in that it transmits short-wave radiation (ultraviolet light) more easily than the long-wave forms. Consequently, the atmosphere allows the entry of short-wave solar radiation but it absorbs a large portion of the outgoing long-wave radiation emitted from Earth. The resultant increase in atmospheric temperatures is called the **greenhouse effect**. If more of the outgoing radiation is absorbed because of increasing carbon dioxide levels in the atmosphere, Earth's surface temperature will increase over time.

Table 14.1 Albedo values of various land surfaces

Land surface	Albedo rating (%)
Dense forest	5
Light vegetation	10
Bare soil	20–30
Snow or cloud	60–75

1. Define the terms 'climate', 'weather' and 'microclimate'.
2. Describe the components of solar radiation.
3. Define the terms 'solar constant' and 'albedo'.
4. Briefly describe the cause of the greenhouse effect.

Net radiation

At the ground surface two forms of radiation can be identified – short-wave incoming radiation from the Sun and long-wave outgoing radiation from Earth. Earth's surface loses approximately equal amounts of radiation through reradiation and the evaporation of water. The losses are almost entirely balanced by incoming radiation. The sum of all radiation energies gained and lost by the surface of the Earth is called **net radiation**. This is a positive figure by day as energy is gained from the Sun, and a negative figure by night as energy is radiated back into space by Earth. Most of the radiation gained or lost by the ground surface is either released as heat energy into the air or stored as heat in the soil.

Surface energy balance

The **surface energy balance** equation is true both for any instant in time and over a period of time. The equation is defined as:

$$\text{Net radiation } (R_n) + \text{Rate of transfer of heat within soil or plants } (S) + \text{Rate of transfer of heat from soil/plant to air } (A) + \text{Heat loss due to evaporation of water } (LE) + \text{Photosynthetic activity } (p) = 0$$

$$R_n + S + A + LE + p = 0$$

By day the net radiation figure is positive and all other factors are negative. By night the net radiation figure is negative, photosynthesis is zero and the other values are positive. Consequently, Earth neither builds up nor runs down environmental energy levels.

The water cycle

Solar radiation is a resource for the light and temperature needs of plants and animals. Radiant energy is also responsible for evaporating water from both free water surfaces and living organisms. Water loss from the ground occurs through evaporation and loss through plant leaves is in a process called **transpiration**. Water vapour cools in the atmosphere and condenses to form clouds. In time, droplets of water form and eventually fall as rain.

Rain will run off unprotected or steep ground surfaces and infiltrate the surface soil layers, which are porous. Water in the soil will either be stored in spaces in the soil, called pores, to be eventually used by plants and other living organisms, or it will move through the soil because of gravity and drain into a permanently saturated area of the soil, called the water table, to form part of an artesian water system. The movement of water between the atmosphere and the ground surface and the events involving water in the atmosphere occur on a global scale to form a cycle. This water cycle, as it is called, produces a water balance between evaporation, run-off, movement of water vapour in the air and rainfall (Fig. 14.2).

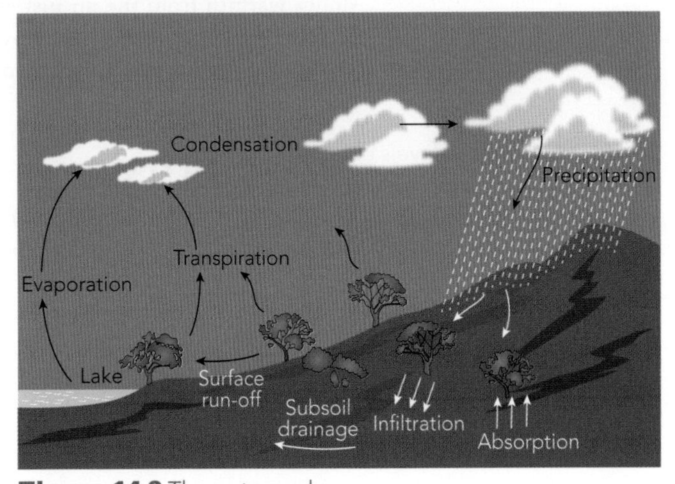

Figure 14.2 The water cycle

ISBN 9780170265560

Temperature

Temperature ranges determine in which part of the country a plant or animal type can be farmed. The approximate temperature range within which plants grow satisfactorily is 0–50°C. Generally, air temperature regulates plant temperature. Table 14.2 shows that there are temperatures below which certain plants cannot grow. Plants such as grapes, peaches and plums become dormant over the winter months to avoid problems associated with low temperatures.

Table 14.2 The minimum temperature requirements for the survival of various crops

Crop	Minimum temperature (°C)
Peas	15–18
Sorghum	18–20
Wheat	– 2–5

All organisms have an optimum temperature range in which they function best. Different functions in plants have different optimum temperatures; for example, potatoes have an optimum temperature of 20°C for **photosynthesis** to occur and 48°C for respiration to operate at optimal levels. Reproductive processes are the first to suffer from heat stress. In situations where animals need to consume more food in winter in order to keep warm, their rate of growth is not as rapid as that of similar groups of animals kept in conditions that best suit them. Growth declines as temperatures increase beyond the optimal range, until death occurs because of:

- enzyme inactivation
- excessive loss of water (dehydration in animals)
- excessive rates of respiration relative to photosynthesis in plants.

Many plants have **thermoperiod** requirements; that is, they grow best under conditions of alternating hot and cold temperatures. For example, tomatoes grow best with a day temperature of 26°C and a night temperature of 19°C. Other plants (e.g. wheat and hyacinth bulbs) require exposure to low temperatures at the seed or bulb stage to induce flowering. This process is called **vernalisation**.

The temperature of the soil has an effect on both plant germination rates and the rate of growth of plants. These in turn influence pasture quality and quantity, which directly affect animal growth and production levels. Soil temperatures are variable, depending upon soil colour, moisture content, depth and composition.

Air temperatures are more uniform than soil temperatures and have significant effects at ground level, depending upon altitude, aspect and landforms.

5 Draw and fully label the water cycle.

6 Define the term 'thermoperiod requirements'.

Fogs

At night the ground surface cools by emitting long-wave radiation. At the same time it draws warmth from the air just above the ground. Providing there is no wind movement, a layer of cold air forms just above the ground and remains trapped under warmer upper layers of air. Moisture droplets condense in this cold air layer, resulting in horizontal fog banks.

Frosts

The probability of frost increases when there are cloudless nights, dry air and a lack of wind. Air above the ground cools to a point where moisture condenses; however, as temperatures continue to drop to below freezing point, frost or ice crystals form on all exposed surfaces and often water freezes in the leaves of plants. The water expands on freezing, causing damage to the delicate cell structures of the plants.

Table 14.3 Relative levels of frost resistance

High	Medium	Low	None
Oats	Cabbage	Maize	Cotton
Barley	Soybeans	Sorghum	Rice

7 How do frosts form and how do they damage plant tissue?

ISBN 9780170265560

Light requirements

Many plants have light requirements for flowering. This mechanism is called **photoperiodism**. The critical period is the length of the dark time period (Fig.14.3). Plants may be divided into three categories on this basis:

1. short-day plants, which flower in conditions of less than eight and a half hours of daylight (e.g. oats and subclover)
2. long-day plants, which flower under conditions of greater than 11 hours daylight (e.g. radishes and maize)
3. day-neutral plants, which are non-specific (e.g. tomatoes)

Under the influence of critical periods of light and darkness, hormone levels in vegetative buds change to produce flowering buds.

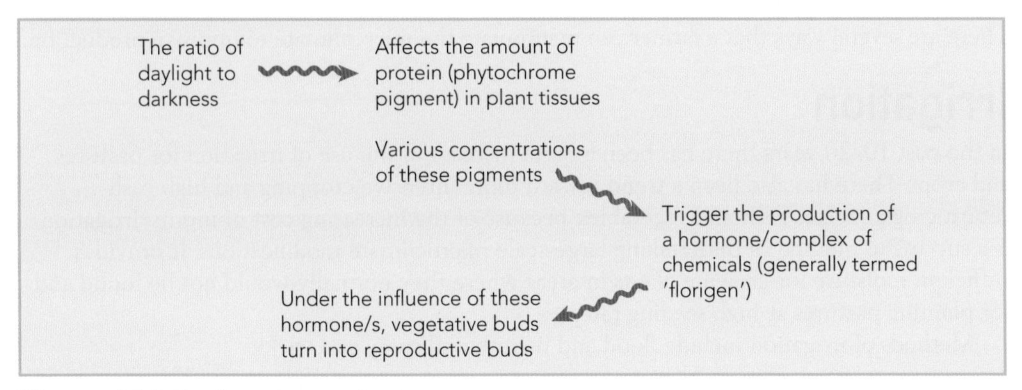

Figure 14.3 The flowering mechanism

connect

Photoperiodism

Review the concepts.

8 Define the term 'photoperiod requirements'.

Animals are also sensitive to day length, which triggers reproductive cycles. Rams become sexually active in situations of decreasing day length, while poultry become more sexually active as day length increases.

Rainfall

Moisture returns to Earth's surface as rain, dew, snow or hail. Farmers are concerned with the following aspects of rainfall:

- the total rainfall, or the amount of rain that falls in an area, which determines if a crop can be grown
- its seasonality, or the time of year that rain can be expected, which determines when a crop can be grown
- its reliability, or how likely that rain will fall, which determines whether or not activities can be planned
- its effectiveness, which is a measure of how much water remains in the soil after evaporation. This occurs when rainfall levels exceed evaporation rates. The months that have effective rainfall are called the **growing season**.

Wind

Uneven heating of Earth's surface by the Sun produces air movements, or currents, which are called winds. Strong winds cause tall crops, such as sunflowers, sudax and sometimes wheat, to fall over or **lodge**, making harvesting difficult. Dry winds cause pollen grains in crops such as maize to dry out, reducing the number of grains per corncob produced.

Humidity

Humidity is a measure of the amount of water vapour present in a sample of air at a certain temperature. Relative humidity refers to the amount of water vapour in the air at a particular temperature in relation to the maximum amount of water vapour that the air could hold at the same temperature. At 100% relative humidity, fog would form. Humidity readings depend on such factors as wind, temperature and moisture levels.

ISBN 9780170265560

During periods of high humidity many vegetable crops suffer from the increased effects of fungal diseases, while animals suffer heat stress, because losing heat by sweating becomes more difficult.

The rate of evaporation increases as the air temperature rises. It also increases during periods of dry, windy weather. A newborn animal with a wet body will die from exposure if it is not protected from wind. Evaporation of water from the animal's body surface causes a rapid cooling down of the animal's body, often resulting in pneumonia. Plants that cannot replace water lost by evaporation through their leaves will collapse or wilt.

connect

Microclimate

Describe how a microclimate can be developed in a polyhouse.

Farm management and the microclimate

There are several ways that a farmer can manipulate the microclimate to improve production.

Irrigation

In the past 10–20 years there has been a rapid increase in the use of irrigation for pastures and crops. There has also been a trend toward more intensive cropping and high cash-return crops, such as fruit and vegetables, because of the increasing cost of inputs. Irrigation is a successful method of undertaking large-scale microclimate modifications. It provides sufficient moisture for growing plants in areas where they normally would not be found and for planting pastures at high seeding rates.

Methods of irrigation include flood and furrow, and spray and trickle.

connect

Irrigation systems

Summarise the main points raised in this video on irrigation systems.

Shade

Provision of shade is an important consideration in agricultural production systems. When stressed because of temperature extremes, animals use much of their body energy to keep either cool or warm. Consequently, their growth rates slow and their production levels fall. Many high-producing animals are more sensitive to temperature extremes than non-producing animals. Every animal has a temperature range in which production is not affected. European cattle should be in a temperature range of 13–18°C, with a relative humidity of 60–70% (Fig. 14.4).

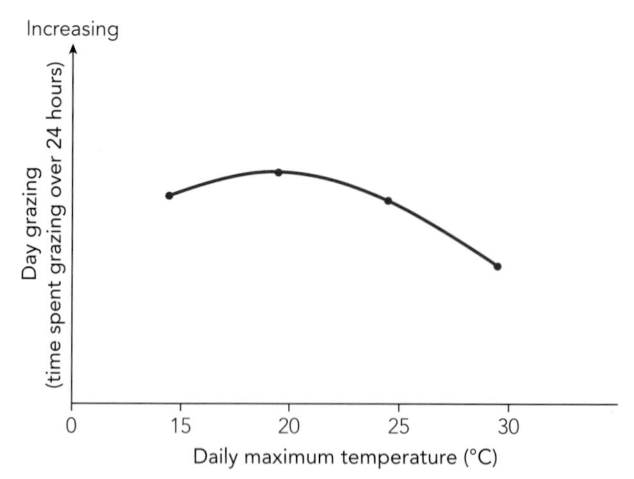

Figure 14.4 The effect of daily maximum temperatures on cattle grazing patterns

Windbreaks

Trees planted near commercial areas as wind breaks, such as orchards and market gardens, help to maintain the quality of the product, provided they are positioned far enough away to not interfere with the growing crop plants.

ISBN 9780170265560

Further information on the use of windbreaks can be found at the Windbreaks for citrus website.

Livestock shelters made from trees, large shrubs or remnant vegetation patches on farms provide places for animals to escape harsh weather and prevailing winds.

connect

Windbreaks for citrus

connect

Climate manipulation

Present a report on climate manipulation in poultry production systems.

Intensive production systems

Intensive animal production systems, such as broiler sheds, intensive piggeries and greenhouses for plants feature specialised methods of microclimate control. Particular management methods are required to control factors such as temperature, humidity and day length, and to maintain ventilation levels. Table 14.4 indicates the desired temperatures for poultry and pig raising.

9 List three ways farmers can manipulate the microclimate.

Table 14.4 Optimum temperatures for pig and poultry production

Production system	Optimum temperature (°C)
Day-old broilers	33
6-week-old broilers	21
Layers	20–28
Newborn pigs	27–35
5-week-old pigs	21

Various types of shadecloths are available to produce ideal daylight and temperature environments for plants. Misting and sprinkler systems may also be used to supply moisture and lower temperature levels. Heating cables are used in nurseries to promote root growth in plants, while heating fans warm up glasshouses. The use of glasshouses or polyhouses is another means of modifying the microclimate to enhance plant growth. In these enclosed environments adequate ventilation is required to avoid excessive humidity and to provide sufficient carbon dioxide levels to sustain plant growth.

The Australian climate

As Australia is a large continent, it experiences a range of climates. The reasons for climatic differences in Australia include:

- changes in latitude
- altitude differences
- proximity to warm or cold ocean currents
- local conditions, such as valleys and mountain ranges
- the effects of world wind circulation systems.

The world wind systems produce a very dry environment across the Australian continent because of the latitudes in which Australia is located. The interior of the country has poor rainfall because of its distance from the sea, the lack of inland lakes, the flat topography and the general shape of the continent.

Rainfall occurs in fairly evident patterns. In the north, rain is received during the summer months, while in the south, rain is received during the winter months. Monsoonal conditions exist north of the Tropic of Capricorn. The rainfall patterns directly affect the type and scale of agricultural production practices across Australia.

connect

Australian climate

Climate change and agricultural production

Recent information gathered by scientists studying the atmosphere suggests that a general warming trend is occurring, because of increases in the level of carbon dioxide and methane in the atmosphere. A rise of only a few degrees Celsius in the temperature of the atmosphere would result in substantial changes in world climate patterns. Increased evaporation rates could occur, which would result in an increase in activity within the water cycle. Humidity levels would be higher and rainfall could increase in some areas and decrease in others. Climate change and its impact is explored in more detail in Chapter 32.

The production levels of plants in an atmosphere with higher temperatures and higher levels of carbon dioxide would increase, as the rate of photosynthesis increased. On the other hand, the incidence of fungal attack, such as rust, and insect activity would increase. The management of agricultural production systems would need to alter to take account of the changing climatic patterns.

connect

Climate change

ISBN 9780170265560

Chapter review

Things to do

1. Visit a farm or irrigation area and record the following measurements for each location.

Location	Temperature	Relative humidity	Light level	Wind
Open area				
Bare soil				
Inside shed				
Shade area				
Glasshouse				
1 m above crop				
Crop surface				
Covered ground surface				
2 cm below bare soil				
2 cm below planted ground				
0.5 m below ground				
0.5 m below planted ground				

What conclusions can you draw from your measurements? What conclusions can you draw from the class measurements?

2. Draw a detailed plan of either an intensive animal production shed (e.g. intensive piggery or broiler production unit) or an intensive plant production system (e.g. a glasshouse or hydroponic unit). On your diagram identify:
   - a all the environmental control systems
   - b the methods of heating and cooling the structure
   - c the methods of waste disposal or drainage of waste
   - d any unique or special aspects of the system you have observed.
3. Visit a windbreak planted on a local property.
   - a What trees or shrubs are commonly used? Why are these used?
   - b What is the planting pattern?
   - c What is the desired height or shape of the windbreak?
   - d What is the critical aspect for the windbreak?
   - e Measure and compare wind intensity at various intervals away from the windbreak.
4. For a particular crop, conduct measurements of the growth rate over 1 week for 3 months of the year. What conclusions can you draw?
5. For a particular form of irrigation system, determine the following:
   - a the wetting pattern
   - b the water delivery rate (amount per time)
   - c the effect of wind on the spray pattern.

ISBN 9780170265560

Things to find out

1. What are the main climatic patterns on Earth?
2. How can growers produce flowering crops (such as chrysanthemums) at specific times of the year?
3. How can farmers manage the water cycle?
4. Define the term 'effective rainfall'. Attempt to determine this factor for your local area.
5. Explain the concept of a growing season, giving examples for different crops.
6. Why do humidity levels vary between the top and bottom of a crop?
7. Describe how climatic patterns influence the distribution of agricultural enterprises in your state. Use two examples: one plant and one animal.

Extended response questions

1. Discuss how farmers have manipulated the microclimate to achieve increased levels of production in your local area.
2. Climate affects many aspects of the agricultural system. Outline the main effects of climate on plant and animal production systems.
3. Explain how a glasshouse environment differs from a natural environment.
4. Examine how possible future climatic changes could affect the management of agricultural production systems.

CHAPTER 15

Outcomes

Students will learn about the following topics.

1. How soil resources influence the distribution of agricultural enterprises in Australia.
2. The effect of soil texture, structure, pH and fertility on plant production.
3. Inorganic and organic fertilisers.
4. Cultivation practices in relation to soil type.
5. The chemical and physical characteristics of a soil.
6. The sources of water on a farm and water management in a farm system.
7. A first-hand investigation project to analyse the physical and chemical characteristics of a soil.

Students will be able to demonstrate their learning by carrying out these actions.

1. Describe the major components of a soil.
2. Define and assess soil texture.
3. Define and assess soil structure.
4. Investigate and describe physical characteristics of the soil, including colour, parent material, soil structure, texture, water holding capacity.
5. Investigate and describe the chemical characteristics of a soil, including pH status and organic matter content.
6. Define the terms 'porosity', 'bulk density', 'coalescence' and 'pulverescence'.
7. Investigate and describe physical characteristics of a soil, including porosity, bulk density, and coalescence and pulverescence.
8. Discuss the factors that develop soil structure.
9. Distinguish between primary and secondary soil minerals.
10. Discuss the role of sand and clay in the soil and compare the properties of sand and clay soils.
11. Discuss the importance of clay in a soil.
12. Describe the influences of soil temperature, water and air on plant growth.
13. Describe the chemical characteristics of a soil, including soil pH, ion exchange capacity, soil carbon (organic matter) and nutrient status.
14. Describe the main effects of soil chemistry on plant production.
15. Discuss the importance of soil organic matter.
16. Outline plant mineral requirements and the role of fertilisers.
17. Identify macro- and micro-nutrients important for plant growth.
18. Select fertilisers appropriate to the soil and crop or pasture requirements.
19. Select appropriate tillage implements and techniques to establish a crop or pasture.
20. Identify the various sources of water and appropriate management of water use on farms.
21. Describe the influence of legislation and government regulations, including licensing, on the availability and use of water for agricultural purposes.
22. Perform a first-hand investigation to analyse and report on the physical and chemical characteristics of a soil.

ISBN 9780170265560

SOILS, NUTRIENTS AND WATER

Words to know

available water the amount of water available to a plant in the soil as measured between the point of field capacity and the permanent wilting point

bulk density the mass of soil solid material relative to the total volume of soil material

C:N ratio the ratio of carbon to nitrogen material in the soil, which determines the effectiveness of micro-organisms in the nitrogen cycle

cation exchange capacity a measure of the amount of exchangeable ions that can be held by a clay particle

clay double layer the layer of tightly and loosely bound ions that surrounds a clay particle

coalescence the degree of moulding or coherence that occurs to a block of soil when subjected to a force at a particular moisture content

colloids an organic or inorganic particle in a soil that is extremely small, but has a very large surface area. Soil colloids may hold plant nutrients on their surface

dispersion the spread of particles in several directions, from areas of high concentration to areas of lower concentration

diurnal rhythm an activity lasting for a day

field capacity the amount of water retained in a soil profile that was saturated after drainage has occurred over a 24- to 48-hour period

inorganic fraction the non-living fraction of a soil, including minerals, soil air and soil water

moisture characteristic the relationship between soil moisture potential and soil moisture content

ped the basic unit of soil structure

permanent wilting point the point beyond which a plant cannot recover from water loss, even if water is applied to it

porosity the percentage of spaces, or voids, in a soil

pulverescence the degree of shattering that occurs when a block of soil is subjected to a force at a particular moisture content

saturated soil a soil in which all pores contain water

soil consistence a measure of the mechanical strength of a soil, based on the force necessary to break a block of soil

soil moisture potential a measure of the energy required by the plant to remove water from the soil at a particular moisture content

soil solution the liquid fraction contained in a soil

soil structure the arrangement of soil particles in a soil

soil texture the percentage of sand, silt and clay in a soil, as determined by particle size

surface scalding the formation of a surface crust on a soil, which is caused by the build-up of salts in the soil

turgid describes a plant cell that is fully expanded owing to the absorption of water; the cell membrane is pressed against the cell wall

weathering the process of rock breakdown to form soil

wilt a condition where the structure of a plant collapses because of lack of water

wilting point the moisture level in the soil when wilting occurs

Introduction

Soils develop over very long periods of time. They result from the breakdown of rock and mineral material on Earth's surface by physical, chemical and biological factors. Chemical processes include the weathering of minerals through the effect of temperature, water, air and acidic solutions. Physical processes include the mechanical breakdown of rock and the transport of particles due to wind, water and wave action. Many living organisms, both plant and animal, combine to assist in the formation of soil from parent rock. Soil provides an anchorage site for plants, an environment for many living organisms, and serves as a source of nutrients for both plants and organisms. **Weathering** is the term used to describe the process of rock breakdown to form soil.

Primary minerals are found in parent rock. Such minerals include feldspars ($KAlSi_3O_8$) and pyroxenes composed of iron and magnesium silicates. These primary minerals are released as parent rocks break down. Secondary minerals are derived from the breakdown of primary minerals. For example, quartz and muscovite (clear) mica weather to form the coarse sand fraction of the soil, while shale, slate and basalt weather to form clay. Granite weathers to form a sand and clay mixture called loam.

Soil is more than just dirt; it is a balanced system, resulting from decomposed rock and the interaction of many physical factors, and chemical and biological systems, and it requires very careful management. Within the soil a dynamic relationship exists between the following three components:

1 solids: the mineral and organic aspects of the soil
2 liquids: the **soil solution**
3 gases: the soil air.

These components influence soil chemistry and temperature, and consequently seed germination rates, plant growth rates and the activities of micro-organisms. Figure 15.1 indicates the general composition of topsoil under normal climatic conditions.

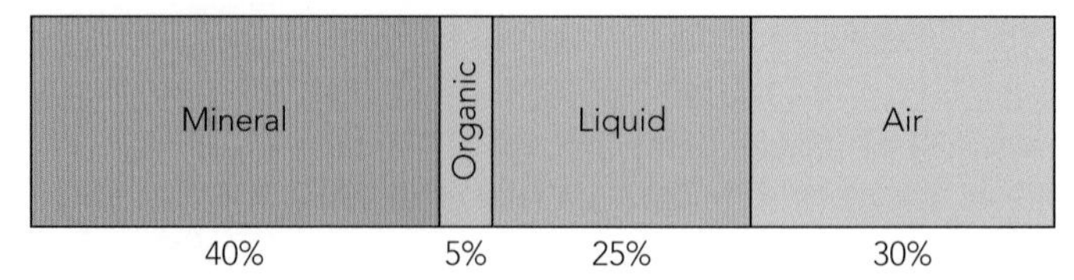

Figure 15.1 The composition of topsoil under normal conditions

Solid phase

Major secondary minerals

As a result of various weathering activities, soils are made up of a number of different particles. Soil particles are classified according to size, which is based on particle diameter. Table 15.1 indicates particle size and classification.

Table 15.1 Soil particle classification, including the organic component humus

Name of particle	Diameter of particle (mm)
Gravel	> 2
Coarse sand	2–0.2
Fine sand	0.2–0.02
Silt	0.02–0.002
Clay*	< 0.002
Humus*	<0.002

*These very small particles are called colloids.

ISBN 9780170265560

Sand (silicon dioxide)

Sand forms an inert structural framework for the soil. Sand particles range in size from 0.02 to 2.0 mm in diameter.

Silt

Silt is very fine particles intermediate in size between sand and clay and often deposited as sediment in floods.

Clay

Clay consists of highly weathered particles with diameters of less than 0.002 mm. Particles of this size are called **colloids**.

1. List the main components of a soil.
2. List five features of a sandy soil.
3. List five features of a clay soil.
4. What is the definition of a colloid?
5. Explain what occurs as organic matter breaks down in a soil.

Organic matter

Organic matter refers to living organisms and their waste products, as well as to the dead organisms in the soil. Humus is the main product yielded from the breakdown of organic matter (Fig. 15.2). Like clay, these particles have the ability to attract and hold both water and minerals. Humus can also cement soil particles together. Humus particles are also colloids.

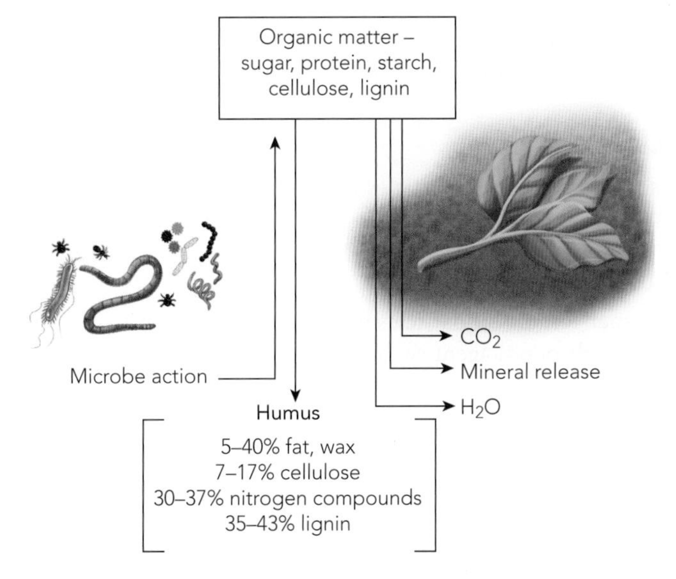

Figure 15.2 The formation of humus from the breakdown of organic matter

Liquid phase

The soil solution contains many elements dissolved in water. Soil water may be either acidic or alkaline and, because of its variable composition, causes changes in soil acidity or alkalinity and mineral availability. The main sources of minerals in the soil are:

- the soil solution
- primary minerals released through weathering
- exchangeable ions from clay and organic matter.

Gas phase

The composition of soil air and the amount of air in the soil vary according to the size of the soil particles produced by the weathering processes. In soils where the amount of oxygen falls to below 10%, there is a decrease in the activity of micro-organisms and a lower rate of root respiration. Consequently, plant growth suffers and processes involved with mineral cycling decline.

ISBN 9780170265560

connect

Who loves soil?

Describe the role of a soil scientist. Outline how to determine soil texture.

Physical properties of soil

Parent material

Parent material is the storehouse of future soil material and soils can be classified into one of two groups according to the source of their parent material. Soils are either sedentary or transported.

- *Sedentary.* This means that they have never moved, and have formed from the solid rock that lies under them.
- *Transported.* This means that the soils have been carried to their present site by gravity, wind or water. They have little relationship to the type of rock found in the area.

The material in the parent rock affects the percentage of sand, silt and clay in a soil, as well as the chemical makeup of a soil and the way soil particles are held together. Sand, for instance, results from the breakdown of sandstone. Rocks such as shale, on the other hand, weather to release many materials, several of which act as cementing agents. Shale also weathers to form a very fine particle called clay that is very sticky and can absorb water.

Soil colour

connect

Soil colour

Soils are often referred to by their colour as this is often the initial observation made about a soil. It is a useful feature for the description and identification of a soil. Refer to the Soil colour website and summarise the features that soil colour can indicate.

Soil texture

Soil texture refers to the size of soil particles in a soil, and allows differentiation of sand, silt and clay fractions in a soil. Soil texture can be determined in a number of ways.

Mechanical analysis

This involves the dispersion of soil particles by violent agitation of a soil sample with a mixture of water and a dispersal agent such as Calgon® or hydrogen peroxide. Particles fall through suspension at a rate proportional to size, which can be calculated according to Stokes' law. The concentration of particles at any one level can be measured by a hydrometer at set time intervals. Once the sample has been shaken and allowed to settle, particles with a diameter greater than 0.2 mm will have all settled after four minutes and 48 seconds, while particles with diameters less than 0.002 mm will not settle for 8 hours.

Texture is a permanent characteristic of the soil, and on the basis of particle diameter measured in millimetres, soil particles are classified according to size (see Table 15.1 on page 203).

Texture triangle

This allows the classification of soils into many textures, based on texture readings, as shown in Figure 15.3. Several soil samples can be plotted on one triangle, allowing easy comparison.

Texture is an important property of soil as it influences the following aspects:

- movement of water into the soil (infiltration)
- movement of water through the soil (drainage, or percolation)
- soil air circulation
- soil structure, to some degree
- the ease with which plant roots penetrate the soil
- the ease with which seeds germinate and emerge from the soil
- the ability of a soil to hold water and minerals.

ISBN 9780170265560

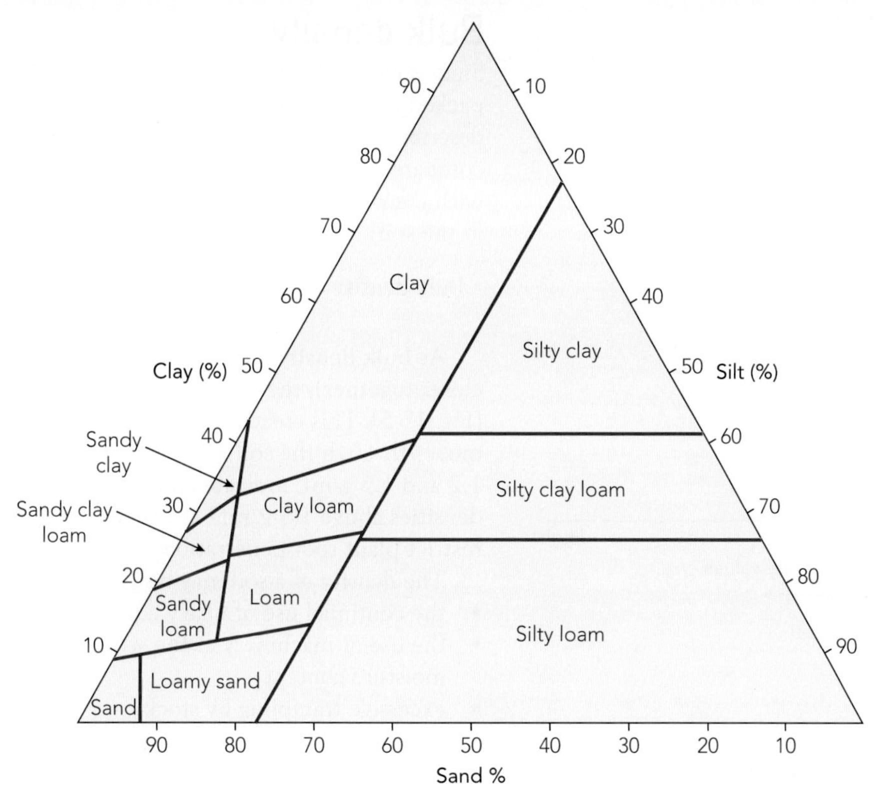

Figure 15.3 The texture triangle

Soil structure

Soil structure refers to the arrangement of soil particles (sand, silt and clay) in a soil. The basic unit of soil structure is called a soil **ped** and forms a natural soil aggregate. Sandy soils, such as beach sand, have no structure because individual particles do not come together to form stable units. These soils are apedal, or structureless.

Structure influences the aeration, water penetration, heat transfer, rates of gaseous exchange, mechanical strength and erosion potential of a soil. This property ultimately affects plant growth and land-use patterns through direct influence on seed emergence and root anchorage.

Soil structure can be easily changed. Mechanical working of the soil when the moisture content is either too high or too low, overgrazing, overcropping and excess cultivation all contribute to a rapid loss or change of soil structural properties. Soil structure can be improved by adding organic matter and reducing the number of times the soil is cultivated.

Soil structure can be determined either visually or by specific measurement. Visual assessment involves recognising the shapes of the various soil peds, as shown in Figure 15.4. Other determinations of structure include the measurement of bulk density and soil porosity.

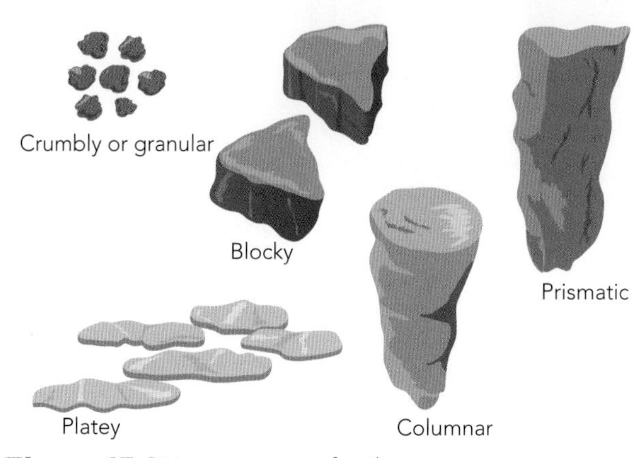

Figure 15.4 Various types of soil aggregate

ISBN 9780170265560

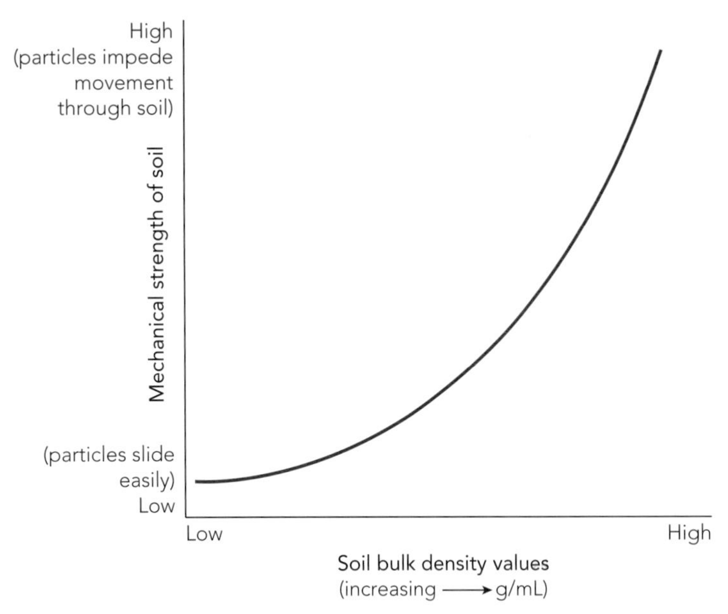

Figure 15.5 Soil bulk density relationships

Bulk density

Since soils can be described as being loosely packed or closely packed, density is a means of describing structure. The bulk density equation compares the percentage of solid particles with the percentage of voids (pores or spaces in the soil):

$$\text{Bulk density} = \frac{\text{mass of soil solids}}{\text{total volume of all constituents}}$$

As **bulk density** increases (particles are closer together), the particles slide less easily (Fig. 15.5). This creates stress on machinery as it moves through the soil. Bulk densities between 1.2 and 1.5 g/mL indicate an average soil, while densities above 1.5 g/mL, as found in clay soils, restrict plant root penetration.

The density of a soil may be altered by:

- the continual use of heavy equipment
- the use of machinery at the wrong soil moisture content
- excessive trampling by stock.

Soil porosity

Pore space relationships are also useful indicators of soil structure. **Porosity** measures structure by examining the percentage of soil spaces, rather than the percentage of solids in the soil. Figure 15.6 illustrates how a suction plate can be used to plot the amount of water draining from a soil sample. The results obtained and shown in Figure 15.7 illustrate how different-sized pores in the soil empty at different rates.

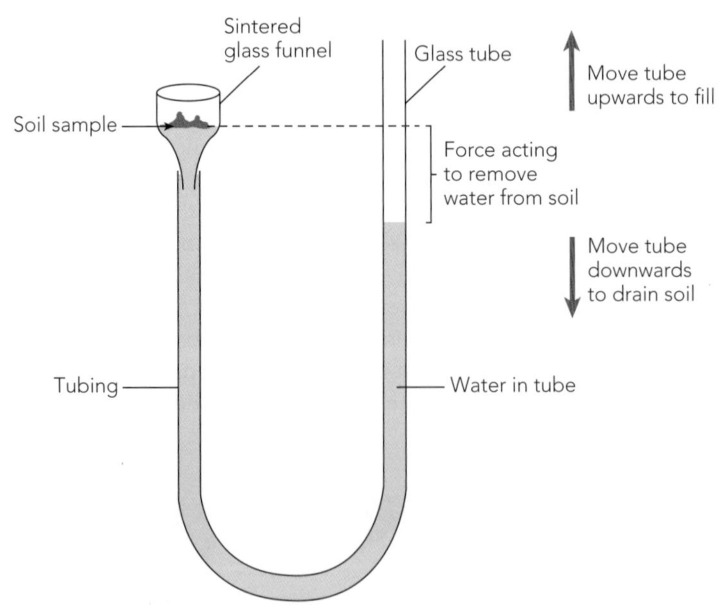

Figure 15.6 Suction-plate equipment used to determine the size of soil pores

ISBN 9780170265560

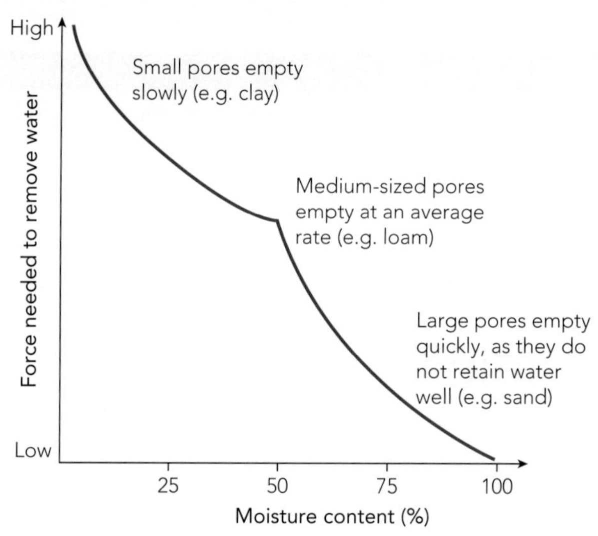

Figure 15.7 Results of soil analysis by the suction-plate method

6 Define the terms 'organic matter', 'soil texture', 'soil structure', 'bulk density', 'porosity' and 'ped'.
7 Describe three ways of determining soil texture.
8 Outline two ways of determining soil structure.
9 Use Figure 15.4 to illustrate various soil peds.

Factors affecting structure and pore size distribution

The ability of a soil to develop a structure can be influenced by:

- alternate freezing and thawing of the soil
- alternate wetting and drying of the soil
- swelling and shrinking in the soil (self-mulching)
- the amount of vegetative cover
- the activity of soil organisms.

The stability of soil structure is influenced by:

- the percentage of organic matter in the soil
- the amount of previous cultivation
- the mineral content of the soil (soils high in monovalent ions, such as sodium, will disperse quickly)
- the type of clay in the soil (some clays swell and contract according to their moisture content)
- the presence (or absence) of chemical cements, such as iron and silicon chelates, in the soil
- the age of the soil (ageing produces stability as an equilibrium situation is achieved).

10 List the factors that tend to develop soil structure.
11 List the factors that influence soil structural stability over time.

Soil consistence

Soil **consistence** is a measure of the mechanical strength of a soil, based on the force necessary to break a block of soil. It depends on the moisture content of the soil, and the manner in which the soil block fails. Block fragments are classified as shattered (**pulverised**) or moulded fragments (**coalesced**). On the basis of the percentage of these fragments, soils may be placed into one of four groups, as shown in Figure 15.8. This is a useful description of how a soil behaves under cultivation, as it includes an assessment of the force necessary to deform particles and the type of particle failure.

The crumbly state is ideal for cultivation as it avoids plastic clod formations or pulverised brittle fragments, which can be easily transported by wind or water. Examples of the types of machinery interactions on the soil are shown in Table 15.2 (page 208).

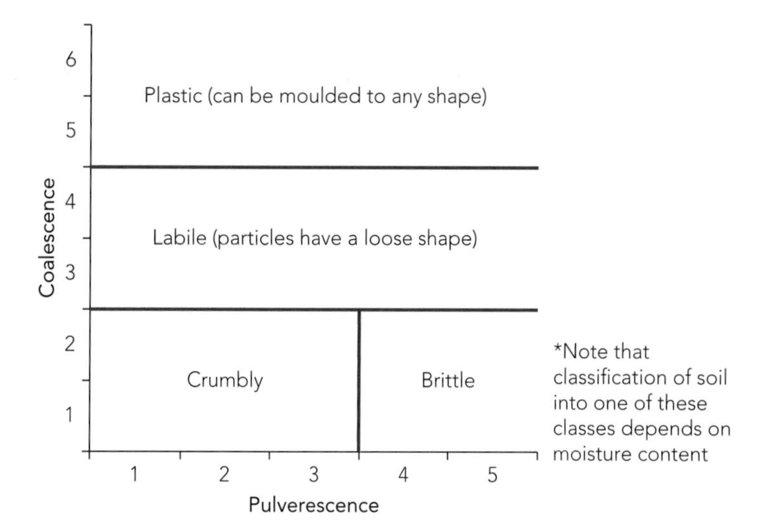

Figure 15.8 Soil consistence classes

ISBN 9780170265560

12 Define the term 'soil consistence'.

13 Describe how machinery affects the physical nature of a soil.

Table 15.2 Soil–machinery interactions

Machine	Effect on soil
Mouldboard	Causes little shattering; compacts soil when wet; may form claypans (compaction areas) in the soil
Disc plough	Stresses soil in three planes; causes some shattering of soil particles; causes less compaction than a mouldboard plough
Rotary hoe	Develops a 'fine' soil structure; produces excess shattering of soil particles in a shallow layer above a more compacted layer

Soil inorganic fraction

The **inorganic fraction** of the soil consists of sand, silt, clay, minerals, air and water. The chemical, or mineral, fraction of the soil is composed of primary minerals and secondary minerals. Primary minerals weather from parent rocks and include the following common components of a soil – quartz, feldspars, pyroxenes, mica and apatite. In terms of resistance to weathering, quartz, muscovite mica and potassium-based feldspars are more resistant than other feldspars, olivine or black mica. Secondary minerals are derived from primary minerals forming the sand, loam and clay fraction of the soil; plus specific minerals (e.g. quartz and muscovite, mica), which form the coarse sand fraction; and other types of mica, which form clay particles.

Major secondary minerals

Sand

Made entirely of silicon dioxide and derived from weathered quartz, sand forms an inert structural framework for soil peds, provided cementing agents are present in the soil. Sand does not absorb water or adhere to other particles, and it is the basis for good drainage and aeration in the soil. Consequently, sandy mixtures are used as the basis of seed-germination mixes, as they allow easy seedling emergence and good drainage. Sand particles have the following characteristics.

- They are physically and chemically inert.
- They do not absorb water.
- They do not stick, or adhere, to each other.
- They warm up readily.
- They are usually of little value as a store of plant nutrient material.

Clay

Clay particles are called colloids, as they have a diameter of less than or equal to 0.002 mm. Clay consists of highly weathered particles, which have the following features:

- a negative surface charge
- a large surface area to volume ratio, which makes clay particles highly reactive in the soil
- the ability to hold water, which causes them to swell on wetting and shrink on drying, which in turn influences soil drainage, aeration and water entry
- the ability to cement other particles together to form soil peds
- the ability to store plant nutrients
- a strong tendency to coalesce if worked when too wet, and pulverise to powder if worked when too dry.

Clay consists of millions of small, three-dimensional cells, called unit crystals (Fig.15.9 on page 209). The structure of each unit crystal is determined by the arrangement of anions (O^{2-}, OH^-) within it. These anions are large compared with cations present in the soil. The visible clay particle could be thought of as being composed of millions of these small three-dimensional units.

ISBN 9780170265560

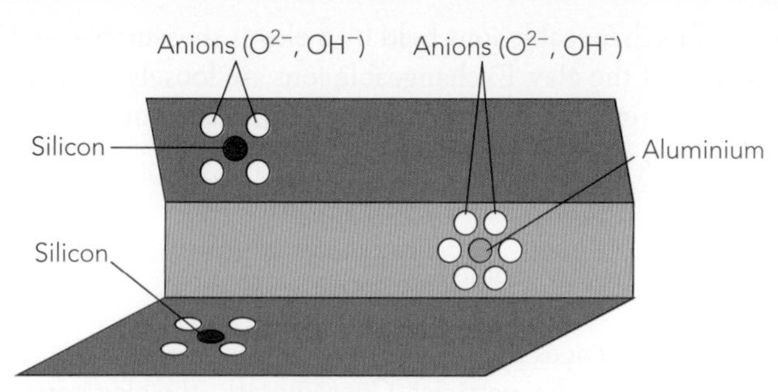

Figure 15.9 A model of a clay unit in a mineral containing two silicon layers to one layer of aluminium

The clay unit is the product of a weathering process. Therefore, the outer section has lost many positive ions, rapidly leading to the formation of a negative surface charge on the particle surface (Fig. 15.10).

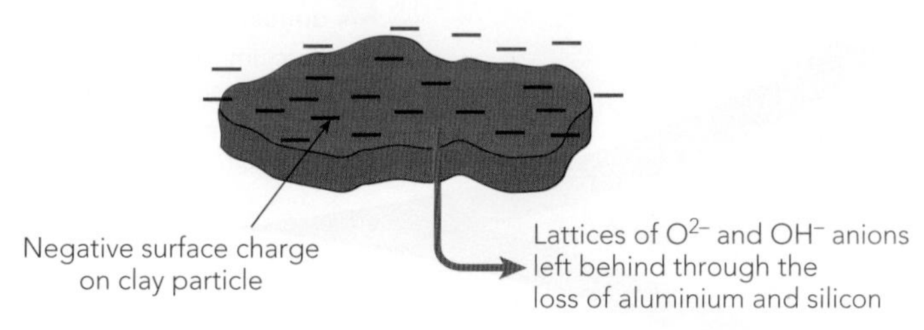

Figure 15.10 The clay surface charge

Further investigation of the surface charge reveals a series of zones, commencing at the surface of the clay particle and moving out, as shown in Figure 15.11a. Positive ions (cations) are held very tightly at the clay surface, as they are attracted to the clay particle by the negative surface charge. Surrounding this build-up of positive ions are negative ions, until an equilibrium is reached some distance from the clay particle. This layer of tightly and loosely bound charged ions is called the **clay double layer**, and it allows the clay particle to develop electroneutrality. The ions closest to the clay particle are not free to take part in chemical reactions in the soil. However, the more loosely bound positive and negative ions around the clay particle are free to participate in soil reactions, and are available to both plant and invertebrate organisms.

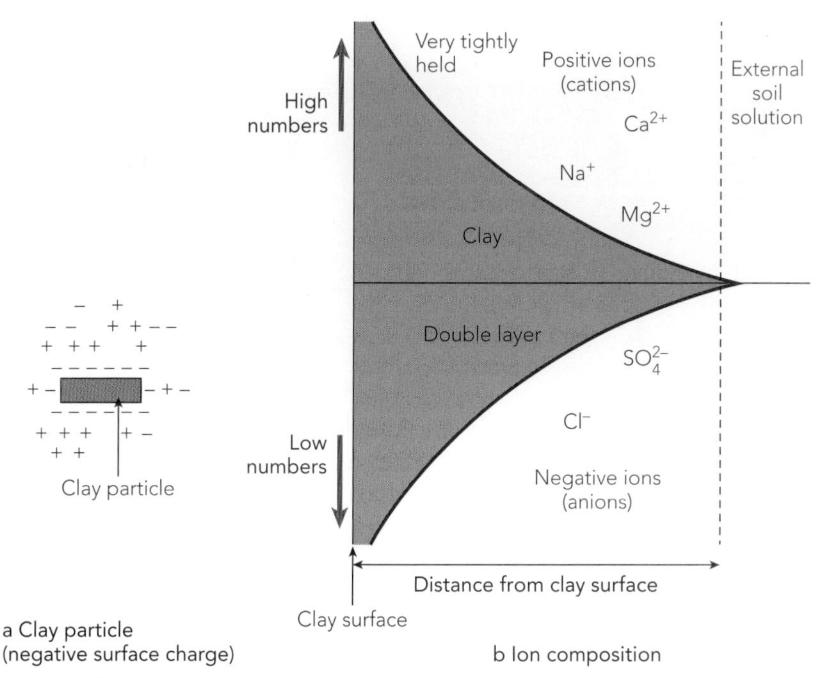

Figure 15.11 The clay double layer

The total amount of exchangeable ions held by a clay is measurable, and is called the **cation exchange capacity** of the clay. Exchangeable ions are loosely attached to the clay double layer and are easily replaced by ions from the external solution found in the soil. The cation exchange capacity of a clay is affected by:

- the charge level on the clay mineral
- the soil pH
- the types of ions located in the double layer
- organic matter.

The clay double layer influences the physical properties of a soil. The larger the double layer, the more buffered are the clay particles. Consequently, the clay particles are less likely to come into contact with one another, producing a situation of instability in a soil. This type of instability occurs in soils with high levels of sodium or potassium ions, increasing hydration of the clay and high soil pH. Figure 15.12 illustrates the effect of various cations on the size of the clay double layer.

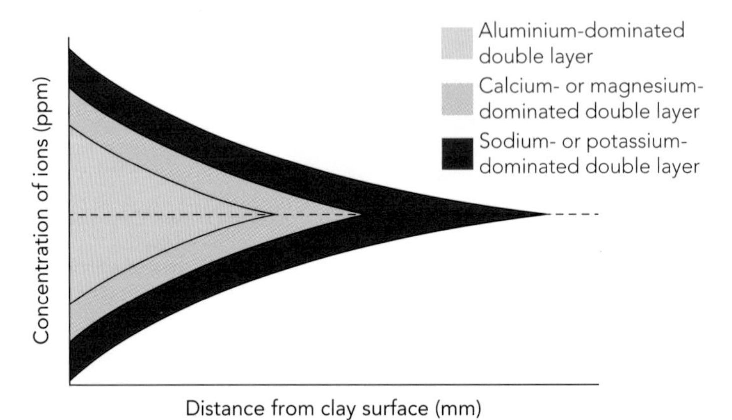

Figure 15.12 Clay double layer size varies with the kind of cation it contains

Plants obtain minerals by a process of active mineral exchange with the surrounding soil particles (Fig. 15.13). Mineral ions are held by clay particles in the soil.

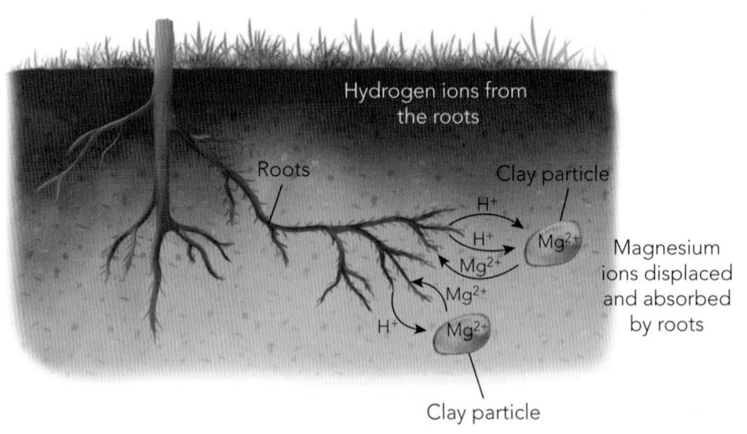

Figure 15.13 Mineral exchange

14 Outline the role of sand in the soil.

15 Outline the role of clay in the soil.

16 Use a diagram to illustrate the main features of a clay particle.

17 Explain how plants obtain minerals from the soil.

At very low pH levels (acidic conditions), there is a dissociation of Fe^{3+} and Al^{3+} cations resulting in Fe (iron) and Al (aluminium) ions moving in and out of the actual cell lattice of the clay particle. The clay becomes unstable and the concentration of iron and aluminium ions in the soil increases to toxic levels for plants.

 ISBN 9780170265560

Table 15.3 indicates the differences between a sand soil and a clay soil.

Table 15.3 Sand and clay soil properties

Sand	Clay
Consists of large angular particles of quartz	Consists of very small particles of various minerals
Has a coarse texture	Has a fine texture
Cannot be moulded	Can be moulded
Is structureless	Particles may form crumbs
Contains few plant nutrients	Is rich in plant nutrients
Contains large individual pore spaces	Contains very small pore spaces
Is well drained and aerated	Has poor drainage and aeration unless well structured
Has a small total pore space (25%)	Has a large total pore space (50%)
Has low water-holding capacity	Has high water-holding capacity
Warms up quickly	Warms up slowly
Produces early soils	Produces late soils
Is easy to work	May be hard to work
Is not affected by wetting and drying	Becomes sticky when wet, may swell and crack on drying
Is improved by adding clay or organic material	Is improved by adding sand or organic material

Soil water

Plants contain between 80% and 90% water. They must replace the water lost through their leaves (by transpiration) by drawing moisture out of the soil through their root system. During this process dissolved nutrients are brought into the plant from the soil solution.

Water is removed from a soil system by the action of gravity, by evaporation, and by transpiration of plants and living organisms. Forces retaining water in the soil are in two categories – physical forces of cohesion and adhesion, which arise because of particle shape and arrangement; and chemical forces arising from the osmotic effects of bore water, fertiliser and the water table.

Soil moisture potential is related to the amount of water in a soil available to plants. Soil moisture potential is a measure of the energy required by the plant to remove water from the soil at a particular moisture content.

connect

Cracking up!

Answer the questions that are posed during the video.

Within a soil the relationship between water content and moisture potential is called the **moisture characteristic** of that soil.

Specific soil types may be identified on the basis of their moisture characteristics, as shown in Figures 15.14 and 15.15 (page 212).

Figure 15.16 (page 212) demonstrates several important soil–water relationships. When all the available pore space is full of water, the soil is said to be **saturated**. This condition can interfere with plant growth as all the air is forced out of the soil. The amount of water remaining in the soil 24–48 hours after saturation by irrigation or rainfall is called **field capacity**. This is the water held in the soil pores by physical and osmotic forces. Plants and organisms will continue to remove water from soil pores until the physical forces holding water in the soil pores prevent this. When a plant can no longer remove water from soil pores it collapses, as plant cells only maintain plant form when they are full of water (**turgid**). When a plant collapses it is said to **wilt**, and the moisture level in the soil when this occurs is

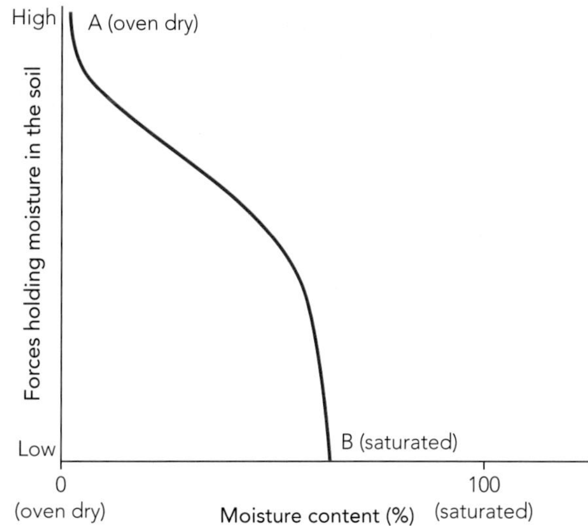

Figure 15.14 Generalised moisture characteristic of a soil

ISBN 9780170265560

called the **wilting point** of the soil. The quantity of water held in a soil between field capacity and wilting point is the total amount of water available to the plant.

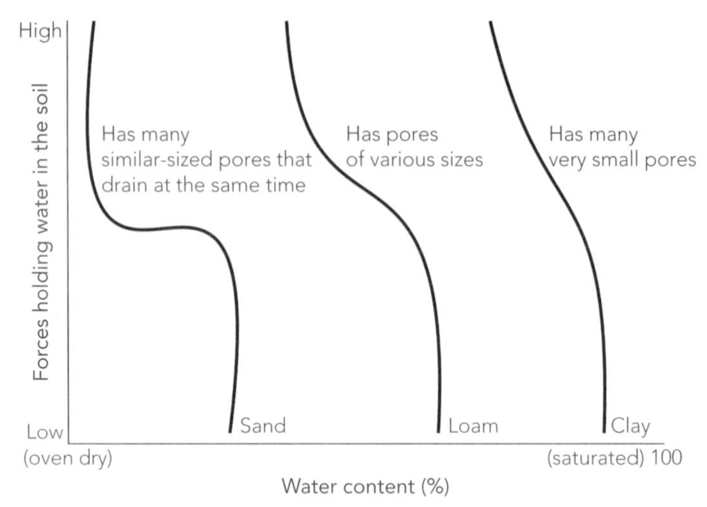

Figure 15.15 Soil moisture characteristics

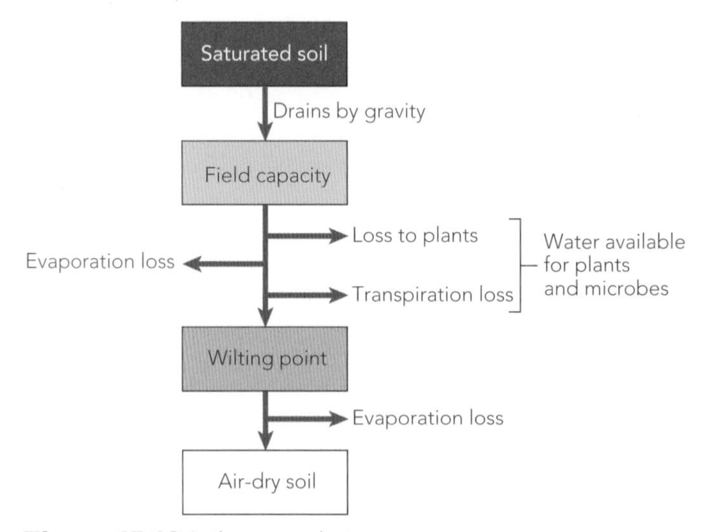

Figure 15.16 Soil–water relations

The soil moisture characteristic is useful for the identification of soil types based on pore size, and in determining the wilting point, field capacity and available water content of a soil sample, as shown in Figure 15.17.

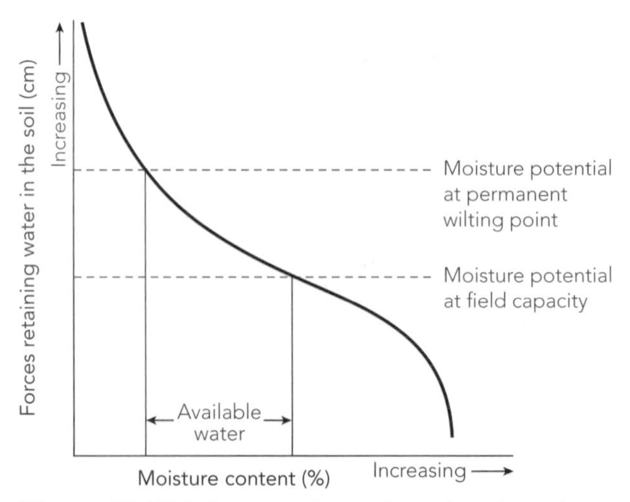

Figure 15.17 Soil–water relations based on the soil moisture characteristic

 ISBN 9780170265560

Water must be drawn out of the soil pores. The smaller the size of the soil pore, the more difficult this task becomes. Note that at the **permanent wilting point** and at saturation, a clay contains more water than a sandy soil. This does not mean that water in a clay is more available; a clay simply contains more water within vast numbers of very small pores. This is because clay soils have a greater total pore space than sandy soils. Water within the clay pores is held extremely tightly to the clay, and is not available to the plant.

18 Define the terms 'moisture potential' and 'moisture characteristic'.

19 Explain how water is retained in a soil.

20 Use Figure 15.16 to describe what happens as a saturated soil dries out until it is air dry.

Water management and sources of water on a farm

There are many sources of water available on farms, such as farm dams, local streams or rivers, and bore water that is obtained from underground artesian water sources. Farmers may have available town water, farm dams or use water permits in irrigation districts to pump quantities of water from local rivers to manage their pastures, crops, commercial orchards and gardens and provide water for livestock. Irrigation provides sufficient moisture to:

- grow plants in areas where they normally would not be found
- plant pastures at high seeding rates, so that the farm can support more livestock. Although competition between plants is high, more plants can be grown because of the availability of extra water.

Most of the water stored for irrigation in Australia is in the area surrounding the Murray–Darling Basin. There are extensive underground water resources, but most of this water is saline. Water is drawn up to the surface by deep-well turbine pumps. Streams and rivers provide the bulk of the water supply for irrigation. There are several methods of irrigation.

Flood and furrow irrigation

These systems are often found in rice and cotton areas. In the flood system, water can be supplied to reasonably flat land where the soil is not very permeable to water (Fig.15.18). A series of parallel banks – called check banks – is built, running down the slope. Each pair of banks encloses a bay, usually graded to remove any irregularities.

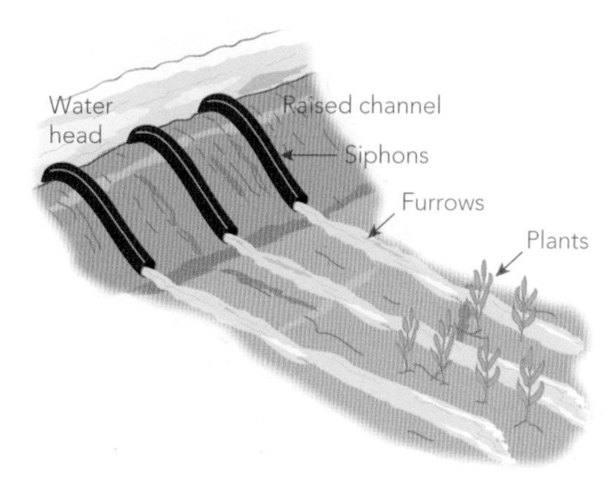

Figure 15.18 A furrow irrigation system

Spray irrigation

The advantages of this type of irrigation are as follows.

- It can be used over a wide range of land surfaces.
- It can be used on slopes.
- There is little waste land (water channels or banks).
- Uniform distribution of water is possible.
- The system is not dependent on the type of soil to be effective.

The disadvantages of this type of irrigation include the following.

- There is a high initial cost to establish the system.
- There is a need for careful planning in order to efficiently cover any area.
- Some time must be spent shifting the lines or sprinkler positions (centre pivot systems and other fixed systems do not need to be moved once they are set up)
- Windy weather can play havoc with the sprays.
- With fine sprinkler systems evaporation may be a problem.

Spray systems all involve the delivery of water under pressure through pipes to a spray nozzle. The system requires a suction pipe with a foot valve and strainer, which is suspended in the water. A pump and power unit is needed to take up the water, which flows down a main line and into side lines placed in particular patterns around the field by the farmer. These pipes may be made of aluminium or PVC. Spray heads are mounted on short lengths called risers. Figure 15.19 is an example of a lateral irrigation system while Figure 15.20 is an example of a central pivot irrigation system (page 214). Also known as circle irrigation systems, this method relies on equipment rotating around a central point (pivot), watering crops with sprinklers. This system reduces labour costs and uses less water than surface irrigation methods, such as flood and furrow irrigation systems. Central pivot systems are easy to set up compared with the lateral aluminium pipe arrangements that move in a straight line.

To work out how much water should be supplied to a crop (the frequency and amount of water needed for each irrigation) the farmer needs to know:

- the amount of water lost from the soil by evaporation each day
- the water needs of the plant per day (e.g. lucerne requires 5 mm of water per day)
- how much water the soil can hold and make available to the plant.

Shutterstock.com/Deyan Georgiev

Figure 15.19 A spray irrigation lateral system

Shutterstock.com/alexmisu

Figure 15.20 Central pivot irrigation system

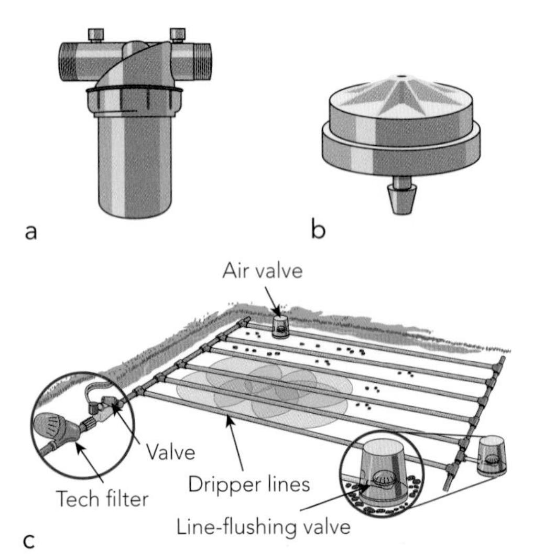

Figure 15.21 Elements of a trickle system **a** in-line screen filter **b** Turbo-SC™ pressure compensating drip emitter **c** a trickle system for vegetables

Trickle irrigation

Trickle irrigation involves a small flow of water at low pressure. This method conserves water, maintains soil structure and uses less energy. Low-pressure polythene pipe, 9–12 mm in diameter, takes water from the main line, with one or two outlets going to each plant. The outlets are either long, thin tubes called spaghetti tubes, or drippers. Water is supplied over a period of several hours in sufficient quantity to water over one-third of the plant root system to a depth of 30 cm. A filter is required in this system, or the delivery tubes and drippers rapidly block up (Fig. 15.21).

Water management

Water is a scarce and increasingly expensive resource. Efficient use of water in a changing climate is essential. Farmers carry out regular maintenance on irrigation equipment, including the metering equipment where allocations are used to determine water use in irrigation areas. Where furrow and flood irrigation practices are used, farmers are covering many of the channels to limit loss of water by evaporation. Where flood irrigation is used (for example, in rice production in the Murrumbidgee irrigation area) laser-levelled rice bays are prepared so that water can be distributed evenly. Farmers can then use the rice bays after harvesting to grow wheat or sow a pasture to maximise land use and fully utilise the subsoil moisture built up after a rice crop.

Vegetable growers and orchardists water early in the morning or late in the day to minimise water loss by evaporation. Many fruit and vegetable growers use trickle irrigation systems.

Technological advances allow farmers to measure soil moisture content using soil moisture probes to detect when moisture levels are low. These signals can then be used to trigger automatic irrigation systems. Consequently, less money is spent on supplying water, and improved efficiency in the delivery of water to plants limits water stress, ultimately enhancing productivity. Problems associated with leaching and salinity are minimised.

connect

Water sustainability

connect

Water use in the Murray–Darling Basin

Complete the activities.

ISBN 9780170265560

Government legislation and farm water management

Government legislation of on-farm water availability and management is designed to restrict the use of water by individual farmers to ensure that everyone has access to the available water sources in a district. The schemes are designed to also provide sufficient water in the local area to maintain stream flows and supply water to local natural ecosystems. More sustainable water use systems prevent water degradation, encourage the protection of existing waterways and help ensure a secure water supply. Use of water on farms is regulated by a system of water licences and water allocations to minimise over-use of water. These regulations encourage farmers to build dams on their properties to intercept run-off and harvest rainwater, and to adopt newer, more efficient technologies in the storage and delivery of water on farms. State and Commonwealth governments have legislation in place to regulate farm water management. In NSW the Department of Primary Industries Office of Water oversees the *Water Management Act 2000*. The Tasmanian Department of Primary Industries, Parks, Water and the Environment grants water licences under the *Tasmanian Water Management Act 1999*. Farmers are required to purchase these licences to irrigate or alternatively purchase farms with licences. Farmers purchase the right to use a specified amount of water (measured in megalitres) for irrigation. They pay an annual fee for these water licences. The licence also specifies the source from which farmers can draw the water to use for crop and livestock enterprises.

Water allocations state the quantity of water (in megalitres) that can be taken in a given time period.

Farmers are subsidised to adopt new technology and equipment that improves the efficiency of water use; for example, for the use of underground pipes instead of open channels. The National Water Initiative was developed to ensure that all significant licenced water users are metered. Water meters enable better monitoring of on-farm water use, allowing farmers to develop more efficient systems of water management between crop and livestock enterprises. Meters also ensure compliance with authorised water usage volumes.

connect
Water management policy in NSW

connect
Murray–Darling

connect
Water buyback

Soil air

In soils where the amount of oxygen available to organisms falls to below 10%, there is a noticeable decrease in the activity of living organisms. Oxygen is required by living organisms for respiration. Plant root growth declines at levels of oxygen concentration less than 10%, although plants can tolerate levels of carbon dioxide concentration greater than 10% in the immediate root environment. In general, there is a trend for carbon dioxide concentrations to increase with soil depth, while oxygen levels decline.

Table 15.4 Comparison of soil air composition with that of the atmosphere

Gas	Soil air composition (%)	Atmosphere composition (%)
Nitrogen	79.15	79.00
Oxygen	20.60	20.97
Carbon dioxide	0.25	0.03

Gases move from areas of high to low concentration in the soil by diffusion. This process is influenced by:

- soil texture
- soil structure (pore size and pore distribution)
- soil water content
- the degree of soil compaction.

The supply of air from the surface of the soil depends on the depth of soil being considered, temperature gradients, pore space relations and the level of water in the soil, and the soil structure.

21 Explain how soil air influences life in the soil.

22 How does the composition of soil air differ from that of the atmosphere?

Soil temperature

Soil temperature is influenced by a number of factors.

- The colour of the soil has an influence on soil temperature, with dark soils absorbing heat.
- Soils with high moisture content require more heat to warm them.

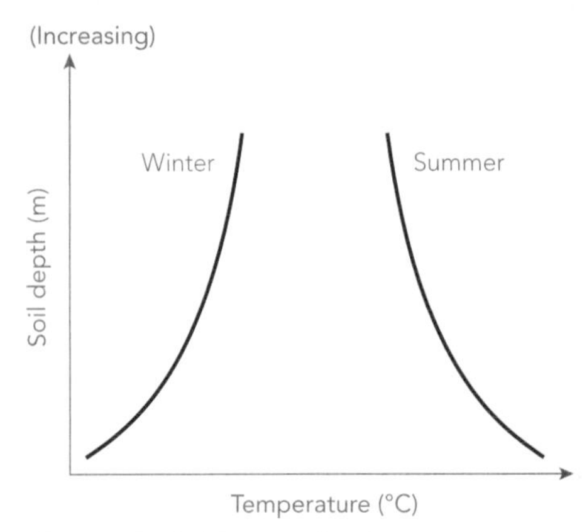

Figure 15.22 Seasonal soil temperature curves

- Soils with high sand content tend to warm up faster than soils high in clay. This is because sandy soils have larger soil pores, which are filled with air. Sandy soils allow earlier planting of crops and more rapid seed germination, and are often called 'early soils' because of these features.
- The amount of solid, liquid and air in the soil determines the soil's thermal capacity, or ability to warm up.
- The rate of flow of heat from hot to cold areas in the soil depends on water content and pore size.

Soil temperature influences the rate of organic matter breakdown in the soil and mineralisation reactions, especially those reactions in the nitrogen, sulphur and iron cycles. The rate of seed germination and plant growth is also directly influenced by soil temperature. Micro-organism activity decreases rapidly as the soil cools, limiting the availability of minerals to plant root systems.

Soils heat up by day and cool by night, a process called **diurnal rhythm**. The degree to which a soil profile is affected by this process depends on the type of soil (sand or clay) and the degree of protection provided by vegetation or surface soil coverings.

23 What factors influence soil temperature?

24 Why is soil temperature an important aspect of soil fertility?

Seasonal temperature variations also occur in a soil profile (Fig. 15.22). There is a net upward movement of heat in the profile in winter as the soil slowly cools, and a downward movement of heat in the summer as the soil mass slowly warms up. Both processes are effective to a depth of 4 m. Seasonal temperature differences affect the time of sowing of crops.

Soil chemistry

Instability in a soil arises from either physical or chemical causes. Aggregate analysis indicates the degree of physical stability of a soil. Physical instability seems to result from air being trapped in the soil aggregates, and expanding outwards as these soil aggregates are immersed rapidly in water.

Chemical instability causing **dispersion**, or loss of structure of soil in water, relates to the nature of the clay mineral, its double layer and the concentration of ions in the solution. The higher the concentration of sodium ions, the higher the degree of soil structural instability. High sodium values may indicate salinity problems. Crops vary in their ability to tolerate high salt levels. Cotton and couch are very tolerant, while apples and citrus are not very tolerant.

Coagulation, or bringing the soil particles together, is encouraged by increasing the concentration of calcium ions in the soil solution by the use of calcium-rich materials, such as gypsum (calcium sulphate). Lowering the water table will assist in maintaining stability. Unstable soils have lower infiltration rates, poor drainage rates, poor conduction rates for air and heat and lower germination rates due to a hard crust forming on the soil surface (**surface scalding**).

Soil nutrient status and pH

The availability of soil nutrients is greatly influenced by soil pH and factors such as temperature, drainage and soil pore size.

Soil pH measurements are dependent on the type of measurement carried out and the ratio of soil and water used in the sample. For any pH measurement of a soil sample, the ratio of soil to water must be stated. Figure 15.23 illustrates how increasing water content gradually dilutes the true pH reading of a soil.

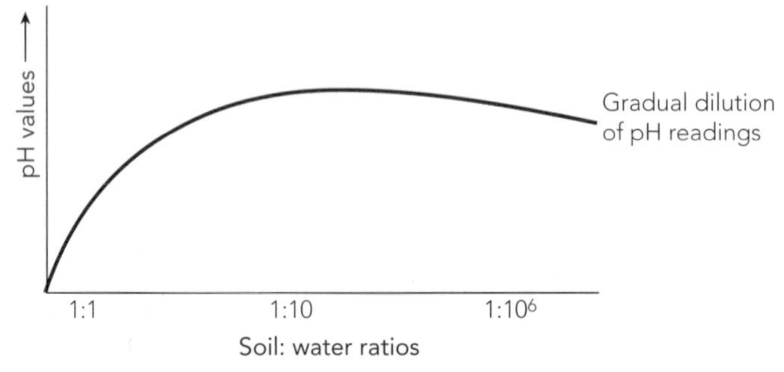

Figure 15.23 Soil pH sample readings

ISBN 9780170265560

The percentage of base minerals in a soil and soil cation exchange values vary with pH. Mineral availability also correlates to soil pH values. In alkaline conditions nitrogen compounds are converted to ammonium salts. However, the bacteria needed to continue the nitrogen cycle to produce nitrate salts are not present in alkaline conditions, and so the nitrogen cycle is limited. Figure 15.24 indicates the availability of minerals at various pH levels.

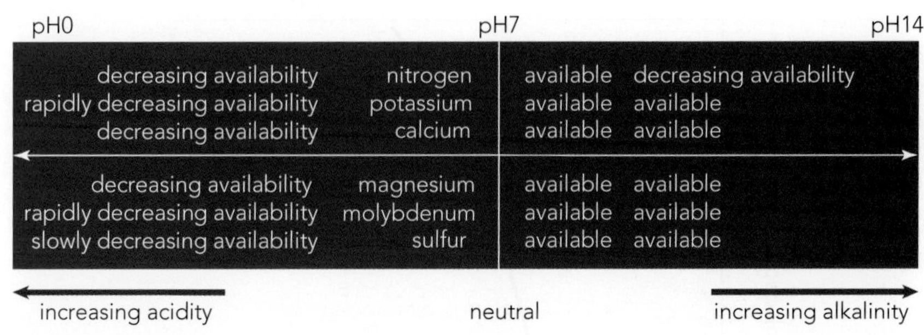

Figure 15.24 Mineral availability and soil pH

Soil pH measurements are made using:

- a pH meter, which reads soil samples against standardised solutions
- a universal indicator on a soil water extract
- a Raupach indicator, which is placed on barium sulfate that has been sprinkled over a moistened soil sample.

Generally, soil pH influences plant growth as indicated by Table 15.5.

25 Why do some soils disperse?

26 Redraw Figure 15.24 in your book and explain the trends in the availability of the named nutrients.

27 Copy Table 15.5 into your book and explain why this information is important.

Table 15.5 Plant growth and soil pH levels

pH value	Response
0–4	Too acidic for plant growth
4–5	Too acidic for rhizobia
5–6	Too acidic for many sensitive plants; e.g. lucerne
6–7	Satisfactory for most plants – optimum range for greatest nutrient uptake
7–8	Becoming too alkaline for some plants
8–14	Too alkaline for plant growth

The soil solution

The minerals within a soil are obtained mainly from:

- the soil solution
- weathering of primary minerals
- exchangeable ions in clay
- the breakdown of organic matter.

The soil solution contains many elements dissolved in soil water. Soil water is either acid or alkaline according to the material dissolved within it. Consequently, chemical behaviour, pH and mineral composition vary. Most salts are of limited solubility and rely on equilibrium reactions for their availability. The amount of nutrients available to a plant at any one time largely depends on the plant's ability to extract minerals through dynamic ion exchange with soil particles, the concentration of ions and the ease with which minerals are released by clay particles.

ISBN 9780170265560

Soil organic matter

Organic matter refers to the living and dead components of the soil and the waste materials that are produced by living organisms. Figure 15.25 illustrates the distribution of organic matter in a typical agricultural soil.

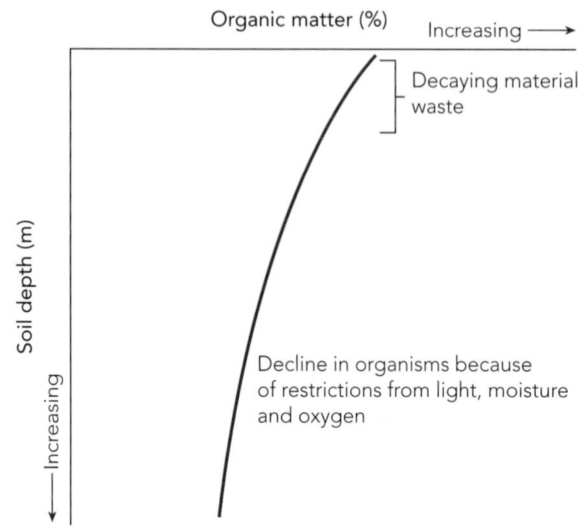

Figure 15.25 Soil organic matter distribution

The undecomposed fraction of organic matter is largely located on the ground surface, and supplies plant nutrients. The decomposed fraction of organic material is colloidal (contains particles with diameters less than 0.002 mm), and is located at depth in a soil.

Humus

The decomposed fraction of organic matter is called humus. It is a colloid and as such is extremely reactive in the soil. It also has a negative surface charge because of the remnants of functional organic complexes left on the particle surface from weathering forces. Negatively charged sites bind to the positively charged ions (cations) in the soil solution, making them more available to the plant by way of ion exchange. While these nutrient cations are accessible to plants, they are held in the soil and are not able to be leached by rain or irrigation. Humus has a cation exchange capacity equivalent to or better than clay minerals. The functional groups form very stable complexes with ions. In combination with aluminium, calcium or magnesium ions, humus forms very stable cementing agents, which glue soil particles together, hence improving soil structure. High levels of organic material in the soil encourage microbial activity. Figure 15.26 indicates the possible products derived from the breakdown of soil organic material.

The ratio of carbon to nitrogen material in the soil is critical in determining the effect of microbial action on soil fertility levels. The **C:N ratio** is determined by the equation:

$$\text{C:N} = \frac{\text{percentage of organic carbon}}{\text{percentage of organic nitrogen}}$$

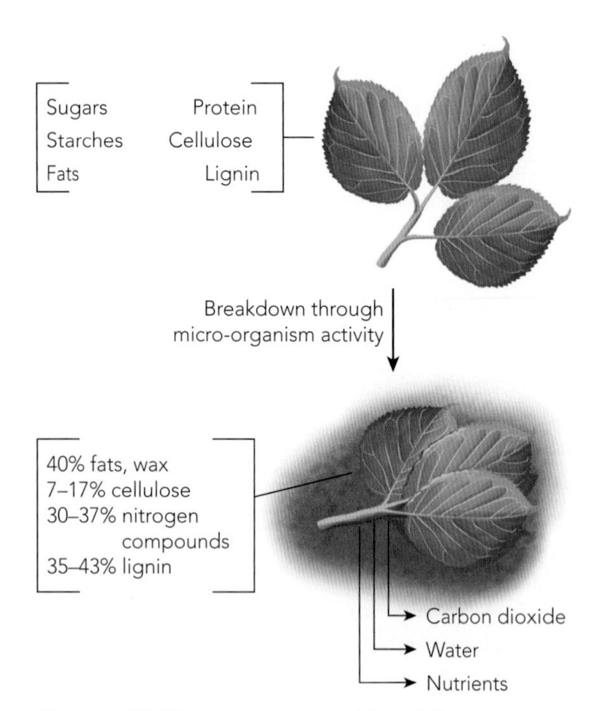

Figure 15.26 Organic material breakdown

ISBN 9780170265560

When the ratio is less than 35:1, micro-organisms attack the nitrogen in the material, releasing nitrates plus ammonium salts into the soil for uptake by plants. Where the ratio is greater than 35:1, the micro-organisms are unable to obtain enough nitrogen from organic material to satisfy their needs. They then rob the soil of its nitrogen reserves, leading to a depletion of nitrogen resources in the soil.

Plant nutrients

Plants require a number of essential nutrients for growth and development. Both the soil and the atmosphere can supply these nutrients. To replace nutrients lost from a farming system, many farmers add nutrients in the form of fertilisers. Figure 15.27 illustrates the minerals required by plants in large amounts (major elements) and those that are essential but only in small amounts (trace elements).

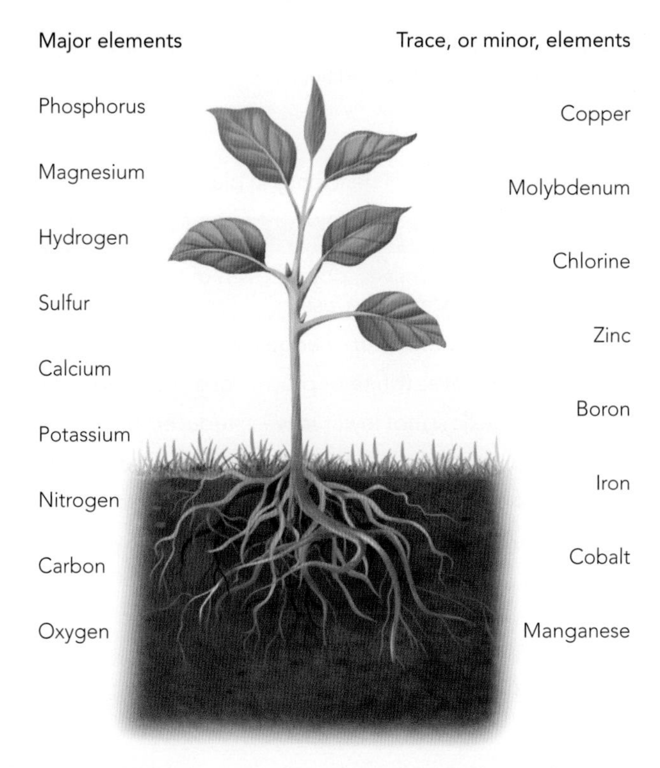

Figure 15.27 Plant mineral requirements

While deficiencies in any of the elements reduce production levels, excess supply can also damage plants. Correct fertiliser application is essential. Excess minerals in the soil are not stored but often leached out of the soil system. Excess minerals also affect soil pH levels and interfere with micro-organism activity. The aim of applying fertiliser is to have nutrients present in the root zone as the plant requires them. There are many types of fertilisers, as indicated in Table 15.6 on page 220.

Farmers should decide what fertiliser to add on the basis of cost effectiveness, availability and how much nutrient plants require. The farmer may base this decision on:

- the visible symptoms of mineral need displayed by the plant, as listed in Table 15.7 on page 220.
- soil analysis
- plant tissue analysis
- pot trials and paddock trials
- field trials.

Local field days and experimental results on crop growth performed by district agronomists are useful guides for the farmer when assessing fertiliser needs.

Table 15.6 Types of fertilisers

Form of fertiliser	General description	Examples
Bulk blended	Mechanically dry mixed	Molybdenum plus superphosphate
Granulated	Hydrated (fertiliser plus water)	Sulphate of ammonia
High analysis	Highly concentrated for particular elements	Urea, high in nitrogen
Composite	Based on nitrogen, phosphorus and potassium ratios (N:P:K)	20:11:0 18:18:0 12:52:0
Liquid	100% soluble	Liquiphos
Organic	Animal based	Manure, urea, blood, bone, bone dust
Gas	Injected into the soil; used in irrigation areas; e.g. cotton, sugarcane	Anhydrous ammonia

Table 15.7 Visible symptoms of mineral deficiency in plants

Mineral	Deficiency symptom
Nitrogen	Pale green to yellow leaves; poor growth of leaves and general vegetative parts of the plant (stems, leaves)
Phosphorus	Leaves blue-green with purple edges; very poor root growth
Potassium	Dead spots on leaves (white or brown); death of leaves from base upward
Sulfur	Upper leaves yellow (not lower leaves); reduced growth of plant
Calcium	Rapid death of growing points; production of several lateral shoots
Iron	Green veins on yellow leaves (in young leaves) or completely bleached leaves
Manganese	Grey to brown spotting of lower leaves; problems with chlorophyll formation, hence leaf yellowing
Molybdenum	Failure of legumes to form nodules; poor lamina development
Boron	Thickened stems, often with a sudden collapse of plant tissue
Zinc	Small leaves in citrus; yellow mottling in other plants

Common fertilisers

There is a large variety of common fertilisers, which can be grouped as follows.

- *Phosphate-based fertilisers.* Phosphate occurs naturally as apatite or guano, a rock composed of calcium phosphate ($Ca_3(PO_4)_2$). When treated with sulfuric acid, superphosphate is produced.

  In the soil, superphosphate reverts to an unavailable form in an equilibrium reaction, to the extent of 80% of the original amount of superphosphate supplied to a crop being unavailable after 18 months. Superphosphate remains readily available to plants for the first 6 weeks.

  Superphosphate is available either as basic superphosphate ($Ca(H_2PO_4)_2H_2O$), which avoids problems in acid environments, or as molybdenised superphosphate (Mo super), which contains the trace element molybdenum, necessary for successful nodulation in clovers.
- *Potassium fertilisers.* Several sources of potassium are available. Muriate of potash (KCl) contains 52% potassium, while potassium sulfate (K_2SO_4) is a common fertiliser. Potassium is readily available in clay soils, being derived from mica and feldspar minerals.
- *Nitrogen fertilisers.* Some of the numerous sources of nitrogen fertilisers are shown in Table 15.8.

 ISBN 9780170265560

Table 15.8 Sources of nitrogen fertiliser

Source	Examples
Nitrates	Sodium nitrate, ammonium nitrate, calcium ammonium nitrate
Ammonia compounds	Sulfate of ammonia, ammonium nitrate, urea
Anhydrous ammonia	Ammonia injected into soil under pressure
Legumes	Pulse crops, green manure crops

The soil–plant environment

The soil–plant environment is an ever-changing, or dynamic, environment in which equilibrium situations are developed between aspects of the soil and the plant. The farmer must manage the soil environment so that the moisture, temperature, air and mineral needs of the plant are met. The physical environment of the soil is important to provide protection and anchorage for plants and organisms. Once again, the farmer must manage this environment so that natural cycles are preserved and the physical aspects of the soil are maintained or improved where possible. Soil fertility depends on the physical and chemical properties of a soil.

28 Describe the role of organic matter in a soil.

29 Explain the term 'C:N ratio'.

30 Redraw Figure 15.27, which shows the mineral needs of plants.

31 List four types of fertilisers.

32 For three major and three trace elements, describe deficiency symptoms.

ISBN 9780170265560

Chapter review

Mandatory soil practical for NSW HSC Agriculture course

Students are to perform a first-hand investigation on the physical and chemical characteristics of a soil. For example, for two soil samples students must assess at least two physical and two chemical characteristics. Suggested practicals are shown below in **Things to do**. Students are then to analyse and report on their results and conclusions.

Things to do

These experiments should be performed on a variety of soils.

Experiment 15.1 Soil components (physical)

Aim: To analyse a typical soil sample

Method

1 Take a sample of garden soil, place it in a measuring cylinder and add water.
2 Shake it thoroughly and allow to stand for 24 hours.
3 Draw the profile as it appears in the measuring cylinder and label the layers.

Questions

1 Why did it take some time for some of the particles to settle?
2 Where would you find the heaviest particles?
3 Where would you find the lightest particles?
4 What is found on top of the water? Why?
5 Why are soils beneficial to farmers?
6 What are the components of a typical soil?

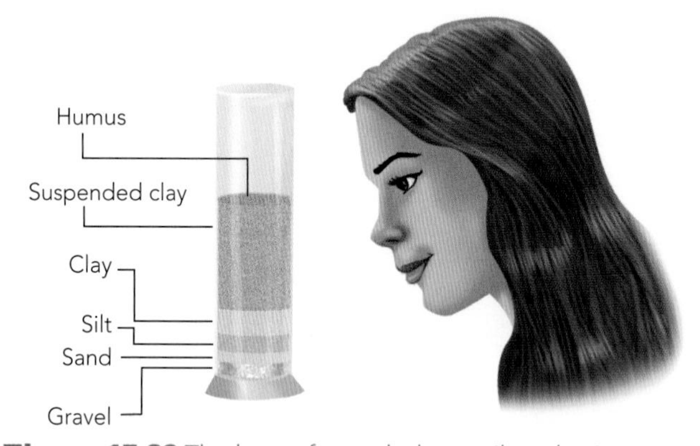

Figure 15.28 The layers formed when soil settles in water

Experiment 15.2 Water-holding capacity of soils (physical)

Aim: To determine how texture affects the water-holding capacity of soils

Method

1 Take three glass funnels and place a piece of muslin over one end of each funnel. Place each funnel in a clamp with a beaker below it.
2 Place 2 cm of soil sample into each tube as follows:
Tube A – sandy soil
Tube B – loamy soil
Tube C – clay.
3 Add 100 mL water to each tube and record the time it takes for the first drop of water to drip from each tube. Also record the amount of water that moves through each sample.

 ISBN 9780170265560

4 Record the results in your books using a copy of the following table.

Sample	Time taken for first drop	Amount of water that moved through
A		
B		
C		

Questions

1. Which sample did the water take the longest time to move through?
2. Which sample retained the most water?
3. What types of soils are best for drainage?
4. Which type of soil contains the most pore space?
5. Which type of soil has the largest pore spaces?
6. Which types of soils have the greatest water-holding capacity? Why?

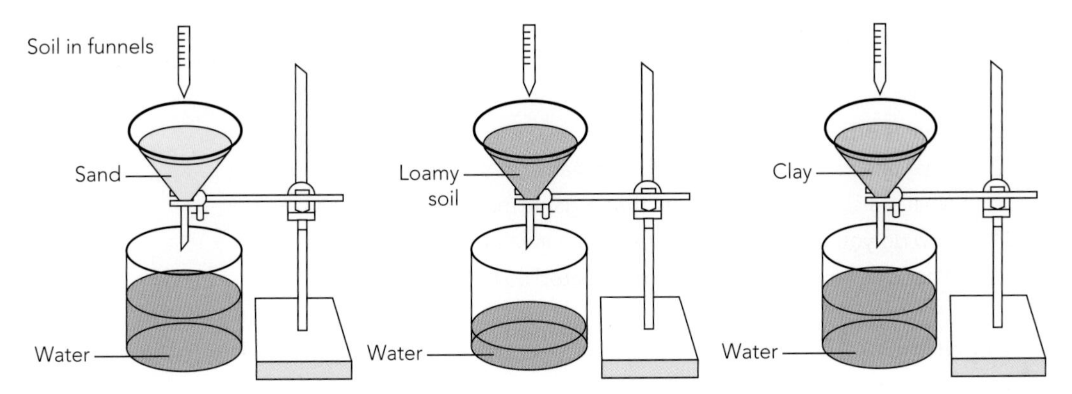

Figure 15.29 Water-holding capacity of three types of soils

Experiment 15.3 Types of organic matter (chemical)

Aim: To examine the types of organic matter found in the soil

Method

1. Take a sheet of newspaper and place on it 500 g of soil rich in organic matter.
2. Using a hand lens and a pair of forceps, pick out any dead or living organic materials and separate them into piles as follows:
   - **a** insects
   - **b** earthworms
   - **c** other animals
   - **d** dead plant material
   - **e** fungi
   - **f** other organic matter.
3. Record the results in your books using a copy of the following table.

Groups	Kinds and numbers of each
Insects	
Earthworms	
Other animals	
Dead plant material	
Fungi	
Other organic matter	

Questions

1. How many groups of living things were found in the soil sample?
2. What kinds of living things were found in the soil?
3. Why is organic matter important to the soil?
4. What is humus?
5. How would you add organic matter to the soil?

Experiment 15.4 Measurement of organic matter (chemical)

Aim: To examine and measure the organic matter in soil samples

Method

1. Collect two soil samples, one rich in organic matter and one poor in organic matter, from two different regions of the school ground.
2. Examine the soil samples carefully with a hand lens or binocular microscope (at magnifications ×40, ×100 and ×400), and describe what you find in the samples.
3. Place a few grams of each sample into separate test tubes of water and shake vigorously. Draw the results in your book.
4. Weigh out 30 g of each sample into separate evaporating basins and heat strongly for 2 minutes. Note in your books any colour changes and smells.
5. Reweigh the samples when cool and assume that any decrease in weight is due to organic matter being burnt. Calculate and record the percentage of organic matter in each sample.

Questions

1. What is organic matter?
2. What affects the amount of organic matter in a soil?
3. What are six important effects that organic matter has on a soil?
4. How do you explain the change in colour when the soils were heated?
5. What does the smell given off the heated soils suggest to you?

Experiment 15.5 Determining soil texture – field determination (physical)

Aim: To practice field determination techniques

Soil texture (the percentage of sand, silt and clay in a soil) can also be classified by feel. A great deal of practice is required to perfect this technique.

Method

1. A sample of soil is just moistened, until it begins to stick to the fingers.
2. The sample is then rubbed between thumb and forefinger to determine a 'feel' for the soil. If the sample feels gritty, it contains a percentage of sand; if it feels smooth and silky, the clay content is high.
3. If the soil sample can be moulded into a ball, which crumbles if an attempt is made to form a ribbon, the sample is a loamy sand. If the sample can be moulded into ribbons, the clay content is high. Clay soils will form long ribbons (indicating they contain over 30% clay). If only a short ribbon of approximately 1 cm is formed, the soil will tend to be a clay loam.

   The field determination technique works well except for soils with high levels of organic matter, which can mask the true texture of a soil.

Questions

1. Define the term 'soil texture'.
2. Why is this soil property basically impossible to change without total replacement of the existing soil?
3. Research the link between the parent material of a soil and its texture
4. Why are seed-raising mixtures composed of a high percentage of sand?

ISBN 9780170265560

Experiment 15.6 Determining soil pH (chemical)

Aim: To determine the pH of a soil sample

Method

1. Place a small amount of soil (20 cent piece size) on a watch glass.
2. Place a few drops of universal indicator on the sample and mix in.
3. Sprinkle the sample with barium sulphate powder.
4. Leave for 2–3 minutes to allow the colour to develop then compare to the colour charts appropriate for a Raupach indicator. Use a commercial soil testing kit.

Questions

1. Why is knowledge of soil pH important?
2. What other methods could be used to determine soil pH?
3. What happens to soil pH values when lime is applied to the soil?
4. What happens to soil pH values when gypsum is applied to the soil?

Things to find out

1. Outline how soil resources influence the distribution of agricultural enterprises in Australia.
2. a What happens to a soil if the amount of organic matter is not maintained?
   b How can a farmer increase the organic material in a soil?
   c A farmer has cropped a paddock for the last 7 years, as he or she does not have enough land to afford 2–3 years of pasture rotation. What method/s of maintaining soil fertility would you recommend?
   d Find out the pH when the greatest number of minerals is available to plants. What minerals are available at this optimum pH?
3. Take two samples of soil and place each into a funnel. Leave one sample untouched and add lime to the other. Pour water through the samples and collect the filtrate. Which sample has the clearer filtrate? Why?
4. Obtain four containers of known volume (e.g. medicine glasses, measuring cylinders or egg cups), and add sand to one, garden soil to another and two known types of clay to the remaining containers. Place the containers into beakers or buckets of water and leave for 3 days. What has occurred?
5. Use a binocular microscope (at magnifications ×40, ×100 and ×400) to observe a sample of soil. Draw the particles and organisms seen, and identify them. What is the texture of the soil? What type of structure does the soil have? Does it contain an adequate level of organic matter?
6. Examine samples of different fertilisers and complete a table showing:
   - name of fertiliser
   - general description
   - list of nutrients.

Extended response questions

1. Why can a soil be described in terms of solid, liquid and gas phases?
2. How can clay affect soil fertility both physically and chemically?
3. Describe several soil processes that are in a dynamic equilibrium in the soil.
4. Describe the farming practices that influence soil temperature, soil pH and soil structure.
5. Describe the influence of legislation and government regulations, including licensing, on the availability and use of water for agricultural purposes.

Outcomes

Students will learn about the following topics.

1. Aboriginal land practices prior to the arrival of Europeans.
2. Changes in the Australian environment that have occurred since the arrival of Europeans.
3. The farming practices that have contributed to soil degradation, such as soil structure decline, loss of soil organic matter, soil erosion, tree decline and dieback, salinity, increased soil acidity and water contamination.
4. Practices that have contributed to changes in water quality and quantity.
5. Management for sustainable production.
6. Sustainable resource management.
7. The role of individual farmers, the broader community and government in reducing the harmful environmental effects of agriculture.
8. The tension between sustainability and short-term profitability approaches in farming systems.

Students will be able to demonstrate their learning by carrying out these actions.

1. Describe the impacts of historical land use practices in the development of Australian land use practices from Aboriginal practices to the present day.
2. Recognise that the problems of land degradation must be tackled in an integrated way, and that sustainable farming is one way of doing this.
3. Recognise that whole farm planning is one way of tackling land degradation and leads towards sustainable farming.
4. Recognise sustainable management practices, including crop rotation, green manuring, minimum tillage and stubble mulching.
5. Sustainable techniques to maintain soil fertility, such as conservation tillage systems, maintenance of soil organic matter (or carbon), organic fertilisers, inorganic fertilisers, pasture ley phase and crop rotations.
6. Explain why land degradation is a major problem facing Australia and Australian agriculture.
7. Explain the farming practices that have contributed to soil degradation, such as soil structure decline, loss of soil organic matter, soil erosion, tree decline and dieback, salinity, increased soil acidity and water contamination.
8. Assess the outcome of these practices on the land and water system.
9. Outline current recommended procedures to alleviate degradation problems.
10. Discuss practices that have contributed to changes in water quality and quantity, including fertiliser use, the effects of stock, effluent management, chemicals, grassed waterways, riparian zones, dam construction and irrigation methods.
11. Describe programs that involve community and government groups working together, such as Total Catchment Management and Landcare.
12. Assess the factors involved in the long-term sustainability of agricultural systems, using Australian land classification/capability and whole farm planning.
13. Identify tensions between sustainability and short-term profitability approaches in farming systems.

ISBN 9780170265560

SUSTAINABLE AGRICULTURAL PRODUCTION

Words to know

acidity of the soil the level of acid in the soil measured by the pH scale, which ranges from 1 (very acidic) to 14 (very alkaline) with a pH of 7 being neutral

catchment area the area of land that collects water for a particular waterway; the term may apply to a small stream in a paddock that only flows when it rains or to a very large river system, such as the Murray–Darling

contour bank a small bank constructed along the contours of the land where it slows running water and prevents it building up speed, and thus prevents erosion

cultivar a particular strain, or variety, of a plant (e.g. Hartog is a cultivar of wheat)

evapotranspiration part of the water cycle in which liquid water is removed from an area with vegetation and moves into the atmosphere by the processes of both transpiration and evaporation

grassed waterway a wide channel that is permanently grassed and rarely grazed, which slows the velocity and carries water away safely

gully erosion a type of erosion (which often starts as rill erosion) that is caused by running water, and is characterised by the formation of deep gullies in the soil

herbicide a chemical used to kill unwanted plants, especially weeds

land degradation the adverse alteration of the land surface and the lowering of the land's capacity to produce; characterised by one or more of the following – soil erosion, increased soil acidity, soil salinity, tree decline and reduced biodiversity

minimum tillage or **reduced tillage** methods of farming that minimise soil damage by reducing cultivation and substituting chemicals for machinery use

natural vegetation any vegetation that has not been cleared

rill erosion the removal of soil caused by small channels of running water

salinity the level of salt in the soil; usually refers to unacceptably high levels

sheet erosion the removal of soil more or less evenly over a wide area of land by wind or running water, usually resulting from heavy downpours

soil erosion the removal of soil by running water or wind

stocking rate the number of animals of a particular type per hectare of land (e.g. 10 sheep per hectare)

tunnel erosion a type of erosion where the saturated subsoil moves and washes away, leaving tunnels

water table the top of the ground water in the soil

zero tillage no tilling takes place to preserve plant cover during the soil preparation phase prior to planting

ISBN 9780170265560

Introduction

Agriculture has contributed greatly to the successful founding and establishment of the Australian nation. Australia's farmers and the technological developments that have supported them have achieved consistent and steady increases in production and efficiency.

The methods employed have produced spectacular results but at considerable cost to the farm land and the environment. The systems of agriculture employed have produced degradation of the land and the environment that, if not arrested, will result in reduced production from agriculture and a decline in the welfare of the nation as a whole. Some indication of the magnitude of the problem was given by Andrew Campbell in his book *Planning for Sustainable Farming* (Lothian, 1991), where he listed the 'impacts of Australian agriculture, mining and forestry on the landscape over the last 150 years'.

- Two-thirds of Australia's forests (40 million hectares) and one-third of all scrub and woodland (63 million hectares) have been cleared.
- 46 mammal species (15% of the total) have become extinct – a world-record rate of extinction.
- More than 500 species have been introduced as **cultivars**, weeds, pests or all three.
- More than half of all cropping and grazing lands require treatment for erosion, salting, soil acidity or soil structure decline.
- Many of the waterways and wetlands on this dry continent have become contaminated by soil run-off and algal blooms caused by fertilisers, pesticides and heavy metals.
- Most irrigation areas are being flooded from beneath by rising saline ground water, due to tree clearing, profligate water use and non-existent or inadequate drainage.

Land degradation includes **soil erosion**, tree decline and dieback, salinity, increased soil acidity, reduced biodiversity and pollution from fertilisers and pesticides. On any particular farm these problems are clearly interrelated, and it would not be wise to attempt to do something about one particular problem without giving consideration to the other problems. A holistic approach is required, which takes into account all the problems at once and puts them in the context of the whole farm and neighbouring farms. This holistic approach will also take account of the fact that the farm needs to be productive enough to provide a comfortable living for the farmer and family.

connect

Science and ancient fire knowledge

Answer the accompanying questions.

connect

Aboriginal fire knowledge

Answer the questions.

Today, the focus is on sustainable farming practices that maintain the quality of a farm's natural resources (water, soil, and animal and plant diversity). Initiatives such as the CSIRO's Sustainable Agriculture Flagship aim 'to reduce the carbon footprint of Australia's land use while achieving the productivity gains needed for prosperous agricultural and forest industries and global food security'. Resources within the natural environment are conserved, greenhouse gas emissions are reduced and recycling of water, minerals and organic matter is optimised. Efficient use of irrigation to avoid problems such as increases in soil salinity, optimal use of fertiliser application through adoption of new technology in relation to soil mineral analysis and fertiliser delivery, and use of wildlife corridors to conserve native wildlife are typical elements of a sustainable farming system. Consequently, sustainable practices are also profitable over time.

Aboriginal land use practices exhibited many principles of a sustainable land use system. The practice of burning vegetation encouraged new growth as the fires were low intensity and attracted animals into the area to feed on regrowth. Minerals were released into the soil from the burnt vegetation and maintained soil fertility. Environmental resources were not overused and due to the people's frequently nomadic existence, overstocking and overcropping problems did not arise. The practice of burning maintained a grassland cover and therefore limited the number of intensive fires experienced in timbered country. Rather than the European view of yearly climate as four seasons, Aboriginal Australians traditionally divide the year up into seasons based on the availability of food and weather features. Usually a year was broken into five or six defined seasons, reflecting a relationship between land management, weather and tribal customs.

1 Write a paragraph defining land degradation.
2 Explain why land degradation is an important issue for Australia and Australian agriculture.
3 List three effects of non-sustainable European farming practices.
4 Define the concept of sustainable farming.
5 List two Aboriginal land use practices and describe their impact on the Australian environment.

Sustainable whole farm planning practices aimed at conserving resources and being profitable include:

- increased emphasis on tree planting for windbreaks, protection from soil erosion and shelter for both animals and plants, especially tall crops liable to be blown over (lodged) by strong winds, making harvesting difficult. Trees with deep root systems will also lower the **water table** and reduce the risk of salinity.

ISBN 9780170265560

- stubble retention after harvest, which reduces the incidence of wind and water erosion
- use of perennial pastures composed of deep rooted plants, such as ryegrass and lucerne, to decrease the frequency of irrigation
- development of riparian zones on farms. A riparian zone or corridor forms a transition zone between the land and a river or watercourse or aquatic environment. These corridors are planted with trees and shrubs to help stabilise river banks, protect water quality by trapping sediment and provide a habitat for other organisms
- land development in accordance within land capability assessments.

6 List three sustainable whole farm planning practices now commonly used in Australian agriculture.

The effects of soil degradation

Soil erosion

Soil erosion is a major problem. The agents of soil erosion are wind and water (Fig. 16.1). These agents provide the means by which soil particles are blown or washed away, resulting in a loss of top soil, reduced nutrient availability to plants and increased fertiliser costs.

Soil erosion is widespread in Australia. A large proportion of the limited amount of land available for cultivation in this country has been affected to some extent by the ravages of soil erosion. Valuable topsoil, which best supports plant growth, has been lost. It is in every Australian's interest that no further erosion takes place and that steps are taken to correct and reclaim land that has already been affected. To prevent erosion, land has been classified according to its potential to erode. Factors taken into consideration when classifying land include the gradient or slope of the land, the climate, the type of soil and the **natural vegetation**. Clearly, some land in high rainfall areas is too steep to be cleared of its covering of trees and to do so would be to court disaster. Other land in low rainfall areas, such as western New South Wales, must be carefully stocked and managed so that the vital vegetative cover is not totally removed.

iStockphoto.com/NeilBradfield

Figure 16.1 Soil erosion

7 Explain why it is in the interest of every Australian to stop soil erosion.

Types of erosion

- **Sheet erosion** describes the removal of layers of topsoil over a wide area. It can be caused by water or wind (Fig. 16.3, page 230).
- **Rill erosion** describes the situation, usually on cultivated land, where small channels approximately 4 cm wide and 2 cm deep are washed in the soil. These rills often develop in the furrows left by cultivation implements (Fig. 16.4, page 230).
- **Gully erosion** is much more severe. Running water that has been concentrated in natural depressions in the land washes away the soil, and deep gullies are formed (Fig. 16.5, page 230).
- **Tunnel erosion** describes the situation where the saturated subsoil moves and washes away. The tunnels that are formed eventually collapse to form gullies (Fig. 16.6, page 230).

Auscape/Wayne Lawler

Figure 16.2 Erosion on very steep land that has been cleared

Not only is the land where the eroded soil comes from degraded, but also the land where the soil is deposited is adversely affected. The extra soil in the running water silts up streams and dams and lowers the quality of water for humans and stock. Windblown soil can be deposited along fence lines, making the fences ineffective in controlling stock movement and eventually causing wire fences to rust through.

8 List the different types of erosion.

ISBN 9780170265560

Figure 16.3 Sheet erosion

Figure 16.4 Rill erosion

Figure 16.5 Gully erosion

Figure 16.6 Tunnel erosion

Prevention and control of erosion

There are several principles used in the management of land that are designed to prevent and control erosion. These principles include:

- maintaining and improving the vegetative cover of the soil
- maintaining and improving soil organic matter content and therefore soil structure
- reducing the speed of run-off water and preventing its concentration in places where it is likely to cause damage
- reducing the speed of the wind at the soil surface.

Maintaining and improving the vegetative cover

It is important that there are always plants covering the soil. The plant stems and leaves break the fall of the raindrops, preventing them from pounding the soil. The plant roots intertwine between the soil particles and aggregates holding them together and prevent them being washed and blown away.

ISBN 9780170265560

The farmer should give attention to the **stocking rate** of the land. If there are too many grazing stock, they will quickly remove the vegetation and expose the soil. The number of stock the land can carry will not remain constant, but will vary with the seasons of the year and general weather conditions. During a drought the carrying capacity of the land will be greatly reduced because pasture growth will be impaired by lack of moisture.

In the seasons of the year when rainfall and temperature favour pasture growth, more stock can be run.

If land is unsuitable for clearing, for example, if it is too steep, it should be left untouched to prevent erosion. Fencing off these uncleared areas and areas along streams that are likely erosion sites will prevent stock from removing the protecting vegetation. Controlling rabbits and feral animals, such as wild pigs and goats, will reduce damage to the vegetative cover and hence reduce erosion.

Shutterstock.com/Janelle Lugge

Figure 16.7 Tree planting

Trees planted in strategic places along waterways and ridges and as windbreaks increase the cover and help to prevent erosion (Fig. 16.7).

Maintaining and improving soil organic matter and soil fertility

Traditional methods used for seedbed preparation when sowing cereal crops included several cultivations using different implements and then leaving the soil exposed to the agents of erosion for months at a time. This exposure led to decreased organic matter in the soil and, consequently, poor soil structure.

In recent years attention has been given to finding new farming methods that maintain and improve the organic matter in the soil, and thus hold it together against the forces of erosion. Such methods involve retaining stubble and reducing the amount of tillage done in preparation for the new crop. **Minimum tillage** (or **reduced tillage**) and **zero tillage** systems preserve plant cover during the soil preparation phase prior to planting. Soil organic matter, structure and nutrient levels are maintained. View the two websites given and compare and contrast the systems.

connect

Minimum tillage systems

connect

No till farming

The aim of stubble retention is to retain the residues (stems and leaves) from the previous crop and incorporate them into the soil. Zero tillage is tillage reduction in its most extreme form, where no cultivating is done except at the time when the seed drill actually sows the crop. Weed control is achieved by the use of **herbicides**. Reduced tillage and minimum tillage use various combinations of cultivation and herbicides to control weeds, prepare the seed bed and give protection from erosion.

Crop and pasture rotation is also important in the control of erosion. The pasture phase (pasture ley) in a crop rotation increases the organic matter levels in the soil and helps improve structure, which, in turn, makes the removal of soil particles by wind and water less likely.

Strip cropping involves strips (20–100 m wide) of the present crop, strips of stubble from the previous crop and strips of fallow that run along the contour of the land (Fig. 16.8, page 232). The strips are at right angles to the flow of water. The growing crop and stubble spread the water out, which prevents it concentrating and causing erosion. Any cultivation is done along the contour of the land.

Mulching is where a natural or artificial layer of material is placed on the surface of the soil. Mulches are applied to conserve moisture, control weeds and improve structure.

By growing then ploughing in leafy legume crops or green manure crops (such as cowpeas, lupins or field peas) soil organic matter reserves are built up. The organic matter improves soil structure by binding soil particles together. When it decomposes, it releases nutrients for plant root uptake. Use of organic fertilisers, such as mushroom compost, Dynamic Lifter® and Bio-Soil (composted sewerage), boosts nutrient and water holding capacity.

ISBN 9780170265560

Shutterstock/Tim Roberts Photography

Figure 16.8 Strip cropping

Reducing the speed of run-off water

The practice of ploughing or cultivating along the contour of the land reduces the speed of run-off water. The ridges left by the cultivating machine are at right angles to the flow of water, and thus stop the water concentrating and running down the slope. This is called contour ploughing.

The strip-cropping approach already described is based on this idea and is applied to gently sloping land (Fig. 16.8). In steeper country the strategy is to build structures that prevent too much water concentrating in natural depressions and gullies where erosion is likely to occur. Contour banks that run along the contour of the land are constructed with graders and dozers (Fig. 16.9). The banks trap the water so that it moves away slowly and soaks into the ground. Dams are made to catch and slow the water down, and to allow soil to sediment out before it is lost. Broad **grassed waterways** may be constructed to safely take water down slopes. These are usually fenced off and not grazed to ensure there is always a vegetative cover. Existing gullies are filled in and the **contour banks** and grassed waterways divert water away from the gully.

9 List the principles employed to prevent and control erosion.

10 For each principle, describe one procedure a farmer could perform to put it into practice.

CSIRO/John Coppi

Figure 16.9 Contour banks

Reducing the speed of wind at the soil surface

The maintenance of a vegetative cover will reduce wind speed at the soil surface. Windbreaks of trees planted at right angles to the prevailing winds not only help reduce wind speed but also provide shelter for livestock.

Total catchment management (TCM)

The successful control of erosion on one farm often depends on what is being done on neighbouring farms. In fact, it often depends on all the land uses in the **catchment area** of the stream in which the farm is located. A typical catchment area is likely to include forests and trees, farms, dams storing water for farm and town use, natural areas, such as parks and bushland, towns, roads and railway lines. Any erosion control plan needs to include the whole catchment area of a stream. The design of the plan and its implementation should involve the cooperation of all the stakeholders and authorities that control the catchment area.

11 Why is it desirable to carry out an erosion control plan in conjunction with neighbouring farms and other land-users?

12 Where can farmers get assistance with their erosion problems?

ISBN 9780170265560

There are many services available to farmers to help them combat erosion. The Soil Conservation Service and the various total catchment management (TCM) authorities provide assistance with drawing up whole farm plans that identify problem areas and possible solutions. It also makes earth-moving equipment and finance available so that plans can be put into action. The Forestry Commission of New South Wales provides advice on trees as well as producing trees for planting. The Department of Primary Industries works in close cooperation with both the aforementioned organisations and landholders.

The underlying theme of the TCM process is the community (farmers and the wider community) and the government agencies work together to achieve sustainable natural resource management.

Alamy/David Wall

Figure 16.10 A windbreak

Soil salinity

Basically, soil **salinity** can be described as the rising of the level of salts in the soil to such concentrations that they adversely affect the growth of plants (Fig. 16.11). Salinity has become a problem in irrigation and dryland areas. The level of salt in the waters of certain streams, in particular the Murray–Darling system, is of concern because it reduces the usefulness of the water for humans, plants and animals.

Many irrigation areas in Australia are constructed on land that was covered by sea millions of years ago. Deep in the soils are large reserves of salt. The use of large quantities of irrigation water for growing crops has resulted in the water tables rising and bringing salt to the surface, where it interferes with plant growth. This is known as irrigation salinity.

Auscape/© Tim Acker

Figure 16.11 Soil salinity

Dryland salinity also results from rising water tables bringing salt to the surface. The cause of the rising water tables is not the addition of more water to the system but the removal of trees that acted as pumps, drawing water from deep in the soil and passing it into the atmosphere by transpiration (Fig. 16.12, page 234). As land was cleared to open it up for cropping and grazing, deep-rooted trees were replaced by shallow-rooted pasture and crop plants. Salinity caused by the removal of tree cover can contribute to soil erosion.

The increased salinity in the river systems, especially the Murray–Darling system, is the result of drainage water that is loaded with salt seeping back into the rivers from irrigation areas.

connect

Salinity

Summarise the information in the report.

ISBN 9780170265560

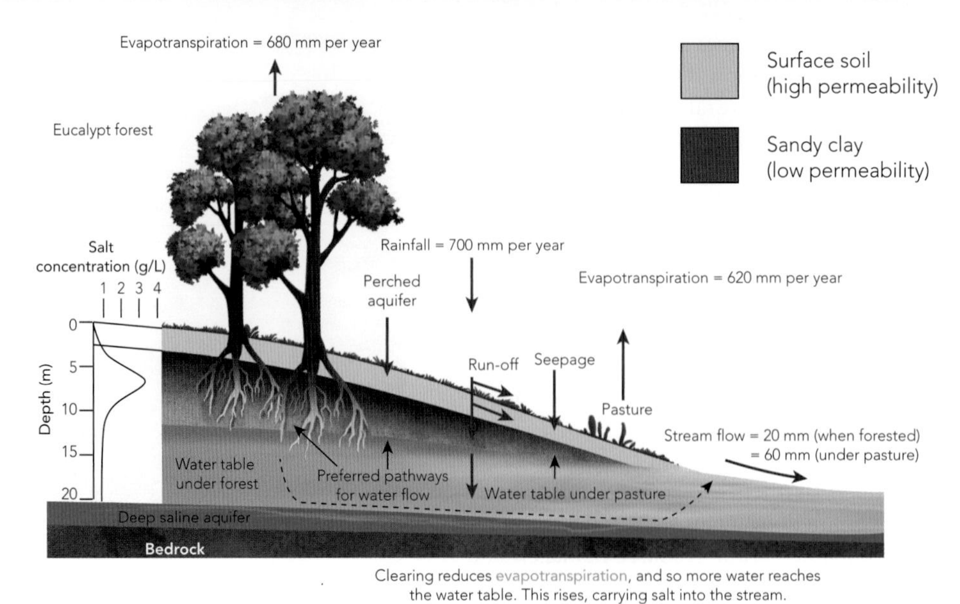

Figure 16.12 How clearing and planting shallow-rooted crops and pastures can lead to salting

Tackling salinity in irrigation areas has involved lowering the water table. This has been done through the installation of drainage systems, by storing salty water in large water basins, through more judicious and efficient use of irrigation water, and by actually pumping water from the water table. Planting trees also has some potential. These measures may be effective for the farm in a particular area but it still leaves the problem of safely disposing of the salt-laden water. Attempts have been made to channel it into salt-retention pans where the water evaporates and the salt is retained. The only absolutely successful method would involve piping the salty drainage water hundreds of kilometres to the sea, but this procedure would be too expensive.

connect

Water use in the Murray–Darling Basin

Summarise the main points raised in the video.

In dryland areas the best approach to tackling problems of salinity seems to include:

- re-establishing trees in the catchment areas to reduce the amount of water filtering into the ground water (the trees remove the water by transpiration)
- fencing off the salt-affected areas and planting them with salt-tolerant pastures and crops
- using reduced stocking rates on the affected areas, are in keeping with the reduced capability of the land.
- catchment planning through the application of technology to detect salt affected areas; for example, using satellite imagery and aerial scanning.

13 What causes the level of salt to rise in the surface soils of irrigation areas?

14 What strategies can be employed to prevent and reduce salt buildup in irrigation areas?

15 What has caused salinity problems in some dryland areas?

16 Suggest an approach to treating a salt-affected area on a farm that does not have irrigation.

The problem of rising salinity in the Murray–Darling system affects the states of Queensland, New South Wales, Victoria and South Australia, and the Australian Capital Territory. Successfully dealing with the problem will have to involve the cooperation of all four state governments and the Commonwealth Government.

Soil acidification

Rising acidity (decreased pH) in soils of higher rainfall areas, such as the tablelands of New South Wales, has led to reduced productivity from grazing and cropping (Fig. 16.13). The widespread use of superphosphate fertiliser in combination with more productive introduced pasture species has given rise to increased production of organic material and, consequently, the return of more organic material to the soil. The breakdown and decay of this organic material produces acids, and hence the pH of the soil has been lowered.

The increase in acidity has caused changes in the availability of plant mineral nutrients, making some less available. It has also caused some mineral elements, such as aluminium, to become soluble. These elements in their soluble (ionic) forms are toxic to plants and contribute to further increasing the **acidity of the soil**.

The application of lime in fairly large quantities appears to be the only currently feasible approach to combating acidity. Unfortunately, this is a reasonably expensive solution.

17 Give an explanation of the rise in acidity in soils of high-rainfall areas.

18 Comment on the effect of decreased soil pH on the availability of plant nutrients and the toxicity of elements such as aluminium.

ISBN 9780170265560

Shutterstock.com/paintings

Figure 16.13 Fertilisers and introduced pasture species contribute to increased soil acidity

Tree decline and dieback

The main reason for tree decline in Australia's agricultural lands is that as the land was opened up, trees were cleared to make way for grazing animals and the growing of crops. (This clearing process is still occurring.) No allowance was made for the regeneration of trees. The trees that were left are now dying out. In some areas they are suffering from a condition known as dieback, where the trees show a sickly appearance with reduced foliage and branches dying back from the ends (Fig. 16.14). It is thought that a combination of factors, including insects, increased soil fertility from the use of fertilisers, and the lack of regeneration, probably attributable to grazing livestock, has brought about the condition of tree dieback.

The obvious way to arrest tree decline is to plant new trees and to preserve existing remnant natural vegetation. Fencing off the newly planted areas is of vital importance if the young trees and seedlings are to be protected from being eaten by livestock. It is also important to fence off existing remnant vegetation to allow regeneration to take place.

19 Explain how planting new trees and fencing them off will assist in reversing tree decline.

Dreamstime.com/Camelotimage

Figure 16.14 Tree decline and dieback

ISBN 9780170265560

Soil pollution

In some places the soil has become contaminated with chemicals and residues from substances that were used as pesticides and herbicides. These contaminants have become a problem because they are present in the products coming from the farm. In particular, there have been problems on certain farms with meat products containing unacceptable levels of pesticides. The farms have not been able to sell meat animals until such time as the levels of contamination have been reduced. This process may take many years, as many of the early pesticides took a long time to break down in the environment.

20 Explain how chemicals used in agriculture have contributed to land degradation.

Fertilisers

No one can deny that fertilisers have facilitated great increases in production from the cropping and pastoral industries. Their use has been, and continues to be, a significant part of Australia's agricultural systems. Problems arise when excess fertiliser is washed or leached from the farms and ends up in streams and waterways. The fertiliser stimulates the growth of algae, which upsets the natural balance in the waterway. Algal blooms can result in the production of toxins and oxygen depletion in the water, which, in turn, results in the death of fish and other water creatures (Fig. 16.15). This is called **eutrophication**.

Excessive use of fertiliser should be avoided. Where possible, natural vegetation along streams should be retained, encouraged and even re-established. This vegetation acts as a filter to water entering the stream.

21 What problems have been created by the use of fertilisers?

Getty images/Universal Images Group

Figure 16.15 Fertilisers can cause algal bloom in waterways

Soil structure decline

This has arisen from excessive cultivation, use of bare soil and fallowing practices, overgrazing and loss of vegetative ground cover. Plant establishment declines in these conditions and yields are reduced. Cultivation becomes both difficult and costly due to soil compaction. There is an increased risk of run-off and erosion.

Irrigation

Irrigation has been regarded as a panacea for agricultural problems in areas where water supplies are limited. Indeed, it has contributed greatly to production. Large areas of land have been opened up to relatively intensive production, which could not otherwise have happened. However, there have been enormous costs in setting up and maintaining the schemes that supply and deliver the irrigation water, and irrigation has brought with it problems of rising water tables, salinity and the safe disposal of salt-laden drainage water. Dam construction has increased water storage for irrigation.

Farming practices that have led to a decline in water quality include:

- excessive fertiliser use resulting in water run-off and seepage carrying fertilisers high in nitrogen and phosphate into waterways and dams.
- poor effluent management that allows seepage into streams and waterways again results in eutrophication of the waterway

 ISBN 9780170265560

- poorly managed livestock access to waterways results in river bank damage and resulting erosion problems; sedement levels also build up in the waterway from damaged stream banks
- excessive pesticide use, resulting in run-off into waterways, killing fish and aquatic plants.

Sustainable farming

As has already been stated, there is only limited benefit in tackling any of the land degradation problems individually. They must be taken on together because they are interrelated.

The farm is a complex system involving the interaction of plants, animals, micro-organisms and invertebrates, the physical environment of topography, climate and soil, the farmer and the farm family, and the wider economic environment. A sustainable farming system is needed. Andrew Campbell's definition in *Planning for Sustainable Farming* helps to clarify what is meant by sustainable farming:

A sustainable farming system is one which is profitable and maintains the productive capacity of the land while minimising energy and resource use and optimising recycling of matter and nutrients.

Note: in this definition there is a balance between making a profit and maintaining and improving the environment where the farming takes place. It is important that the farmer makes a comfortable living and it is important to maintain the farmed land in such a state that it will continue to support people for many future generations.

connect

Who is Landcare?

connect

Landcare

Many programs across Australia involve Landcare groups. Study this website and use the information obtained to answer question 22.

connect

Case studies

Sustainable resource management – Landcare

Landcare groups form the basis for active community involvement in resource management. Individuals, community groups, business and government agencies have a role in caring for the environment through this organisation. Activities implemented under Landcare are consistent with the strategies developed by catchment management committees. Read the information on the role of Landcare groups contained in their websites and summarise the information. The websites outline how members of the community and organisations exchange information, and discuss management options and action programs. Landcare encourages land users on a group basis to take responsibility for local problems.

22 From the Landcare website or from secondary sources:
- identify the location and describe the type of Landcare groups
- describe the problems that the Landcare group seeks to solve
- describe the programs on which the group is working
- assess the success of two of their programs.

Whole farm planning

Planning is required to run a farm in a sustainable way. The concept of whole farm planning is an effective approach to taking a farm towards sustainability. Andrew Campbell lists the four principles on which whole farm planning is based.

1. Farms should be subdivided into homogeneous land units based on natural features, such as soil type, slope, drainage and vegetation, rather than on arbitrary boundaries made by people.
2. Each land unit should be managed according to its potential and its limitations (potential to degrade), with an understanding of the ecological processes in operation both within the farm boundary and over the land system and catchment in which the farm is located.
3. Farm improvements, such as water supply, drainage, crop and pasture improvements, access roads and revegetation, should not be looked at in isolation, but rather integrated into a plan which considers the farm as a whole, not a collection of discrete parcels of land.
4. Farm management should aim to incorporate the elements of existing natural systems that convey the stability, resilience and ability to recover from disturbance that characterise a sustainable system. These elements include structural and species diversity within populations of plants, animals and micro-organisms, efficiency in the use of energy and resources, optimal turnover and recycling of matter and nutrients, efficient energy flow, adaptation of crop and animal components to the environment, and conservation of the renewable resource.

One of the first steps in applying these principles would be to take a fresh look at the farm in question using an aerial photographic map (Fig. 16.16), and to start planning the farm from scratch as though no improvements already existed. Plans should be made and put into practice as finance and labour become available. Plans will take at least 10 years to fully implement, and once conceived will not remain static as the influence of new knowledge, technology and economic forces impinges on the farming operation.

There are three main benefits of whole farm planning.

1 It helps farmers recognise problem areas on their property.
2 It assists farmers in developing solutions to problems on their property. These solutions may take several years to develop.
3 It develops improved farming practices to achieve sustainable production on their property.

23 What is the function of whole farm planning in achieving sustainable farming?

Figure 16.16 A whole farm plan based on an aerial photographic map

Land classification and capability systems

Land resource planners have developed a five-point and an eight-point land capability classification system to indicate the most suitable use for land. This ensures that land will remain productive by minimising practices that would lead to land degradation. For each category, recommendations are given for the management of this resource.

In summary, for the five-point system, land designated Class 1 and 2 are suitable for cultivation, Class 3 and Class 4 for grazing, and Class 5 is land unsuitable for agriculture. With the eight-point system Classes 1–3 refer to land suitable for cultivation, Classes 4–6 is grazing land, Class 7 is land suitable for the establishment of trees and Class 8 is unsuitable land for agricultural activity. Management recommendations are outlined for each land capability class that ensure sustained production and protection to natural resources, such as the soil.

These systems are designed to support sustained production over time and may cause tension with approaches to land management that seek short-term profitability outcomes.

connect

Classify rural lands

An in-depth look at both five-point and eight-point land classification systems.

 ISBN 9780170265560

Chapter review

Things to do

1. Obtain an aerial photographic map of your school farm or another farm with which you are familiar, and draw up a new development plan using the principles of farm planning set out in this chapter. You will need to spend quite a lot of time on the farm, becoming familiar with its soils, drainage patterns, existing vegetation and topography.
2. Grow some plants in pots and water half the pots with salty water and the other half with water from the tap. Observe the effects of salt on plant growth.
3. Obtain some acid soil and measure its pH using a universal indicator. Add lime to the soil a little at a time and measure the pH after each addition is made. Keep adding lime until the pH has been brought up to the range of 6.0–6.5, which is suitable for most plants. By carefully measuring the lime added you should be able to estimate how much lime per hectare would need to be applied to bring the soil to a pH more suitable for crops and pastures.
4. Set up a stream tray and fill it evenly with sand. Halfway along the tray, sow a 10 cm band of grass seed (e.g. ryegrass) across the tray. Water the tray carefully, and when the grass has germinated and established, tilt the tray and use a watering can or hose with a sprinkler to apply rain on the slope above the band of grass. Observe and record the effect the grass has on the flow of water down the slope, and how it holds the sand particles together.
5. Conduct an experiment to observe the effect of raindrops on moving sand particles. Place a teaspoon of sand in the middle of a large piece of newspaper on the floor. Now use an eye-dropper to drop water from a height of 1 m. Measure how far the sand particles are moved by the bombarding raindrops. What happens if the rain falls from a lower or higher?
6. Survey your school grounds and farm, if it has one, for signs of erosion. Things to look for include the build-up of soil at the bottom of fences, small stones on little piles of earth after rain and bare patches of earth where vegetation has been removed. Mark the problem areas on a map of the school or farm and make suggestions for what might be done to improve the erosion situation. If possible, put the suggestions into practice.

Things to find out

1. How does adding lime to an acid soil reduce acidity?
2. By what mechanisms does a high level of salt in the soil adversely affect plant growth?
3. What levels of pesticides are considered dangerous in farm produce (meat, milk and vegetables), and how are their levels monitored? What are the consequences for the farm that has high levels of pesticide in its products?
4. Where is the nearest Soil Conservation Service or catchment management authority (CMA) office located, and what are its main projects at the moment?
5. Identify how unsustainable farming practices in the short term might generate large profits, but in the long term might damage the environment and, consequently, degrade the farm and lower profitability.
6. What is the actual extent of the erosion problem in Australia?
7. Research a catchment management group and:
   - identify the roles of government and community members
   - describe and assess the success of three management programs.
8. For a farm that you have studied describe the changes in its layout and management that have arisen from the development of a whole farm plan. Outline the expected benefits from implementing the whole farm plan.
9. For a farm that you have studied describe a land classification/capability system available to farmers. In your description:
   - outline the characteristics that would be used to give the farm a land use class number and state the class number
   - outline the recommended uses and landcare practices for this farm.
10. For a farm you have visited, outline what sustainable farming practices could be developed to generate increasing profits over time.

connect

Catchment management authorities

This website may assist you in answering Question 7.

Extended response questions

1. Write an article for your local newspaper, alerting its readers to the problem of soil erosion in your area and explaining what can be done about it.
2. Discuss the advantages of whole farm planning to the farmer, the farm and the broader ecosystem.
3. Outline strategies and educational programs that could be used to change attitudes and promote sustainable land use.
4. For one soil degradation problem discussed in this chapter:
   a. describe the farming practices that have contributed to the problem
   b. describe the impact of this problem on farm soils and water
   c. describe the current recommended procedures/practices to alleviate the problem.
5. Outline the role and responsibility of government and agencies in encouraging responsible land and resource use.
6. Research and summarise existing legislation controlling land resource use. Include information about environmental protection from the Environmental Protection Authority (EPA), water licensing and the *Native Conservation Act 2003.*
7. Describe techniques used to manage soil fertility: conservation tillage (minimum tillage), maintenance of soil organic matter (or carbon), crop rotations, organic fertilisers, inorganic fertilisers and pasture ley phase.
8. Discuss the arguments you would put to a group of farmers to convince them that adopting the sustainable farming concept is a good idea and in their best interests.

ISBN 9780170265560

UNIT 4
PLANT PRODUCTION

Shutterstock.com/Grey Carnation

CHAPTER 17

Outcomes

Students will learn about the following topics.

1. Regionally significant plants.
2. Constraints imposed by environmental factors.
3. Competition in plant communities.
4. Managing the constraints on plant growth and development to maximise production.

Students will be able to demonstrate their learning by carrying out these actions.

1. Identify a range of regionally significant plants.
2. Describe the influence of climate, especially rainfall, temperature and day length, on plant production.
3. Outline how soil fertility, both physical and chemical, influences plant production.
4. Describe the adverse effects of pests and disease on plant production.
5. Describe sources of competition in plant communities.
6. Explain how interference from the crop plants themselves affects plant production.
7. Explain how interference from weeds influences plant production.
8. Explain how the genotype or variety of a plant influences its production.
9. Describe how management decisions influence production.
10. Outline how soil is prepared for sowing a crop.
11. Outline what may have to be done to prepare a site for growing a crop.
12. Explain that when planting a crop the aim is to sow the seed at the correct depth and at a spacing so as to achieve a suitable plant density.
13. Investigate how farmers manage plant competition through plant density and weed control strategies.
14. Perform a first-hand investigation to determine the effects of planting density on plant growth/yield.
15. Explain that caring for a growing crop involves providing moisture and plant nutrients and controlling diseases, pests and weeds.
16. Explain that harvesting a crop involves taking the crop off at the right time and preparing the product for market.
17. Know that products of the plant production system have to be transported from the farm to a market or some other place for processing.
18. Appreciate the role played by management in the conduct of a successful plant production system.

ISBN 9780170265560

FACTORS INFLUENCING PLANT PRODUCTION

Words to know

allelopathy where one organism produces one or more biological chemicals that influences the growth, survival or reproductive capacity of another organism

coleoptile a protective cone of tissue over the terminal bud in the monocot embryo, which gives the bud protection until it reaches the soil surface

crop one kind of plant cultivated to produce some particular product (e.g. wheat is grown to produce grain for bread)

density (of plants) the number of plants per hectare

fallowing a farming system in which land is left without a crop for extended periods to accumulate soil moisture

fodder crop a crop that is grown to produce feed for grazing animals (e.g. lucerne)

growth habit the way a species of plant grows (e.g. kikuyu grass sends out stems horizontal to the ground surface and is said to have a creeping habit)

harvesting gathering a product from the plants being grown

hydroponics growing plants without soil in carefully balanced nutrient solutions

integrated a combination of practices

long-day plant a plant that appears to flower in response to a long photoperiod, but in fact responds to the short dark period (i.e. spring or summer)

mineral nutrients elements that the plant requires in ionic form for healthy growth (e.g. potassium (K^+) and nitrate (NO_3^-))

morphology the form and structure of a plant

mulch material spread over the soil surface to reduce loss of water by evaporation (e.g. crop residues, black plastic, wood chips and compost)

optimum temperature ideal temperature

photoperiod the length of the daylight period in a 24-hour day

post-emergent herbicide a herbicide that is applied to the crop after it has emerged from the soil

pre-emergent herbicide a herbicide that is applied before or at sowing, before the sown crop emerges from the soil

selective herbicides herbicides that target particular weed species but have little or no effect on the desired crop plants

short-day plant a plant that appears to flower in response to a short photoperiod, but in fact flowers in response to the long dark period (i.e. autumn or winter)

soil fertility a soil's ability to support plant growth

weed a plant growing where it interferes with other plants or that, because of its characteristics, causes harm to humans or grazing livestock, or degrades the value of animal or plant products

ISBN 9780170265560

Introduction

Plants produce a wide range of products, including:
- cereal grains, such as wheat, oats, barley, sorghum, rice, triticale and maize
- oilseeds, such as canola, soybean, safflower and linseed
- legume grains, such as chickpeas, lupins and field peas
- vegetables, such as potatoes, tomatoes, cabbages, cauliflowers, carrots, beans, peas and onions
- fruit, such as oranges, apples, pears, peaches, bananas, apricots, cherries and avocados
- fibres, such as cotton, flax, jute and hemp
- herbage or feed for livestock from pastures.

How well an individual plant, or a crop of plants, produces depends on the interaction between the plant's **genotype** (see page 59) and its environment.

The plant's genotype, or genetic makeup, sets the upper limit for production and is determined by the set of genes inherited from the parent plants.

The environment of the plant or crop includes the climate, the soil, pests, disease, interference from plants of the same species and other species (such as **weeds**), and the management practices carried out by the farmer. The environment determines how far production reaches towards the limit set by the genotype. All the environmental factors listed above can individually or collectively limit plant growth and production. The things that are necessary for plant life include water, carbon dioxide, **mineral nutrients**, oxygen, the right temperature, light, and freedom from pests, disease and interference by weeds. Management plays an important role here – first, in selecting the type and variety of plant to be grown, and second, in determining how well the growing crop or pasture is supplied with the requirements for growth.

1 Explain how the production of a crop is determined by the crop plant's genotype and environment.

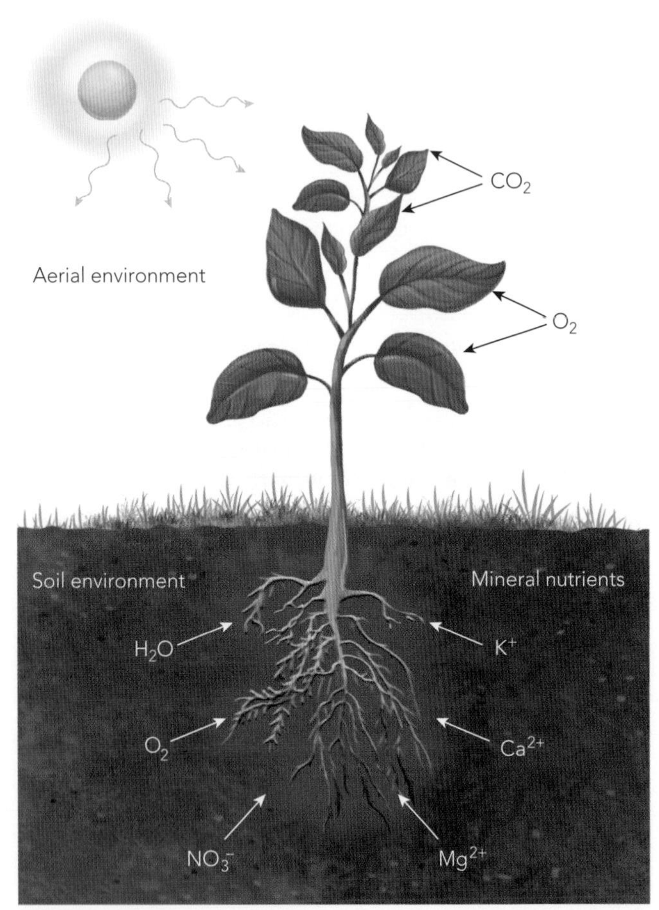

Figure 17.1 The source of a plant's requirements from the aerial and soil environments

The plant environment

The plant has a stem, leaves, flowers and fruits in an aerial environment, and roots in the soil environment. From the air the plant obtains carbon dioxide, solar energy (sunlight) and oxygen, and from the soil it obtains water, mineral nutrients and oxygen (Fig. 17.1). The plant must also have a certain temperature regime if it is to grow and produce well.

The aerial environment

The climate, or aerial environment, needs to supply carbon dioxide, sunlight and the right temperature to a plant. The carbon dioxide concentration in the air is approximately 0.03%. The level of carbon dioxide in the air in the canopy of a rapidly growing crop can become so reduced by absorption by the crop plants that it is limiting to plant growth.

The amount of sunlight is rarely limiting to plant growth, although in very cloudy situations crop yields may be reduced. The **photoperiod**, or the number of hours of sunlight in a day, has important effects on the initiation of flowering in plants. In fact, plants can be classified according to their day-length requirements for flowering. **Long-day plants** flower in response to what appears to be long periods of daylight. They actually respond to the short dark periods. **Short-day plants** appear to respond to short periods of light, whereas they actually respond to the long dark periods. Day-neutral plants flower irrespective of what day-length regime they are exposed to.

ISBN 9780170265560

Temperature has a very important effect on plant growth and production. For each species of plant there is a maximum temperature above which it will not grow, and a minimum temperature below which it will not grow. Somewhere in between there is a small range of temperatures over which the plant grows best, and this is the **optimum temperature** range for that species (Table 17.1).

Certain plants require exposure to fairly specific temperature regimes before flowering is initiated. For example, winter wheat and tulip bulbs need a certain amount of exposure to very low temperatures.

Table 17.1 The minimum, optimum and maximum temperatures for the germination of several crops

Species	Minimum (°C)	Optimum (°C)	Maximum (°C)
Lucerne	1	30	38
Wheat	4	25	32
Maize	9	33	42
Rice	11	32	38

2 List the components of the plant's aerial environment and briefly explain how each component affects plant growth.

The soil environment

The soil environment is very important, and the term 'soil fertility' is often used to rate a soil's ability to support plant growth. **Soil fertility** describes how well a soil provides the plant with the necessities of life. Soils are said to be fertile or infertile. A fertile soil provides everything a plant needs, while an infertile soil is one that lacks at least one of the things the plant requires, and so limits plant growth and production. Soil fertility has both physical and chemical aspects. Physically, the soil provides a store for water, oxygen, warmth and actual physical support.

A physically fertile soil will hold sufficient water and air in its pores at the same time to give maximum growth, be at the right temperature and give the roots something in which to anchor the plant. Chemically, the soil provides air, water and mineral nutrients that can be easily absorbed by the plant, and has a certain pH (acidity or alkalinity). A chemically fertile soil has a large supply of available mineral nutrients, air and water and is at a pH that is suitable to the plants growing in it.

Rainfall is one aspect of the climate that directly affects the amount of water in the soil and therefore available to the plant roots. If rainfall is inadequate, there will be insufficient soil moisture to support plant growth and production will decrease. Conversely, excessive rainfall, especially over extended periods, will reduce plant growth because there is insufficient oxygen in the soil.

3 Describe the features of a fertile soil.

Pests and diseases

Pests and diseases limit plant growth. Pests and disease organisms reduce the plant's capacity to grow and produce by one or more of the following methods:

- by eating or damaging the leaves, thus reducing the photosynthetic capability (e.g. the cabbage white butterfly and its larvae eat the leaves of cabbages, cauliflowers and broccoli)
- by sucking the juice out of the plant, thus using up photosynthetic products that would have been used for growth (e.g. aphids attack lucerne plants)
- by damaging the roots or stems so that water and mineral nutrients cannot be obtained in sufficient quantities, or the translocation of these is interrupted and the products of photosynthesis are reduced (e.g. the take-all fungus disease attacks the roots of wheat).

Production can also be reduced by pests and diseases attacking the harvested products. For example, weevils can attack stored grain and the fungus brown rot can destroy harvested stone fruit.

4 List three ways in which pests and disease reduce plant yield.

ISBN 9780170265560

Interference

A plant usually grows in fairly close proximity to other plants, either of the same species or of different species. A **crop** is almost entirely made up of plants of the same species, and usually the same variety of that species. Weeds are plants of a different species from the crop species. A pasture is made up of several different species growing together, usually legumes (such as clovers) and grasses. The growing plants interact or interfere with other plants in several ways.

- There is competition between plants for resources in the environment, such as mineral nutrients, water, space, sunlight and carbon dioxide. Competition occurs between the crop plants as well as the weed plants. In fact, some competition is necessary if yield per hectare is to be maximised.
- Many plants release certain chemicals into their immediate surroundings that affect the growth of neighbouring plants. The suppression or enhancement of growth in some plants by chemicals released by another plant is known as **allelopathy**. The allelopathy may be either positive or negative and may be produced by dead plant residues as well as living plants. For example, the weed thornapple produces chemicals that adversely affect the germination and growth of wheat seeds.
- By acting as alternate hosts to pests and disease organisms, plants cause interference to other plants. For example, weed plants growing near a vegetable crop can be infected with a virus and infested with aphids. The aphids transfer the viral infection to the neighbouring crop plants.
- Differences in **growth habit** and **morphology** of plants may result in one species modifying the **microclimate** (see page 192) of another so that growth and production is reduced. For example, the weed Paterson's Curse begins its growth by spreading its leaves out in a rosette horizontal to the surface of the ground, thus shading out other plants that are close to it.

5 Which one of the components of plant interference is likely to be of most significance in its effect on production? Why?

Farmers can manage plant competition by:

- selecting the optimum sowing rate to control plant density
- using weed control measures, which include cultivation and herbicide use.

Density

The decision about the number of plants per hectare (the **density**) of plants in the final crop or pasture will influence the yield per hectare significantly. It has been shown, as Figure 17.2 illustrates, that with increasing plant density the vegetative yield per hectare (leaves and stems) increases to a certain maximum level. After that, any further increase in plant density does not result in any further vegetative yield increase.

Figure 17.3 shows how plant density affects the reproductive yield per hectare. That is, the material that is harvested from the reproductive parts of the plant, such as grain and seed, increases with increasing plant density until a maximum is reached. Further increases in plant density then result in a decrease in reproductive yield per hectare.

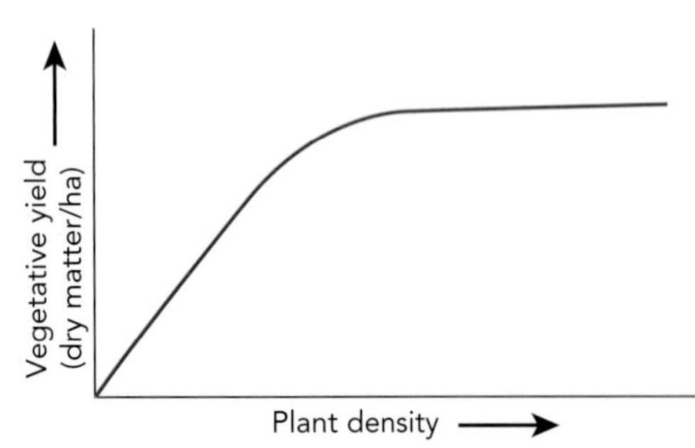

Figure 17.2 The effect of increasing plant density on vegetative yield per hectare

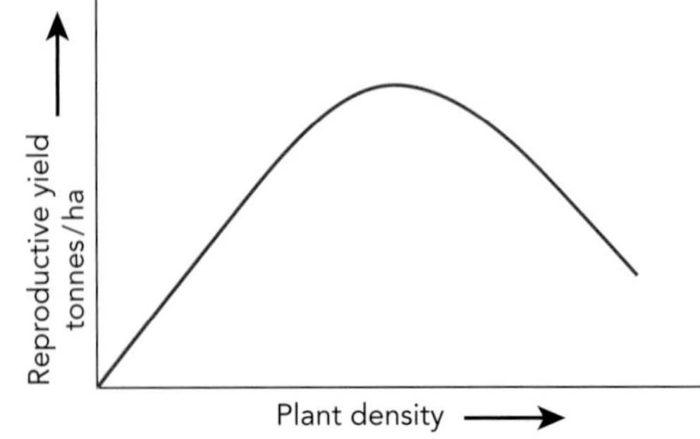

Figure 17.3 The effect of increasing plant density on reproductive yield per hectare

ISBN 9780170265560

The density chosen will be determined to some extent by the end use of the crop. If vegetative yield is required, say in a **fodder crop** of oats, a density should be chosen that just ensures the maximum vegetative yield per hectare. Any density higher than this will not increase vegetative yield. If the farmer wishes to produce oat grain, a density should be chosen that maximises the yield of grain per hectare, and there is only a very narrow range of densities that will achieve this.

Fertiliser use

The farmer also has to decide which fertiliser to apply and at what rate. The fertiliser chosen will depend on two factors:

1 which particular nutrients are deficient – in Australian soils phosphorus and nitrogen are likely to be deficient
2 what kind of production is desired – in general, vegetative growth is stimulated by nitrogen, and reproductive production is stimulated by phosphorus.

The rate of application will depend on the cost of the fertiliser and the expected return from the product.

Weed and pest control

The farmer has a number of options when deciding how to control weeds in the crop or pasture. Traditional preparation for sowing crops involved long fallows, where the soil was left ploughed for several months and worked after rain to kill the germinating weeds. **Herbicides** (see page 231) have been used increasingly in recent times. They have formed an important part of the **reduced tillage** (see page 231) programs, where such programs have replaced cultivation as the major method of weed control. The land is less exposed to the possibility of erosion with these techniques. The herbicides are applied prior to the crop emerging (**pre-emergent herbicides**), or after the crop is established (**post-emergent herbicides**). Biological methods that use insects and diseases to attack the weeds have also been introduced. Skeleton weed, a major weed of southern wheat crops in Australia, has been controlled to some extent by this method. In pastures weeds are controlled by adopting grazing strategies, such as strip grazing, where the animals eat everything including weeds very quickly, slashing after grazing to cut down weeds, and introducing vigorous pasture species that compete successfully with weeds.

Decisions on pest and disease control will depend on the likelihood of attacks occurring. Control measures should only be carried out where their cost is well and truly covered by the returns from the crop.

6 Explain why a farmer may have made the following decisions while growing a crop.

a A disease-resistant crop variety was chosen.
b Superphosphate fertiliser was applied at planting.
c A post-emergent herbicide spray was used.
d The crop was irrigated several times.
e The crop was sprayed with insecticide.

Watering

Where irrigation is available, decisions have to be made as to when to apply the water and how much to apply. Ideally, enough water should be applied to wet the root zone of the growing crop. When to irrigate will depend on rainfall, evaporation rates and soil type.

Steps in crop production

In general, there are several steps that must be taken in producing a crop:

- preparing the soil and site
- planting the crop
- caring for the growing crop
- harvesting the crop
- transporting the product to market.

Preparing the soil and site

Preparation of the soil has traditionally involved ploughing the soil in some way, and to a large extent this is still the first step. This initial ploughing might be done with a mouldboard plough, disc plough or chisel plough (Fig. 17.4). It is carried out some months before sowing the crop and is followed up by one or two cultivations with offset disc harrows or scarifiers. This extended resting time between cultivations is known as **fallowing**.

Shutterstock.com/Chukov

Figure 17.4 A chisel plough might be used to prepare the soil

Ploughing and cultivating the soil have the effects of:

- reducing the number of other plants (weeds) that can compete with the crop plants once they are planted
- aerating the soil and thus stimulating micro-organism activity, hastening the release of plant mineral nutrients from organic material in the soil and making them available for the crop
- reducing loss of water by transpiration through other plants, so that more water is retained in the soil for the crop.

A large proportion of farmers now use **minimum tillage** (see page 231) or reduced tillage technology, which has substituted the use of herbicides for fallowing. Advantages are that the soil is not exposed to the agents of erosion (wind and water) for any length of time and that the level of organic matter in the soil is increased. The cost is approximately the same as that of traditional fallowing, because the cost of herbicides is approximately equal to the cost of the fuel used for cultivation and ploughing. The development of sensitive spray equipment that applies herbicide only to weeds when they are detected will save a lot of the cost of herbicides, because the amount used will be reduced by approximately 90% (Fig. 17.5).

7 Compare traditional fallowing and minimum tillage as methods of preparing the soil for planting a crop.

Shutterstock.com/Federico Rostagno

Figure 17.5 Spraying with herbicide has replaced ploughing and cultivation in minimum and reduced tillage systems

Site preparation might well involve other techniques, such as laser levelling the land so that an efficient irrigation system can be installed, or constructing contour banks and grassed waterways so that the land is protected from erosion while the crop is being grown. It might also involve clearing the land of some of the timber and scrub that are present.

Note that not all crops are grown in soil. Some horticultural crops, such as flowers and vegetables, are grown using **hydroponics**, with all their mineral nutrient requirements supplied dissolved in the water supply.

ISBN 9780170265560

Planting the crop

Most crops are grown from seed. Planting them involves placing the seeds in the soil so that they can germinate and become established as crop plants. For a crop to grow and produce well, three things must be considered at planting:

1 the depth at which the seed is to be sown
2 the desired density of plants in the growing crop to achieve maximum yield
3 the timing of planting.

Depth

The aim is to place the seed in the soil deep enough for there to be sufficient moisture for germination, yet close enough to the surface for the seedling to be able to establish itself before running out of food reserves. Seed planted too close to the surface is in danger of having insufficient moisture because the soil there dries out quickly. Seed planted too deep is at risk of running out of seed food reserves before the germinating seedling reaches the surface and can produce its own food through photosynthesis.

Density

As discussed earlier in this chapter, the density in a crop has significant effects on the yield. Whether the crop is being grown for its vegetative or reproductive yield is critical to the choice of density when planting the crop.

To achieve the correct sowing depth and the desirable plant density in a crop, sophisticated precision-planting machines have been developed (Fig. 17.6). These machines accurately place the seed at predetermined intervals in the row and at a specified depth and are used for crops such a cotton and soybean. Combined seed and fertiliser drills are still widely used for sowing crops such as wheat. These machines are not quite as precise with spacing and planting depth.

Timing

The timing of planting is important. To a large extent it depends on environmental influences. The aim is to plant the crop at such a time that the climate best suits it while it is growing. For example, a wheat crop planted too early in autumn is more at risk of being damaged by frosts at flowering in early spring than one planted later (Fig. 17.7). However, wheat planted too late in winter has little time for vegetative growth before flowering begins. This reduces yield.

8 Why is it important to plant seed at the correct depth?
9 Why is the density of plants important for a crop?
10 How can the timing of planting influence the production of a crop?

Corbis/Richard Hamilton Smith

Figure 17.6 A precision planter gives accurate sowing depth and precise placement of seeds in the row, which result in a more even crop

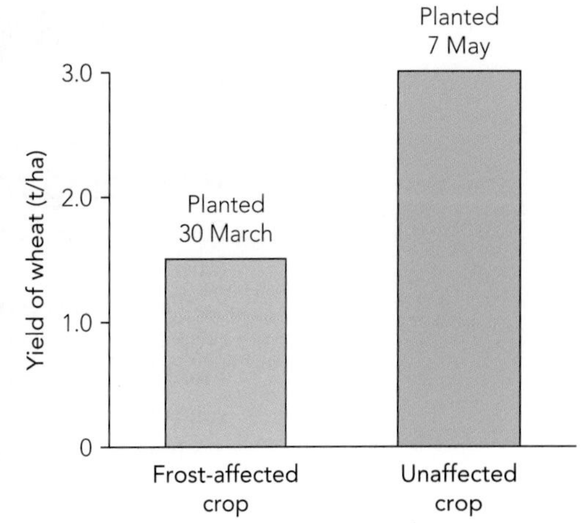

Figure 17.7 Frost damage at flowering reduces yield and results from planting too early

ISBN 9780170265560

Caring for the growing crop

Caring for the growing crop involves providing nutrients and moisture and successfully controlling pests, diseases and weeds.

Providing nutrients

Nearly all Australian soils are deficient in nitrogen and phosphorus, and other plant nutrients as well are deficient in particular areas. Therefore these elements must be provided for the growing crop. The use of fertilisers such as superphosphate and sulfate of ammonia is one common way of overcoming these deficiencies. Fertilisers are usually applied at the time of planting, using the same machine that sows the seed.

Rhizobium bacteria, found in nodules on the roots of legume plants, are able to fix nitrogen into forms that can be used by plants, and this ability has been used extensively to boost nitrogen supplies to crops. Legume seed is inoculated with *Rhizobium*, so that the resulting legume crop achieves its own supply of nitrogen. Pastures containing legumes and legume crops (Fig. 17.8) are used in rotation to build up nitrogen levels in the soil, for use by subsequent crops. Using legumes to supply nitrogen is far cheaper than using nitrogen fertilisers.

11 Describe two methods of providing nitrogen for a crop.

Figure 17.8 Legumes fix nitrogen and provide a cheap source of nitrogen for crops

Providing moisture

For most crops farmers rely on natural rainfall to supply all the moisture for the growing crop.

Fallowing the ground (cultivating the soil and leaving it bare) for a long period before the crop is grown allows the level of moisture in the soil to build up, because there is little or no loss of water through transpiration from plants during this time. This reserve of moisture in the soil is then available to the crop as it grows.

Irrigation can supply the growing crop with adequate water (Fig. 17.9, page 251). There are many methods of irrigating crops, and the method adopted should be the one that achieves the desired results cheaply. Efficient use of water for irrigation is one of the major issues faced by agriculture as demands for water outstrip supply. (Water management and irrigation are explained in more detail in Chapter 15.)

Other techniques to improve the availability of water to the crop include increasing the organic matter content of the soil, which increases the ability of the soil to store water, and using **mulches**, which decreases the loss of water from the soil by evaporation.

12 List the methods that can be used to increase the availability of water to the growing crop.

ISBN 9780170265560

Shutterstock.com/prudkov

Figure 17.9 Irrigation ensures adequate supplies of water for the growing crop

Disease and pest control

Diseases and pests have the potential to reduce the yield of a crop drastically; therefore it is prudent to carry out measures to control them. The idea these days is to use a range of control measures as part of an integrated pest management (IPM) program against a particular pest or disease. As far as possible, chemicals should not be seen as the first or only approach to controlling diseases and pests. The use of biological and cultural control measures, such as growing resistant varieties and carrying out good hygiene, should be encouraged wherever possible. Using seed certified to be disease free is a good start.

13 What is an integrated pest management (IPM) program for the control of pests and diseases? Provide an example.

Weed control

Weed control is essential to give good yields (see Table 17.2). Weeds compete for environmental resources, including space, water, plant nutrients, sunlight and carbon dioxide. They can harbour pests and disease and modify the environment of nearby crop plants. They can also have adverse effects on the growth of crop plants through the chemicals that they produce and release into their immediate environment. This is known as **allelopathy**.

Control measures include fallowing to reduce the weed population before planting and, in some crops, cultivating between the rows after the crop has emerged and become established. Herbicides are commonly used and have proved to be very effective. Pre-emergent herbicides are applied prior to the crop's emerging from the soil; they kill weeds as they germinate. Post-emergent herbicides are applied after the crop plants have become established. These herbicides have to be **selective**; they must kill the weeds but have little or no effect on the desired crop plants.

14 List the four ways weeds can adversely affect a crop.

Table 17.2 Results of a student experiment to demonstrate the adverse effect of weeds on the yield of potatoes. The experiment was replicated four times.

	Yield of potatoes (kg/10 m²)	
Replicate	**Weeded**	**Unweeded**
1	51.1	11.7
2	55.7	16.9
3	29.9	14.7
4	35.6	30.1
Average	43.0	18.6

ISBN 9780170265560

Harvesting the crop

The timing of **harvesting** is particularly important. Taking the crop off at just the right time is vital to the quality and quantity of the product. Harvesting too early can give products that do not have enough of a particular constituent, such as sugar or protein, or the total quantity might not be as much as it should be. For example, wine grapes harvested too early will not have sufficient sugar content to make good wine. Harvesting too late might lead to products that are not of high quality because their constituents have been changed or lost in some way. For example, the protein content of wheat is lowered when mature grain is exposed to rain before harvesting.

Machinery plays a big role in modern day harvesting (Fig. 17.10). The efficiency of these machines means that crops can be harvested at just the right time with far less labour than was previously required.

Alamy/Rick Dalton - Ag

Figure 17.10 A modern harvesting machine harvesting wheat

Once the crop is harvested, it has to be prepared for market. For some crops, such as fruit, this means packing into boxes and cartons for transport to the markets. For other crops, such as wheat, this just involves storing the grain in silos or bins until it can be transported to the grain storage at the railway.

15 Why is the timing of harvesting important?

Transporting the product to market

Trucks are the means by which crop products are carried off the farm. Increasing attention is being given to the careful handling and transport of crop products, so that they arrive at the processor or consumer in as good a condition as possible. For example, refrigerated trucks are now used to transport fresh fruit and vegetables to the markets (Fig. 17.11). The produce might go directly to the markets, as with fruit and vegetables, or it might go to some other place for further processing before it is sold. Here are some examples: cotton is taken to the cotton gin, where the seed is separated from the lint; wheat is delivered to railway silos, from which it is further transported to ports and loaded onto ships for export; wine grapes go to a winery for processing into wine; and soybean grain is taken to a mill, where its oil is extracted.

16 Why is transport important to a plant production system?

iStockphoto.com/volkansengor

Figure 17.11 A refrigerated truck taking fresh fruit to market

ISBN 9780170265560

Management

Management plays a large part in the successful running of a plant production system. The manager exercises influence at all stages of the system. Decisions are made as to what crop and which variety will be grown (in effect, choosing the genotype), what inputs will be used and at what rates, when activities such as sowing the crop and applying fertilisers and herbicides will be done, when harvesting will take place and how the crop will be sold, to mention just a few. In dryland farming, planting time to a great extent is determined by the incidence of rainfall. Crops cannot be planted until after rainfall.

The level of skill and expertise that the manager brings can be seen as human input into the system. The manager receives a large amount of feedback information from observing the processes going on in the system and the products being produced. Information also comes from outside the plant production system in the form of information about new technologies, such as new varieties or new harvesting machinery, financial information, such as the product price, interest rates and weather forecasts for the coming cropping season. The manager uses all this information to assess and make adjustments to the plant production system.

The pattern of interactions is illustrated in Figure 17.12.

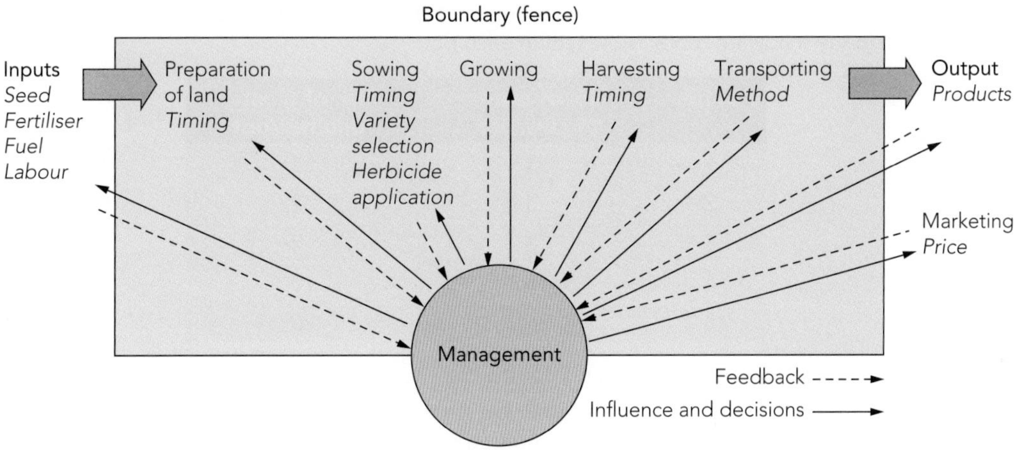

Figure 17.12 A plant production system showing the role of the manager

Chapter review

Mandatory plant density practical for NSW HSC Agriculture course

Students are to perform a first-hand investigation to determine the effects of planting density on plant growth or plant yield. **Things to do** Question 1 is an example of a plant density trial.

Things to do

Experiment 17.1 Effects of plant density on plant growth and yield

Aim: To observe the effects of plant density on plant growth and yield

Method

1 Set up five large pots (approximately 30 cm in diameter) and label them 1 to 5. In the pots, sow the seeds of a cereal grain, such as wheat, using the sowing rates shown in the table below. Ensure the seeds are spaced evenly in the pot. When the seedlings are established, remove any excess seedlings (thin them out) so that the required number of plants, as shown in the following table, is present in each pot.

Pot no.	No. of seeds to sow/pot	No. of plants required/pot
1	3	1
2	5	2
3	8	5
4	15	10
5	25	20

2 Make the following observations and use the data collected to carry out the calculations.

- a Determine the number of tillers (new side stems; see Fig. 18.3 on page 259) in each pot. This is best done towards the end of the vegetative phase of growth. From this data calculate the average number of tillers per plant at each plant density.
- b Determine the height of each plant in each pot. Measure to the top of the seed head. This is best done at maturity. From this data calculate the average height at each plant density.
- c Determine the number of seed heads in each pot. This is best done at maturity. From this data calculate the average number of seed heads per plant at each plant density.
- d Determine the weight of grain produced by each pot. This is best done at maturity. From this data calculate the average weight of grain produced per plant at each plant density.

Questions

1 Calculate and compare means and standard deviations for each plant density (see Chapter 2) for number of tillers, height, seeds per head or weight of grain produced per plant at each density.

2 Which of these measurements is most important for the farmer ?

3 Use the data to construct the following graphs. In each case use the horizontal axis for plant density (the number of plants per pot).

- a Graph the number of tillers per pot versus plant density. On the same axes plot the number of tillers per plant against plant density.
- b Graph the average plant height versus plant density.
- c Graph the number of seed heads per pot versus plant density. On the same axes plot the number of seed heads per plant versus plant density.
- d Graph the weight of grain per pot versus plant density. On the same axes plot the weight of grain per plant versus plant density.

ISBN 9780170265560

4 Use the graphs to answer the following questions.

  a What conclusion can be drawn about the relationship of plant density to vegetative yield?

  b What conclusion can be drawn about the relationship of plant density to reproductive yield?

  c What could be done to the design of this experiment to improve the reliability of the results? (Refer to Chapter 2)

The experiments in Questions 2–4 explore other factors affecting plant yield.

2 Design and carry out an experiment to explore what effect weeds have on the yield of a crop such as potatoes. Try to employ the principles of good experimental design. Describe the meaning of each principle and outline why each is used.

3 Test a weed plant for allelopathic effects on crop or pasture plants, by carrying out germination trials with a 'grass' plant such as wheat, and with a broad-leaf plant such as clover, lucerne, radish or carrot. Place the leaves from the weed plant you wish to test for allelopathic effects in a Petri dish, cover them with filter paper and then put the seeds of the plant you wish to grow on top of the filter paper. Add sufficient water to keep the seeds moist but not prevent air reaching them. For a control you will have to set up Petri dishes that do not have any of the weed leaves in them.

  After germination, record the percentage germination and measure the length of the primary root or the **coleoptile** of the germinating seedlings (see Fig. 18.16, page 268).

  Does the weed you tested have any allelopathic effects?

4 Grow a vegetable crop suited to your locality, and keep careful records of all the inputs you use, the processes you carry out and the output you obtain. How could you improve your system?

Things to find out

1 For a crop that is commonly grown in your area, find out:

  a what are considered to be the most troublesome weeds

  b why they are considered to be weeds

  c what methods are used to control them

  d what is a major pest and what is the recommended approach to controlling it

  e what is a major disease and what is the recommended approach to controlling it.

2 Find out the temperature, rainfall and day length requirements of a crop plant commonly grown in your region. Can these requirements be met by the natural climate of your region? How have the conditions been modified to ensure the crop produces at a satisfactory level?

3 Identify a range of significant plants grown in your region.

4 Trace all the steps and processes that the product of a crop goes through before it actually reaches the consumer. For example, trace the steps between wheat leaving the farm, and bread being bought by the consumer.

5 If fertilisers are used when growing a crop in your area, which fertilisers are used, at what rates are they applied and what mineral nutrients are they supplying?

6 Research the growth requirements of a crop grown in a region that has an entirely different climate from the one you are in.

Extended response questions

1 Explain why the yield of a crop is determined by the interaction of the crop's genotype and environment.

2 For a named plant enterprise:

  a briefly explain the environmental factors that limit crop plant production

  b discuss strategies that farmers might adopt to overcome these limitations.

3 Competition occurs between plants for environmental resources.

  a Explain how crop yield is affected by plant competition.

  b Describe how farmers manage plant competition through plant density and weed control strategies.

CHAPTER 18

Outcomes

Students will learn about the following topics.

1 The basic morphology and function of leaves, stems, roots, flowers, seeds and fruits.
2 The process of photosynthesis.

Students will be able to demonstrate their learning by carrying out these actions.

1 List the major parts of a plant.
2 Distinguish between monocotyledonous and dicotyledonous plants.
3 Describe the functions of the stem.
4 Describe the external and internal structure of monocot and dicot stems.
5 Describe the functions of the roots.
6 Describe the external and internal structure of monocot and dicot roots.
7 Distinguish between tap and fibrous root systems.
8 Distinguish between the seminal and nodal root systems of monocots.
9 Describe the functions of the leaves.
10 Describe the external and internal structure of monocot and dicot leaves.
11 Describe the functions of flowers, fruits and seeds.
12 Describe the structure of monocot and dicot flowers.
13 Explain how pollination and fertilisation take place.
14 Describe the different types of fruit that plants produce.
15 Describe the structure of monocot and dicot seeds.
16 Explain how the light and dark phases of photosynthesis take place.
17 Distinguish between C3 and C4 plants.
18 Describe the environmental (abiotic) factors that affect growth, development and production in plants.
19 Explain how farmers can manage plant production systems to overcome environmental constraints.
20 Perform a first-hand investigation to determine the effect of light on plant growth.

ISBN 9780170265560

PLANT STRUCTURE AND FUNCTION

Words to know

adventitious roots roots that arise from stems, usually at nodes

annual plant a plant that completes its lifecycle within 1 year

anther or pollen sac the terminal portion of the stamen, containing pollen in sac-like structures

cambium a layer of meristematic tissue found in stems and roots between xylem and phloem

chloroplast a discrete membrane-bound part, within a cell, that contains chlorophyll and is capable of photosynthesis

coleorhiza a protective cone of tissue over the radicle in the monocot embryo

cross-pollination pollination between flowers of different plants of the same species; in fruit trees, of one variety with another

cuticle a waxy layer covering the epidermis of plants, particularly leaves and herbaceous stems

dicotyledon a plant that has two seed leaves or cotyledons in its seed

embryo the developing organism in early pregnancy; in plants, the small immature plant found in the seed

endodermis a layer of cells in the root between the cortex and the vascular tissue that controls the movement of water into the vascular tissue

endosperm the food reserve in the monocot seed

epidermis a layer of cells covering the outside of young roots, stems and leaves

fibrous roots roots of monocot plants, all about the same size

herbaceous plants or plant parts that are fleshy as opposed to woody

inflorescence the flowers of a plant borne in a particular way on one stalk (the arrangement of flowers on a plant)

meristem or meristematic tissue tissue consisting of cells capable of cell division and thus of producing new cells

monocotyledon a plant that has only one seed leaf or cotyledon in its seed

phloem living conductive tissue consisting of sieve tubes and companion cells, through which the products of photosynthesis translocate throughout the plant

photosynthesis the process where carbon dioxide and water are combined using light energy to form high-energy sugar compounds and oxygen

pollination the transfer of pollen grains from the anther to the stigma

root cap protective tissue that surrounds the tip of the root, protecting it as it grows through the soil

root hair a finger-like projection of the epidermal cells of the root, responsible for absorbing water and mineral nutrients from the soil

scutellum the cotyledon in monocot seeds, which absorbs food from the endosperm during germination

self-pollination movement of pollen from the anther to the stigma of the same flower

stamen the male part of the flower, consisting of the filament (stalk) and anther (pollen sac)

stigma the receptive tip of the female part of the plant

stomata (singular: **stoma**) small pores or holes in the epidermis of a leaf that allow the diffusion of oxygen and water vapour out of the leaf and the diffusion of carbon dioxide into the leaf; most often occur in the lower epidermis but are found in the upper epidermis of some species

style tube connecting stigma with ovary

tap root the primary root of a dicot plant, which is larger than the others and grows straight down

testa the coat of a seed

tiller a secondary stem in a monocot plant

vascular bundle a bundle of tissue in roots and stems, consisting of phloem and xylem

xylem dead conductive tissue consisting of tracheids and vessels, through which water and dissolved mineral salts move from the roots throughout the plant

ISBN 9780170265560

Introduction

Broadly speaking, a plant can be said to consist of these major parts: stems, leaves, roots and reproductive parts, which include flowers, seeds and fruits, as shown in Figure 18.1. It is important from the outset to relate the structure of the various parts of a plant to the functions that each part performs. When studying plant structure we should ask continually, 'How does the structure of this part of the plant allow it to carry out its function in the plant?' Plants can be divided into two groups:

1 monocotyledons
2 dicotyledons.

Monocotyledons have only one seed leaf or cotyledon in their seeds. They include a great number of agriculturally important plants, such as the cereal grain crops (wheat, rice, maize, oats, sorghum and barley), sugar cane, and all the grasses that are vital to grazing animals. **Dicotyledons** have two seed leaves or cotyledons in their seeds. They also include many agriculturally important plants, such as the legumes, lucerne and soybean, cotton, canola, and fruit crops, such as oranges and peaches.

1 List the major parts of a plant.

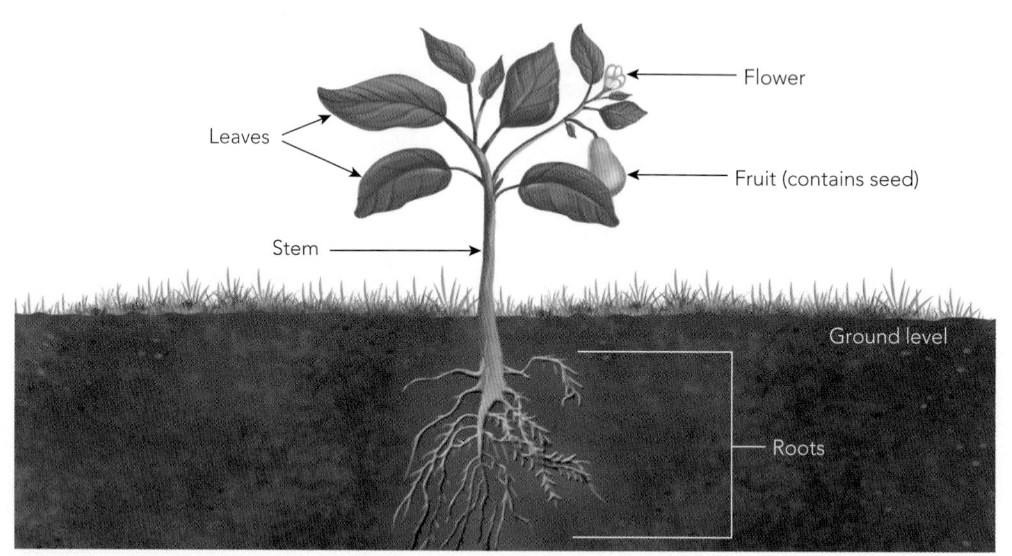

Figure 18.1 Major parts of a plant

The stem

The stem forms the framework of the above-ground part of the plant. It has several functions:

1 It holds the leaves up in the air where they can intercept sunlight, absorb carbon dioxide, lose oxygen and transpire water.
2 It moves water and dissolved plant mineral nutrients from the root system to the leaves.
3 It translocates food material (the products of photosynthesis) from the leaves to the root system.
4 It stores excess food material.
5 It carries out photosynthesis in some cases.

External structure

Dicotyledons

A dicotyledonous plant usually has a main stem with branches or laterals leading off it. The parts of the stem are shown in Figure 18.2. At the end of each stem is a bud known as the terminal bud. It is from here that growth in length occurs. The terminal bud contains **meristematic tissue**, which is capable of cell division, thus producing new cells and growth.

ISBN 9780170265560

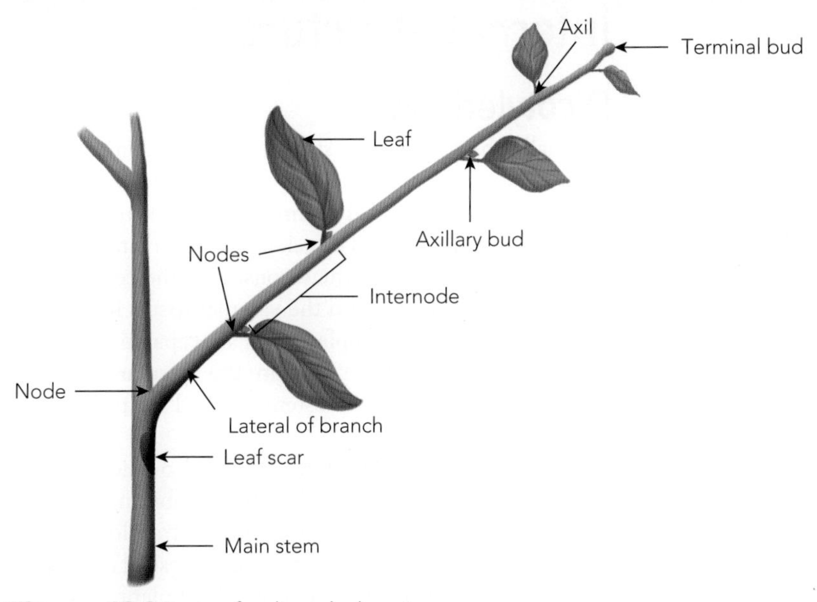

Figure. 18.2 Parts of a dicotyledon stem

Careful examination of a bud will show that it contains undeveloped leaves and stem. At intervals down the stem are thickenings called nodes. At each node a leaf is attached to the stem (or was attached, leaving a leaf scar). In the angle between the leaf and the stem (the axil) a bud is found. The buds at the nodes are called axillary buds and will grow and develop into branches or flowers. The parts of the stem between nodes are the internodes.

Monocotyledons

The monocotyledonous plant has a slightly different stem, as shown in Figure 18.3. There is still the terminal bud, which produces new leaves, and axillary buds, which produce new stems known as **tillers**, but these buds do not produce elongation of the stem. For much of the life of the monocotyledon the stem remains very short, approximately 17.5 cm high, and cannot be seen for the leaves. Elongation of the stem occurs when the terminal bud starts to produce reproductive parts (an **inflorescence**) and is brought about by the activity of meristems situated just above each node producing new tissue.

2 List the functions of stems.
3 What are the similarities and differences between the external structures of monocot and dicot stems?

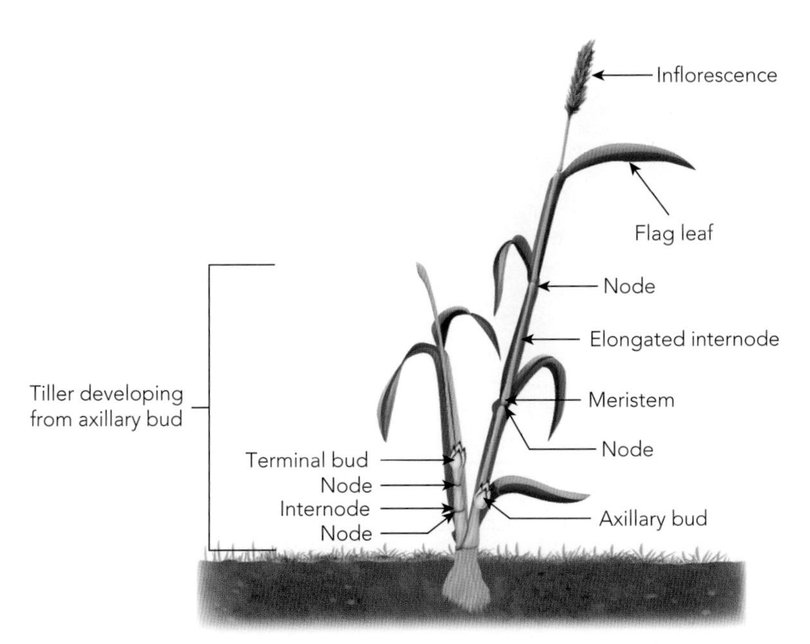

Figure 18.3 Parts of a moncotyledon stem. Half of each leaf has been removed to expose the stem structure

ISBN 9780170265560

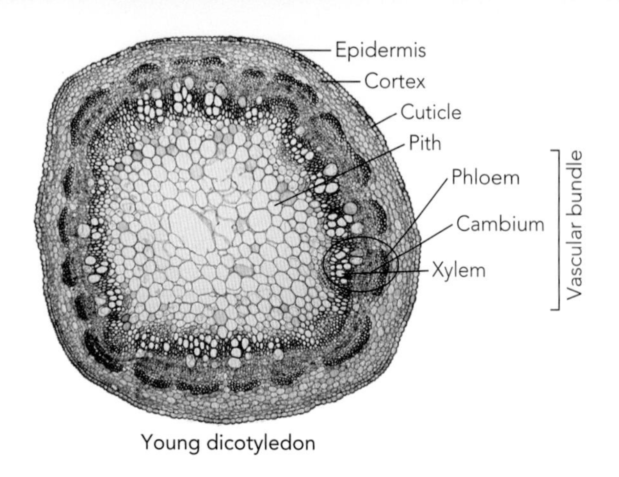

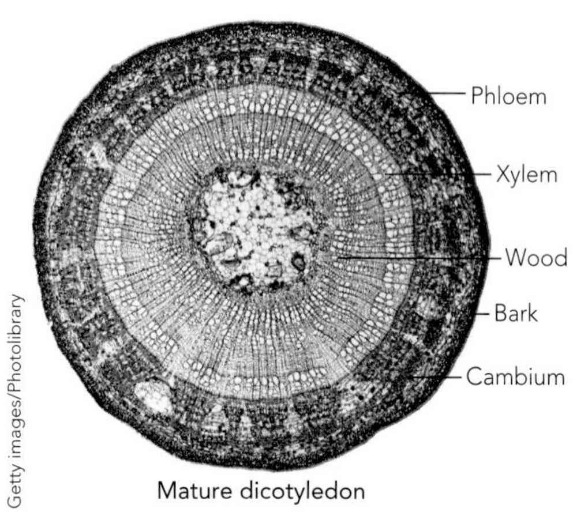

Getty images/Photolibrary

Figure 18.4 Cross-sections of dicotyledon stems

Internal structure

Dicotyledons

Internally, the dicotyledon stem has the structure shown in cross-section in Figure 18.4 and described as follows. The outer layer of cells, called the **epidermis**, secretes a **cuticle**, which helps to prevent water loss and attack by micro-organisms. Inside the epidermis is the cortex, which consists of the cells filling the space between the epidermis and the vascular tissue. In some plants these cells contain **chloroplasts** and are capable of photosynthesis. Towards the centre of the stem are the **vascular bundles**, arranged in a circle. The centre of the stem is called the pith and is made up of thin-walled cells.

Each vascular bundle, going from the outside in, has phloem tissue, meristematic tissue called the cambium, and xylem tissue. The **phloem** tissue is responsible for the translocation of the sugars and other products of photosynthesis. The **xylem** tissue allows movement of water and dissolved plant mineral nutrients. The **cambium** makes new phloem and xylem and causes the stem to grow in diameter.

Herbaceous plants and **annual plants** may not have the cambium. **Perennial** (see page 143) plants do, and the activity of the cambium produces phloem on the outside and xylem on the inside. The vascular bundles join up and form a cylinder, and in mature woody stems all the tissues outside the vascular cambium form the bark and everything inside forms the wood. This explains why ringbarking kills trees: it cuts the phloem and therefore the supply of food to the roots.

Monocotyledons

Monocotyledon plants have an epidermis, and their vascular bundles are situated near the surface of the stem. The vascular bundles are similar to those in dicotyledons except that they do not have a cambium. A sheath of thickened fibres surrounds each vascular bundle and provides rigidity to the stem. Figure 18.5 shows a cross-section of a monocotyledon stem.

4 Draw diagrams that show the differences in internal structure between monocot and dicot stems.

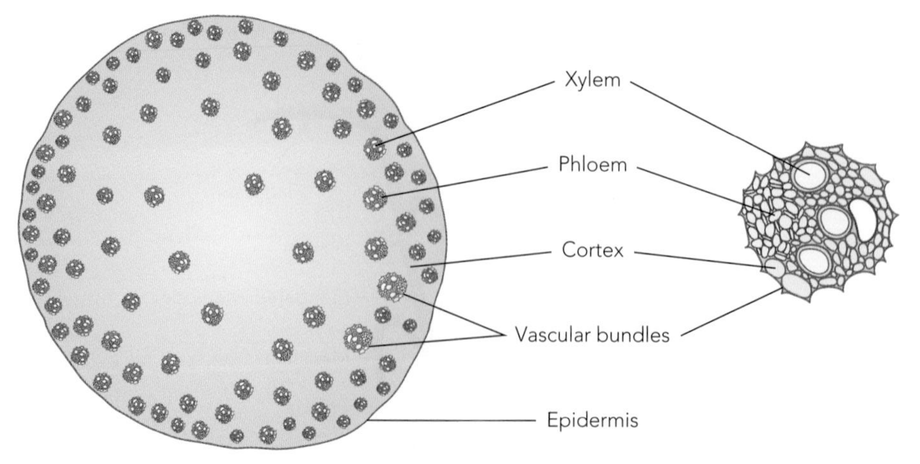

Figure 18.5 A cross-section of a monocotyledon stem. In the stems of the majority of monocotyledons many vascular bundles are present and lie scattered throughout the tissue

ISBN 9780170265560

Roots

The roots form the underground parts of the plant, and their extent is not generally appreciated. There is as much weight of root material as there is stem and leaf material. Roots have these functions:

1 absorbing water from the soil
2 absorbing mineral plant nutrients from the soil
3 anchoring the plant in the soil
4 storing the products of photosynthesis.

External structure

Dicotyledons

In dicotyledons the root system is known as a tap root system and consists of a major root that grows straight down, called the **tap root** (Fig. 18.6). This can penetrate very deeply into the soil; for example, the tap root of lucerne goes down several metres.

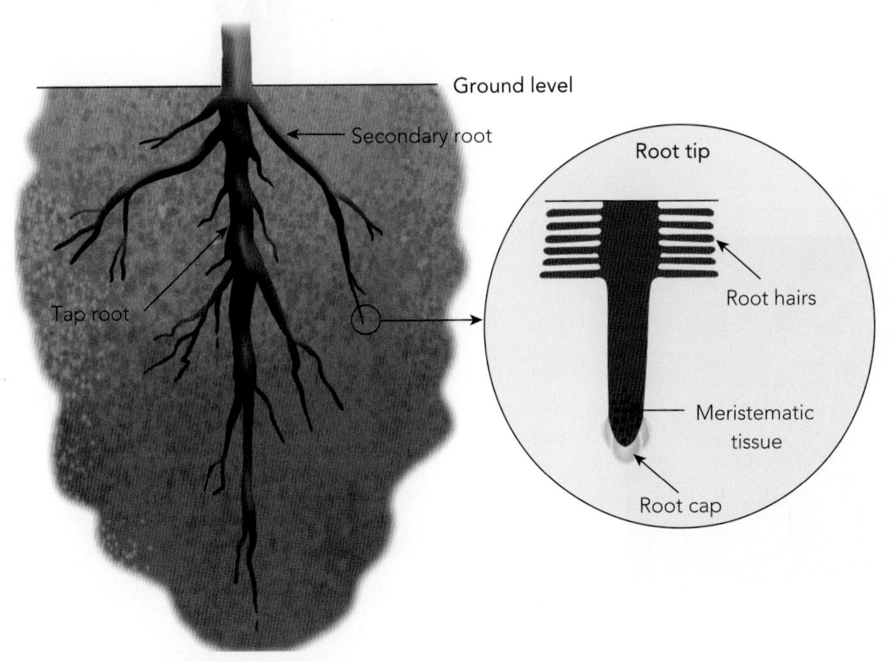

Figure 18.6 The dicotyledon tap root system

Thinner secondary roots branch from the tap root, and from these further branching takes place, so that the soil is penetrated in all directions by roots.

The end of each root is covered by a **root cap**, which is several cells thick and protects the root as it is pushed through the soil. Just behind the root cap is meristematic tissue, which is responsible for growth in length. A little further up the root is a zone where the epidermal cells have developed long projections called **root hairs**. These root hairs absorb water and mineral plant nutrients from the soil. They greatly increase the surface area of the roots, allowing them to absorb the mineral plant nutrients and large quantities of water that the plant needs.

Monocotyledons

Monocotyledons do not have a definite tap root; all the roots are approximately the same diameter. There are many of these, and the root system that they form is known as a **fibrous root** system (Fig. 18.7). These fibrous roots do not go down as far as tap roots but extensively penetrate the surface layers of the soil. It has been estimated that the roots on a single wheat plant, put end to end, would have a length of approximately 70 km and an absorbing area of 400 m^2.

Careful observation of the growth of monocotyledons, such as wheat or maize, after germination will show that the first roots arise from the **embryo** in the seed. These roots form the seminal or seed root system. As further development takes place, roots grow from the nodes of the stem and form the bulk of the root system of the plant. These are known as **adventitious roots** because they arise from the stem and are collectively called the nodal root system. These root systems are shown in Figure 18.8.

The ends of the roots of monocotyledons are the same as those of dicotyledons, having a root cap, a meristem and a root hair zone.

5 List the functions of the roots of a plant.

6 Distinguish between:
   a nodal and seminal roots
   b a tap root system and a fibrous root system.

Figure 18.7 The monocotyledon fibrous root system

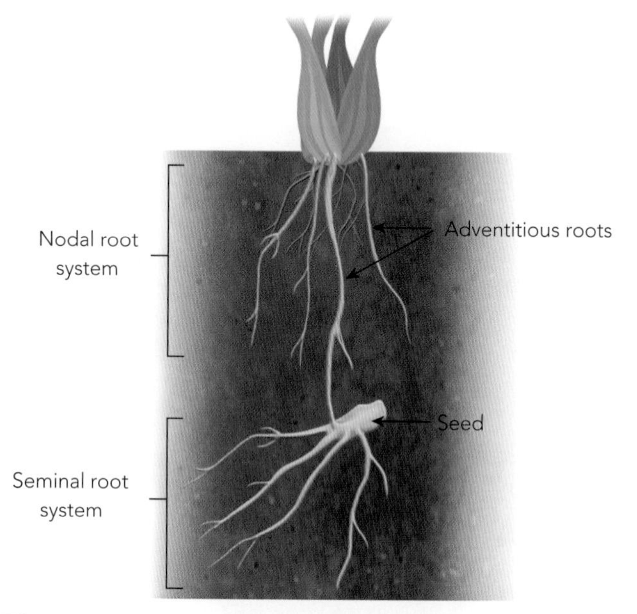

Figure 18.8 The seminal and nodal root systems of the monocotyledon

Internal structure

The internal structure of roots is composed of several different tissues, as shown in Figure 18.9. The epidermis is the outer layer of cells. It does not have any cuticle because it functions as an absorbing surface. The surface area of the epidermis is enlarged by the presence of root hairs, which are finger-like outgrowths of epidermal cells. Inside the epidermis is the **cortex**, composed of thin-walled cells. The cortex is used for storage and has become greatly enlarged in some species harvested for food, such as the carrot. Inside the cortex is a ring of cells called the **endodermis**, which controls the movement of absorbed materials into the central vascular cylinder or stele.

The tissue of the vascular cylinder is made up of xylem and phloem. The xylem consists of relatively large xylem vessels which conduct water and dissolved plant mineral nutrients to the stem and leaves. The phloem consists of sieve tubes and companion cells and carries the products of photosynthesis from the leaves and stem to the roots.

ISBN 9780170265560

In dicotyledons a **cambium** develops between the xylem and phloem. New xylem and phloem are produced, and the root grows in diameter. Mature roots (Fig. 18.10) are covered with bark and are no longer able to absorb water and plant mineral nutrients. No such growth in diameter occurs in monocotyledons, which just produce more roots instead.

7 Explain the significance of the root hairs that occur just behind the growing root tip.

8 List the tissues found in a root, and write the function of each.

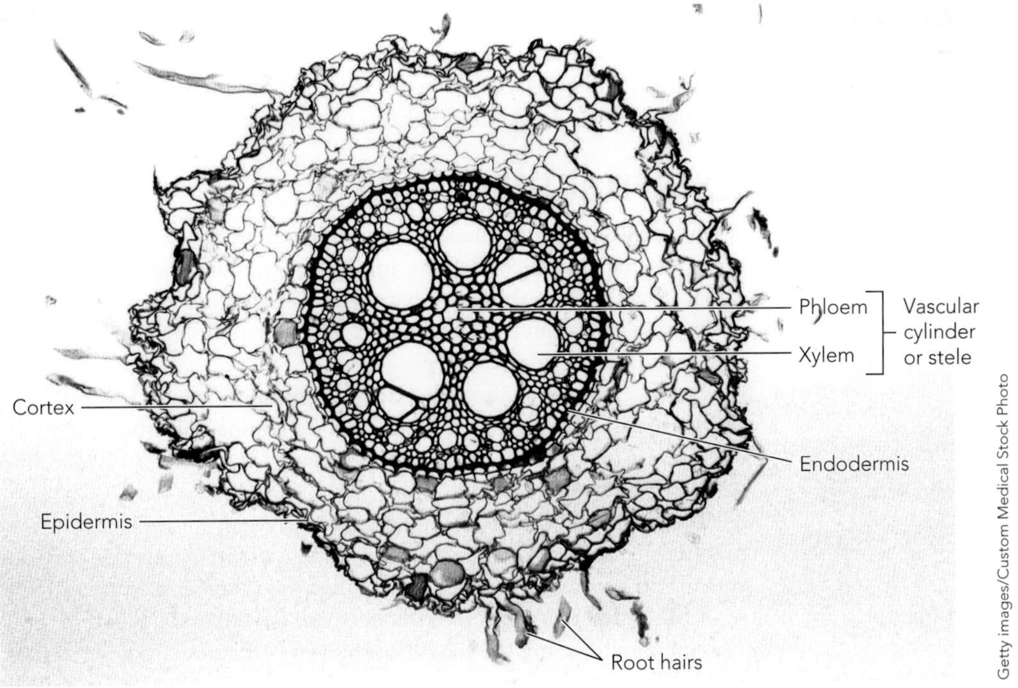

Phloem
Xylem
Vascular cylinder or stele
Cortex
Endodermis
Epidermis
Root hairs
Getty images/Custom Medical Stock Photo

Figure 18.9 A cross-section of a young root showing the internal tissues

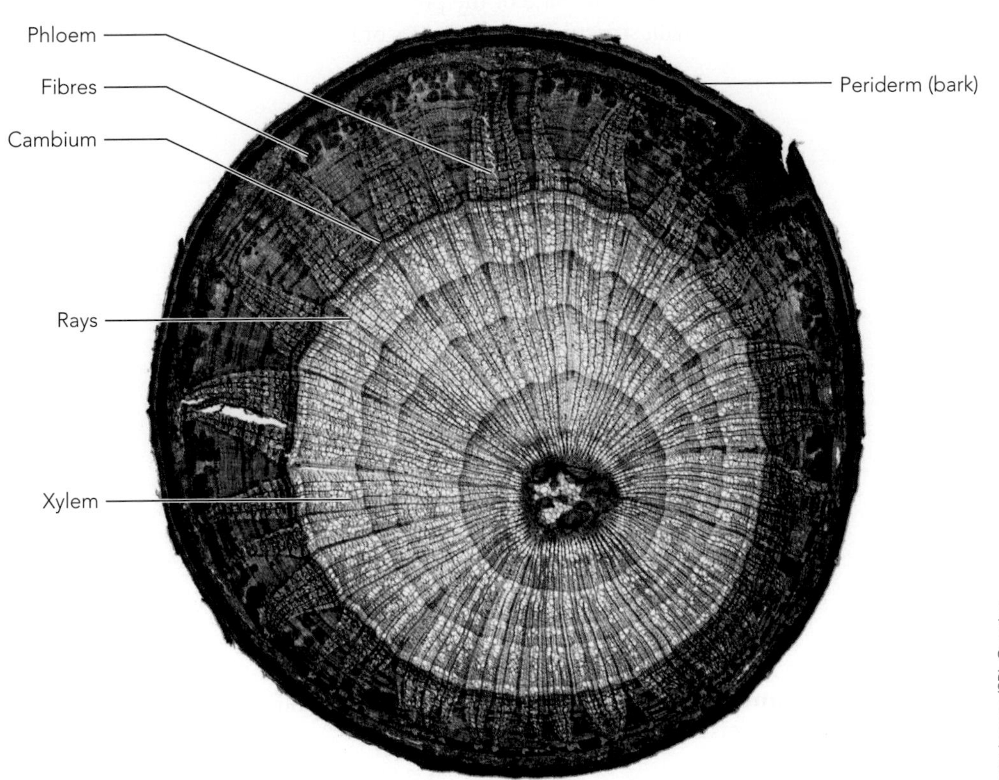

Phloem
Fibres
Cambium
Rays
Xylem
Periderm (bark)
Getty images/SPL Creative

Figure 18.10 Internal tissues of a mature root of a dicotyledon

ISBN 9780170265560

Leaves

Leaves are the powerhouse of the whole plant, for here the energy coming from the Sun as light is converted into chemical energy that can then be used by the plant. The functions of leaves are:

1 to carry out photosynthesis
2 to carry out transpiration
3 to store the products of photosynthesis, in some cases.

The process of photosynthesis will be discussed in detail later. Suffice to say that carbon dioxide and water are combined using energy from light to form sugars and oxygen.

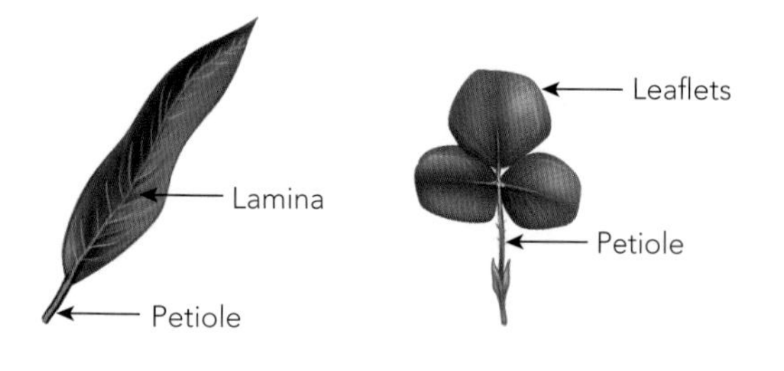

Figure 18.11 A simple leaf (peach) and a compound leaf (white clover)

External structure

A leaf consists of a leaf blade (or lamina) and a stalk (or petiole), as is the case in most dicotyledons. In monocotyledons the petiole is replaced with a sheathing leaf base. The leaf always joins a stem at a node, and there is an axillary bud between the petiole or sheathing leaf base and the stem.

Leaves can be classified into two large groups: simple leaves and compound leaves. In simple leaves the lamina consists of one continuous piece, while in compound leaves the lamina is divided into smaller leaflets (Fig. 18.11).

The veins in leaves contain xylem and phloem for transporting materials to and from all parts of the leaf. In dicotyledons the veins form a network pattern attached to the main vein or midrib running down the centre of the leaf and are described as network or reticulate venation. In monocotyledons the veins run parallel along the length of the leaf blade and are described as parallel venation.

The ligule in the monocot is a membranous piece of tissue that prevents dirt and moisture from running down between the stem and the sheathing leaf base. Auricles are outgrowths that occur at the junction of the sheath and blade of leaves of some grasses. Stipules are leaf-like outgrowths of the base of the petiole.

Figure 18.12 shows the main structures of the leaves of both monocots and dicots.

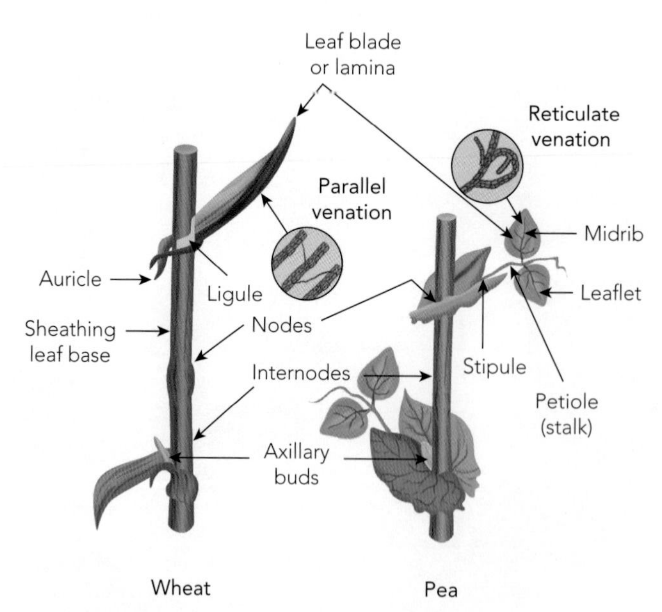

Figure 18.12 Main structures of the leaves of both monocots and dicots

ISBN 9780170265560

Internal structure

The internal structure of the leaf is best described by looking at a section through the leaf blade, as shown in Figure 18.13. At the top is a waxy coating called the cuticle, the function of which is to prevent loss of water and entry of pathogenic organisms, such as bacteria and fungi. The cuticle is secreted by the epidermis, which consists of a single layer of cells. The upper epidermal cells contain no chloroplasts and therefore do not photosynthesise, but they allow light to pass through to the chloroplast-containing cells below. Under the upper epidermis are the palisade mesophyll cells. These elongated cells are packed fairly closely together in one or two layers with their axes perpendicular to the surface of the leaf. They contain chloroplasts and carry out most of the photosynthesis in the plant. Under the palisade mesophyll is the spongy mesophyll, which consists of loosely packed, irregularly shaped cells containing chloroplasts with interconnecting air spaces between them. The air spaces allow for the movement of carbon dioxide, oxygen and water vapour. The bottom layer of cells is the lower epidermis and is similar to the upper epidermis, having no chloroplasts. Below this is another cuticle.

In the middle of the leaf the vascular bundles or veins are found. They have xylem tissue on top and phloem tissue underneath and are surrounded by a layer of cells forming the vein sheath. The veins connect to the vascular tissue of the stem through the petiole or sheathing leaf base.

Small pores or holes called **stomata** (singular: stoma) are found in the epidermis. They most often occur in the lower epidermis but are found in the upper epidermis of some species. A stoma is formed by two kidney-shaped guard cells, which contain chloroplasts. The guard cells open and close the stoma. An open stoma allows the diffusion of oxygen and water vapour out of the leaf and the diffusion of carbon dioxide into the leaf. A closed stoma prevents loss of water by transpiration. The stomata open into the intercellular air spaces between the spongy mesophyll cells.

Leaves are flat and thin so that they can intercept sunlight and allow gases to diffuse into and out of the leaf easily: carbon dioxide from the air to the photosynthetic mesophyll cells inside, and oxygen and water vapour from the mesophyll to the atmosphere outside.

9 List the functions of leaves.
10 Draw up a table comparing dicot and monocot leaves.
11 List the parts of a section through a leaf, starting at the top with the cuticle above the upper epidermis. Write the function of each part.

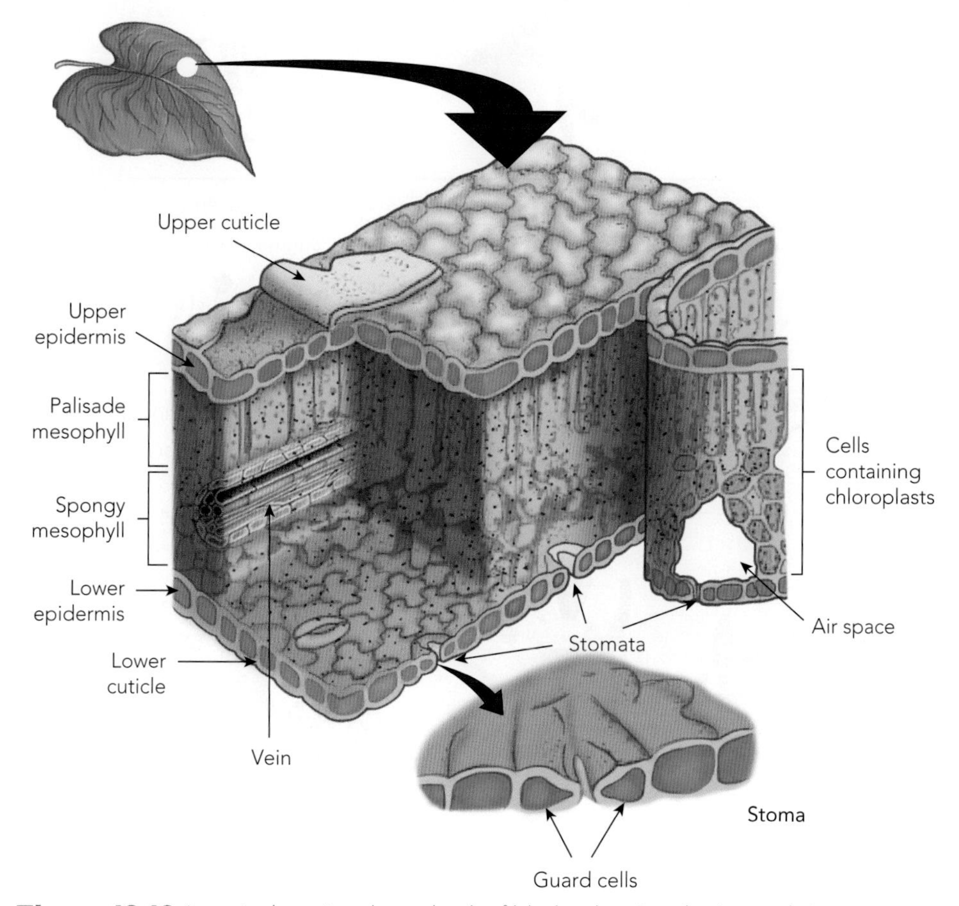

Figure 18.13 A vertical section through a leaf blade, showing the internal tissues

ISBN 9780170265560

Flowers, fruits and seeds

Flowers are the reproductive organs of the plant. The flowers produce seeds, enclosed in fruits, and thus provide for the plant's survival, multiplication and dispersal. The main function of flowers is sexual reproduction, while fruits are involved in dispersal and seeds give survival.

Flowers

Flowers are usually borne in groups on the stem and are called an **inflorescence**. There is wide variation in inflorescences between species. The simplest form of inflorescence has just one flower on a stalk or **peduncle**, such as the tulip. In most plants there are a number of flowers on each peduncle, making up the inflorescence, such as wheat or canola.

In the centre of a typical flower (Fig. 18.14) the female parts known as the pistil are found. The pistil usually consists of one carpel, but in some species there are more than one. The carpel is bottle shaped. The top part is the **stigma**, and its function is to catch pollen that might be blowing in the wind or be on the body of a visiting insect. Some stigmas are sticky and some are feathery, as in the grasses. The stalk that the stigma sits on is the **style**, which makes it more accessible to the pollen. The broader lower part of the pistil is the **ovary**. The ovary contains one or more ovules, and each ovule contains a female sex cell.

Surrounding the pistil is a ring of **stamens**, the male part of the flower. A **stamen** is made up of a fine stalk called the filament, which holds aloft the pollen sac or **anther**. The anther contains the pollen grains, and each pollen grain contains a male sex cell. The number of stamens varies from species to species. There are 10 in lucerne flowers, while wheat flowers have only three.

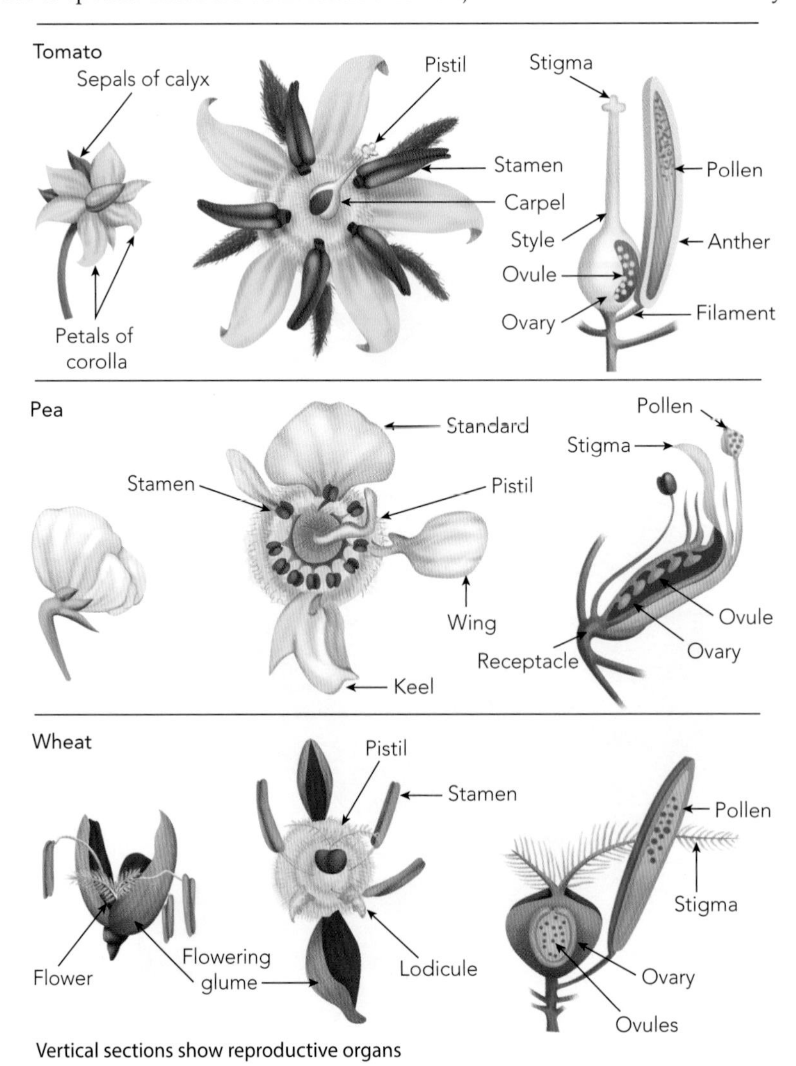

Figure 18.14 The structure of typical dicot flowers (tomato, pea) and a grass (monocot) flower (wheat)

 ISBN 9780170265560

Outside the stamens the petals are attached, and are known collectively as the corolla. Under the corolla, leaf-like structures called sepals are attached, known collectively as the calyx.

All the parts of the flower are attached to a base, known as the receptacle.

Legume flowers (e.g. peas) have five petals. The standard pea flower has a large petal, and on either side of this are two wing petals. The keel is made of two petals that surround the pistil.

In wheat flowers there are no petals as such, but leaf-like structures called glumes. The lodicules are structures at the base of the ovary that swell up, causing the flower to open and expose its anthers and stigma so that pollination can take place.

12 What is the purpose of flowers, fruits and seeds?

13 Show that you understand the difference between a flower and an inflorescence.

14 List all the parts of a flower, and write the function of each.

Fruits

Pollination occurs when pollen grains are transferred from the anther to the stigma. This can be done by the wind or by insects. **Self-pollination** occurs when the pollen from the anther lands on the stigma of the same flower, or flowers on the same plant. **Cross-pollination** occurs when the pollen comes from the anther of a flower on a different plant. Once the pollen grain lands on the stigma it germinates and sends a pollen tube down through the style into the ovary and into the ovule. The pollen tube carries with it the male sex cell. Fertilisation occurs when the male sex cell from the pollen grain unites with the female sex cell in the ovule. The fertilised ovule develops into a seed. The ovary wall develops into a fruit and encloses the seed.

There are many types of fruit, some of which are illustrated in Figure 18.15.

1 Simple fruits form a single, simple or compound ovary. They include:
   a simple, dry, indehiscent fruits (e.g. maize grain, wheat grain, hazel nut, sunflower seed); indehiscent fruit does not open along a definite seam when ripe
   b simple, dry, dehiscent fruits (e.g. bean, pea, soybean), which split along a seam when ripe, releasing the seeds
   c simple, succulent fruits, which have a fleshy ovary wall (e.g. plum, peach, tomato, orange, banana, mango).

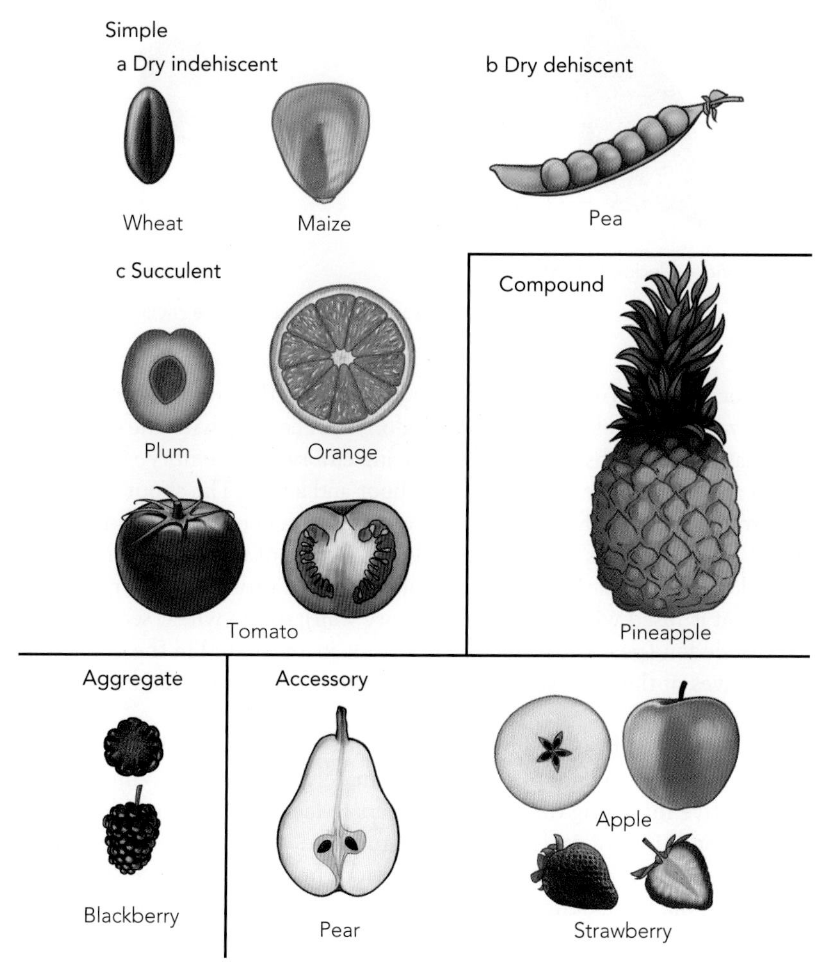

Figure 18.15 Various kinds of fruits

ISBN 9780170265560

15 What is the difference between pollination and fertilisation?

16 For each of the following fruits, state from which part of the flower the 'fruit' developed: apple, wheat, tomato, blackberry.

2 Aggregate fruits form from a single flower that has several carpels and are a collection of simple fruits (e.g. raspberry, blackberry).
3 Compound fruits form from a group of flowers or an inflorescence (e.g. fig, pineapple, mulberry).
4 Accessory fruits form when other parts of the flower, such as the receptacle, develop with the ovary to make up the fruit (e.g. apple, pear, strawberry).

Seeds

The seed consists of a seed coat or **testa**, a food reserve and an embryo or tiny plant. Figure 18.16 shows the structure of a wheat seed (monocotyledon) and a pea seed (dicotyledon).

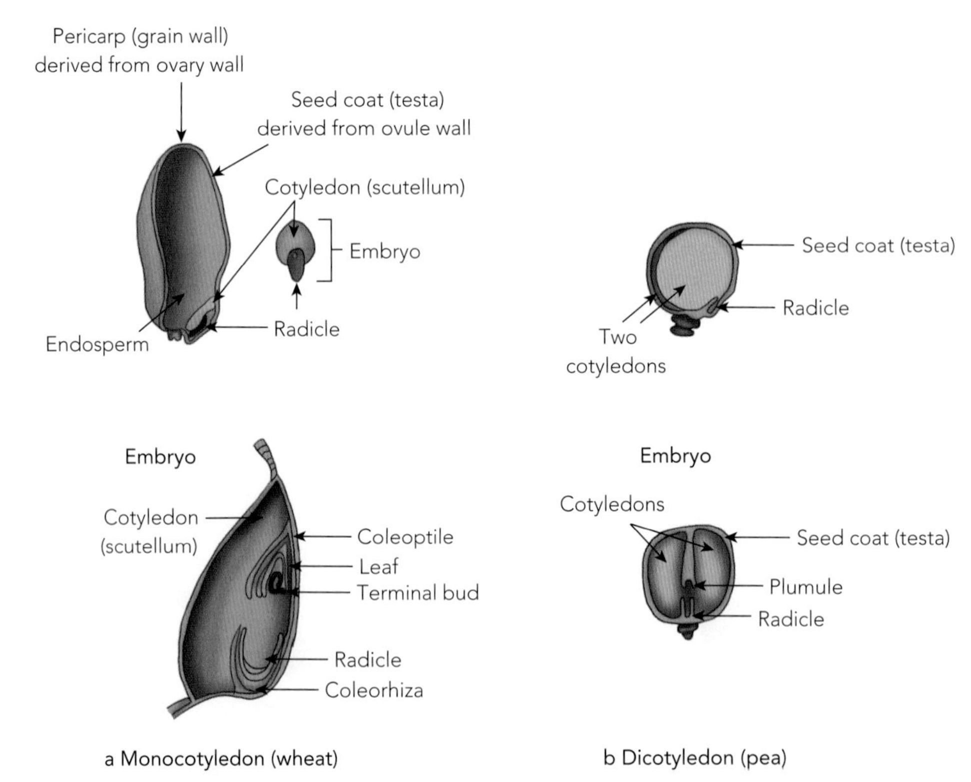

Figure 18.16 The structure of a monocotyledon seed (wheat) and a dicotyledon seed (pea)

In the monocotyledon (wheat seed) the food reserve is the **endosperm** and is separate from the embryo. The **scutellum** (single **cotyledon**) is in contact with the endosperm, and during germination it absorbs food for the growing embryo. The embryo is attached to the scutellum and is a minute plant, having a shoot and a root. The shoot consists of the coleoptile, terminal bud and first leaves. The coleoptile is a cone of tissue protecting the growing point of the shoot as it grows towards the surface of the soil. The root consists of the coleorhiza and the radicle, or first root. The **coleorhiza** is a protective layer for the radicle. The testa or seed coat is fused with the pericarp (ovary wall); so the wheat seed is in fact a fruit.

In the dicotyledon (pea seed) the food reserve is incorporated in the cotyledons, which are modified leaves of the embryo itself. The embryo has a plumule and a radicle. The testa forms a protective coat.

When a seed is ripe it has dried out and contains approximately 10% water. (An actively growing crop or pasture plant is approximately 80% water.) While it remains dry the seed can survive for a long time – several years in fact – and be capable of germination when the right conditions of moisture, temperature and oxygen supply are present. Some seeds remain dormant even with the right conditions because they have some factor that prevents germination. For example, in a sample of legume pasture seed some seeds are known as 'hard seeded'. The testa of these seeds remains impervious to water and thus prevents their germination. Germination occurs only when the testa is physically broken

 ISBN 9780170265560

in some way, allowing water to enter. Some other plant seeds absorb water but still do not germinate because they contain some chemical that inhibits them. Germination occurs only when the inhibiting chemical is leached out or breaks down.

17 A seed consists of an embryo, food reserve and a protective coat. Draw diagrams of a monocot seed and a dicot seed, showing these structures.

Photosynthesis

Photosynthesis is the fundamental process on which all living organisms depend for their food supply. It is where they get the material they are made from and their energy. In photosynthesis, light energy is trapped by chlorophyll, converted into chemical energy and used to combine carbon dioxide and water to form simple sugars. It can be summarised in the following equation:

$$6CO_2 + 12H_2O \xrightarrow[\text{chlorophyll}]{\text{light energy}} C_6H_{12}O_6 + 6O_2 + 6H_2O$$

carbon dioxide + water → simple sugar (high in energy) + oxygen + water

This is an oversimplification of what really occurs, and the description that follows also is a simplification of the reactions involved.

Process of photosynthesis

There are many steps in the process of photosynthesis, which can be divided into two phases: one that requires light, and one that does not.

Light phase

Here the chlorophyll in the chloroplasts traps light energy. This energy is used to split water molecules into hydrogen and oxygen. The oxygen is released and escapes from the leaf to the atmosphere. The energy allows hydrogen to be picked up by an NADP molecule to form $NADPH_2$, and it is also used to make a high-energy ATP molecule by adding a phosphate (P) to a low-energy ADP molecule (see Table 18.1 for the full chemical names for these abbreviations).

Table 18.1 Abbreviations for the chemicals involved in photosynthesis

Abbreviation	Chemical
P	Phosphate
ADP	Adenosine diphosphate
ATP	Adenosine triphosphate
NADP	Nicotinamide adenine dinucleotide phosphate
$NADPH_2$	Reduced nicotinamide adenine dinucleotide phosphate

The inputs to the light phase are water and light energy; the outputs are oxygen, energy-rich $NADPH_2$ and ATP.

Dark phase

The dark phase also takes place in the chloroplasts. The energy in ATP and $NADPH_2$ is used in a complex set of reactions to combine carbon dioxide with other molecules and form simple sugars. No light is required; hence it is called the dark phase. The ATP has a phosphate removed and so becomes ADP and phosphate again, and the hydrogen is removed from $NADPH_2$ so that it becomes NADP again. The energy released by these two changes is used to combine carbon dioxide in a cyclic process with other carbon compounds to form simple sugars. ADP, phosphate and NADP are returned for reuse in the light phase.

$NADPH_2$ and ATP are used in the dark phase to fix carbon dioxide. The inputs to the dark phase are carbon dioxide, $NADPH_2$ and ATP; the outputs are ADP, phosphate, NADP and high-energy simple sugars from which all the substances found in a plant can be synthesised. Figure 18.17 (page 270) outlines the reactions of photosynthesis, showing the light and dark phases. It also shows the cycling between the two of ATP, ADP, NADP and $NADPH_2$.

18 From the summary equation of photosynthesis, list the inputs and the outputs.

19 Draw black box system models for the light and dark phases of photosynthesis.

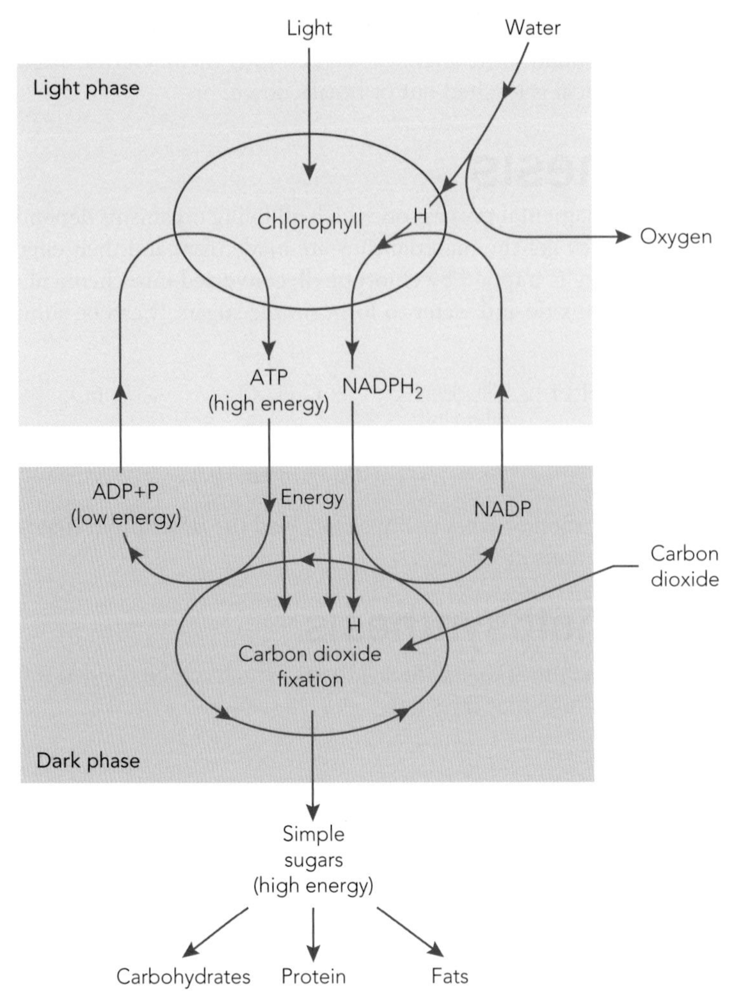

Figure 18.17 Photosynthesis, showing the relationship between the light and dark phases

C3 and C4 plants

Plants can be divided into two groups according to differences in their photosynthetic systems. One group has a three-carbon system and the other a four-carbon system, and they are known as C3 and C4 plants respectively. Under high temperatures the photosynthesis of C4 plants is much more efficient than that of C3 plants. Many tropical plants (e.g. sugar cane) have the C4 system, as do some weed species.

Factors that affect plant growth, development and production

The rate of photosynthesis is affected by environmental factors, such as temperature, light, water, carbon dioxide concentration and mineral nutrients, and by the genetic potential of the plant.

Light

As the intensity of light is increased, the rate of photosynthesis increases, until a point is reached where there is no further increase (see Fig. 18.18). From there on the rate remains constant. Different plants have different light requirements, and their maximum rates of photosynthesis occur at different light intensities. Certain ferns grow best when they are in shady situations because they are suited to lower light intensities, while other plants do well only in full sunlight. In the horticultural industries, shadecloth and other devices are used to create the light intensities preferred by the plants being cultivated.

 ISBN 9780170265560

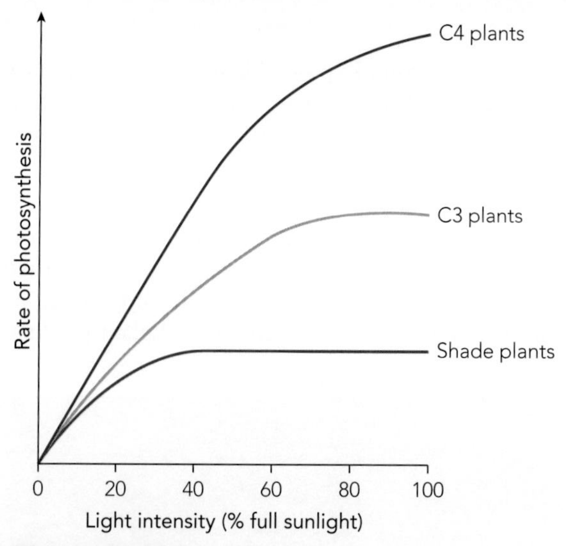

Figure 18.18 The response of photosynthesis to increasing light intensity in C3, C4 and shade plants

Temperature

The effect of temperature on the rate of photosynthesis is shown in Figure 18.19. In C3 plants, which tend to be temperate species such as wheat and soybean, photosynthesis reaches a maximum at 20–25°C and then quickly declines with further rises in temperature. C4 plants, which tend to be tropical species, reach their maximum at much higher temperatures. Tropical plants have a greater potential for maximum growth than temperate plants because of their more efficient photosynthetic system.

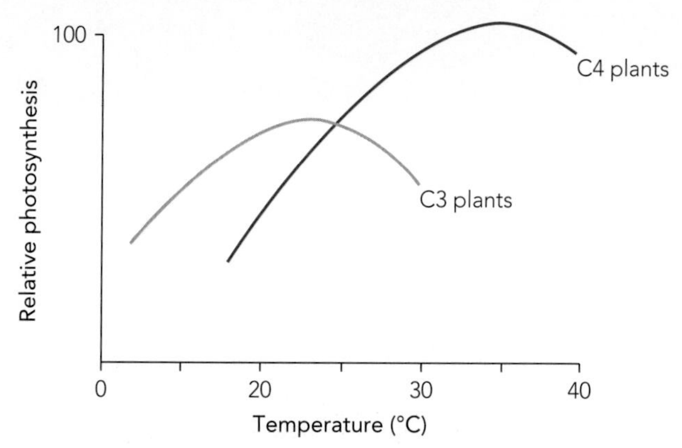

Figure 18.19 The response of the rate of photosynthesis to temperature in C3 and C4 plants

Carbon dioxide

As the concentration of carbon dioxide in the air surrounding the plant increases, the rate of photosynthesis also increases, as shown in Figure 18.20. The normal concentration of carbon dioxide in the air is 0.03%. Carbon dioxide concentrations above this give increases in the rate of photosynthesis, and in some glasshouses carbon dioxide is added to the air to achieve greater growth.

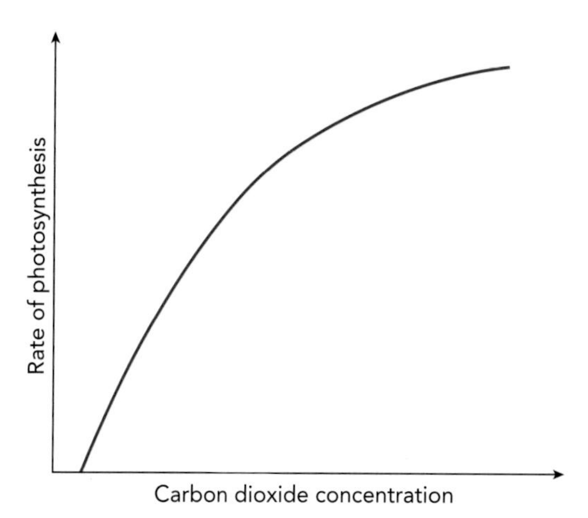

Figure 18.20 The response of the rate of photosynthesis to increasing CO_2 concentration

ISBN 9780170265560

Wind

Hot dry winds can damage grain crops. Exposed filaments and anthers can be damaged, lowering successful rates of pollination and hence yield. Hot winds also increase **evapotranspiration** (see page 227), so less water is available for photosynthesis.

Water

Water is one of the important inputs of photosynthesis and therefore a shortage of water will result in a decreased photosynthetic rate. Complete lack of water will result in the death of the plant. Irrigation is a common practice in farming to supplement natural rainfall, to ensure that lack of water will not limit photosynthesis and plant production.

© Dreamstime.com/Ekays

Figure 18.21 Irrigating young sugar cane: lack of water limits photosynthesis and plant production

20 View the *What do plants need to grow?* video on the website below, and answer the following questions.
  a What are the essential ingredients for plant growth?
  b Why do plants need the following minerals: nitrates, phosphates, potassium and magnesium?

connect

What do plants need to grow?

21 Draw three graphs showing that the rate of photosynthesis depends on the availability of carbon dioxide, light and water. Put the rate of photosynthesis on the vertical axis.

22 Explain why the maximum rate of photosynthesis for C3 and C4 plants occurs at different temperatures.

Mineral nutrients

Phosphorus is involved in the capture and transfer of energy in the photosynthetic process, and magnesium is an important part of the chlorophyll molecule that initially traps the light energy. Deficiency of these mineral nutrients will therefore result in decreased photosynthesis and growth. Fertilisers are commonly used to ensure that lack of mineral nutrients will not limit growth and plant production.

Genetic potential

Species vary in their ability to carry out photosynthesis. This can be attributed to the various species being adapted to the environments in which they developed. Wheat is suited to temperate climatic conditions, for example, while sugar cane is suited to the warmer tropics. Variation in photosynthetic capacity also occurs between varieties within a species. For example, certain varieties of soybean show a difference of up to 20% in photosynthetic rate under the same environmental conditions of light intensity and carbon dioxide concentration.

ISBN 9780170265560

Chapter review

Mandatory practical for New South Wales HSC Agriculture course

Students are to perform a first-hand investigation to determine the effect of light on plant growth. Refer to **Things to do** Question 3 below.

Things to do

Experiment Effects of light on plant growth

Aim: To design an experiment to observe the effects of different light intensities on plant growth

Method

1. Examine some dicot and monocot plants, and identify as many of the structures and parts of roots, stems, leaves, flowers, fruits and seeds listed in this chapter as possible.
2. Put some wheat and bean seeds on filter paper in Petri dishes, and add some water.
   a. Once the seeds have absorbed water and softened, carefully dissect some and identify the parts of each seed. For wheat, identify the pericarp/testa, endosperm, scutellum and embryo. For beans, identify the testa, cotyledons, plumule and radicle.
   b. Let the remaining seeds germinate. Identify the coleoptile and seminal roots of wheat and the radicle of the bean.
   c. Carefully place a root on a microscope slide, and use low magnification (×40) to identify the root hairs and root cap.
3. Set up an experiment to test the effect of different light intensities on the production of pasture grasses or legumes. This can be done by growing the plants under different grades of shadecloth (full sun, medium shade, high shade) and comparing their production. Production can be measured by the amount of dry matter produced.

Things to find out

1. Analyse what wavelengths and colours of light are important for photosynthesis.
2. For a particular species of crop plant, list the current recommended varieties. What are the differences between the varieties? Are they suited to different climates and localities?
3. Discuss what makes the photosynthesis of C4 plants more efficient than that of C3 plants at higher temperatures.

Extended response questions

1. You have conducted a first-hand investigation into the effect of light on plant growth. State the conclusion of the experiment and suggest one improvement for the experimental method.
2. Describe how a farmer applies knowledge of the factors that affect the rate of photosynthesis when growing a particular crop.
3. Outline two environmental factors that limit production of a named crop that you have studied, and for each discuss a practice that farmers use to reduce this limitation or environmental constraint.
4. For a named crop you have studied evaluate two management practices that have overcome environmental constraints.

CHAPTER 19

Outcomes

Students will learn about the following topics.

1. Propagation techniques.
2. Growth and development in plants.
3. Processes of respiration, photosynthesis, net assimilation rate, water and nutrient uptake.
4. The role of plant hormones in the growth and development of plants.
5. The phases of growth of one monocot and one dicot used in agriculture.

Students will be able to demonstrate their learning by carrying out these actions.

1. Propagate plants by sexual and asexual methods.
2. Distinguish between growth and development in plants.
3. Describe how plant growth can be measured.
4. Explain that growth can occur only when the rate of photosynthesis exceeds the rate of respiration.
5. List and explain the factors that affect plant growth.
6. Explain the difference between the vegetative and reproductive phases of plant development.
7. Grow and monitor a crop/pasture from planting through to harvest.
8. Describe the phases of development of the wheat plant.
9. Describe the phases of development of the potato plant.
10. Identify temperature and day length or photoperiod as factors that affect the onset of the reproductive phase in plants.
11. Outline the effects of plant hormones, including auxins, gibberellins, cytokinins, ethylene and abscisic acid.
12. Explain how plant hormones have been used to manipulate plants to our advantage.

ISBN 9780170265560

PLANT GROWTH AND DEVELOPMENT

Words to know

abscisic acid a plant growth substance that functions chiefly as a growth inhibitor

anthesis the stage of development of a plant when the anthers become prominent and pollen is shed; particularly applies to cereals (e.g. wheat)

apical dominance a state where the growth of all the axillary buds on a stem is suppressed, apparently by the growing terminal bud

auxins plant hormones causing cell elongation, secondary thickening of stems and roots, fruit development and apical dominance

biennial plant a plant that completes its lifecycle within 2 years

cytokinins plant hormones that promote fruit ripening and cell reproduction, and initiate the production of roots and shoots

dormancy a state of a seed or plant where it remains living but does not grow or germinate

ethylene a plant hormone that promotes fruit ripening

flag leaf the last leaf of a monocot

gibberellins plant hormones causing stem growth and flowering; they break seed dormancy

herbicide a chemical used to kill unwanted plants, especially weeds

reproductive phase the development of reproductive structures in a plant (i.e. flowers, seed, fruit)

respiration a process in all living cells where complex organic molecules are oxidised, releasing energy necessary for life

tuber an underground stem swollen with materials stored by the plant (e.g. a potato)

vegetative phase the stage of plant development where roots, stems and leaves grow

viable (seed) living and capable of germination

Introduction

Plant growth can be distinguished from plant development in that growth involves an increase in size whereas development is the sequence of changes that occur in the life of a plant as it grows. After a seed is planted it germinates, grows stems and leaves, produces flowers, fruits and seeds, and then dies.

Plant growth

The growth of a plant can be measured simply by measuring the height at, say, weekly intervals. It may also be measured by determining the amount of dry matter production each week. This is done by taking a sample of plants each week, drying it in the oven and weighing the dry matter that remains. Figure 19.1 shows how the weight of dry matter in a plant changes from fertilisation through to maturity. Note that the dry weight actually declines as the seed germinates. This is because the **respiration** of the germinating seed uses up some of the food reserves stored in the seed. As the seedling becomes established and starts to photosynthesise, its dry weight begins to increase. At first the growth rate is slow, but as the amount of tissue capable of **photosynthesis** (see page 269) increases, so does the growth rate. The growth rate then slows down as the proportion of non-photosynthesising material (stems, flowers, ageing and shaded leaves) increases.

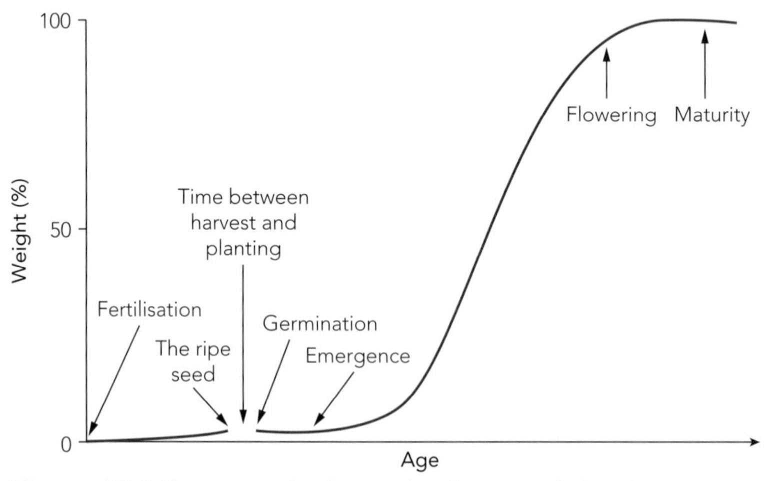

Figure 19.1 Changes in the dry weight of an annual plant from fertilisation to maturity

For a plant to grow, the rate of production of material by photosynthesis must exceed the rate at which this material is used up by the plant in respiration. The material not used in respiration is used to produce new cells and hence, growth. Respiration is the process that occurs in all living things whereby complex organic molecules are systematically broken down, releasing the energy that they contain to meet the energy needs of the organism.

Figure 19.2 shows how respiration and photosynthesis change through a 24-hour period. Respiration goes on all the time, but photosynthesis occurs only when there is sufficient light (between 6 a.m. and 6 p.m. in the case shown). It is only when photosynthesis exceeds respiration that growth can occur.

1 How is growth different from development in plants?
2 How can the growth of plants be measured?
3 Explain what the relationship between photosynthesis and respiration must be for a plant to grow.

ISBN 9780170265560

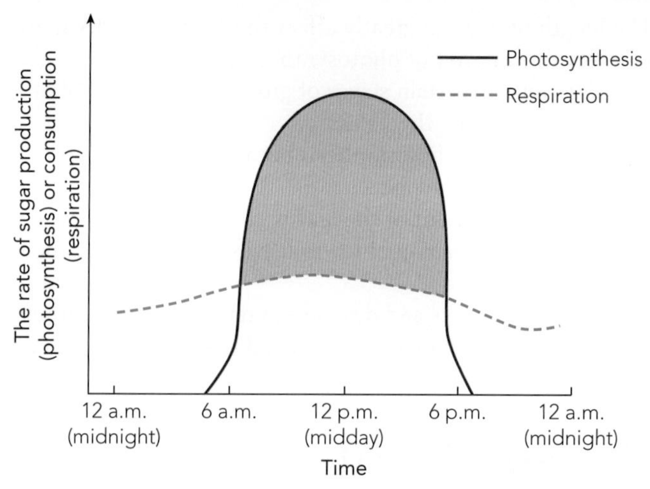

Figure 19.2 Changes in the rates of respiration and photosynthesis over a 24-hour period

Net assimilation rate

Net assimilation rate (NAR) is one measure of the growth of plants resulting from the difference between photosynthesis and respiration. It gives an important indication of the growth of a crop or pasture. It can be defined as the average increase in dry matter per unit area of leaf material in a given period. To calculate NAR, samples of the crop or pasture are cut at the beginning and end of a time interval, such as 1 week, and the leaf area and weight of dry matter at each time are determined. These data are then processed in the following formula:

$$\text{NAR} = \frac{W_2 - W_1}{T_2 - T_1} \times \frac{1}{\frac{1}{2}(A_1 + A_2)}$$

where:

W_1 = dry weight at the beginning of the sampling period
W_2 = dry weight at the end of the sampling period
T_1 = time the first sample was taken
T_2 = time the second sample was taken
A_1 = area of leaf in the first sample
A_2 = area of leaf in the second sample.

The units most often used for NAR are grams of dry matter per square decimetre of leaf per week ($g/dm^2/week^2$). NAR then is a measure of the dry matter gain of the plant over a period.

NAR is affected by photosynthesis, respiration and a number of other factors. The factors that affect NAR (plant growth) are listed here.

- *Rate of photosynthesis.* This is actually controlled by the interaction of factors such as light intensity, carbon dioxide concentration near the leaves, availability of water and the temperature around the plant.
- *Rate of respiration.* This is influenced by factors such as the availability of complex organic molecules (sugars) within the plant, the availability of oxygen and the temperature.
- *Availability of water.* Plants need water to carry out the many metabolic reactions, such as photosynthesis, cellular respiration, enzyme production and hormone production, that occur in their cells. Water is also needed for cell division and elongation. Low water levels within and around plant cells can lead to slow metabolism rates, reduced growth, wilting and death.
- *Availability of inorganic nutrients.* Nutrients such as nitrogen, phosphorus, sulphur and potassium are necessary for the synthesis of the different tissues in the plant. Phosphate is required to form the energy compound ATP. Nitrate ions are needed for protein synthesis, potassium for enzyme production, magnesium for chlorophyll production.
- *Proportion of non-photosynthetic tissue.* The more roots, stems, flowers and fruits there are in the plant, the more respiration will take place.

ISBN 9780170265560

- *Photoperiod*. The length of day can greatly affect the amount of dry matter produced, because it will affect the amount of photosynthesis in a day.
- *Stage of plant development*. At certain stages of growth the rates of photosynthesis and respiration will change, affecting the NAR.
- *Leaf area*. This directly affects photosynthesis because, in general, the greater the leaf area the more photosynthesis can take place.
- *Canopy structure*. The arrangement of the leaves on the plant can influence the NAR, especially if this leads to the shading of lower leaves, thus reducing their photosynthetic capacity.
- *Insect and disease damage*. Insects and disease can damage photosynthetic tissue by chewing holes in leaves or producing areas of dead tissue (necrosis) and this will decrease plant photosynthetic activity and therefore NAR.

4 The rates of photosynthesis and respiration are both temperature dependent; that is, an increase in temperature results in an increase in the rate of each.
  a What temperature regime or pattern over a 24-hour period would result in maximum dry matter gain by a plant?
  b Provide an explanation for why this occurs.

Plant development

Plant development is the sequence of definable phases that a plant goes through as it grows, from the time it is planted as a seed until it reaches maturity. As a plant develops, various tissues and organs, such as roots, leaves, stems, flowers, fruits and seeds, grow at different times in its lifecycle.

In annual plants the lifecycle is complete within 1 year. The development of these plants can be divided into two phases.

1 **Vegetative phase**: in this phase the seed germinates and the plant becomes established. It continues to produce roots, stems and leaves. The plant is building its capacity to make food in preparation for the reproductive phase.
2 **Reproductive phase**: during this phase flowers form, fertilisation takes place, seeds and fruit develop, and the plant matures and dies.

Annual plants, such as wheat, rice, maize, barley, sorghum and oats, are very important because they make up a large proportion of the world's food crops.

In **perennial** plants (see page 143) the growth and development of the vegetative material may continue throughout the life of the plant, which can be up to thousands of years for some trees. Reproductive development occurs at regular intervals, most often yearly, for a short period of that year. The plant flowers and produces seeds and fruit.

In **biennial plants** the lifecycle takes 2 years. The vegetative phase takes place in one year, and the reproductive phase occurs in the following year. In the first year, food reserves are built up, and in the second year they are used to supply the reproductive phase.

5 Briefly describe what happens in the vegetative and reproductive phases of development of a plant.

Development of the wheat plant

The wheat plant is an annual and has the following developmental phases, illustrated in Figure 19.3.

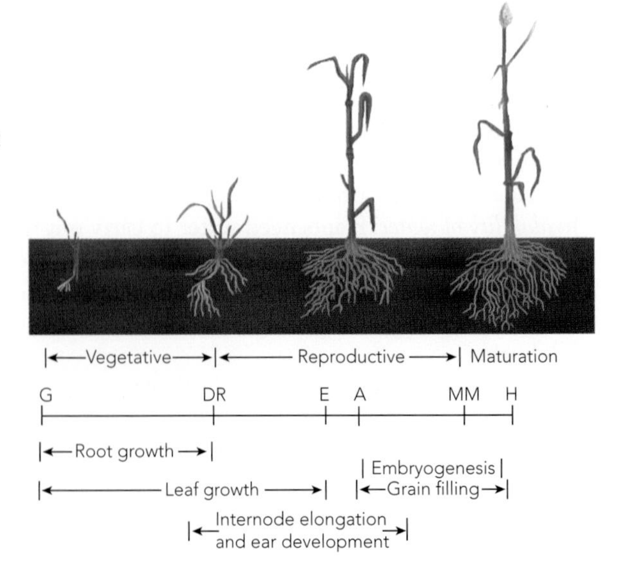

Figure 19.3 Phases in the development of the wheat plant from germination to maturity

Key
G, germination
DR, double ridge
E, ear emergence
A, anthesis
MM, maturation of ears started
H, harvest

 ISBN 9780170265560

Vegetative phase

The first process in this phase is germination. For the seed to germinate it must be **viable** and have a supply of water and air (oxygen) and the correct temperature. The seed absorbs water and swells, and then the first root emerges. This is closely followed by the emergence of the coleoptile (a protective sheath around the shoot) and further roots from the seed. These roots that arise from the seed are seminal roots, which grow to form the seminal root system. The coleoptile grows to the surface of the soil, and through it emerges the first leaf or plumule. This leaf is followed by several other leaves. Very near the surface of the soil, the crown node forms. From this node adventitious roots grow and form the major root system of the plant, known as the nodal root system. Development continues for the rest of the vegetative phase, with the growth of more stems or tillers from the axillary buds of the leaves. During this time the actual stems remain very short, being less than 2 cm above the soil surface. The plants appear to be much higher because the leaves grow up.

Reproductive phase

The change from vegetative to reproductive phase is marked by a change in the terminal bud or growing point on the stem. During the vegetative phase it was a rounded dome producing new leaves as it grew. Now it changes to a double ridge and goes on to produce an inflorescence. The stem now grows up, caused by the growth and elongation of the internodes. The inflorescence is pushed up; and when it is enclosed by the last leaf (**flag leaf**), the plant is said to be in the 'boot' stage. After the inflorescence emerges from the flag leaf, the anthers become visible. This is known as **anthesis**. Pollen is shed, fertilisation takes place, and the seed containing the embryo and endosperm develops. The occurrence of frosts at anthesis can adversely affect the amount of grain set and thus the final yield of the crop. The endosperm or food storage organ of the seed fills with starch. The flag leaf and the inflorescence itself are responsible for most of the photosynthesis that produces the material for storage in the seed. Once the seeds have completed filling, the plant begins to die and dry out. The seeds also dry out and harden and are then ready for harvest, their moisture content being 10–12%.

6 List the events that occur in the vegetative and reproductive phases of the development of the wheat plant.

Development of the potato plant

The potato is grown as an annual crop, but unlike wheat it does not grow from a seed but from a **tuber**. The tuber is the 'potato' itself and is a swollen stem. It has leaf scars and buds, which we know as the 'eyes' of the potato (Fig. 19.4). The development of the potato plant (Fig. 19.5, page 280) can be considered in two phases.

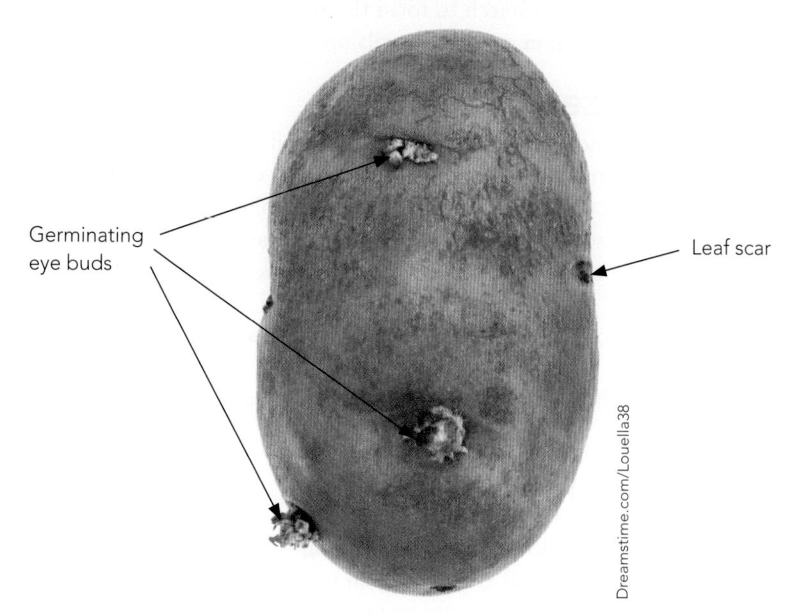

Figure 19.4 A potato with eyes

ISBN 9780170265560

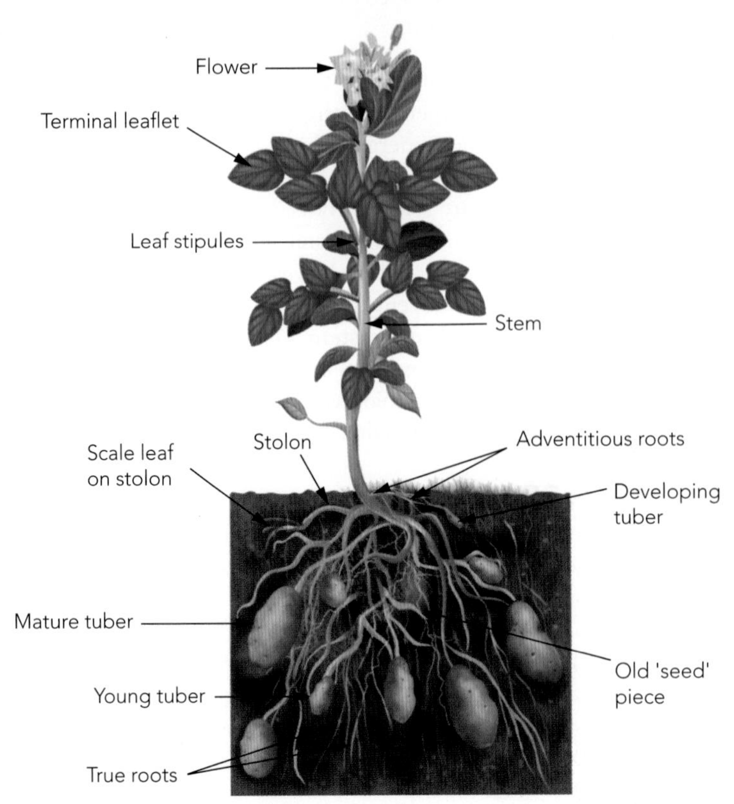

Figure 19.5 A potato plant approaching maturity

Vegetative phase

The tuber is planted in the ground in much the same way as the seed of other crops would be. The buds on the tuber sprout and produce shoots using the food stored in the tuber. Once the shoots break through the surface of the soil, they begin to photosynthesise and supply their own food requirements. Adventitious roots develop from the stems under the ground and grow through the soil, seeking moisture and mineral nutrients. The aerial stems continue to grow and produce leaves. Approximately 10 days after the emergence of the stems above the ground, the axillary buds of the underground part of the stems start to send out horizontal stems called stolons. These stolons grow out from the plant through the soil. The ends of them swell up, storing starch to form the tubers that are finally harvested. No material is stored in the 'seed' tuber, which shrivels up and dies.

Reproductive phase

As the above-ground stem and leaf material reach their full size, flowers are produced. Their colour varies from white to mauve to blue-purple. In many varieties the flowers fall off at an early stage. Some flowers that remain on the plant are pollinated and after fertilisation produce small fruits. These are not unlike tiny green tomatoes in appearance and are only as big as a pea. Since potatoes are grown from tubers, the seeds are important only to those breeding new varieties.

The tubers under the ground continue to grow, storing starch. The above-ground parts of the plant eventually die off, leaving the tubers in the ground. In practice the above-ground parts are killed off with **herbicide** or by slashing before they die, to enable digging and harvesting of the tubers.

7 List the events in the development of the potato plant.

What triggers the reproductive phase?

Several factors provide the stimulus for a plant to change from the vegetative phase to the reproductive phase. The chief stimuli are changes in temperature and photoperiod (day length). For a plant to flower, it must have been exposed to a combination of a certain temperature for some time and a certain day length or photoperiod. Wheat, for example, requires temperatures from 2–7°C for at least 1 month, and to some extent, long days before

 ISBN 9780170265560

it will flower. Citrus trees, being subtropical plants, require at least 1 month of temperatures from 10–15°C before they will flower. Meeting the requirement for a period of cold temperatures before a plant will flower is known as **vernalisation** (see page 191). Table 19.1 categorises plants by how their flowering responds to day length.

Plants show wide variation in their response to these stimuli. Some are very precise and will not flower unless a particular requirement for day length or temperature is met, while in others flowering is just enhanced by these conditions.

8 Describe how photoperiod and temperature act as stimuli for the initiation of the reproductive phase in plants.

Table 19.1 Responses of plants to day length (photoperiod)

Plant group	Flowering
Short-day plants (e.g. maize, rice, soybean)	Short-day plants that appear to flower in response to a short photoperiod, but in fact flower in response to the long dark period
Long-day plants (e.g. wheat, oats, ryegrass)	Long-day plants that appear to flower in response to a long photoperiod, but in fact respond to the short dark period
Day-neutral plants (e.g. tomato)	Day-neutral plants flower irrespective of what day-length regime they are exposed to.

Plant hormones

Plant hormones are organic substances produced by various parts of the plant that, in very small amounts, affect the growth and development of other parts of the plant. They have a role in the coordination of plant growth and development. Their action depends on their concentration. For a particular hormone a low concentration may promote growth and a higher concentration may inhibit growth. A particular concentration may promote growth of roots and inhibit leaf growth. Plant hormones include the following.

- **Auxins** are produced in young leaves, germinating seeds, meristematic tissue (all buds, root tips and cambium), and developing flowers and fruits. They cause cell elongation, secondary thickening of stems and roots, fruit development and dominance of the apical (terminal) bud. The response of roots, buds and stems to one auxin is shown in Figure 19.6.
- **Gibberellins** are produced by germinating seeds and young stems. They cause stem cell elongation and flowering and have a role in breaking seed **dormancy**.
- **Cytokinins** are produced in the growing roots, in developing seeds and cambial tissue. They affect cell division and root growth.
- **Ethylene** is particularly produced in young growing tissues and ageing leaves and ripening fruit. Ethylene promotes the ripening of fruit.
- **Abscisic acid** reduces cell division, inhibits growth and closes plant stomata when there is a severe water shortage.

connect

Role of auxin in plant response mechanisms

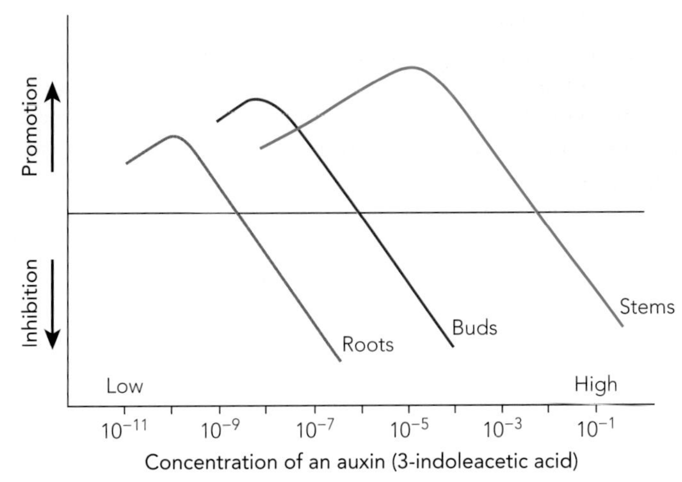

Figure 19.6 Response of roots, buds and stems to change in concentration of an auxin

ISBN 9780170265560

Plant hormone applications in agriculture

As knowledge of plant hormones and their effects has increased, hormones have been put to use in various ways to help manipulate plants to our advantage. Their application often involves producing them synthetically. They have been used for various purposes.

- Hormone preparations are used as herbicides, where excess of the hormone results in abnormal functioning of the unwanted plant and its subsequent death (Fig. 19.7).
- In tissue culture, a bud from the parent plant is taken and grown on media containing the hormones. By changing the relative concentrations of the hormones, the mass of tissue that grows is caused to differentiate into many new plants, all genetically identical to the parent (Fig. 19.8).

Department of Agriculture and Food, Western Australia

Figure 19.7 Chlorsulfuron, a herbicide, can cause stunted and spiky wheat plants with pruned roots

Shutterstock.com/jaboo2foto

Figure 19.8 Plants being propagated using tissue culture

- To prompt root growth on cuttings, freshly taken cuttings are dipped in a hormone preparation, and the formation of roots is enhanced.
- Where a lot of fruit is set on one tree, the result is a lot of small, less desirable fruit. A hormone spray reduces the number of fruit on the tree (this is called thinning the crop). The remaining fruit will be larger and more marketable.
- Hormones can be used to ripen fruit. Bananas are treated with ethylene gas to ripen them. This means that fruit can be harvested and transported to market before ripening. The advantage is that the fruit is less likely to be bruised in transport.

9 Use an example to explain how knowledge of plant hormones has been applied in farming to improve production.

Plant propagation

The propagation of plants by asexual methods has been practiced for centuries. It enables plants with identical genotypes to be reproduced, which is impossible in sexual reproduction. Asexual techniques do not involve the exchange of gametes, so new plants are identical to parent plants. Sexual propagation of plants involves the production of seeds; asexual (vegetative) plant propagation uses cuttings, budding and grafting, etc. to produce new plant material.

connect

How to grow plants from cuttings

 ISBN 9780170265560

Chapter review

Things to do

1. Construct a table to outline the role of plant hormones in plant growth and development. Use the following headings to construct your table.

Name of hormone	Where hormone is produced in plant	Action of hormone

2. Measure the growth rate of a pasture. Every 2 weeks cut the same square metre of pasture to approximately 3 cm above the ground. Weigh the cut material, oven-dry it, and weigh the dry matter.
3. Plant a plot of wheat or potatoes. Monitor the crop weekly. Dig up a representative plant and carefully record the stages of development as they occur (from planting through to harvest).
4. Carry out an experiment to demonstrate that water and air are required before seeds will germinate.
5. Design and carry out an experiment to determine the optimum temperature for germination of a particular crop.
6. Carry out an experiment to demonstrate the phenomenon of **apical dominance**. Grow two plants in separate pots. Carefully remove the terminal bud from one plant. Write down what you expect to happen to the axillary buds of each plant. Observe what happens to the axillary buds on both plants in 2 weeks' time. Write an explanation for your observations.
7. Take some cuttings of a plant commonly propagated by cuttings. Treat half the cuttings with a commercial hormone treatment to promote root development, and leave the other half untreated. Compare the rates at which roots develop in each group.
8. Propagate plants through stem and leaf cuttings.

Things to find out

1. What are selective herbicides, and what is their role in crop production?
2. Discuss the difference between pre-emergent and post-emergent herbicides.
3. Explain how budding and grafting are used in fruit crops to control the size of the trees and improve their resistance to disease.
4. For the crops grown in your area, find out whether the varieties used have any particular requirements of temperature and day length for stimulating flowering.
5. What are the active ingredient chemicals in several different herbicides? How do these chemicals affect the target plants? Are these chemicals derived from plant hormones?

Extended response questions

1. Describe the phases of growth of one named agricultural dicot and one monocot crop.
2. For a crop that is grown in your region:
   a. list the management practices carried out during the growth of the crop in the order in which they would be carried out
   b. describe each of the management practices.
3. Explain:
   a. how plant hormones might be used to manage farm operations, such as killing weeds and ripening fruit
   b. the role of two named hormones in the manipulation of plant production.

ISBN 9780170265560

CHAPTER 20

Outcomes

Students will learn about the following topics.

1 The interaction of genotype, environment and management.
2 The aims of plant breeding programs.

Students will be able to demonstrate their learning by carrying out these actions.

1 Explain that existing crop and pasture plants are the result of selection down through the ages and of scientific selection and breeding in the last century.
2 List the general aims of plant breeders in selecting and breeding plants.
3 Explain that many of the characteristics that plant breeders endeavour to improve, such as yield, are quantitative characteristics controlled by many genes acting together.
4 Explain that the phenotype of a plant is the result of the interaction of its genotype and its environment.
5 Describe the heritability of a characteristic as the proportion of variation in that character due to genotype.
6 Explain that there is considerable variation for any particular characteristic within a species of plant throughout the world.
7 Describe the basic steps that a plant breeder takes in developing a new variety.
8 Describe the modes of reproduction of plants as self-pollinating, cross-pollinating and vegetative.
9 Explain how the mode of reproduction is taken into account in plant-breeding programs.
10 Explain how hybrid varieties are produced from self-pollinating and cross-pollinating species.
11 Explain how the new technologies of wide crosses, tissue culture and genetic engineering have been applied to plant-breeding programs.
12 Explain that the sources of genetic variation are:
    a those already existing naturally in the population of a species
    b those arising from mutations, both natural and induced
    c those arising from transferring desirable genes from one species to another using genetic engineering.

ISBN 9780170265560

PLANT BREEDING AND SELECTION

Words to know

breeding the process of crossing and selecting plants to produce new types or varieties

characteristic a distinguishing feature of a plant, resulting from the interaction of a gene or genes with the environment

clones organisms that are genetically identical

lodging occurs when wind causes tall plants to fall over, making harvesting difficult

mutations a change in the amount or structure of DNA in the chromosomes of an organism

quantitative characteristic a characteristic that can be measured in some way (e.g. yield of wheat in tonnes per hectare)

selection choosing which plants will be used as parents to produce the next generation

species plants of similar kind that can be crossed and produce fertile offspring

variety a type of plant or animal with a particular genotype within a species (e.g. wheat varieties Eagle, Gatcher and Timgalen); in plants, also known as a cultivar

ISBN 9780170265560

Introduction

The selection of plants for food crops has gone on ever since humans ceased to be hunters and gatherers and started to settle in relatively stationary communities, growing crops for food. The seed for next year's crop was kept from the existing crop. The seed kept was consciously or unconsciously selected because it was better or simply because it came from plants that had survived and produced seed. Through this process of selection, plants very slowly, over hundreds of years, became better adapted to their particular environment, and yields very gradually increased. It was this process that brought about the crops we presently grow, such as wheat, rice, maize, sorghum, sugar cane, potatoes and many vegetables.

Since 1900, when the work of Gregor Mendel on crossing garden peas was rediscovered, breeding and selection have resulted in spectacular improvements in the yields, adaptation and quality of many crop and pasture plants. This progress is now also being enhanced by the process of **genetic engineering** (see page 131). The researchers who carry out the work of plant selection and breeding are known as plant breeders.

1 What has brought about the increase in production from plants since 1900?

Aims

In general, the aims of plant selection and breeding are to:

- improve the yield of the crop or pasture plants
- improve the quality of the product
- increase the adaptation of the plants to the physical environment
- increase the plants' resistance to the ravages of diseases and pests.

Genetic principles

The genetic principles that enable plant breeders to make improvements to crops and pastures are the same as those that apply to animals, which have been described and discussed in Chapter 9.

Quantitative characteristics

Plant breeders are interested in improving characteristics, such as yield (e.g. in wheat), plant height (barley), cob size (maize), staple length (cotton), sugar content (sugar cane) and ability to make great chips (potatoes). Such characteristics are not controlled by single genes but are the result of the action of a large number of genes acting together. They are known as quantitative characteristics.

Let us consider one such quantitative characteristic: the yield of a crop, of say, tomatoes. The yield of fruit from this crop is determined by the interaction of the genotype with the environment. The **genotype** (see page 59), or genetic makeup, of the plant sets the potential yield that can be achieved – an upper limit to the yield.

The environment includes all those factors that affect the yield of tomato fruit but are not genetic, such as the mineral nutrients available in the soil, available moisture, soil acidity/alkalinity (pH), level of interference, temperature and incidence of disease and pests. This idea can be summarised in the following equation:

$P = G + E$

where P = phenotype of the crop (its resulting appearance or yield)

G = genotype of the crop (its genetic makeup)

E = environment that the crop grows in.

2 List three quantitative characteristics of plants.

3 The yield of a wheat crop is 1.8 tonnes per hectare. Explain how this yield is the result of the genotype and environment of the wheat crop.

Heritability

When plant breeders are selecting for improvement in a characteristic such as yield, it would be very useful to know what proportion of the observed yield is due to the genotype and what is due to the environment. The more that is due to genotype, the faster progress can be made in improving the characteristic by selection and breeding. The heritability of a characteristic is a measure of the proportion of a characteristic that is due to genotype. A characteristic with a heritability of 1.0 would be entirely due to the plant's genotype. A characteristic with a heritability of 0.33 would be 0.33 due to genotype and 0.67 due to environment.

ISBN 9780170265560

Variation within a species

For improvement in a characteristic to be made, the plant breeder must combine into one genotype as many as possible of the **genes** (see page 131) that will contribute most to this characteristic. Within any one **species** of plant there is wide variation in any particular quantitative characteristic. A species has many genes, and different forms (**alleles**, see page 131) of genes, which contribute to the characteristic spread throughout the species all around the world. There tends to be most variation in regions where the particular species originated. For example, potatoes originated in Peru, South America, and the largest range of varieties of potatoes is found there.

To preserve the variation in plants and make it available to plant breeders, various seed banks and collections of plants have been set up and maintained around the world, including the Australian Seed Bank Partnership. In New South Wales, the Agricultural Research Centre at Tamworth maintains a collection of more than 3000 wheat varieties. They are stored as seed at low temperature and low humidity. They provide a bank of genetic variation from which the wheat breeders of Australia can draw for their breeding programs. Periodically, each **variety** must be grown to give a fresh supply of seed.

connect

Australian Seed Bank Partnership (ABSP)

The ABSP's mission is to conserve the diversity of Australia's native plants through seed collection, banking and research, and the sharing of knowledge.

It is an online resource of the seed collections held in Australia for conservation purposes. Experiment with the search feature to locate different seeds.

Basic steps of plant breeding

The basic steps in plant breeding are:

1 collecting and selecting varieties that between them have the characteristics required in the final crop or pasture
2 **crossbreeding** (see page 131) the collected varieties to combine the desired characteristics
3 selecting among the offspring of the crosses those individuals that do have the desired characteristics
4 evaluating the selected offspring for the presence of the desired characteristics under field conditions.

This process takes a long time. It takes 10–15 years to produce a new wheat variety.

4 Why is it important that collections of the different varieties of a crop plant, such as wheat, be maintained?
5 How does a plant breeder make use of these collections of plant varieties?
6 Briefly list the basic steps in breeding a new variety of plant.

Mode of reproduction

The mode of reproduction has some bearing on the methods adopted in breeding programs. Most plants are naturally cross-pollinating, and some are self-pollinating. Vegetative reproduction is also important as a means of reproducing desirable genotypes. Table 20.1 lists some important self-pollinating and cross-pollinating crop and pasture plants.

Table 20.1 Important crop and pasture species classified by their mode of sexual reproduction

Self-pollinating	Cross-pollinating
Wheat	Maize
Rice	Phalaris
Barley	Rye
Oats	Lucerne
Soybean	Red clover
Subterranean clover	White clover
Lettuce	Sunflower

Self-pollinating species

Self-pollinating plants naturally pollinate their own flowers and are in fact inbreeding. To cross-pollinate two varieties of a self-pollinating species, the plant breeder must:

- remove the anthers from the flowers of the plant that is to be the female parent
- take anthers containing pollen from the flowers of the plant that is to be the male parent
- transfer them to the flowers of the plant that is to be the female parent.

ISBN 9780170265560

The plants that grow from the seed produced – the F1 generation – are allowed to self-pollinate. This self-pollination goes on for 8–10 generations, by which time the plants are true breeding because they are virtually **homozygous** (see page 131) for all their genes; that is, the two genes for a particular character trait are the same. (By contrast, **heterozygous** (see page 131) plants would not breed true, because the two genes for each particular character trait are not the same.) Figure 20.1 shows that homozygosity very closely approaches 100% after 8–10 generations of self-fertilisation. During this time, selection takes place for various characteristics.

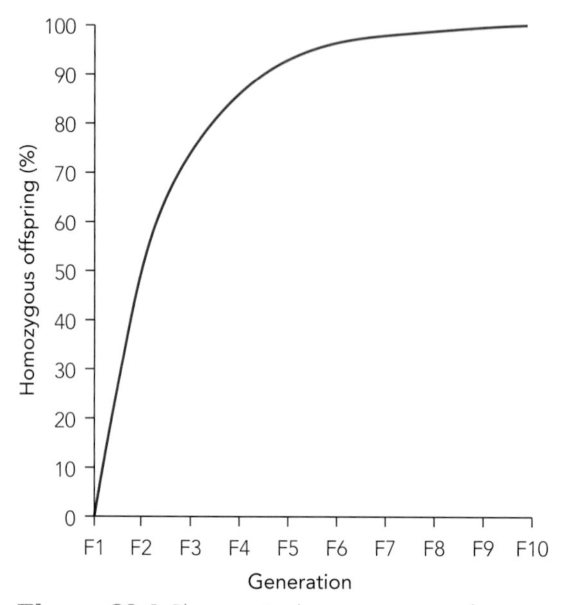

Figure 20.1 Changes in the percentage of homozygous offspring resulting from continuous self-fertilisation

Now the seeds from individual plants can be grown, and the plants can be tested in the field for their suitability to the target environment. The plants from one individual plant at this stage are known as a pure line, because they are homozygous and have the same genotype. The result of these field trials should be at least one new variety of plant that meets the requirements that the plant breeder was looking for.

Wheat-breeding programs in Australia follow this breeding technique. Breeders are continually trying to produce new varieties that are resistant to the important disease black stem rust. The rust fungus changes genetically and overcomes the resistance in existing varieties; so new varieties resistant to the new strains of rust are constantly required. As well, the new varieties of wheat must have other qualities equal to or better than the varieties that they are replacing. These qualities include protein content, yield and baking quality.

Commercial **hybrid** (see page 37) varieties of rice and wheat that are self-pollinating have been developed. Pure lines or varieties are produced as explained previously and are crossed to give F1 hybrids. Selection is based on the performance of these F1 hybrids. The pair of lines that gives the best F1 performance is chosen, and these are then used as parents to produce the commercial seed.

Cross-pollinating species

Cross-pollinating plants naturally pollinate other plants of the same species and frequently have mechanisms to prevent self-pollination. For example, maize and pumpkins have separate male and female flowers on the same plant; kiwi fruit has separate male and female plants; and other plants have built-in incompatibility that prevents fertilisation, even when pollination with pollen from the same plant does take place. Three avenues are open to the plant breeder for breeding these species: open pollination, production of hybrid varieties, and cross-pollination combined with vegetative reproduction.

 ISBN 9780170265560

Open pollination

Open-pollinated varieties of species, such as lucerne and rye, are produced by selecting a number of plants that have the desired characteristics and allowing them to cross-pollinate.

Selection is carried out on the offspring of the first crosses, and then these plants are allowed to cross-pollinate. This process is repeated until a population of plants is achieved with the desired characteristics.

The production of Hunterfield – an aphid-resistant variety of lucerne – by the South Australian Department of Agriculture is an example of this process. It was produced by three cycles of selecting and crossing within the existing variety Hunter River.

1 A collection of seed was made from across the lucerne-growing areas of Australia, and 12 600 seedlings were grown. They were subjected to the spotted alfalfa aphid, and 28 tolerant plants were found. A further 3000 seedlings originating in South Australia were tested with blue-green aphid, and 26 tolerant plants were found. The two groups of aphid-tolerant plants were crossed by hand to produce the next generation.
2 In this generation 28 800 plants were tested for tolerance to both aphids, and 147 tolerant plants were found. These plants were allowed to intercross randomly, using honey bees as the pollinators, to produce the next generation.
3 In the next generation, tolerance to the pea aphid also was tested, and 500 plants resistant to the blue-green aphid, spotted alfalfa aphid and the pea aphid were selected. Honey bees were again used to cross-pollinate these plants randomly. The resulting seed was then used as the basis for the new variety Hunterfield.

7 Calculate the proportion of heterozygous plants in each of 10 generations of self-pollination of a first generation (F1) produced by crossing two varieties of rice, using the following formula (assume that the F1 is heterozygous for all genes):

$p = (1/2)^{n-1}$

where

p = proportion of heterozygous plants

n = number of generations of self-pollination.

8 Lupins are a self-pollinating plant. A lupin breeder crosses a high-yielding disease-susceptible variety (M) with a low-yielding disease-resistant variety (H), producing an F1 generation. Describe the steps that the plant breeder would then take to produce a new variety that combined high yield and disease resistance.

Production of hybrid varieties

Hybrid varieties of crops such as maize are produced by the following steps, illustrated in Figure 20.2.

1 Selected maize plants are inbred by self-pollination for 8–10 generations to give true-breeding inbred lines. These inbred lines are less vigorous and have lower yields than the open-pollinated varieties that they were derived from. Many lines actually die out, as inbreeding brings detrimental and lethal recessive genes together in homozygosity.

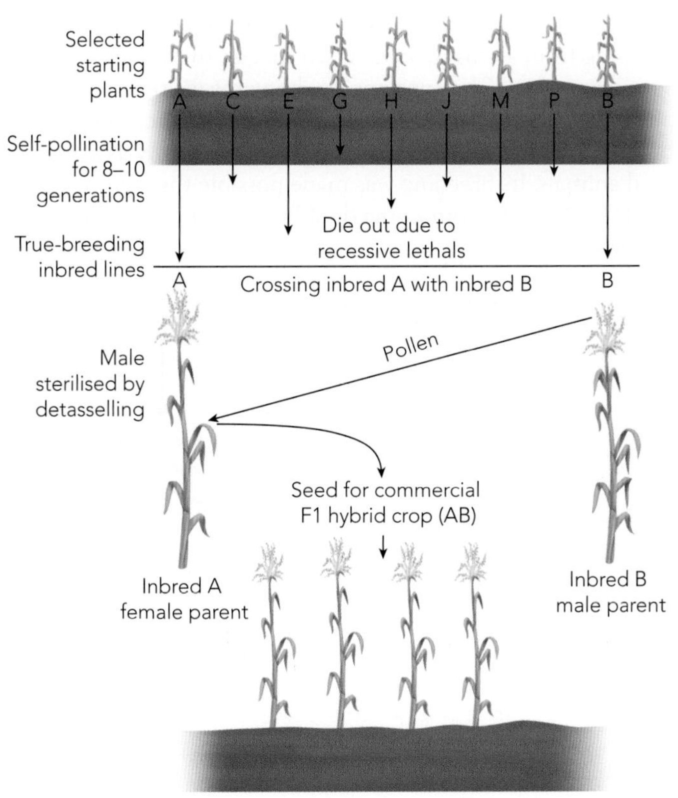

Figure 20.2 The production of a single cross-hybrid maize variety

2 Individual inbred lines are crossed with each other, and the performances of the F1 offspring are tested. The F1 offspring show a phenomenon known as **heterosis**, or **hybrid vigour** (see page 59), where the performance of the F1 offspring is superior to that of their parents and even of the open-pollinated varieties. The cross between the individual inbred lines that produces the best F1 offspring performance makes up the new variety.

To produce commercial seed of this new variety, the two parent inbred lines must be crossed. This is done on a fairly large scale, with the female and male inbred lines being planted in alternate rows. The female line is rendered male-infertile by mechanically detasselling (cutting off the tassels), cytoplasmic male sterility (a gene that makes the pollen infertile) or treatment with a male gametocide (chemical that kills the pollen). Crossing occurs as pollen is then available only from the nearby male parent.

connect

Production of new varieties of maize

Cross-pollination and vegetative reproduction

Many crops that cross-pollinate are propagated vegetatively (e.g. bananas, potatoes, grapes, sugarcane). Cross-pollination is carried out, and the resulting offspring are further reproduced by vegetative means, such as cuttings, tubers or tissue culture. This is **cloning**. Testing can then be carried out. A selected desirable genotype can be multiplied rapidly by cloning and put into commercial use.

Application of new technology

Technologies with the potential to have a substantial impact on plant breeding include wide crosses, tissue culture and genetic engineering.

Wide crosses

Wide crosses are crosses between different species of plant. In the past such crosses have been made, producing plants, but the offspring of these plants were never fertile. Infertility of the offspring resulted from the production of gametes (i.e. reproductive cells) with varying numbers of chromosomes, because pairing did not occur during the meiotic cell division that formed the gametes.

Technology now exists that allows the doubling of the chromosomes of the zygote (i.e. fertilised gamete) arising from a cross between two species so that fertile offspring have been produced, at least in some cases. Chromosome doubling means that chromosome pairing can occur in meiosis, forming gametes with a full complement of chromosomes. Triticale is the result of a cross between wheat and rye and is in use commercially as a grain crop, producing feed for humans and animals. Its breeding was made possible through the chromosome-doubling technique (Fig. 20.3). Chromosome doubling is induced by the application of chemicals such as colchicine.

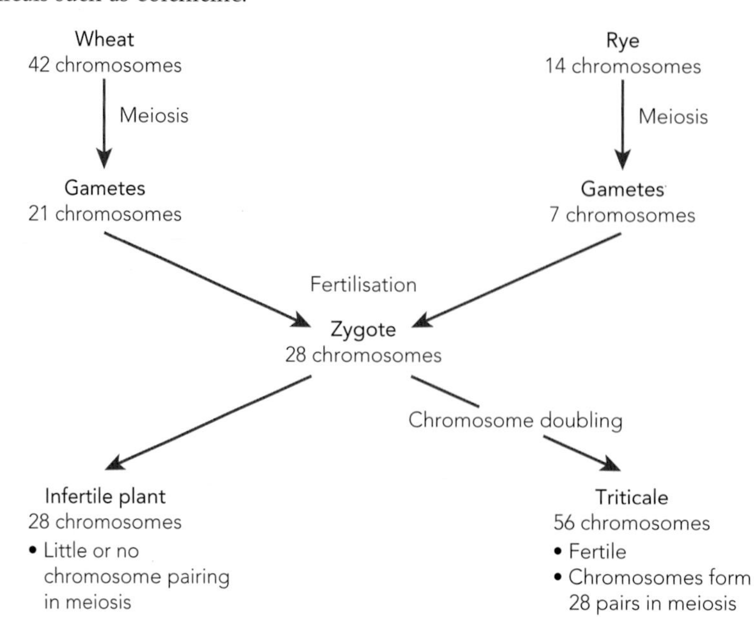

Figure 20.3 How the new plant triticale was produced by crossing wheat with rye using chromosome doubling

ISBN 9780170265560

Tissue culture

Tissue culture is the technique of taking a small part of a plant, such as a terminal bud, and growing it on sterile agar medium (a jelly-like material containing nutrients and hormones). Bud tissue is grown; and later, by changing the concentration of plant hormones in the medium, the tissue can be caused to develop into new plants. These plants are **clones**, having genetic makeup identical to the plant from which the original bud was taken. The advantages for the plant breeder are that a genotype can be multiplied rapidly and that its testing and evaluation can take place much faster. Once a new genotype proves itself, it can be rapidly multiplied in commercial production by the tissue culture technology.

Genetic engineering

Genetic engineering is perhaps the most exciting of the new technologies, because it appears to offer the possibility of making rapid genetic gains that would take years with conventional breeding programs. Genetic engineering involves four steps.

1 *Identification*. The segments of DNA that code for a particular gene responsible for a desired feature are identified.
2 *Isolation*. The DNA segment is isolated.
3 *Cloning*. The DNA segment is cloned by introducing it into a host, such as bacteria or yeast. As the host multiplies, so does the introduced DNA segment.
4 *Transference*. The cloned DNA can then be transferred to other members of the species that it came from or to another species where it produces the desired effect.

Some success has been achieved in this field. The CSIRO field tested two virus-resistant potato varieties. These were bred by adding one gene, giving resistance to the potato leaf-roll virus (Fig. 20.4). The resistant gene was made by taking a gene from the virus itself – the gene that codes for the protein coat of the virus. The research meant that less pesticide is used in controlling the aphids that acted as vectors for the virus.

Almost 100% of Australia's cotton crop is grown with **transgenic** (see page 131) varieties. Bollgard II®, a genetically engineered variety of cotton, contains two genes from the naturally occurring soil bacteria *Bacillus thuringenesis*. The genes produce a chemical that kills the *Helicoverpa* moth larvae (a major pest of cotton) as they feed on the cotton plant. Again, a reduction in the use of pesticides has been achieved. Australian cotton growers reduced their insecticide use by approximately 80%, with some crops not sprayed for insects at all.

Two cotton varieties in use, Roundup Ready Flex® (with genes from the soil bacterium called *Agrobacterium tumefaciens*), and Liberty Link® (with genes from the soil micro-organism *Streptomyces hygroscopicus*) were engineered to be unaffected by the herbicide Roundup® early in their lives. This means Roundup® can be used to control weeds early in the life of this crop without damaging the cotton plants.

9 Name three new technologies that have been applied to plant breeding programs.

connect

Cotton biotechnology

More information on the benefits of transgenic cotton

Alamy/Nigel Cattlin

Figure 20.4 The two potato plants have been exposed to the potato leaf roll virus. The plant on the left has been genetically engineered to possess a virus-resistance gene, and is healthy. The plant on the right is infected.

connect

Risks of gene technology

View this video and answer the questions.

ISBN 9780170265560

Sources of genetic variation

There is wide variation within one species, considered on a worldwide basis. Variation is important for the continuation of genetic progress in breeding programs, producing new varieties that are better adapted to their target environments, resistant to disease and pests, and improved in yield and quality.

Mutations are a source of genetic variation. A mutation is the failure of a gene to replicate itself exactly in the process of meiosis. There is a change in the DNA, which causes a change in the expression of the gene; that is, a new allele is formed. Most mutations under current circumstances are detrimental, resulting in reduced fitness. However, there have been some useful mutations. The gene for short straw in wheat (dwarf wheat) was a mutation discovered in Mexico. Because the wheat plants were not subject to **lodging** (i.e. falling over when blown by wind and rain), more wheat could be harvested. This played a part in changing Mexico from a wheat-importing country to a wheat-exporting county.

Genetic engineering opens up the availability of genes from almost any organism. These genes can be transferred to plants, giving them new attributes that could not be achieved by normal within-species breeding programs. For example, there is no known gene for resistance to the pest pea weevil in field peas. The French bean does have a gene for resistance to this pest and it has been transferred to the field pea plant, giving it resistance.

ISBN 9780170265560

Chapter review

Things to do

1. Buy some hybrid sweet corn seed and grow a crop. Make careful records of characteristics, such as the height of plants, the size of cobs, the number of cobs per plant and the time from planting to harvesting. Keep some of the seed produced, sow it next season, and observe whether there is more variation in this generation than in the previous one. Careful records of means and their standard deviations in each year will be necessary. Make sure that the same planting density is used each year. A photographic record may help in making comparisons between the two generations.
2. Carry out an experiment to compare the yield of a hybrid variety of a crop with an open-pollinated variety. A vegetable crop, such as tomatoes or zucchini, could be used.

Things to find out

1. Explain the term 'genetic engineering'. Describe how this is carried out with plants.
2. Explain one example of genetic engineering or associated technology that has resulted in improved plant production.
3. Describe the process a plant breeder would use to produce a new variety for a self-pollinating species.
4. Describe how a new variety of plant is produced for a cross-pollinating species (e.g. maize) by a plant breeder.
5. Discuss the need for the development of plant seed banks around the world.
6. Find out where breeding programs are being carried out in Australia for the major crop grown in your region, and what the aims and priorities of these programs are.
7. Research and outline the genetic basis of the following methods used in plant breeding: selective breeding, hybridisation, genetic engineering
   a. Describe each method.
   b. Outline the genetic basis for each method.
   c. Describe how each method has improved production for quality, yield and environmental adaptation.
8. Evaluate arguments for and against the use of food products by people from genetically modified plants.

Extended response questions

1. Describe the role of the plant breeder in improving productivity in plant production systems, and explain the basic genetic principles applied in plant breeding.
2. Plant breeding systems have developed to incorporate new genetic techniques aimed at improving product quality, yield and environmental adaptation. Discuss the implications of this development on Australian agricultural systems.
3. One of the results of genetic engineering of crop plants is food that is processed from them. Some people are calling for food produced from genetically engineered plants to be labelled as such.
   a. Discuss the arguments for labelling food from genetically engineered plants.
   b. Discuss the arguments saying that such labelling is unnecessary.
   c. State what your opinion is on this issue and the reasons for your opinion.
4. Explain how tissue culture is performed and assess its use in commercial plant production systems.

CHAPTER 21

Outcomes

Students will learn about the following topics.

1. The nature and impact of microbes, invertebrates and pests on plant production systems.
2. Responsible and strategic use of chemicals.
3. Integrated pest management.

Students will be able to demonstrate their learning by carrying out these actions.

1. Describe the effects of disease on plants.
2. List the types of disease that can occur in plants as microbial, metabolic and metazoal, and give examples of each.
3. Outline one important disease and one pest for a selected crop/pasture.
4. Evaluate methods that can be used to control and prevent plant pests and diseases.
5. Describe how infection of a plant by a pathogen and the establishment of disease can take place.
6. Explain the interrelationship between the pathogen, host and environment in causing disease.
7. Describe the mechanisms of resistance in plants to micro-organisms, insects and mites.
8. Interpret an agricultural pesticide label and relate it to safe practice and correct usage.
9. Define integrated pest management (IPM).
10. Outline IPM's ability to reduce the problems of pesticides and chemical resistance in target organisms.
11. Explain how pests and diseases are controlled using these:
    - a eradication
    - b quarantine
    - c management control
    - d genetic control
    - e biological control
    - f induced sterility
    - g pheromones
    - h chemical control
    - i integrated control.
12. Describe an IPM program, naming the target organism and plant host.

ISBN 9780170265560

PLANT PESTS AND DISEASES

Words to know

antibiosis the adverse effects of the plant on insects or mites after they have consumed at least some of the plant

chlorosis the loss of chlorophyll in a plant

companion planting growing different crops in rows next to each other; usually one type of plant has factors to ward off pests that would normally attack the other plant

contact or **stomach poisons** poisons that are absorbed by some part of the insect's body

crop rotation the process of growing different crops, including pasture, from one year to the next on the same piece of land

damping off where young seedlings wilt and die due to fungal attack

deficiency the lack of a substance important for normal growth, especially mineral nutrients such as nitrogen and phosphorus

dieback applies to trees and woody plants and is characterised by the progressive death from younger branches to the base of the plant; causal agent is root pathogens

fungicide chemicals used to control fungal diseases

galls abnormal swellings and outgrowths found on plants caused by some pathogens and insect activity

haustoria specialised hyphae found in fungi for extracting nutrients from inside cells

hyphae thread-like structures that are part of the structure of fungi

induced sterility sterilisation of male pests (e.g. flies) by chemical or radiation means; the flies are then released into the wild population to breed and no offspring are produced from these matings

inoculum the infective units of a pathogen that start the infection process

insecticide a chemical substance used to kill insects

long fallowing the practice of ploughing the land for a crop several months before planting (to accumulate soil moisture) and then cultivating it after rain

micro-organism an organism that cannot be seen with the naked eye but can be seen with the aid of a microscope; in general, organisms with a diameter of less than 1 mm

miticide chemical used to control mites

necrosis the death of tissue

pheromones a chemical substance released by an animal that influences the development or behaviour of other animals of the same species

rhizosphere the area of soil in contact with, and under the influence of, the root system of a plant

salinity the level of salt in the soil; usually refers to unacceptably high levels

toxicity the effects of the oversupply of a substance, usually a mineral nutrient

toxin a chemical produced by a micro-organism that has an adverse effect on the life or function of another organism

volunteer (plants) that are not intentionally planted (e.g. wheat plants growing by the side of the road or around silos)

withholding time the period of time after the application of the pesticide before the product can be harvested and used for human consumption or processing

Introduction

Constraints on plant production include environmental factors, competition, plant genetic makeup, management practices and pests and diseases. Plants used in agricultural production systems are affected by diseases caused by a wide range of pathogens, including bacteria, viruses, fungi, insects, arachnids and nematodes. The most common diseases are caused by fungi, bacteria and viruses. Some common plant diseases are shown in Figure 21.1a–d.

a Bacterial canker of tomato (caused by a bacterium). Leaflets wilt and turn brown. Cavities develop in the stem. Raised brown fruit spots are surrounded by a white halo.

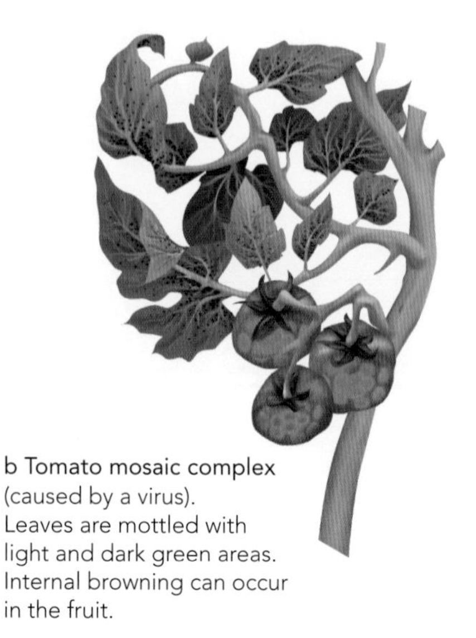

b Tomato mosaic complex (caused by a virus). Leaves are mottled with light and dark green areas. Internal browning can occur in the fruit.

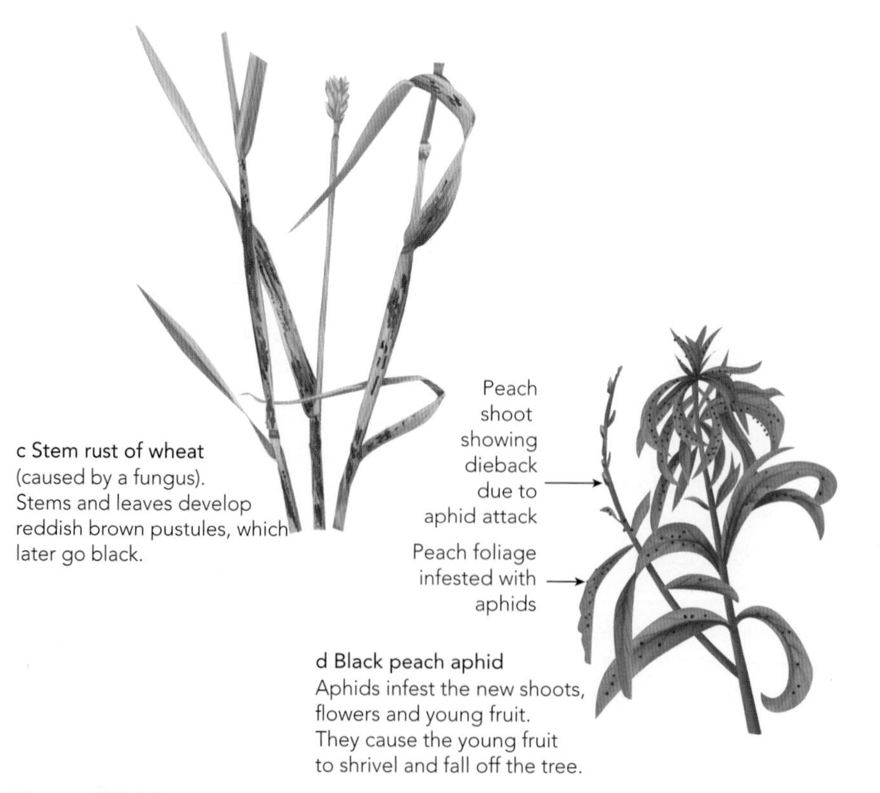

c Stem rust of wheat (caused by a fungus). Stems and leaves develop reddish brown pustules, which later go black.

d Black peach aphid
Aphids infest the new shoots, flowers and young fruit. They cause the young fruit to shrivel and fall off the tree.

Figure 21.1 Some common plant pests and diseases

ISBN 9780170265560

Non-infectious diseases of plants may be caused by mineral deficiencies, mineral toxicities, temperature extremes, water excesses or deficiencies, hail or farm implements.

Disease in plants can be considered as any change from the normal growth of a plant. This is a very wide definition and includes the effects of pests, **micro-organisms** and physical factors such as **mineral nutrient** (see page 243) deficiencies and toxicities, water shortage or oversupply, and actual physical damage. Micro-organisms that cause disease are called pathogens. A pest is any insect or organism that injures, irritates or damages plant products, and can adversely affect productivity.

Effects of disease on plants

Disease may affect only one part of a plant – the leaves, stem, roots or reproductive structures (flowers and fruit) – or several parts at once, or the entire plant.

Symptoms of disease

There are many signs of disease (symptoms).

Stunting and reduced yield

Often caused by specific pathogens or in association with other conditions (e.g. drought, root-rot), stunting and reduced yield are nearly always associated with disease. They can be caused by mineral nutrient **toxicity** or **deficiency**, drought or waterlogging, as well as by micro-organisms and pests. To gauge whether stunting and reduced yield have occurred, it is helpful to have some healthy plants for comparison (Fig. 21.2). Performance in previous years might be a guide here. Black stem rust of wheat causes reduced yield, as does a deficiency of phosphorus or nitrogen.

Necrosis

Necrosis, or death of tissue, might be localised in one particular organ, or it can be general to the whole plant (Fig. 21.3). Again, it can be caused by particular micro-organisms or by mineral nutrient deficiencies and toxicities. Shot-hole disease of stone fruit leaves is caused by a fungus that invades the leaves, causing small round pieces of the leaf to die and drop out. The resulting appearance of the leaves is as though they have been peppered with shotgun pellets. Potassium deficiency causes the death of the edges of leaves in bananas and apples, and boron toxicity causes the tips of orange leaves to become necrotic.

Figure 21.2 An alfalfa mosaic virus-infected (AMV) lentil plant showing yellowing and stunting of plant on the left compared with the healthy plant on the right. Note that the plant on the left shows considerable stunting of the leaves.

Figure 21.3 Leaf with necrosis

Chlorosis

Yellowing, or **chlorosis**, is the changing of leaves from green to yellow, or their not ever turning green as they should when they develop. Chlorosis is a very common symptom of disease. Its origin can be a nutrient deficiency (e.g. nitrogen) or micro-organisms, such as bacteria, fungi, viruses and mycoplasmas. Chlorophyll is either destroyed by the invading pathogens, or its formation is disrupted (e.g. by lack of nitrogen).

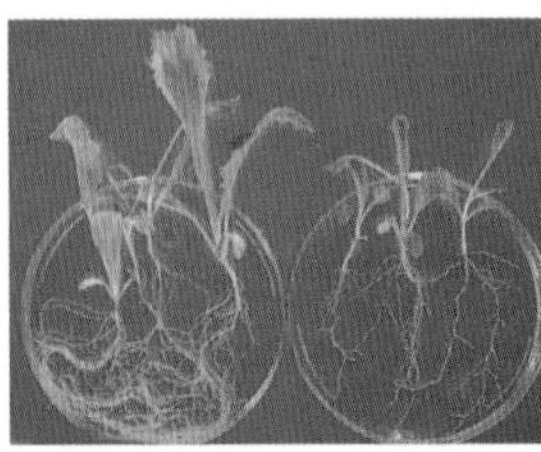

School of Biological Sciences, University of Sydney

Figure 21.4 Damping off symptoms shown by the seedling on the right

Damping off

Damping off is where young seedlings in particular wilt and die due to attack by the fungi pythium and rhizoctonia on their root systems and stem bases, as shown in Figure 21.4.

Wilting

Wilting (see page 201) applies to more mature plants and is the result of their not being able to obtain sufficient moisture. This might be simply because there is insufficient available water in the soil, or it might be because pathogens have damaged the root system or **vascular tissue** of the plant. The vascular wilts group of diseases damage the xylem (see page 257) tissue and cause wilting.

Alamy/kris Mercer

Figure 21.5 The two peach leaves centre and left are affected by peach leaf curl fungus

Dieback

Dieback applies to trees and woody plants and is characterised by progressive death from the younger branches to the base. It is caused by pathogens that affect the roots and vascular tissue. Damage from insect species may also contribute to dieback.

Colour deviation

Deviation from normal colour is where the plant leaves or flowers have different colours from normal. These changes are often associated with virus infections. Tobacco mosaic virus in leaves shows a mosaic pattern of varying shades of green. White streaks in the flowers of some plants are the result of virus infection. A deficiency of phosphorus is characterised by leaves of a dull bluish green with purplish tints.

Galls

Swellings and outgrowths are caused by some pathogens and the activities of some insects. These abnormal growths are called **galls**. Figure 21.5 shows how peach leaves are distorted by the peach leaf curl fungus.

Rot

May result from death of cells or breakdown of cell walls following the action of enzymes

Disease syndromes

Often one particular disease will have several of these signs or symptoms, and the combination of these symptoms is known as a syndrome. The virus disease necrotic yellows of lettuce, as the name suggests, is a combination of chlorosis and necrosis.

1 List the symptoms of disease in plants.
2 What is meant by a 'syndrome' of a particular disease?

Types of disease

There are three types of disease in plants: microbial, metabolic and metazoal.

Microbial diseases

Microbial diseases are caused by pathogens. These diseases are most commonly caused by fungi and viruses, though some diseases are caused by bacteria, including mycoplasmas. Some examples of microbial diseases are as follows:

1 *Fungal diseases*. Black stem rust of wheat, powdery mildew (which affects many plants), late blight of potato, yellow stripe rust of wheat, take-all of wheat, and black spot of roses, to name just a few (Fig 21.6).

 ISBN 9780170265560

2 *Viral diseases*. Bunchy top of bananas, tobacco mosaic, potato leaf-roll virus (Fig. 21.7), broad bean wilt, necrotic yellows of lettuce, and mottly dwarf virus of carrots
3 *Bacterial diseases*. Black rot of cabbage, halo blight of beans, bacterial wilt of potato and crown gall of stone fruit
4 *Mycoplasmal diseases*. Big-bud of tomatoes, summer death of beans and yellow dwarf of tobacco.

Alamy/Nigel Cattlin

Alamy/Nigel Cattlin

Figure 21.7 The potato leaves on the left are infected with potato leaf roll virus and show characteristic rolling. The leaves on the right are normal.

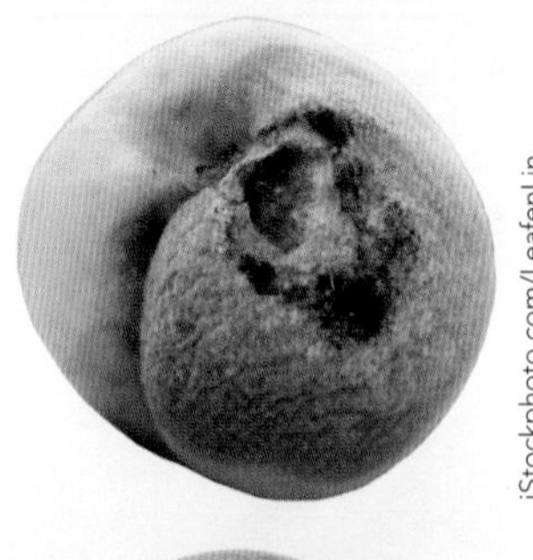
iStockphoto.com/LeafenLin

Shutterstock.com/bonchan

Figure 21.6 At the top, the effects of the fungal disease brown rot on peaches. The peach below is normal.

3 List the types of pathogens that cause disease in plants.

Metabolic diseases

Metabolic diseases are caused by mineral nutrient deficiency or toxicity.

Mineral nutrient deficiencies

The visible symptoms of mineral nutrient deficiencies are listed in Table 15.7, page 220. The symptoms of the three most likely deficiencies are:

1 *Nitrogen*. Growth is stunted; shoots are dwarfed; production of laterals is restricted; leaves are sparse, pale green to yellow; lower leaves die early.
2 *Phosphorus*. Growth is stunted; shoots are dwarfed; leaves are bluish green, often with purplish tints; leaves die progressively from the base of the plant upwards; root growth is very restricted.
3 *Potassium*. Growth is stunted; shoots are dwarfed; there is excessive production of lateral shoots; leaves develop small white or brown dead spots, which gradually extend to produce scorching on their margins; leaves die from the base of the plant upwards. There is variation in the expression of these symptoms.

Nutrient toxicities

Nutrient toxicities show themselves in various ways.

Salinity is an extreme form of toxicity, and is likely to be an excess of several nutrients. Salinity restricts plant growth and in severe cases results in plant death.

The symptoms of chloride toxicity are scorching or burning of the ends and edges of the leaves. There may also be yellowing of the leaves.

Metazoal diseases

Metazoal diseases are caused by organisms that are visible to the naked eye. These organisms include:

1 *Roundworms or nematodes*. Nematodes attack the stems, leaves, roots and other underground parts of plants, such as tubers. An example is the root knot nematode.
2 *Arachnids*. Arachnids, especially mites, cause considerable damage to plants. Examples include the red-legged earth mite, red spider and two-spotted mite.

ISBN 9780170265560

4 Describe the difference between microbial and metazoal diseases.
5 Give an example of a metabolic disease in plants.

3 *Insects*. Many species of insect attack crops, pastures and stored plant products. Some examples are the cotton boll worm, cabbage white butterfly (Fig. 21.8), cut worms, Queensland fruit fly, aphids (Fig. 21.9) and rice weevil.
4 *Parasitic plants*. Examples are dodder and mistletoe. Dodder lacks chloroplasts and is entirely parasitic on lucerne plants.

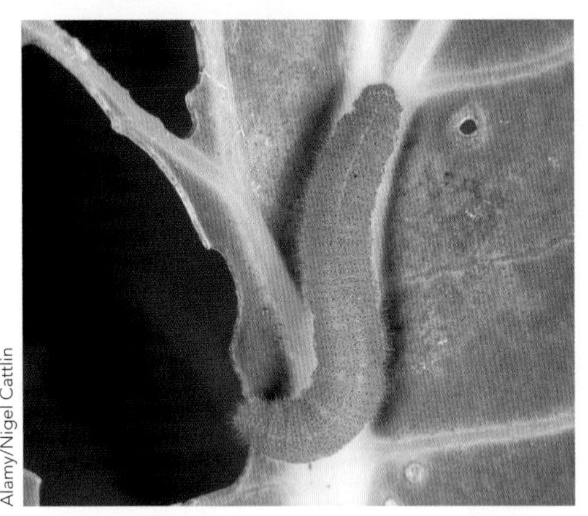
Alamy/Nigel Cattlin

Figure 21.8 The larva of a cabbage white butterfly in a characteristic position on the midvein of a cauliflower leaf

Alamy/Nigel Cattlin

Figure 21.9 Aphids feeding on young tissue and leaves at the end of a citrus branch. Aphids can act as vectors of viral diseases, transferring virus particles from infected plants to uninfected plants

Infection and disease

Micro-organisms cause disease by entering a host, multiplying and leaving the host. Once inside the plant host, the extent of damage is determined by the growth of the organism or enzymes it produces that are able to decompose host tissue. The severity of the disease can also be influenced by the environment. If the environment is unfavourable, the disease will not occur.

This interrelationship, or interaction, is shown in Figure 21.10 and is known as the **disease triangle** (see page 143). Plants affected by the activities of micro-organisms show various symptoms according to the particular interaction of host, pathogen or parasite, and environment.

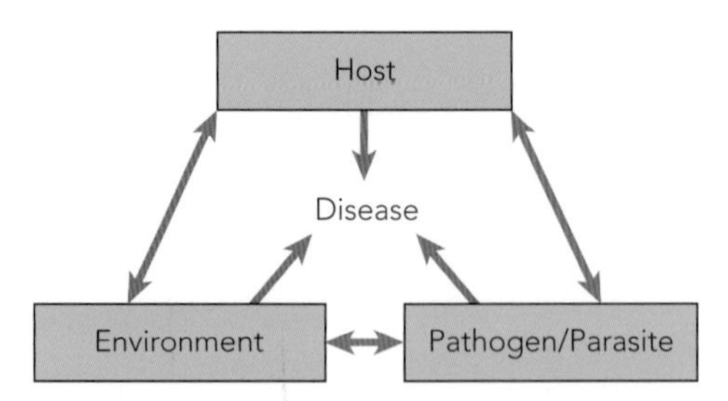

Figure 21.10 Disease triangle

The interactions between host, pathogen and environment may result in little disease with minor economic loss in one season, but severe disease and major losses in another season. For example, some diseases result in major losses in wet conditions, but not during dry conditions. Downy mildew can only infect grape vines in wet conditions.

The infection process depends on the type of pathogen. An infective unit of a pathogen that start the infection process is referred to as an **inoculum** (e.g. fungal spore, virus particle). Plants can be infected by a number of pathogens at the same time.

Plant pathogens may enter their hosts in a number of ways. Some pathogens enter by penetrating the **cuticle** (see page 257). For example, powdery mildew fungus penetrates the cuticle and epidermal wall and then produces a number of structures that feed in the epidermal cell.

ISBN 9780170265560

Some pathogens, especially bacteria, enter their hosts through wounds (e.g. punctures), which can be caused by feeding insects, pruning, agricultural implements, hail, rain and wind.

Other pathogens enter through natural openings, such as stomata. The bacterium that causes soybean blight enters the leaves through the stomata and usually causes localised spots.

Figure 21.11 a and b illustrate the disease triangle relationships for infection of cabbage by an insect pest, the cabbage white butterfly, and for a bacterial infection called black rot. Black rot, caused by the bacterium *Xanthomonas campestris* pv. *Campestris* (Xcc), is a significant disease of cabbage and other crucifer crops worldwide.

6 What factor plays a significant role in plant disease development?

7 List the ways that pathogens may enter a plant.

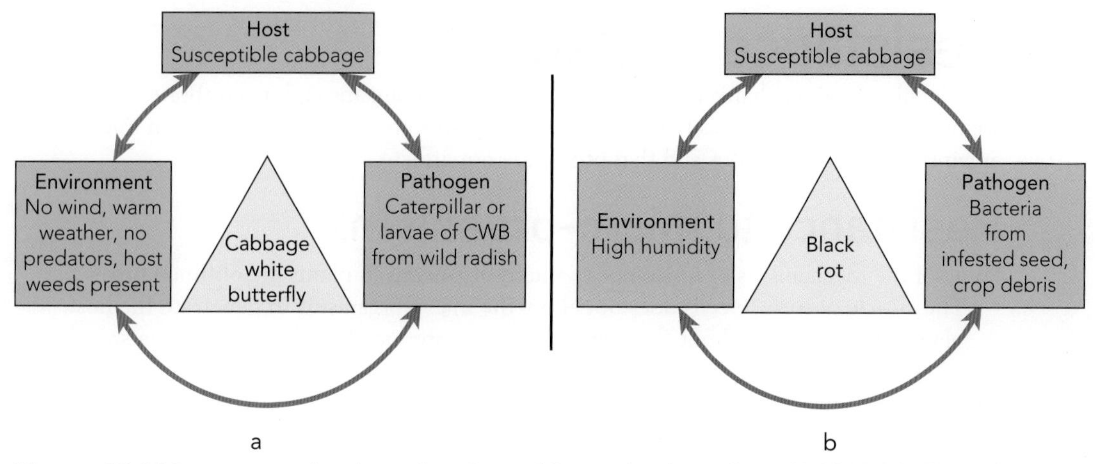

Figure 21.11 Disease triangle relationships for **a** cabbage white butterfly and **b** black rot infections of cabbage

Black stem rust

Let us consider one particular disease: black stem rust of wheat. The inoculum for this disease in Australia is uredospores (rust spores) from **volunteer** (i.e. not intentionally planted) wheat plants growing nearby in old wheat paddocks, around headlands, around silos, in railway yards, and along railway lines and roadways. As these spores are light, they are wind-blown and land on the wheat plant. There they germinate and infect the plant by their **hyphae** growing through the stomata. The fungal mycelium grows throughout the tissue of the plant, feeding from the cells by means of **haustoria** (specialised hyphae for extracting nutrients from inside cells). The fungus then produces masses of rust-coloured uredospores, which burst through the epidermis and are blown off in the wind to infect other plants (Fig. 21.12).

The fungus produces disease in two ways:

1 by using up the plant's nutrients
2 by reducing the photosynthetic capacity of the plant as the spores are produced.

Alamy/Nigel Cattlin

Figure 21.12 Black stem rust

Extent of damage

The extent of damage within the host plant depends on one or more of:

1 the growth of the organism
2 the production of **toxins**
3 the production of enzymes that decompose host tissue.

Disease control

For disease to occur in plants, three conditions must be present. There must be:

1 a host plant
2 a pathogen
3 most importantly, an environment (moisture, temperature, chemical factors) that suits the pathogen.

The interrelationship between these three conditions or factors is shown in Figure 21.10.

8 What must happen for a plant to become infected with a pathogen and become diseased?

9 Why is the relationship between the host, pathogen and environment important in disease control?

ISBN 9780170265560

Control of disease often relies on removing the effects of one of these conditions. A host plant resistant to the pathogen can be grown (e.g. wheat varieties that are resistant to black stem rust).

The pathogen can be kept out by quarantine – the reason, at least in part, for Australia's strict quarantine regulations preventing the entry of plant material or soil that could bring with it pathogens and pests not already present in this country. The environment can be changed so that it does not favour the pathogen; for example, changing from overhead spray irrigation to furrow or trickle irrigation, thus reducing humidity and wetness in the crop canopy and making it less favourable for fungi that attack the leaves.

Resistance in plants

The details of the mechanisms of resistance to micro-organisms and invertebrates in plants are not well understood. It is known that certain plants are susceptible to certain micro-organisms and invertebrates and that others are not affected at all.

Resistance to micro-organisms

Some of the mechanisms of resistance to micro-organisms in plants are outlined below.

- The cuticle of surface cells does not allow the micro-organisms to penetrate the host plant cells, with or without the help of enzymes produced by the micro-organisms that digest the cuticle.
- The waxy substance called **suberin**, which impregnates cork cells in bark, making them impermeable to water and resistant to decay, forms a natural barrier to micro-organisms.
- The lack of nutrients and energy sources for the micro-organisms on the surface of the plant does not allow them to multiply or grow.
- Chemicals exuded onto leaf surfaces or into the **rhizosphere** of the plant (i.e. the region of the soil in contact with its roots) are toxic to or inhibit the growth and multiplication of the micro-organisms.

In some cases the micro-organism succeeds in penetrating the plant but fails to become established. Reasons for this include the following.

- The micro-organism is unable to obtain nutrients and energy from the host because it cannot produce the enzymes necessary to break down host tissue.
- Chemicals produced by, or already in the plant, inactivate or neutralise the enzymes produced by the invading micro-organism.
- Chemicals produced by, or already present in the host plant, are directly toxic to the micro-organism. These chemicals are known as phytoalexins (*phyton* = plant, *alexin* = warding-off compound). The chemicals produced during the death and breakdown of cells penetrated by micro-organisms in some plants are inhibitory or toxic to the micro-organism and so prevent further invasion.
- Special cells that sustain the micro-organism in susceptible plants are not formed. For example, developing nematodes feed on giant cells that develop in susceptible plants, but in resistant plants these giant cells are not formed and the nematodes cannot feed.
- Layers of cork cells are formed, or extra cell wall thickening is laid down by the plant around the penetrating micro-organisms, thus restricting their further penetration.

10 List the mechanisms that prevent micro-organisms from penetrating the potential host plant.

11 What mechanisms prevent a micro-organism from becoming established once it has penetrated a potential host plant?

Resistance to insects and mites

The resistance of plants to insects and mites involves:

1 the non-preference of insects and mites for certain plants
2 **antibiosis** – the adverse effects of the plant on the insects and mites after they have consumed at least some of the plant.

Non-preference

In non-preference, resistant plants are not chosen by the pest for feeding or egg laying, or for both feeding and egg laying. On some plants on which eggs are laid the hatching larvae feed only slightly or not at all. Other plants are not selected at all for egg laying.

It is likely that chemical factors are involved in non-preference. These chemicals may be volatile repellents formed by the plant, or non-volatile taste factors that have their effect

 ISBN 9780170265560

after feeding commences. Physical factors also may be involved, such as the colour of the plant surface (both infrared and visible), the nature of the plant surface (e.g. its hairiness or waxiness), and the shape or outline of the whole plant or parts of the plant (e.g. leaves).

As an interesting aside, some of these factors are used by certain plants to attract insects so that pollination can take place. For example, some orchids have flowers whose shapes mimic those of female insects. The male insects 'mate' with the flowers as though they are female insects and in so doing transfer pollen from one flower to another. The daisy plant's flower 'perfume' attracts flies, which aid in its pollination.

Antibiosis

Antibiosis effects are shown after the insects and mites have started to feed. They can be the result of toxic chemicals in the plant or of the plant's not providing an adequate diet for the pests. Effects include:

- Insects or mites die, and if they reach the adult stage they are smaller and less vigorous than they would be had they been feeding on susceptible plants.
- Fewer and less fertile eggs are laid than if they were feeding on susceptible plants.
- Feeding time in the juvenile stages is longer, resulting in a reduced number of generations in a season or year than if feeding had been on susceptible plants.

12 Show how non-preference and antibiosis are thought to act to stop the attack of insects on plants.

Management and control

Pests and disease take a considerable toll on plant production in any one year, and losses would be much higher if control measures were not put into practice. There is always the overriding factor of economics with the use of any control measure; that is, the extra returns generated by the use of the control must exceed the cost of its implementation.

Control measures include: eradication, quarantine, management control, genetic control, biological control, induced sterility, pheromones, chemical control (pesticides) and integrated pest management.

Eradication

In the horticultural industries, eradication practices, such as soil sterilisation, pasteurisation and fumigation using heat and chemicals, are carried out to rid the soil of pathogens and pests. These processes have a high cost.

Quarantine

Quarantine measures can apply at the national level, as previously mentioned, where the importation of plant material (including seeds) and soil into Australia is strictly monitored and controlled. Quarantine is also used to prevent pests already present in one area of the country from entering other areas. For example, fruit is not allowed to be taken into the Murrumbidgee Irrigation Area in New South Wales, to prevent the entry of the Queensland fruit fly (Fig. 21.13). The aim is to keep out diseases and pests that are potentially devastating to plant production.

Alamy/Graphic Science

Figure 21.13 The Queensland fruit fly

Quarantine is on the whole successful, but it depends on the vigilance of its enforcement. Some breaches do occur, as shown by the accidental introduction of lucerne aphids in 1976 and their subsequent devastation of the Australian lucerne crop. This does not mean that quarantine is not worthwhile. Its cost is well and truly justified by the potential expense of dealing with the many exotic pests and diseases that could be introduced, and the loss of export markets, if it did not exist.

13 Why are the extensive Australian quarantine regulations justified?

Management control

Management control relies on taking positive steps to disrupt the lifecycles of pests and diseases and thus stop their build-up to levels where production is adversely affected.

- *Crop rotation*. The idea in **crop rotation** is to grow different, but generally unrelated, crops and pastures in succession on the same piece of land, so that a disease or pest of one crop does not build up to damaging levels. If a pest or disease is present and affects one crop, it will not be able to attack the next crop because this is generally so different. The population of the pest or disease will decrease and be very low when the original crop is grown again. Legume crops (e.g. soybean, lupins, field peas) are grown in rotation with wheat. Other crops (e.g. oats, barley, sorghum, sunflower, canola) are often included, as are legume-based pastures. Crop rotation is important in the control of soil-borne diseases, such as take-all in wheat.
- *Long fallowing*. The practice of **long fallowing** means ploughing the land for a crop several months before planting and then cultivating it after rain. Materials such as old crop residues and weeds, which could harbour disease inoculum or pests, decay or are destroyed. When the crop is finally sown, there is less chance of disease occurring.
- *Changing the climate*. Climate cannot be changed on a large scale, but it can be modified in situations such as glasshouses. For their establishment, some pathogenic fungi like high humidity and even free water on leaves, so the adoption of a watering system that does not wet the leaves or raise the humidity is necessary. The removal of weeds in field crops also can assist in reducing pathogenic fungi because there is less chance of a build-up of high humidity in the crop canopy.
- *Sanitation*. Sanitation involves removing and destroying any potential sources of disease. It ranges from cleaning and disinfecting equipment, pots and the insides of glasshouses, to removing volunteer plants that may be a source of disease or pests from near a growing crop. Grain harvesting, handling and storage equipment should be thoroughly cleaned to remove all grain that could harbour any grain insect pests. Individual potato plants that are stunted or show any other signs that they may be infected with a virus disease should be dug up and burnt, so that insects such as aphids cannot transfer the virus to other plants in the crop. Prunings from fruit trees often contain diseased material, and so they should be burnt to destroy potential sources of pathogen inoculum.
- *Disease-free planting material*. When planting crops are propagated asexually, such as with strawberries and potatoes, it is advisable to use material that has been declared free from disease. This planting material, such as tubers, bulbs or stolons, comes from farms where the growing of the crop that produced it has been carefully monitored and inspected by an appropriate independent authority that can certify that the crop is disease free. These schemes are particularly important in preventing the spread of viral diseases in potatoes and strawberries. In the horticultural industry the rootstocks chosen for budding and grafting are often selected because of their resistance to certain soil-borne diseases. For example, in citrus, *Trifoliata* rootstock is resistant to root-rot.
- *Companion planting*. In **companion planting**, different crop plants are grown in rows next to each other. One kind of plant has factors that ward off pests that would attack the other kind of plant. For example, it is claimed that the herb basil, grown with tomatoes, reduces the incidence of insect pests in the tomato crop.

14 Explain how each of the management control measures discussed aims to disrupt the lifecycles of pests and disease.

Genetic control

Genetic control involves breeding and selecting varieties of plants that are resistant to disease and pests. It is a very effective method of control. It has been particularly important in the continuing success of the Australian wheat industry, where there are continuous breeding programs to produce varieties that are resistant to the constantly changing black stem rust and other fungi. Lucerne varieties have been bred to resist lucerne aphids (see page 289), and growing them has replaced spraying with insecticides as a means of control. Unfortunately, resistance may not be found for all pests and diseases, and it takes a long time to develop resistant cultivars. The new technology of genetic engineering has potential for

 ISBN 9780170265560

speeding up the process and providing opportunities in directions that were previously only dreams for plant breeders. For example, genes from a bacterium that naturally produces chemicals that act against the heliothis caterpillar (a major pest of cotton) have been introduced into the genome of cotton plants. With this gene incorporated, the cotton plant has a built-in supply of 'insecticide'. Also, the CSIRO has developed potato varieties that are resistant to the potato leaf-roll virus. A synthetic gene has been added to the plant's genetic makeup. The gene came from the virus itself and codes for the protein coat of the virus. It has taken approximately one-fifth the time to develop the genetically engineered varieties as it would have taken to develop a new variety using conventional breeding methods.

15 Give two examples where breeding resistance has been successful in helping to control pests and disease.

16 Why is genetic engineering thought to offer great potential in developing plants resistant to pests and disease?

Biological control

Biological control is an intriguing form of control where the natural enemies (predators, parasites or diseases) of pests are used to keep their numbers at acceptable levels. There have been some spectacular successes with biological control. Here are some examples.

1 *Cottony-cushion scale*. In California the citrus industry was almost wiped out by the introduction of the cottony-cushion scale insect. The Australian vedalia ladybird beetle, which feeds on the cottony-cushion scale, was imported from South Australia and released. So effective was the ladybird that the scale insect has never again become an important pest in California.
2 *Green vegetable bug*. The green vegetable bug is a pest of vegetable crops, including tomatoes and beans. It is controlled by a parasitic wasp that lays its eggs in the eggs of the green vegetable bug. The wasp's eggs hatch, and the larvae feed on the green vegetable bug's eggs, destroying them and preventing them from hatching.
3 *Cabbage white butterfly*. The cabbage white butterfly is a very destructive pest of crops belonging to the family Cruciferae, such as cabbage, cauliflower and broccoli. Its larvae literally eat away the leaves, leaving only the major veins. It can be controlled by spraying the bacterium *Bacillus thuringiensis* onto the plants, which when ingested by the larvae cause them to become diseased and die. The bacterium is harmless to humans and does not remain in the environment, which means that it has to be applied when required.

Where biological control has failed there has been failure to test adequately the control organism for its effects on other non-target species. The cane toad is one such organism. It was introduced to the Queensland cane fields from Hawaii to control the cane beetle. It ate everything else but the cane beetle and has become a pest in itself. It has no predators and is poisonous to animals that make the mistake of eating it. The cane toad has an adverse effect on many Australian native fauna, such as reptiles.

Three conditions are necessary for biological control to be successful.

1 The predator or parasite must be species specific for the pest to be controlled.
2 The predator or parasite must become established in its new environment so that it survives in sufficient numbers to control the pest.
3 The predator or parasite must be easily spread in the target pest population.

Clearly, much research is involved in finding successful biological controls, but they are attractive because they are self-sustainable, once established, and hence their continued cost is limited.

17 Using an example, explain how biological control works.

18 Why is biological control a very attractive form of control?

19 What precautions must scientists take before releasing a new biological control agent into the environment?

Induced sterility

In the **induced sterility** method of control, large numbers of the pest are produced in the laboratory, and the males are sterilised with radiation or chemicals. These males are then released into the wild population, where they mate with females. No offspring are produced from these matings, and so the population is decreased. This method relies on females mating only once in a lifetime. It has been successfully used against the screw worm fly in the southern United States and Mexico. Research has been done in Australia using this method to control the Queensland fruit fly, but the vastness of the country has meant that it has not been effective.

Sterility may also be introduced into a pest population by means of recessive genes.

20 Explain how induced sterility can control a pest.

Pheromones

Pheromones are chemicals produced by animals of one sex to attract members of the other sex. The female oriental peach moth produces a pheromone that attracts males. This pheromone has been produced artificially and is used in two ways.

1 Most importantly, it is used to disorientate male oriental peach moths in the peach orchard, as shown in Figure 21.14. Small strips (very much like the ties used on plastic bags) impregnated with the artificial pheromone are tied to trees throughout the orchard to be protected. The male moths become confused with all the pheromone in the air, limiting their chance of finding a real female, and so the population is kept down.
2 The pheromone can be placed in traps to which the males are attracted and where they are eliminated, thus reducing the rate of reproduction.

21 Explain how the oriental peach moth has been controlled using artificially produced pheromones.

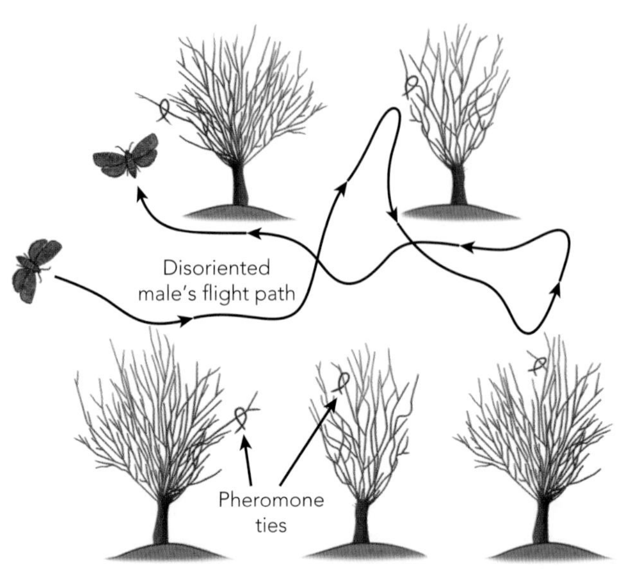

Figure 21.14 Synthetic pheromones are used to confuse the male oriental peach moth in search of a female. Pheromone-impregnated ties, distributed throughout the orchard, cause the males to become disorientated

Chemical control

Chemicals used to control pests and diseases are known as pesticides. Pesticides have been, and indeed continue to be, very important in the control of pests and diseases. They are grouped according to the organisms that they are used against. Fungicides are used to control fungal diseases, **insecticides** are used against insects and **miticides** against mites.

Fungicides

Fungicides are of two kinds.

1 *External.* Some fungicides are applied to the outside of the plant and protect the plant by killing the fungal spores as they germinate on the plant. They have to be applied at regular intervals, because any new growth by the plant that occurs after the fungicide was applied will not be protected. Examples of this kind of fungicide include Bordeaux mixture, copper oxychloride, zineb and mancozeb.
2 *Systemic.* More modern systemic fungicides are transported throughout the plant and thus provide protection continuously. Examples include benomyl (Benlate), triadimefon and pyrazophos.

Insecticides

Insecticides are poisons, either **contact** or **stomach poisons**. A wide range of chemicals serve as insecticides, some of which are shown along with their characteristics in Table 21.1.

 ISBN 9780170265560

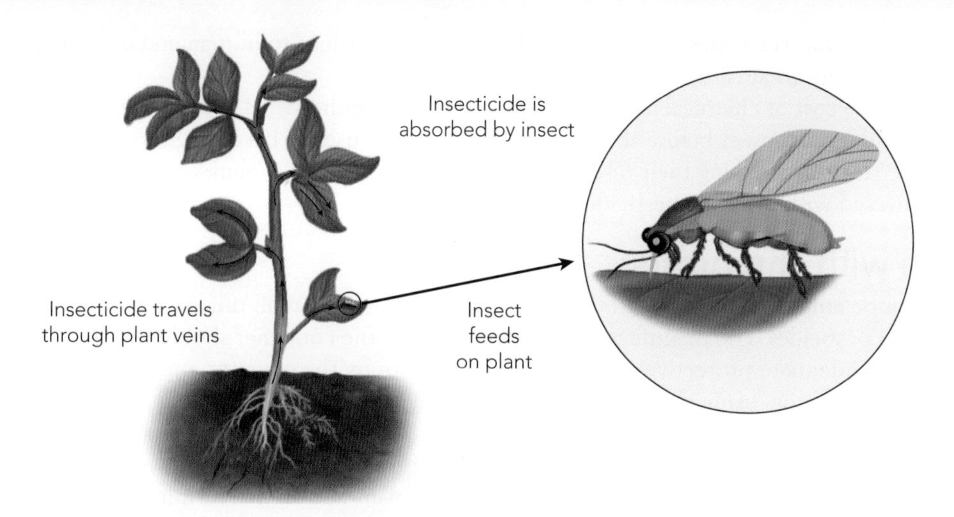

Figure 21.15 Operation of a systemic insecticide

In general, they have been very successful and have contributed greatly to the increased quantity and quality of plant production. Figure 21.15 shows the use of a systemic insecticide.

Problems with pesticides

The extensive use of **pesticides** has brought with it a number of problems.

1 *Resistance*. When a chemical is applied to a population of pests, there are always some that are not killed. These pests are able to cope with the pesticide and survive to reproduce. The capacity to deal with the pesticide is passed on to the next generation, and the result is a population of pests no longer affected by that pesticide.

Table 21.1 Characteristics of insecticides used in agriculture

Class	Examples	First used	Current use	Persistence*	Mammalian toxicity
Botanicals	Rotenone, pyrethrin	1800	Low	Short	Low
Carbamates	Propoxur, carbaryl, methomyl	1950	High	Short to medium	Moderate to high
Organophosphates	Fenitrothion, malathion	1950	High	Short to medium	Moderate to high
Synthetic pyrethroids	Permethrin, bioresmethrin	1970	Medium	Short to long	Low to moderate
Miscellaneous	Dimilin, chlordimeform	1970	Medium	Short to medium	Low

* Length of biological activity after application
Source: K. O. Campbell and J. W. Bowyer (eds) (1988), *The Scientific Basis of Modern Agriculture*, Sydney University Press, Sydney.

2 *Non-specificity*. The pesticide always kills a wider number of species than just the target organism. In so doing, natural enemies of the pest can be destroyed. This was particularly the case with the earlier broad spectrum insecticides. The natural enemies that kept another species under control also may be effectively wiped out, thus allowing this other species to become another pest.

3 *Residues*. Residues of the pesticide build up in the food chains of the environment and may reach levels in the wildlife at the end of the chain that adversely affect their reproduction and survival. Humans are effectively at the end of the food chain and potentially can have high levels of pesticide in their bodies. Strict regulations regarding **withholding time** and pesticide use now are designed to protect people from this hazard.

4 *Direct hazard*. Pesticides as poisons can be directly hazardous to human and animal life, necessitating great care in their use.
5 *Cost*. The cost of chemical control is increasing. The requirements that potential pesticides must meet before they can be registered for use are becoming more and more stringent, making their research and development more expensive. These costs are recovered with profit by pesticide manufacturers.

22 What are the two kinds of fungicide, and how do they differ?

23 The chlorinated hydrocarbons were very effective insecticides when first introduced. Use the information in Table 21.1 to explain why their use has been virtually phased out.

24 One problem that has arisen with the use of pesticides is that in many cases the target organisms have developed resistance to the pesticides used against them. Explain how this resistance could have arisen.

25 What are the general steps involved in the development of a new chemical pesticide for use in agriculture?

26 Why is insect resistance a worry in agriculture?

Care with chemicals

It cannot be emphasised too much that care must be exercised at all times when handling and using pesticides. The manufacturer's instructions on the container should be followed absolutely. Adequate protective clothing should be worn by the operator applying the chemical. Operators using chemicals must be suitably trained in the handling and application of chemicals.

Australian companies specialising in the manufacture of agricultural and veterinary chemicals produce a wide range of products for both plant and animal farming systems. The Australian Pesticides and Veterinary Medicines Authority (APVMA) issues permits for the use of sprays and registers chemicals for use in agriculture.

Chemical products (either synthetic or natural) are only released for general farm use after a long period of research and testing. Many chemicals are tested in laboratories for 1–2 years for their effect on pests and diseases. Substances that show promise are then tested in the laboratory for up to another 5 years. During this time the dosage rates, toxicity levels and the effects of the chemical on non-target insects and animals are assessed. One very important measurement is the chemical's LD 50 rating – a measure of how much of the chemical is needed to cause death to 50% of the test animals. This is tested both by application to the skin (dermal basis) and by ingestion (oral basis). The lower the figure obtained, the more toxic the chemical. The final testing stage (taking another 3–4 years) includes testing the product in the field to establish safe usage levels and residue levels. After 10 years the new chemical may then be registered for use.

Chemical safety

Exposure of the user and friendly or non-target organisms to chemicals can be prevented or reduced if chemicals are used in a safe manner; that is, with personal protective equipment (Fig. 21.16) and with an understanding of the chemicals being used. When handling chemicals, follow these rules.

1 *Read the label carefully.* The toxic ingredient and any other chemical ingredient must be listed on the label. The label must have warning signage (poison schedule, dangerous goods classification). Information about storage and mixing instructions must be shown.
2 *Never use a chemical without looking at the directions.* This includes the first aid directions, crop registrations, application rates and withholding periods.
3 *Wear personal protective equipment* (PPE). This includes a hard hat, washable hat, goggles, respirator, rubber boots, thick rubber gloves and spray shields over the helmet when handling a concentrated product.
4 *Wear protective clothing.* Most poisons are absorbed through the skin. Poisoning can occur by inhalation of vapours or by ingestion as well.
5 *Take care not to spill any chemical.*
6 *Only mix up enough of the chemical to use immediately.*
7 *Avoid spray drift.*
8 *Carefully clean all clothing and equipment after use.*
9 *Wash up with soap and water afterwards.*
10 *Maintain records of chemical use*. Records can be kept using chemical data forms available from state agriculture departments.

Material Safety Data Sheets (MSDS) contain a thorough coverage of the physical, chemical and toxic properties of any active ingredients in a chemical product. Information also includes health effects, precautions to follow and emergency procedures. These data sheets are available at the point of purchase of a pesticide.

ISBN 9780170265560

Farmers also need to ensure that they have any necessary licences; for example, if you are using ground equipment to spray herbicides on another person's property you will need a Commercial Operator's licence.

When storing chemicals, follow these rules.

1 Keep the area clean and tidy.
2 Always lock the chemicals away in a cupboard out of the reach of children.
3 Work in well ventilated areas.
4 Never store excess chemical in an unlabelled bottle or container.

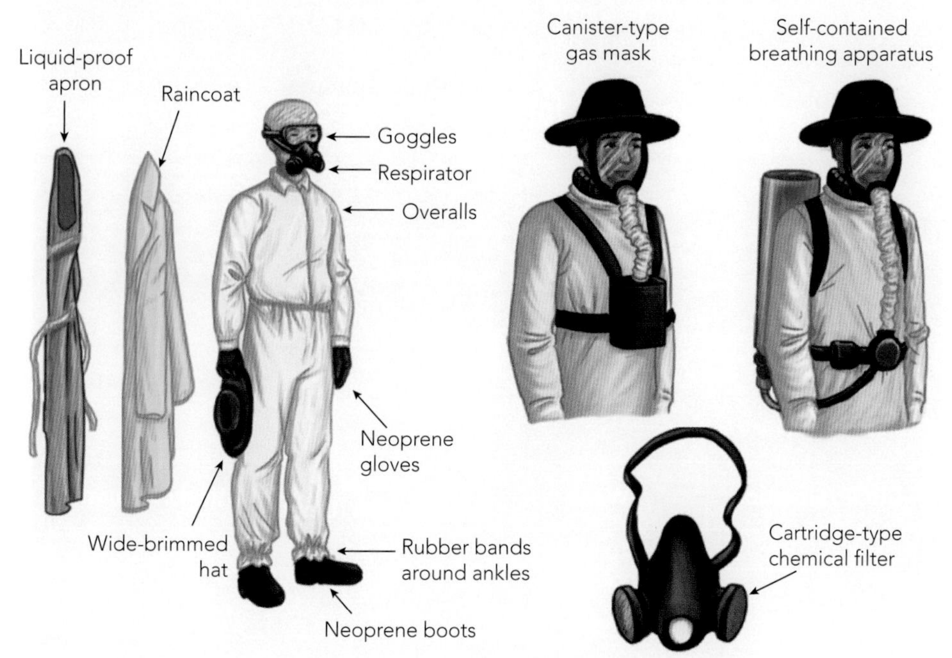

Figure 21.16 Protective clothing for handling chemicals

Spillage and disposal of chemicals

Unwanted containers and surplus chemicals are often disposed of carelessly. Empty containers should be thoroughly washed several times (at least three times), then punctured and crushed if they are metal, or smashed if glass. Remember: do not puncture any aerosol can. The containers can then be buried or taken to a waste disposal site. Unused chemicals must be disposed of at a toxic waste dump. Strict precautions are also needed against contamination of ground and surface water by leaking chemicals.

27 What rules should be followed for these activities?
- **a** Handling chemicals
- **b** Storing chemicals
- **c** Disposing of chemicals

Chemical residues

Chemicals applied to production systems remain active for a period of time. During the withholding period the product that was sprayed cannot be harvested. This protects the consumer from buying poisoned food. Each pesticide has a known withholding period during which time it remains active. This also applies to chemicals sprayed on pastures, since it is possible to contaminate livestock if withholding periods are not met.

The effect of toxic chemicals may be seen immediately (acute toxicity) or chemicals may accumulate in the soil or in a living system, increasing in concentration until the health of animals is affected (chronic toxicity). This residual effect has led to the poisoning of many non-target animals, including livestock and humans. Care should thus be taken to prevent the build-up of chemicals that do not easily break down in the soil.

28 Define the following terms in relation to chemicals.
- **a** Residual effect
- **b** LD 50 rating
- **c** Withholding period
- **d** Material Safety Data Sheets
- **e** Chemical data forms

Pesticide safety and labels

The most important source of information is the pesticide label. Even experienced users still need to read the label for any changes in application rates, safety and disposal instructions. Refer to Figure 21.17 for a sample product label indicating the generic information all labels carry.

ISBN 9780170265560

Sample product label information

Adapted from information provided by the APVMA

*The **signal heading** shows the level of risk of using the pesticide and provides safety guidance*

Trade name for marketing

***Active constituents**, solvents and scheduled ingredients are important for first-aid instructions and identification*

***Mode-of-action group** guide for resistance management guide in agriculture*

***Purpose of use** – the registered use*

***Manufacturer** and emergency contact*

***The directions for use** table provides instructions on how to use the product and the rates to use*

***General instructions** offer advice on use of the product and how to prevent problems*

*Follow the **withholding period** indicated to prevent unacceptable residues in produce*

***Safety directions** explain how to protect yourself when preparing and using the pesticide*

***First-aid** instructions indicate what to do in the event of poisoning*

***Material safety data sheet** (MSDS) contains extra hazard information and is available from the supplier*

*Instruction on how to avoid unintended off-target damage to **crops, environment or livestock***

***Dangerous goods** information assists safe transport and storage*

Not all pesticides are classed as dangerous goods; dangerous goods are identified by a class diamond and UN number. Packing group indicates the level of danger.

***APVMA Approval Number** shows the pesticide has been assessed and registered by the APVMA (the last 4 digits are the date of the last assessment).*

*The **batch number** and **manufacturing date or expiry date** allow tracking of product to prevent use of out of date product.*

CAUTION
KEEP OUT OF REACH OF CHILDREN
READ SAFETY DIRECTIONS BEFORE OPENING OR USING

NO-WEED HERBICIDE 500

CONTENTS: 20 L

Active constituent: 500 g/L glyphosate
Solvent: 200 g/L liquid hydrocarbon
GROUP M HERBICIDE

For weed control in fallow and non-crop situations
Full directions for use are contained in the attached booklet

Manufactured by the Joe Bloggs Chemical Company, 101 Newcastle St, Sydney, NSW, 2000 Emergency contact no: 1800 ....

Directions for use

Restraints

- Do not apply when rain is expected within 4 hours
- Do not apply to plants that are stressed
- Not to be used for any purpose, or in any manner, contrary to this label unless authorised under appropriate legislation

CROP/Situation	Weed	Rate of application (L/ha)	Critical comments
Fallow	Refer to attached booklet	1.2–2.4	Apply in a minimum of 100 L water per hectare when weeds are actively growing.
Non-crop situations		1.5–3.0	

General instructions

- **Withholding period**: Do not cut or graze stock food for 7 days after spraying.
- **Resistant weed warning**: This herbicide belongs to the Group M mode of action group and is subject to a herbicide resistance prevention strategy.
- **Storage**: Store in original container in a cool, dry, well-ventilated area, out of sunlight.
- **Disposal**: This is a drum MUSTER eligible container. Container must be triple rinsed before disposal. Do not contaminate waterways with rinsate.
- **Precautions**: Do not apply under temperature inversion conditions or when drift is likely.
- **Mixing**: Mix in a well-ventilated area. DO NOT mix with phenoxy herbicides.

Safety directions

- Will irritate eyes and skin.
- When opening the containers and preparing the spray, wear face shield or goggles, overalls and impervious gloves. Wash hands after use.
- **First aid**: In the event of poisoning, contact a doctor or the Poisons Information Centre on 13 11 26.
- **MSDS**: Additional information is listed in the MSDS, which is available from ...
- **Protection of wildlife, fish, crustaceans and environment**: Do not allow chemical or used containers to contaminate streams or waterways.

DANGEROUS GOODS INFORMATION

Pesticide, flammable, n.o.s.
UN Number 1234
APVMA Approval number: 12345/0215
HAZCHEM 3WE
PACKING GROUP III
Batch number: A54321 DOM: 060115

Figure 21.17 Reading a pesticide label

ISBN 9780170265560

Integrated pest management (IPM)

Integrated pest management involves the application of a number of different control measures in a systematic and effective way to prevent a pest from causing severe economic loss. IPM has the ability to reduce the problems of pesticides and chemical resistance in target organisms.

An IPM approach to the control of black rot in cabbages might include the following actions.

- *Sterilising seedbeds*
- *Not placing seedbeds on land used previously to grow cruciferous crops*
- *Rotating crops*. Do not grow crucifers on the same land more than once in every 3 or 4 years.
- *Sanitising crops*. Examine seedlings at transplanting and destroy diseased ones. Collect and burn all diseased material as soon as it is noticed. Bury deeply all crop residues at the end of the season.
- *Controlling insects*. Biting and sucking insects need to be controlled with an insecticide as they can carry bacteria from diseased to healthy plants.

There are several advantages of using IPM to target an insect pest or disease.

- *Reduced pesticide use*. This leads to less pesticide resistance, less environmental contamination (that is, less run-off, affecting fish, crustaceans, aquatic plants), less death of beneficial organisms (including predators, pollinators, soil organisms). Less broad spectrum chemicals, less risk to farm operators, less risk of pesticide residues on produce (leading to improved customer satisfaction).
- *More effective control*. Targeting the pest or disease using different methods increases the chance of effective control.

connect

Pest management

Answer the questions in relation to use of IPM strategies in controlling tomato leaf miner infestations.

connect

Principles of IPM

ISBN 9780170265560

Chapter review

Things to do

1. What causes yellowing or chlorosis of the leaves in plants?
2. Carry out an experiment to test the validity of the claims of companion planting. Make sure that you have a control (i.e. a plot planted with the crop but no companion plants).
3. Read the labels on some pesticide containers. Record the active ingredient (chemical), what pests or diseases it is recommended for, what precautions should be observed when using it, whether it has a withholding period, and if it does, how long the withholding period is.
4. Interpret an agricultural pesticide label and explain what it tells you about the following.
   a Safe work practices for the safety of the user, safe practices for the protection of the environment and safe practices for the safety of the consumer; for example, describe advice given about the withholding period.
   b Correct usage for maximum effectiveness.
5. Survey your school farm for pests and diseases. Try to identify as many as possible.

Things to find out

1. Describe the interaction between problem organisms (pathogenic microbe or invertebrate), the host and the environment for one plant disease that you have studied.
2. One of the oldest fungicides is Bordeaux mixture, and the story of its discovery is fascinating. What is in the mixture? How was it discovered?
3. For a plant production system that you have studied:
   a name one important insect pest
   b describe the type of damage that the pest causes
   c describe the lifecycle of the pest
   d name the season(s) when the pest is a problem
   e outline the main methods of control.
4. Outline the effects that conservation tillage practices have on the levels of pests and disease in crops. What has been done to overcome these problems?
5. Describe a major plant pest or disease in your area? Name the crop or pasture that is affected by this pest or disease. What methods are employed to control the pest or disease? Is any research being done to find new ways to control it?

Extended response questions

1. For a crop grown in your region:
   a list the pests and diseases that occur in the crop
   b outline the control program carried out over a 12-month period
   c for one major pest of the crop, discuss whether or not the control measures used against it constitute an example of integrated pest management.
2. a Name a pest or disease of a plant or crop that you have studied.
   b Describe an integrated pest management program for its control.
   c Explain how integrated pest management programs can reduce the problems of pesticides and chemical resistance in pest and disease organisms.

ISBN 9780170265560

3. The interactions between host, pathogen and environment may result in disease.
   a. Name a plant disease that you have studied.
   b. For this disease, describe the interaction between the pathogen or parasite and the host.
   c. Describe how knowledge of this interaction would enable farmers to prevent or control this disease.
4. Write two letters to the editor of a newspaper. In the first, present a case for the continued use of pesticides in plant production, and in the second, present a case for the discontinuation of their use.
5. a. Define integrated pest management (IPM).
   b. Evaluate an IPM program for a named pest or disease that you have studied. In your answer, include examples of management strategies used in this program, and name the target organism and the plant host.

ISBN 9780170265560

CHAPTER 22

Outcomes

Students will learn about the following topics.

1 Various grazing practices.
2 Cultivation practices.
3 Grazing management strategies.
4 Pasture production systems.

Students will be able to demonstrate their learning by carrying out these actions.

1 Describe the role of pastures in animal production, soil improvement, erosion and weed control.
2 Describe the pasture establishment techniques of prepared seedbed, direct drilling and broadcasting.
3 Describe the grazing management strategies that can be used for pastures: rotational grazing, strip grazing, cell grazing, deferred grazing and zero grazing.
4 Identify native, improved and introduced pasture species and describe their roles in pasture production systems.
5 Explain the significance of a diverse pasture mix.
6 List the factors to be considered in selecting species for improved pastures.
7 Explain how and why inoculation and lime pelleting of legume pasture seed are carried out.
8 Describe the effects of fertiliser application on improved pastures.
9 Describe how sowing rate and timing of sowing influence pasture improvement.
10 Describe the effects of pasture on grazing animals.
11 Describe the effects of grazing animals on pasture.
12 Describe the relationship between stocking rate and production per head and per hectare of pasture.
13 Describe the patterns of production from various pasture species, and strategies that can be employed to overcome shortages of feed during the year.

ISBN 9780170265560

PASTURES

Words to know

bloat an unusual build-up of gas in the rumen, which can result in the death of an animal; occurs in cattle grazing pastures containing predominantly lucerne or clover

effective rainfall where mean monthly rainfall exceeds the monthly evaporation rate

improved pasture introduced species of pasture grasses or legumes, or a combination of both, to enhance pasture production

native pasture contains plants originating in that particular country (e.g. Australian native plants originating in Australia)

natural pasture contains plants of both native and naturalised origins

sodseeding sowing seed directly into the soil of an existing pasture or crop stubble without any prior ploughing or cultivation and using a specially designed sodseeding machine

Introduction

In a very general sense, a pasture is a plant community grazed by animals such as sheep, cattle and goats. It consists of several species growing together, and ideally it is a balanced mixture of grasses and legume species. A farm must adhere to certain regulations before clearing native vegetation to use land for agricultural purposes.

Role of pastures

Pastures have a number of roles and uses.

- *Animal production.* The primary role of pastures is to provide the feed for grazing animals that produce meat, milk and wool. This is by far the most important use of pastures. Australia's extensive sheep and cattle industries depend on pasture, which is the cheapest form of feed.
- *Soil improvement.* Growing pastures improves soil by adding organic matter. Organic matter improves soil structure, which gives better aeration and drainage, and it increases the soil's capacity to hold water and mineral nutrients. If legumes are included, the chemical fertility of the soil also will be improved by the increased levels of available nitrogen in the soil. A 3–4 year pasture phase is often included in rotation with crops, with the express purpose of improving the soil's fertility. This is called ley farming.
- *Erosion control.* Pasture plays an important role in minimising the risk of erosion, by reducing the impact of raindrops on the soil and slowing water down as it runs across the surface.
- *Weed control.* An established vigorous perennial pasture (e.g. perennial ryegrass or phalaris) provides competition for weeds and thus restricts their germination and growth.

1 What are the roles played by pastures? Arrange them in order of importance.

Types of pasture

Pastures can be thought of in three different categories, depending on the origin and adaption of the species involved: native, natural and improved pastures.

Native pastures

Native pastures are made up of species indigenous to Australia. These species were here before the arrival of the British in 1788 and include wallaby grass (*Austrodanthonia*), kangaroo grass (*Themeda triandra*), Mitchell grass (*Astrebla lappacea*) and saltbush (*Atriplex*). They are adapted to the climatic conditions and the soil, which is generally low in nitrogen and phosphorus. Their production of feed for introduced livestock is low, with poor palatability and lower digestibility. They show little response to fertilisers, but should not be disregarded.

Currently, there is renewed interest in native pastures because of the enormous benefit they provide in being persistent through periodic dry times. Several selection and domestication programs are being conducted, with the aim of developing **cultivars** (page 227) that can be sown in the same way as improved pastures. Native pastures are particularly valuable in the marginal areas of Australia, where few introduced species survive.

connect

Native pastures

Natural pastures

Natural pastures consist of native and introduced species. These introduced species have not necessarily been deliberately sown. They have just invaded and grown in an area for a long time and become naturalised; that is, they grow as though they were natives. In coastal areas of New South Wales, paspalum (*Paspalum dilatatum*) and kikuyu (*Pennisetum clandestinum*) are examples of natural pastures, as well as plains grass (*Austrostipa aristiglumis*) and subterranean clover (*Trifolium subterraneum*). Paspalum originated in South America and kikuyu in Africa.

 ISBN 9780170265560

Improved pastures

Improved pastures contain species that have the capacity to produce large amounts of palatable, digestible feed for livestock over a much longer period than native pastures do. These species have been introduced into Australia from other countries and include phalaris (*Phalaris aquatic*), ryegrass (*Lolium perenne, Lolium rigidum*), cocksfoot (*Dactylis glomerata*), white clover (*Trifolium repens*) (Fig. 22.1) and subterranean clover. They require soil that is more fertile, so their success in Australia depends on the use of fertilisers (superphosphate in particular) to supply phosphorus and legumes to supply nitrogen.

Shutterstock.com/Lu Mikhaylova

Figure 22.1 An improved pasture consisting of white clover and ryegrass

The use of improved pastures has allowed the carrying capacity of land to increase, and consequently the production of animal products such as milk, meat and wool has increased. Soil fertility has been improved not only by the use of fertiliser and the nitrogen-fixing ability of legumes, but also by the increase in the amount of organic matter added to the soil. The amount of production from improved pastures at certain times of the year far exceeds the requirements of animals. This excess can be conserved as hay and silage and used at other times in the year when production from pasture is low (e.g. in winter), as shown in Figure 22.2.

2 A farmer observes that a particular grass is growing all over the farm. It is readily grazed by livestock. A botanist tells the farmer that the grass is a native of the Middle East. What group of pastures does this grass belong to?

Shutterstock.com/Taina Sohlman

Figure 22.2 Excess pasture has been stored as silage in plastic bags for use in winter when pasture growth is reduced

ISBN 9780170265560

Characteristics of pastures

Pastures provide cheap feed for grazing animals that produce meat, milk and wool. They add organic matter and improve soil structure, aeration, drainage, and water- and mineral-holding capacity. Legumes planted in pastures also add nitrogen.

Pastures also assist in erosion control by slowing down run-off and in weed control through increased competition for light, water and minerals, which restricts weed germination and growth.

The following characteristics are desirable in pasture plants. They must be:

- palatable
- digestible
- nutritious
- persistent under extremes of climate
- vigorous in both growth and regrowth
- resistant to grazing pressures
- prolifically self-seeding
- possessed of a high leaf-to-stem ratio.

Table 22.1 Comparing native and introduced pastures

	Native pasture	Introduced pasture
Characteristics	• Adapted to low fertility soils, drought conditions, cold and frosty winters. • Responds quickly to rain. • Fertiliser ineffective. • Used in tablelands, where introduced pastures cannot be sown; in western plains, where soil fertility is poor, temperatures hot and rainfall low (e.g. Mitchell, wallaby, weeping). • Found in marginal areas. • Won't withstand heavy grazing. • Unpalatable.	• Most productive in places that do not endure extremes. • Rotating with crops such as wheat replenishes soil and gives it a rest. • Highly productive pastures used to maximise production. • Conserved as hay. • Increases carrying capacity. • Includes legumes, such as white clover, red clover and lucerne, and grasses, such as ryegrass and kikuyu. • More leaf and quantity of feed. • Palatable, nutritious and digestible. • Maximises animal production.
Sustainability	• Can grow well without many inputs such as fertilisers. • Well adapted to low fertility soils. • Adapted to low rainfall and persist during dry spells. • Free seeding.	• Legumes in pasture fix nitrogen in the soil, reducing the fertiliser needed to be top-dressed. • Can withstand heavy grazing. • Highly productive and boosts animal production.
Unsustainability	• Do not survive heavy grazing. • Don't produce quantity and quality of feed that introduced pastures do, therefore animal production is poor. • Often unpalatable with limited nutritional value.	• To gain maximum benefit, input costs are high; e.g. fertilisers, irrigation, herbicides, insecticides, cultivation for sowing. • Do not survive under extremes of climate. • Productive life is short. • Easily invaded by weeds, so need to be resown regularly.

ISBN 9780170265560

Pasture improvement

Pasture improvement involves sowing selected improved species of legumes and grasses into existing natural or native pastures, with the addition of fertiliser. Careful planning and management are required for success in establishing a more productive pasture. Attention must be given to a number of activities, now described.

Species selection

It is usual to select both legumes and grasses for a pasture. The legumes, with their relationship with *Rhizobium* bacteria (see page 355), will build up the level of nitrogen in the soil and provide protein for grazing livestock. The grasses will utilise the increased available soil nitrogen and grow faster, providing more feed, competing better with weeds and reducing the risk of bloat. An ideal pasture mix contains 60% grass species and 40% legumes; for example, a mixture of ryegrass and oats for the spring/autumn and winter period and white clover and kikuyu over spring/summer period. This brings advantages: legumes are high protein and nutritious and grass contributes carbohydrates, adding bulk. As the grass gets taller, quality deteriorates into more complex carbohydrates and less protein. Legumes fix nitrogen into a form that plants can use; that is, nitrates. When legumes decompose they release proteins that are high in nitrogen. However, disadvantages can be that the proportion of grasses to legumes may change over time. For example, legumes are more palatable so they are selectively grazed by animals. Legumes can also be shaded out by tall vigorous grasses, so good management is needed to maintain the ratio of grasses to legumes by strip grazing and reseeding. Legumes can cause bloat, a condition in the rumen where foam and bubbles can suffocate the animal that has eaten too much legume material.

When selecting species and cultivars for a pasture the following factors need to be considered:

- compatibility (i.e. grow well together)
- suitability to the local climate
- ability to provide feed all year round
- ability to provide nutrition to livestock
- palatability and digestibility
- persistence
- response to fertiliser
- harmlessness to livestock
- ability to produce large amounts of seed
- ability to produce vigorous seedlings and compete with weeds.

3 List the factors that should be considered before selecting a species or cultivar to use in an improved pasture.

Establishment techniques

There are three common methods for establishing improved pastures: prepared seedbed, direct drilling or sodseeding, and broadcasting or aerial sowing.

Prepared seedbed

The area to be sown is ploughed and cultivated several times to produce a fine seedbed that is ideal for sowing small seeds. The soil aggregates are small crumbs, and weed competition is reduced as much as possible. A seed drill is used to plant the seeds just below the soil surface and apply fertiliser near the seed. A rubber roller is often used to firm the soil down around the seed after sowing, improving the contact of the soil with the seed and thus increasing the chance of effective germination and establishment.

Advantages of a prepared seedbed are that the best rate of establishment is achieved using the least amount of seed, and pasture species are evenly established. Disadvantages are that it is the most costly method, it takes the most time, and the paddock being improved is out of production for the longest period.

ISBN 9780170265560

Direct drilling or sodseeding

A special **sodseeding** machine is used to sow the pasture seed and place fertiliser directly into the existing pasture without any prior cultivation (Fig. 22.3). Competition and existing vegetation or crop residue are reduced prior to planting by heavy grazing or using non-selective herbicides.

Sodseeding has the advantages of requiring only a short break in grazing and involving less cost and time. Disadvantages are that establishment rates are lower and more seed is required.

iStockphoto.com/marekp

Figure 22.3 Direct drilling or sodseeding pasture seed into an existing pasture

Broadcasting or aerial sowing

Pasture seeds are spread over the surface of the area to be sown using a fertiliser spreader or aeroplane. Grazing and herbicides may be used to reduce competition prior to sowing. Broadcasting, especially using aeroplanes, has application in rugged country that cannot be traversed with conventional sowing equipment.

Advantages of aerial sowing are that large areas can be covered quickly, inaccessible areas can be sown, and less cost and time are involved. Disadvantages are a lower establishment rate, the requirement for more seed and the expense of aeroplanes. Also, seed lying unprotected on the ground surface can be stolen by ants.

4 Write the methods used for pasture improvement, and list the advantages and disadvantages of each.

Inoculation and lime pelleting

It is recommended that whenever legumes are sown they should be inoculated with the correct strain of *Rhizobium* bacteria and lime pelleted. As the legume seed germinates, the *Rhizobium* bacteria invade the roots through the root hairs and form nodules that are effective at fixing nitrogen. In acid soils the lime creates a neutral condition around the seed, which favours the *Rhizobium* bacteria and thus aids in their invasion of the roots and subsequent nodulation. In neutral and alkaline soils the lime protects the *Rhizobium* and allows germination only when sufficient moisture is available.

The inoculation and lime-pelleting process involves three steps.

1. The *Rhizobium* inoculum, which comes in a peat medium, is mixed with a liquid adhesive, such as gum arabic or methyl cellulose.
2. This mixture is then poured onto the seed in a cement mixer, where mixing continues until all the seed is coated with the adhesive containing the inoculum.

connect

Seed inoculation and lime pelleting

ISBN 9780170265560

3 Microfined lime (finely ground limestone) is now added, and the final result is that each seed is coated with *Rhizobium* inoculum and a layer of lime.

Recent technological developments have enabled legume seed to be inoculated and lime pelleted by the seed merchant and treated seed can last up to 6 months before sowing.

5 Explain why legume seed is inoculated with *Rhizobium* bacteria and then lime pelleted.

Fertiliser application

Australian soils in general are deficient in nitrogen, phosphorus and to some extent sulphur. Superphosphate is the most widely used fertiliser for pastures. It supplies phosphorus and sulphur and is applied at the rate of approximately 125 kg per hectare. Where the trace element molybdenum is deficient, molybdenised superphosphate is used once every 5 years. Molybdenum is essential for effective nodulation to occur.

In pastures, legume symbiosis (through the bacteria, *Rhizobium)* is relied on to provide the increased amounts of available nitrogen required by introduced species. The decaying residues of the legume plants release nitrogen in forms available to plants (i.e. ammonium and nitrate). Since introduced species have the capacity to respond to higher levels of available nutrients, the use of fertiliser favours their growth. The use of high levels of superphosphate favours the transition from legume to grass dominance of the pasture. Figures 22.4 and 22.5 show how the pasture improvement process occurs, starting with native pasture.

In many situations the phosphate in superphosphate is quickly turned into unavailable forms after application. It is subsequently made available only very slowly. Once approximately 2.5 tonnes per hectare have been applied, after a number of years phosphate is available at adequate rates and subsequent applications of superphosphate do not have to be as high.

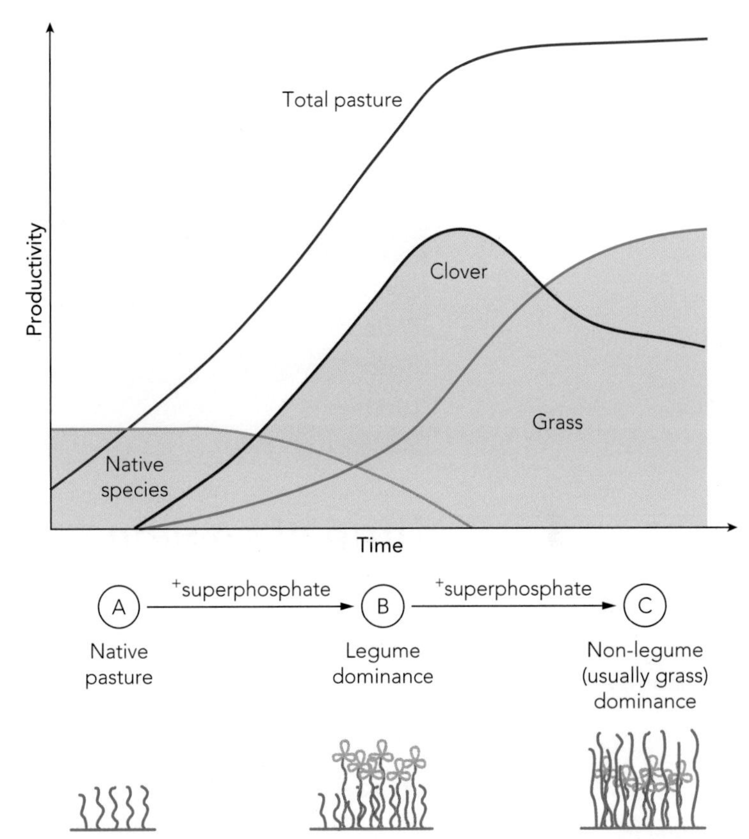

Figure 22.4 The improvement of native pastures. A generalised representation of the relative productivity of the various pasture components over time. Depending on the relative success of establishing improved grass and legume species and the annual rate of superphosphate application, the progression from native pasture to grass dominance will take approximately 3–10 years.

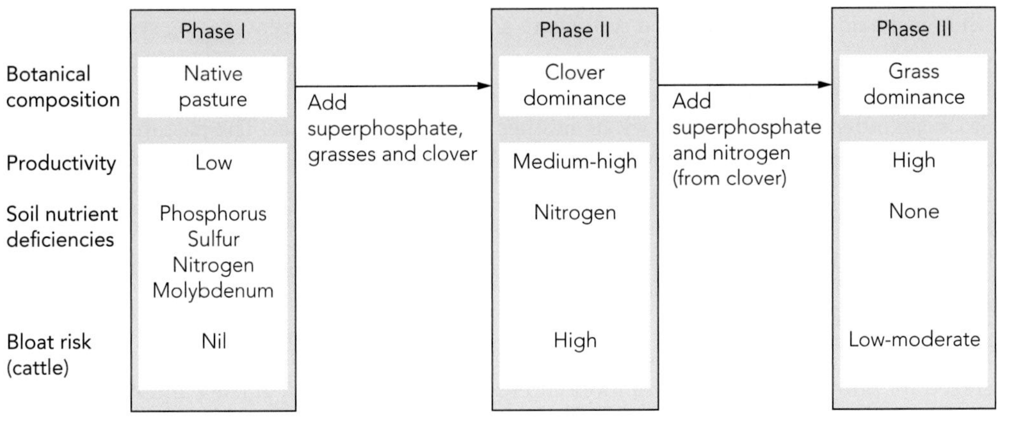

Figure 22.5 Phases of development of a perennial grass-based tableland pasture

6 Explain the roles that superphosphate and legumes play in pasture improvement.

7 Why is molybdenised superphosphate applied to pastures every 5 years in some areas?

ISBN 9780170265560

Sowing rate

Almost always a mixture of pasture species is sown. In selecting seeding rates there needs to be a balance between species to get sufficient plants established, provide optimum production, but not cause strong competition between the faster- and slower-establishing species. Recommended seeding rate varies with the weight of seed of each species, as shown in Table 22.2. Sowing rate is smaller for small-seeded species. Lower rates also apply to pasture species sown in mixtures. The greater the number of different species included, the smaller the sowing rate.

Table 22.2 Relationship between seed size, number of seeds per kilogram and sowing rate

Species	Number of seeds (per kg)	Recommended sowing rate (kg/ha)
White clover	1650000	1–2
Phalaris	700000	1–3
Perennial ryegrass	550000	2–10
Lucerne	440000	2–10
Subterranean clover	150000	2–8

8 What is the relationship between weight of seed and sowing rate?

9 What is the major factor that determines when pasture is sown in a region?

Time of sowing

The time of sowing is really determined by the availability of moisture for the germination and establishment of the pasture plants in the first few months. In southern Australia, with its winter-dominant rainfall, the best time to sow is in autumn. In northern Australia, with a summer-dominant rainfall, the best time is mid-spring to early summer. Temperatures will be more suitable at this time, too, for tropical species.

connect

Grazing guidelines for winter forage crops

Write a short summary.

10 Why is it important to leave a pasture for as long as possible after sowing, before grazing it for the first time?

Grazing management

The difficulty is to leave sufficient time between sowing a pasture and grazing it, to allow for the plants to become established so that they will remain productive for a number of years. In general, the more prepared the seedbed, the shorter the time between sowing and grazing. This might be 6 months for a prepared seedbed and over a year for sod-seeded or aerial-sown pastures. Care must be taken with overall stocking rates on the property, so that newly sown pastures are not put under grazing pressure until they are well established. Grazing in subsequent years is influenced by the need to allow flowering and seed set for the pasture and to supply feed to grazing animals at high demand periods, such as calving and lambing.

The aim of grazing management is to obtain the best return possible from animal products while maintaining the pasture and soil. There are a number of different grazing systems.

Rotational grazing

Stock are moved from one paddock to another in a regular sequence. The pasture is grazed down, the stock are moved on to the next paddock, and the pasture is then allowed to recover before being grazed again. A 6–8 week rotation has been found to be effective when grazing lucerne. This allows reserves to build up in the tap root of the plant to enable it to recover from grazing.

11 What is the rationale behind rotational and strip grazing?

Strip grazing

The pasture is grazed in strips using moveable electric fencing (Fig. 22.6). Each strip is grazed for only 1–2 days. This is a more intense form of rotational grazing and is often used on dairy farms. It is thought that the pasture is better utilised as it is more uniformly eaten down.

ISBN 9780170265560

Alamy/FLPA

Figure 22.6 Dairy cattle strip grazing pasture

Cell grazing

In this form of rotational grazing, the pasture on a farm is divided up into a large number of small paddocks with groups of these paddocks connected by a central hub so that stock can be quickly and efficiently moved from one paddock to another. Large mobs of sheep or herds of cattle graze a paddock for 2–3 days. The paddock is then spelled for 3–4 months before being grazed again. The system encourages beneficial species, particularly natives, improves pasture composition and soil structure, and reduces the use of fertiliser, such as superphosphate.

Deferred grazing

Stock are held off a pasture until it is required for a particular need or to allow the pasture to seed or to enhance its survival. A pasture may be put aside in autumn or spring when its production is at a peak, and then grazed in winter or at a time to finish stock for market, or to provide feed to stock with young at foot.

Zero grazing

Stock never actually go onto the paddocks where the pasture is growing. It is harvested with forage-harvesting machines and carted to the stock, which are stall-fed. The idea is that the pasture is uniformly utilised and that there is less loss through trampling and pastures being covered with dung. Costs include the expense of harvesting and transporting the pasture, and also the cost of replacing the soil nutrients carried away in the cut material.

Interaction between grazing animals and pasture

The interaction between grazing animals and pasture can be considered from two points of view: the effect of the pasture on the grazing animals, and the effect of the grazing animals on the pasture.

ISBN 9780170265560

Pasture effects on animals

Pasture affects grazing animals in a number of ways.

- *Nutrition.* Pasture is the source of nutrition for the animal, providing the energy, protein, vitamins and minerals required for the growth and production of products such as milk, meat and wool. As pasture ages, the bulk of material increases, but its digestibility and protein content decrease. The levels of fibre and indigestible lignin increase with age. The ideal pasture has a composition that is 40% legumes and 60% grasses.
- *Toxicity.* Certain pasture plants contain substances that are detrimental to animals. Some cultivars of phalaris cause a condition known as phalaris staggers. Young plants in the sorghum family contain dangerous levels of cyanide (which causes prussic acid poisoning), and certain pasture weed species (e.g. fireweed) can poison stock.
- *Bloat.* Bloat is likely to be a problem on pastures containing a high proportion of legumes, such as clovers and lucerne.
- *Product contamination.* Burr and seeds from pastures and weeds are responsible for vegetable matter contamination of wool, which reduces its value. Milk can be tainted by animals eating weeds in pastures, such as hexham scent (*Melilotus indicus*) and St John's wort (*Hypericum perforatum*).

12 Outline reasons why zero grazing is sometimes employed on dairy farms.

13 What is the ideal composition of a pasture mix?

Animal effects on pasture

Grazing animals affect pasture in a number of ways.

- *Defoliation and selective grazing.* Sheep and cattle are selective grazers, choosing diets high in digestibility, metabolisable energy and protein. They usually choose legumes before grasses, green material before dead material, and leaves before stems. Sheep are more selective than cattle, and the introduction of cattle into a grazing system with sheep can result in better utilisation of the available pasture and less likelihood that undesirable species may become dominant.
- *Nutrient transfer.* Some of the mineral nutrients eaten by livestock are lost from the system because they are taken out in products, such as meat, milk and wool. However, most of the nutrients are in fact returned in the form of dung and urine. The problem is that they are not uniformly distributed over the pasture. Nutrients might become concentrated in areas where animals camp (Fig. 22.7), around watering points, and around dairy sheds/yards where cows are brought twice a day for milking.

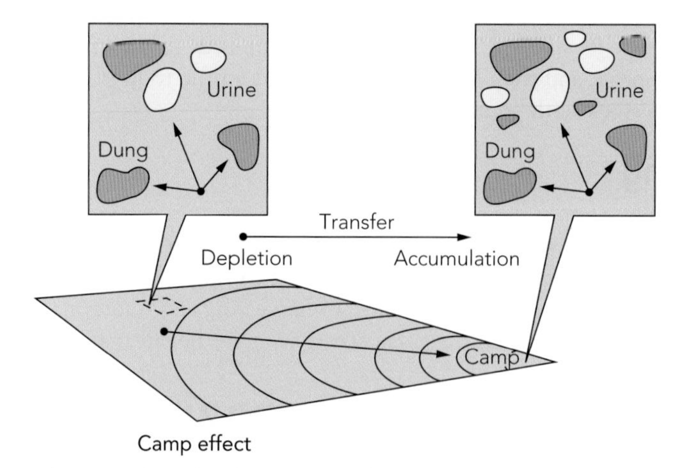

Figure 22.7 The distribution of dung and urine and the effect of a stock camp on nutrient transfer in a paddock

ISBN 9780170265560

- *Seed dissemination.* Animals are responsible for dispersing seeds by carrying them around in their hooves, coats and wool. Some seed is also transferred by being carried in the digestive tract. Some seed (e.g. subterranean clover) can pass through the digestive tract without being damaged.
- *Treading or trampling.* Pasture that is trodden on is lost and unavailable to stock. The number of stock per hectare will influence the damage done by trampling, with higher stocking rates resulting in more loss. The trampling of soil by hard-hoofed animals can also have a detrimental effect on soil structure, and this effect is worse if the soil is wet.

14 List and briefly describe the effects of animals on pasture.

Stocking rate and animal production

There is a general relationship between stocking rate and production per head and production per hectare, which is illustrated in Figure 22.8. As stocking rate increases, there is a steady decline in production from individual animals. This is explained by competition between animals for the available pasture. Production per hectare increases with increased stocking rate and reaches a maximum. Shortly after this point is reached, production per hectare drops dramatically. This is known as the crash point.

When there is a plentiful supply of high quality feed, production per head will be at a maximum. Any restriction of quality or quantity of feed will cause reduced production per head. Restriction of quantity can be brought about by increased stocking rate.

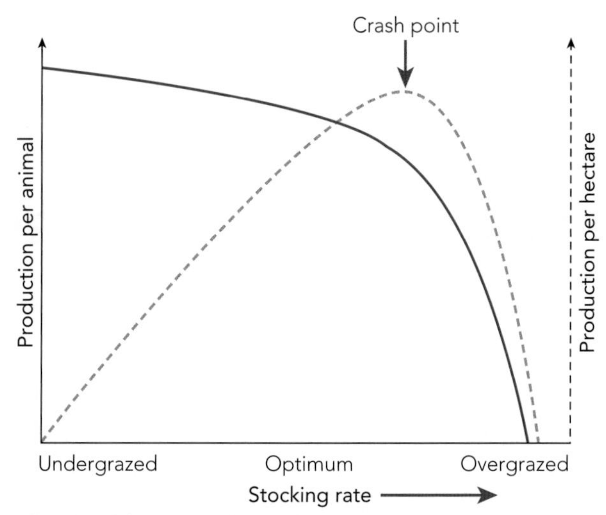

Figure 22.8 The relationship between stocking rate and production per head and production per hectare

Wool production

As stocking rate increases, the weight of wool cut per head decreases slightly, but production per hectare increases steadily. The quality of wool is little affected. At higher stocking rates there is an increase in the proportion of tender fleeces.

Prime lamb production

Increases on low stocking rates produce some increases in production per hectare. Further increases do not produce the increases of production per hectare that occur in wool production. This is because reproduction, growth and the laying down of fat are adversely affected by reduced intake of feed, which occurs at higher stocking rates.

ISBN 9780170265560

Beef production

Production of beef per hectare follows a similar pattern to that of prime lamb production as stocking rate increases. This is illustrated in Figure 22.9.

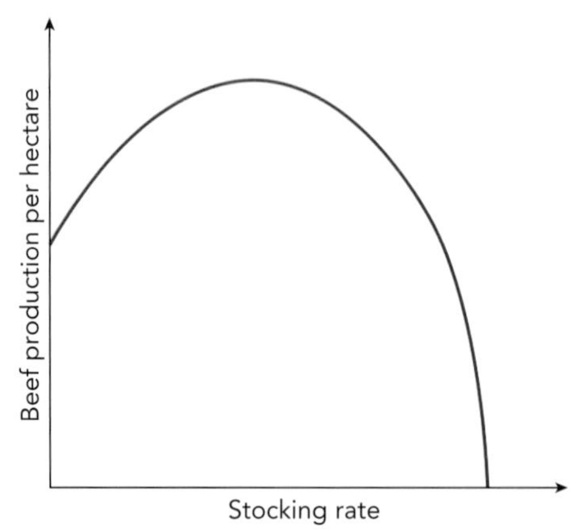

Figure 22.9 The relationship between beef production per hectare and stocking rate

15 Draw a graph that shows the general relationship between stocking rate, production per animal and production per hectare of a pasture.

Milk production

There is an increase in production per hectare as stocking rate is increased, and an associated decrease in production per animal. If stocking rate is increased to high levels, crash point will be reached and production will be severely reduced.

Patterns of growth and the feed year

It would be very neat if pasture kept producing feed at a constant rate all year round. This clearly is not the case. The growth rate of pastures depends on temperature and **effective rainfall** (i.e. rainfall that exceeds evaporation). During the winter months many species stop growing, and those that do grow are slowed down. Lack of effective rainfall also reduces growth, and this is often seasonal. In southern Australia rainfall is winter dominant, while in the north it is summer dominant, and in the central parts there is no pronounced seasonal dominance.

Pasture species can be classified according to their growth patterns in spring, summer, autumn and winter. Subtropical species (e.g. paspalum, lucerne and white clover) have their highest production in summer and virtually no growth in winter. Temperate species (e.g. ryegrass and subterranean clover) have peaks of growth in autumn and spring, with some growth in the winter and little in summer (none in the case of subterranean clover because it is an annual and dies off in late spring).

Figure 22.10 shows the production patterns of pasture species at Camden, and Figure 22.11 shows them for Tamworth, in northern New South Wales. At Camden there is a shortage of feed in the winter, which can be filled by growing fodder crops, such as oats and forage ryegrass. These crops provide feed that supplements the pasture. Another alternative to provide for the winter feed gap is to harvest the surplus pasture in the summer, store it as hay or silage, and feed it out during the winter. Tamworth also has a shortage of feed in the winter. Again this can be filled by growing winter forage crops (e.g. oats) and by storing hay (especially lucerne) in the summer months.

The planting of improved pastures is important in improving productivity in Australia's tropical northern regions. There is increasing pressure to replace native pastures with improved species to increase animal productivity due to increasing production costs on farms. There are pasture species adapted to tropical climates, such as buffel grass, Rhodes grass, glycine and siratro and greenleaf desmodium. Further information on tropical pastures can be located on the websites given.

connect
Pastures

connect
Plants

ISBN 9780170265560

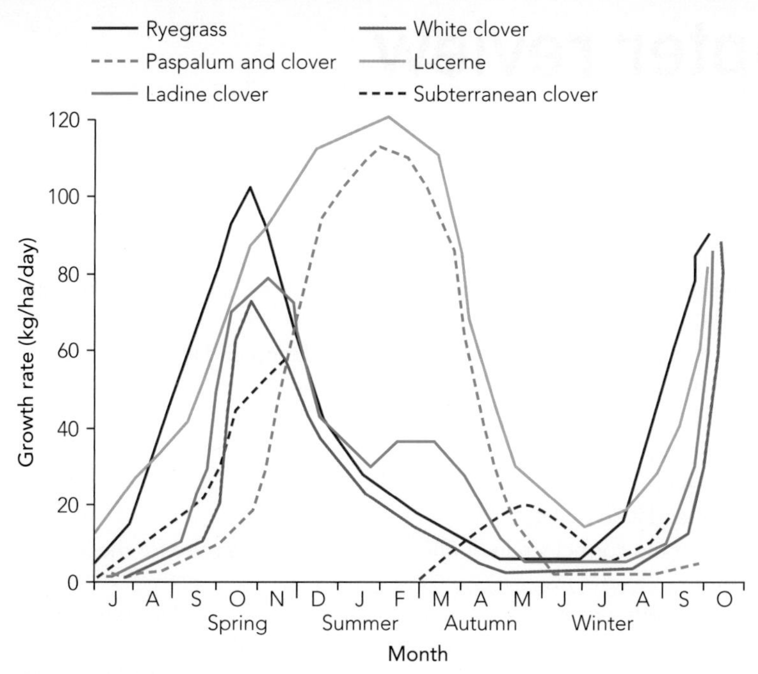

Figure 22.10 Production patterns of temperate and subtropical species at Camden

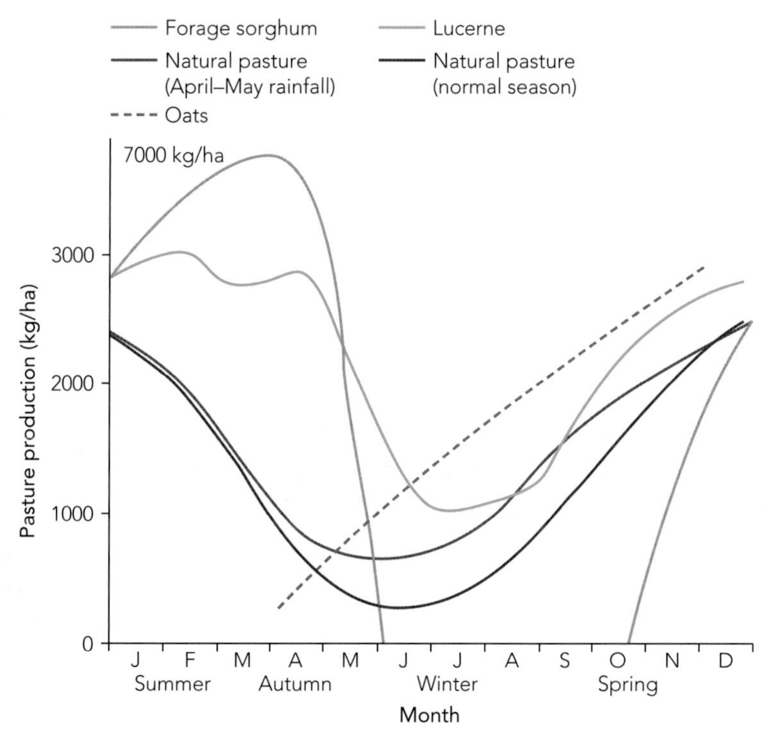

Figure 22.11 Production patterns of pasture species and forage crops at Tamworth

connect

Land clearance

A discussion on the legislation and guidelines to conserve native vegetation

ISBN 9780170265560

Chapter review

Things to do

1. Grow trial plots of individual pasture species and cultivars recommended for your region. Grow a trial plot of the recommended mixture. Carefully observe and record the growth of each plot throughout a 12-month period.
2. Collect the seed of native grasses that grow in your area, and try to germinate and grow them in plots.
3. Carry out an experiment to test the effect of inoculating and lime pelleting pasture legume seed. Use these seed treatments:
   a no inoculation or lime pelleting (control)
   b inoculate only
   c lime pellet only
   d inoculate and lime pellet.

   Make sure that the correct strain of *Rhizobium* inoculum is used for the legume species being grown.
4. Conduct an experiment in pots where four different strains of *Rhizobium* inoculum are tested with the four different legume species they are suited to. This will be a four-by-four experimental design, with each species being treated with each *Rhizobium* strain. The effectiveness of the *Rhizobium* bacteria can be judged by assessing the number of nodules formed and by determining their effectiveness at fixing nitrogen. Effective nodules are salmon-pink inside when cut open.
5. Estimate the amount of dry matter in a pasture. Randomly select five different square metres in the pasture.
   a For each square metre, do the following.
      i Cut the pasture to a level that it would be grazed to, say 3 cm above the ground surface, and collect the cut material.
      ii Weigh the cut material.
      iii Take a small sample of the cut material, and weigh it.
      iv Dry the sample in an oven at 105°C, and weigh it after 2–3 days.
      v Calculate the dry matter percentage of the sample and the amount of dry matter.
   b Next, carry out these calculations.
      i Calculate the average dry matter using the results from the five square metres.
      ii Calculate the amount of dry matter per hectare and if the area of the paddock is known, calculate how much dry matter there is on the paddock.
   c If a cow requires 20 kg of dry matter per day and the stocking rate is two cows per hectare, how long will this pasture provide feed for the cows?

Things to find out

1. For the district in which you live or for the farm case study that you have studied, research the range of pasture species that you could use to ensure feed is available all year round.
2. For an improved pasture mixture recommended for your region by the NSW Department of Primary Industries, find out the following.
   a What species and cultivars are included?
   b Describe their role in pasture production systems.
   c What sowing rate is recommended for each species and cultivar?
   d At what time of the year is sowing carried out?
   e What method of sowing is recommended?
   f What fertiliser is recommended, and at what rate should it be applied?
3. Outline the native pasture species that grow in your area and the role of each in sustainable pasture management.
4. Discuss the reasons why legislation has been introduced to manage native vegetation protection.

ISBN 9780170265560

Extended response questions

1. Explain the significance of a diverse pasture mix.
2. For the pasture species grown in your region, do the following.
   a. Describe their production patterns.
   b. Explain the relationship between the production patterns and climate (rainfall and temperature).
   c. Describe management strategies that farmers use to overcome limitations, imposed by the production patterns of pasture species, to production from grazing animals.
3. Explain how the management of a major livestock enterprise of the region is integrated with the pasture production over a 12-month period.
4. a. Describe the following types of grazing systems and grazing practices used in animal production:
      - paddock rotation
      - strip grazing
      - set stocking
      - zero grazing.
   b. Discuss grazing systems or practices used in your region or local district.
   c. Explain the impact of overstocking and overgrazing on plant production systems. Include the effects on both plants and soil.

ISBN 9780170265560

CHAPTER 23

Outcomes

Students will learn about the following topics.

1. Consumer and market requirements for commercial plant products.
2. The use of technology in the production of plant products.
3. The use of technology in the marketing of plant products.
4. Recent research findings that contribute to plant production systems.
5. The role of research in plant production systems.

Students will be able to demonstrate their learning by carrying out these actions.

1. Recognise the features of plant products that are important to consumers.
2. Use a range of sources to gather information about a specific agricultural problem or situation in plant production systems.
3. Outline the impact of research on agricultural production systems.

ISBN 9780170265560

PLANT PRODUCTION: MARKET REQUIREMENTS, TECHNOLOGY AND RESEARCH

Words to know

chemical ploughing the use of Roundup® or similar herbicides to eliminate weeds and plant competition prior to sowing a pasture or crop

hydroponics the growing of plants without the use of soil in carefully balanced nutrient solutions

tissue culture a method of propagating plants by asexual means

tramlining techniques controlled traffic farming when working soil

Introduction

A number of studies that involve plant breeding and related research aimed at improving productivity in plant production systems are currently being conducted.

Consumer and market requirements

Consumer preferences or tastes are an increasingly important marketing factor as producers and retailers seek to diversify markets for their products. A visit to any supermarket will allow you to identify various marketing strategies. In the marketing of plant products, examples of the following requirements can be seen:

- a uniform product
- a disease-free product
- a blemish-free product
- a desirable size or shape for the product meeting consumer preferences for colour and flavour
- absence of pests and pest excreta
- reasonable price
- freshness
- the particular kind of production chain; for example, organic, sustainable, low energy.

Technology

The work of individuals and agricultural scientists is not the only reason for Australia's increased capacity to produce food. New technologies or innovations in industry have also been important.

Improving production

connect

Tissue culture

Create a summary of the information.

connect

Plant hormones

connect

How to breed tomatoes

Identify the steps to produce a cross-pollinated plant.

connect

Gene technology

Outline the benefits of gene technology.

There are a number of studies of technological innovations which are aimed at improving productivity (quantity and quality).

The following technological innovations are used to increase production and reduce costs.

- *Tissue culture* (Chapter 20). **Tissue culture** has emerged as a powerful technique for asexually producing plants.
- *Plant breeding techniques* (Chapter 20).
- *Hormones* (Chapter 19). As knowledge of plant hormones has increased, new ways to use them to manipulate plant production have been discovered, such as in the manufacture of herbicides, and their use in tissue culture, fruit ripening and the encouragement of root development from plant cuttings.
- *Hybrid cultivars* (Chapter 20). Hybridisation is the crossing of different varieties of plants and is carried out by plant breeders in order to create new varieties that have characteristics that are more desirable. The methods used by plant breeders depend on the mode of reproduction of the plant species – self-pollinating or cross-pollinating.
- *Genetic engineering* (Chapter 20). This is the manipulation of an organism's genetic material. Food products from genetically modified plants are meeting some resistance from consumers because of possible environmental and health impacts. On the other hand, companies producing them are trumpeting their advantages of increased production and decreased use of environmentally damaging chemicals, such as weedicides and insecticides.
- *Improved potting mixes and fertilisers.*
- *Improved machinery and equipment.* The adoption of mechanical innovations (e.g. computerised irrigation systems, minimum or conservation tillage equipment, modern harvesting equipment and broadacre farming equipment) does require farm resource reorganisation, and therefore takes time to implement.
- *Hydroponics.* In **hydroponic** systems, plants are grown without soil in carefully balanced nutrient solutions.
- *Glasshouses.*

ISBN 9780170265560

- *Computers*. Computers are used for record keeping, data analysis, environmental monitoring through the use of sensors linked to computers, and to control machinery (e.g. to deliver precise amounts of fertiliser).

Marketing the products is an important phase in plant production systems. The markets for plant products are both domestic and international or export. These markets have specific requirements and these change over time because of such things as the whims of fashion and the influence of advertising.

1 Name three technological innovations used to increase production of plant products.
2 Describe one of these technological innovations in greater detail.

Improving marketing

The following innovations are used to improve the marketing of plant products.

- *Product grading*. Product grading uses lasers to scan and grade vegetables (e.g. tomatoes) by size.
- *Waxing*. Wax is applied to fruit to prevent the growth of fungus.
- *Ethylene ripening of fruit*. Bananas are treated with ethylene gas to ripen them. This means that fruit can be harvested and transported to market green before ripening. The advantage is that the fruit is less likely to be bruised in transport.
- *Labelling*. Foods that have been genetically modified (GMOs) are identified by label.
- *Convenience packaging of vegetables*. Convenience packaging refers to small packages in polystyrene trays that are plastic wrapped.
- *Targeted marketing campaigns*.

3 Identify three innovations used to improve the marketing of plant products.
4 Describe one of these in greater detail.

Researching sustainable plant production systems

The sustainability of plant production systems is constantly being improved. Some approaches and innovations are described here.

- *Salt-tolerant plants and new varieties of suitable pasture plants*. In salt-affected soils pasture plants, such as lucerne, often struggle to grow.
- *Tramlining techniques or controlled traffic farming when working soil*. **Tramlining techniques** mean that within a paddock there are set wheel tracks which are always used whenever a tractor moves to carry out soil preparation (e.g. ploughing), seed sowing, crop management and spraying with insecticides or herbicides. Tramlining confines heavy tractor movement to less than 10% of the soil in the paddock. The remainder of the paddock, over 90%, is not compacted by machinery and has greatly reduced soil compaction or soil structure decline. Therefore this large area remains well-aerated and allows water to easily infiltrate into the soil. This procedure results in greatly increased crop yields.
- *Chemical ploughing*. **Chemical ploughing** uses Roundup Ready Flex® or similar herbicides to eliminate weeds and plant competition prior to sowing a pasture. This technique reduces the number of cultivations required to prepare the seedbed for sowing. The advantages of this method include:
  - time saved because the number of cultivations is reduced
  - reduced costs related to fuel savings from reducing the number of cultivations
  - less damage to soil structure
  - less compaction of the soil.
- *Genetically engineered crops*. Crops can be genetically engineered to resist pests (for example, Bt cotton) and to be unaffected by commercial herbicide treatments (for example, Roundup Ready® Flex). Cotton plants were under attack from two insects, the cotton budworm and the native budworm, which destroy crops by eating the leaves and buds. Parathyroid sprays were not effective in controlling these two pests as both pests had developed high levels of resistance to all chemicals. Genetically modified cotton plants that would reduce the need for aerial spraying offered considerable promise. These transgenic cotton plants contain an extra gene inserted into their genetic information so they produce a substance that is poisonous to the budworm. This toxic protein is normally produced by the bacterium *Bacillus thuringiensis* (Bt). Scientists have transferred the gene that produces the toxin from the bacteria to the cotton plant. Bt cotton is more sustainable for these reasons:
  - less chemical is needed to control insect pests
  - less resistance to insecticides develops in the pest population

connect

Salt tolerant plants

Learn more about salt-tolerant plants, and how and why they are being developed to improve pasture production.

ISBN 9780170265560

  - less money can be spent on chemicals by the farmer (more economically sustainable)
  - less pesticide runs off into rivers and surrounding water systems (more environmental benefits).
- *Minimising herbicide resistance in weeds.* Herbicide resistance is a risk or problem for farmers practising minimum tillage cropping, but its development can be prevented by effective agronomic techniques and not relying on herbicides as the only method of weed control. Management of resistance is critical to ensuring the future sustainability of conservation farming methods. Preventing resistance can be achieved by:
  - maintaining accurate paddock records of weed species, herbicide application rates and levels of control
  - rotating herbicides that have differing modes of action; this practice minimises the development of herbicide resistance.
- *Partial root zone drying (PRD).* Usually grape vines are drip line irrigated. Partial root zone drying involves placing a drip line on either side of the row of grape vines or citrus trees (Fig. 23.1). With PRD, one side of the vines is irrigated four to six times over a 2-week period. Half of the root system is kept dry while the other half of the root system is irrigated for 2 weeks. This results in a 40–50% reduction in water use, because reduced leaf growth means there is reduced water loss through stomata. No significant reduction in yield occurs. The fruit quality (colour, flavour) was also improved. Further trials include above and below ground level use of drip lines, different irrigation schedules and experimenting with the PRD technique on oranges, pear and peach trees.

iStockphoto.com/sarasang

Figure 23.1 Partial root zone drying

connect

Monitoring technologies

Learn more about the enviroscan system.

5 Identify three research findings that contribute to sustainable plant production systems.

6 Describe in detail one current research finding that contributes to sustainable plant production systems.

- *Enviroscan.* This is a soil-water monitoring probe. Throughout a vineyard, onion or potato crop, a number of probes are placed in the soil to measure soil water levels. These probes record water levels at regular time intervals (for example, every 10 minutes). The irrigation system can be switched on automatically based on soil-water levels. With this system it has been possible to reduce water use by 10%, consequently reducing the cost of irrigation. Improved quality has been attained in grapes, onions and potatoes, and better growth of crops recorded as moisture stress is avoided.

 ISBN 9780170265560

Chapter review

Things to do

1. Describe the process a plant breeder would use to produce a new variety for a self-pollinating species.
2. Describe how a new variety is produced for a cross-pollinating species by a plant breeder.
3. Explain the terms 'tramlining' and 'chemical ploughing'.

Things to find out

1. Explain how tissue culture is carried out.
2. Explain the term genetic engineering and describe how it is carried out in plants.
3. For one innovative technology aimed at improving productivity (quantity and quality):
   - a name the innovative technology
   - b explain its benefits to production and its potential use in plant production systems
   - c identify impacts of this innovative technology on:
     - plants
     - products
     - cost of production, and profitability
     - environment
     - other species
     - health and safety or workers and consumers.
4. Describe two innovative technologies used to improve the marketing of named products in the plant industry.
5. Describe the post-harvest techniques and handling processes that are carried out on two named products to help meet market specifications.
6. Briefly explain one example of genetic engineering or associated technology that has resulted in improved plant production.
7. A number of recent developments in plant production have contributed to sustainable plant production. For one named finding or development, briefly describe how the finding has contributed to sustainable plant production.
8. Outline the impact of research on plant production systems.

Extended response questions

1. There are people who support the use of genetically engineered plants in production systems and those who oppose it. Put forward arguments supporting both points of view and come to a reasoned conclusion to decide your own opinion on this issue.
2. For a named example of an agricultural production or marketing technology:
   - a describe this technological innovation
   - b assess the impact of this innovation.

CHAPTER 24

Outcomes

Students will learn about the following topics.

1. The farm as a production unit and the enterprises on the farm.
2. The subsystems operating on a farm.
3. The goals of the farmer and why these have been established.
4. The routine management procedures for production and explain why each exists.
5. The ways products from the farm are marketed.
6. The relationship between inputs and outputs in farm production.
7. Technology used in management and production on the farm.
8. Safe work practices employed in agricultural work places.
9. The economic performance of this production unit (farm enterprise).

Students will be able to demonstrate their learning by carrying out these actions.

1. Define the terms 'input', 'output', 'processes' and 'boundary'.
2. Observe, collect and record information on the physical and biological resources of the farm.
3. Construct a calendar of operations for an enterprise production cycle.
4. Describe methods of agricultural recording.
5. Identify various measures of performance including gross margins.
6. Identify problems associated with production on the farm.
7. Identify factors impacting on farm management decisions.
8. Describe the effect of demand and the role of consumer trends on farm production.
9. Identify management practices used to address environmental sustainability.
10. Identify marketing strategies.
11. Identify technologies used on the farm.
12. Gather data using appropriate instruments to measure resources, including weather and soils.
13. Explain ways in which technology is used in farm management and production.
14. Identify potential safety hazards in agricultural work places.
15. Outline work, health and safety (WHS) legislative requirements that affect a farm.

ISBN 9780170265560

CASE STUDY: A WHEAT AND SHEEP FARM

Words to know

backlining pouring a chemical in a line along the sheep's back to kill lice

cover crop a crop that is grown to protect another tender species until it is established

direct drilling sowing technique in which seed is placed into the previously undisturbed surface of an existing pasture

hogget a shorn or unshorn sheep from approximately 12 to 15 months of age

mulesing surgically removing skin from the breech/rear area of the sheep so wool will not grow there

sheep year a yearly roster of routine husbandry operations or activities for sheep

static display model indicates inputs, outputs, boundaries, subsystems and the interactions between subsystems on a farm

stubble crop remains (stalks) left after grain has been harvested

Introduction

Ken James's wheat and sheep farm Rockdale, at Parkes, is typical of the well-managed family farms in the central west of New South Wales. The farm consists of 1000 hectares of land owned by the James family.

Physical features

Climate

The climate provides an ideal environment for wheat and sheep. Figure 24.1 shows the average monthly rainfall and temperature for Rockdale.

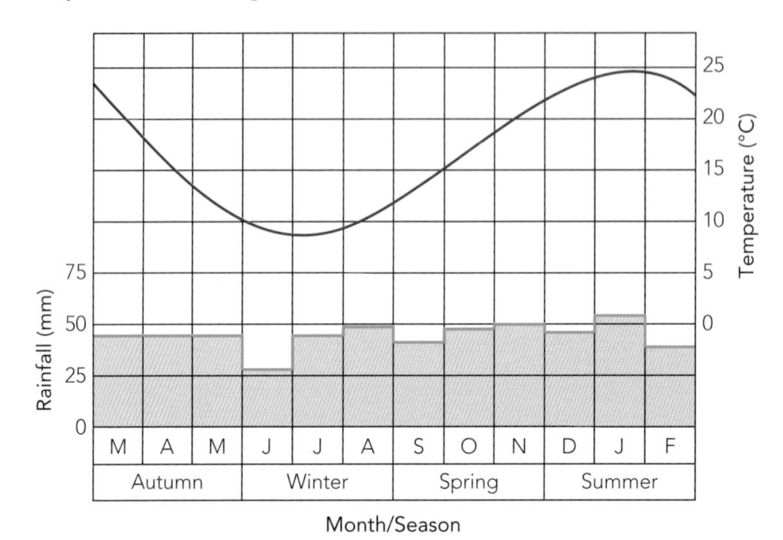

Figure 24.1 Average monthly rainfall and temperature for Rockdale

Figure 24.2 Round bales of hay used to feed sheep

Rainfall is a major limiting factor in wheat growing. When the rain falls is more important than the total amount of rainfall. Rain is particularly important throughout the growing season. Rain is needed from May to June to germinate the seed and in September and October for flowering and head filling. Conserving water from thunderstorms is achieved by spraying and killing summer weeds. Annual precipitation at Rockdale is approximately 525 mm. Summer rain comes from the north and winter rain from the south. Often summers are dry, and sheep have to be hand-fed with hay or grain stored during the previous year or years (Fig. 24.2).

Winters are mild enough to avoid heavy frosts which would damage the young wheat crop. Avoiding frost damage at flowering is very important also. Frosts would cause the flowers not to be fertilised, and this would result in a loss in seed set.

Most of the farm is gently undulating to flat. A small portion is hilly.

1 Name the two critical times during the growth of wheat when rainfall is important.

Water supply

The main source of water is a number of dams located around the property. Water is pumped from one large dam (20000 m^3 capacity) to a tank on a hill, from which it flows to water troughs in most of the paddocks. A bore is also connected to the tank. The bore runs for several hours each day or as required. There are other dams around the property where good run-off occurs.

Soils

Red-brown earths cover most of the property. They are well drained and, because of light seasonal rain, not heavily leached. Their good structure and loamy texture make them suitable for cultivation. These soils tend to be low in phosphorus and nitrogen. Soil is stonier on the hilly country.

2 What is the main source of water on the farm?

3 Why are red-brown earths a suitable soil for cultivation?

ISBN 9780170265560

Farm operations

Pastures

On the hilly country there are a number of native pasture species, including corkscrew, spear grass, kangaroo grass and barley grass. Burr medic (a legume) also can be found. On the arable country there are a number of sown pasture species, including lucerne, subterranean clover, tetila ryegrass, white clover and Persian or balansa clover (which likes wet or damp conditions). Lucerne provides green feed in summer when other species have died off and can also be used for hay. Vetch and oat crops can be grazed and then cut for hay. Topdressing of pastures with single superphosphate is done during late summer, so that the pastures can take full advantage of autumn rains.

4 Name the main sown pasture species on Rockdale.

Sheep

The property carries a self-replacing Merino flock of ewes and wethers. Ewes are classed when they are **hoggets**. The top 70% are joined with Merino rams. The bottom 30% (culls) are crossed with Border Leicester rams to produce first-cross lambs. All mating is usually completed within 5 weeks during February to March. Two rams are joined with 100 ewes (Fig. 24.3). Because prices are higher for finer quality wool, replacement Merino rams purchased over the last few years have been from studs with a reputation for fine wool and fleece weight. The aim is to increase the fineness and therefore the value of the wool produced on the property. However, in recent years the cost/benefit of the fine wool has diminished when compared with medium wool, so now fleece weight is very important.

Shutterstock.com/Lee Torrens

Figure 24.3 Sheep

Sheep are kept in their different mobs. Each year the property carries mobs consisting of:

- 1000 Merino ewes
- 500 Merino culled ewes
- 500 young Merino wethers
- 1000 Merino lambs
- 500 crossbred lambs.

The different mobs are moved from one paddock to the next depending on the availability of pasture. This is called rotational grazing. In summer the sheep can be put onto the wheat **stubble**. If there is a shortage of feed, it might be necessary to hand feed them with grain or hay (Fig. 24.4). Urea blocks can be used to make greater use of dry feed.

Shutterstock.com/Phillip Minnis

Figure 24.4 Silos used to store grain for sheep

ISBN 9780170265560

Pasture availability is matched to flock requirements. Merino ewes are pregnancy tested before lambing (using ultrasound) to detect twin-bearing ewes, which are put into a different group/paddock and are then fed a supplement to prevent pregnancy toxaemia. Merino lambs are given better pasture than Merino wethers.

Alpacas (four) are run with the lambing ewes to protect them from foxes.

The main pests and diseases affecting sheep are worms, lice, flystrike, footrot and clostridial diseases such as pulpy kidney (also called enterotoxaemia), caused by bacteria.

- Worms are controlled by drenching sheep once in spring and moving them regularly to a clean or worm-free paddock.
- Lice are controlled by off-shears **backlining** with an insecticide.
- Flystrike is common in humid conditions. It can be prevented by applying, for example, Click insecticide, which is absorbed into the wool and protects the sheep from flystrike for 6 months.
- Footrot (although not a problem at the moment due to many years of drought) is controlled by making sheep walk through a footbath, foot trimming and disposing of affected animals.
- Bacterial infections are controlled by vaccinations such as Glanvac® 6 in six vaccinations, which controls six bacterial diseases. Ovine Johne's Disease (OJD) is prevented by buying rams from studs that are accredited OJD free and by vaccinating lambs.

connect

More information on OJD

5 What breeds of sheep are crossed to produce a prime lamb?

6 How are the Merino ewes pregnancy tested?

7 What are the main pests and diseases that affect sheep?

8 Name 10 husbandry operations or activities carried out in the yearly management of a sheep enterprise.

During the year many operations are carried out in the management of the sheep enterprise. Ken uses two sheep dogs (kelpies) to help move the sheep. They help to push the sheep into one large mob, direct them towards a gate, catch one sheep for closer inspection, or in the sheep yards they help to move the sheep forward or up into the shearing shed. Routine husbandry operations or activities (the **sheep year**) include:

- shearing
- dipping
- drenching
- sale of cast-for-age (culled) sheep
- joining or mating
- pregnancy testing ewes (ultrasound)
- lambing
- lamb marking, including vaccination with Glanvac® 6
- crutching
- mulesing
- weaning, including vaccination with Glanvac® 6
- stubble grazing.

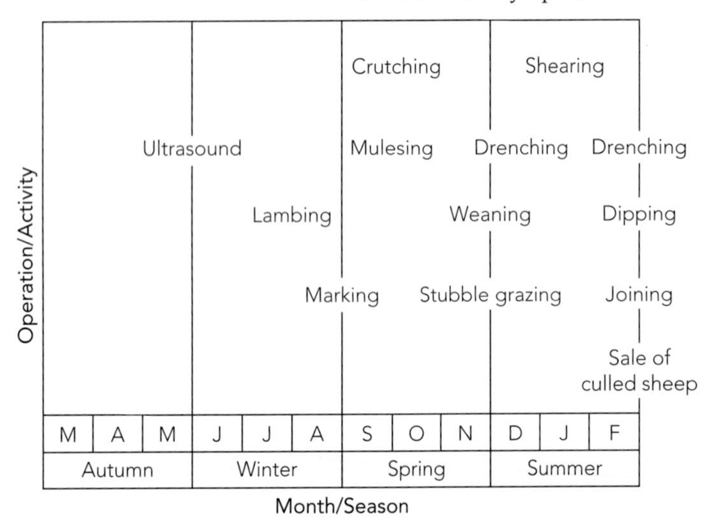

Figure 24.5 Yearly calendar of operations associated with the sheep enterprise at Rockdale

There are products on the market that can be used to drench and vaccinate sheep in one procedure – this can save time and labour.

Merino sheep are **mulesed** at lamb marking (at approximately 8 weeks of age) to keep the sheep's breech/rear clean and prevent flystrike. They are treated with pain relief immediately after mulesing and before they are released from the yards back into the paddock with their mothers. In the long-term it is planned to avoid the need to mules by using good ram selection.

Figure 24.5 shows a yearly calendar of operations associated with the sheep enterprise at Rockdale.

Crops

Each year, 350 hectares are sown with crops. Wheat occupies 100 hectares; oats, 30; barley, 100; canola, 50; lupins, 20; and grazing oats, 50. Ken James has made trial plantings of new crops such as grazing brassicae (a winter crop) in recent years. The actual paddocks sown to the crop each season vary with the needs of the property's rotation.

ISBN 9780170265560

An 8-year rotation is used. This involves cropping the paddock for 3 years and then sowing it to pasture for 5 years. The crop sown in the first year is canola and second year is wheat. In the third year, barley is sown and in the fourth year, pasture seed. The pasture seed includes lucerne, subterranean clover, annual ryegrass and white clover. Some farmers use **cover crops** but Ken James prefers to grow pasture by itself as this has proved to be a successful method at Rockdale. All legume seed is inoculated (i.e. coated) with the correct strain of *Rhizobium* bacteria. Oats and vetch are grazed and cut for hay. Vetch, a legume, is a good source of protein for sheep.

Traditionally, the first step in crop growing is to prepare the seedbed. A disc or chisel plough is used to break up the soil, and this is followed by several cultivations with implements such as scarifiers. However, minimum tillage is preferred. Fallows are commenced using herbicides and sometimes an off-set disc in the first year and after the first year **direct drilling** is practised.

Seed is sown with a combine or trash seeder. The combine simultaneously places fertiliser next to the seed, so that when the seed germinates it immediately has a supply of nutrients. Compound fertilisers, which contain nitrogen and phosphorus, are used in this way. Many farmers are using air seeders, but at Rockdale Ken James prefers to use a trash seeder due to the smaller scale of his operation.

The main weeds that can reduce crop yields are black oats, saffron thistle, silver grass, fleabane and spiney emix. These are controlled with herbicides, and occasionally cultivation.

A number of diseases can affect cereal crops. Diseases caused by fungi are the most important, including black stem rust and stripe rust, crown rot and take-all. Rust infection is controlled by the use of rust-resistant varieties of wheat. Seed is dusted before sowing to prevent crown rot and take-all and both these diseases are controlled by good crop rotation and by removing their grass hosts, such as barley grass and silver grass, 10 months before sowing. Canola and lupins are an important part of the rotation as they enable root diseases to be controlled.

Insects and mites can attack cereal crops by eating either the plants or plant products (e.g. grain). Red-legged earth mites attack pastures by sucking the juice out of the leaves and stems, especially when they are at seedling stage. Oat mites also eat oat plants when they are at seedling stage. Heliothis moth attacks lupins and canola by eating into the seed pods. Weevils attack stored grain. To prevent this, grain is treated before going into storage. In general, control of insects and mites is achieved by careful use of pesticides, crop rotation and using varieties of crops resistant to pests.

9 What crops are sown each year at Rockdale?

10 What are the main pests and diseases that affect cereal crops?

11 Name four operations or activities carried out in the yearly management of a crop enterprise.

The cereal crops are harvested using a header, which cuts the crop and separates the grain from the chaff.

During any one year, three crop cycles are in operation.

1 The stubble from the previous crop is grazed through summer and sprayed to control weeds. Off-set discs are used during March to incorporate any remaining stubble (if it is too heavy to go through the minimum tillage machine) into the soil. Only very occasionally is the stubble burned off – if it is very heavy.
2 The final spraying of the ground for the current crop takes place in April. The crop is sown in April–May, weeds are sprayed in July, and the crop is harvested in November–December. Canola may have to be sprayed in October or November for heliothis moth and/or aphids.
3 Meanwhile, preparation for the next crop is underway, with initial grass removal in June–July and ploughing or spraying in August or September.

During any one year, the management operations or activities include:

- ground preparation
- sowing
- spraying weeds
- harvesting.

Figure 24.6 shows a yearly calendar of operations associated with the wheat enterprise at Rockdale.

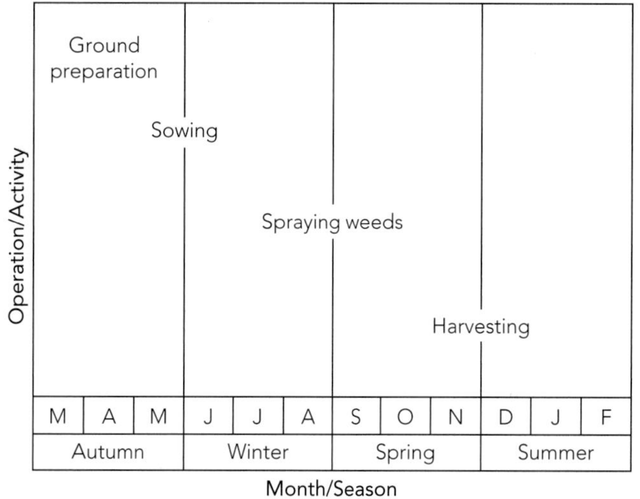

Figure 24.6 Yearly calendar of operations associated with the wheat enterprise at Rockdale

ISBN 9780170265560

Farm management

The seasonal routine on Rockdale is largely determined by the activities associated with the two main enterprises of sheep grazing and growing winter crops, already described. Priority is given to sowing and harvesting.

Other farm activities that need to be carried out during the year include:

- fence repairs
- new fencing
- water trough repairs
- dam dredging
- maintenance of vehicles, plant and equipment.

These activities are fitted between other sheep or crop activities. A computer is used to track the farm business and obtain information on the internet, such as weather and commodity prices for wheat, lamb and wool. A mapping program on the computer has aerial photos, with the area of each paddock. This can be used for making decisions regarding weed spraying, harvesting and hay making where the exact area of a paddock is needed. Contractors can also be given a farm map that identifies the paddock (and size) so it can be sprayed for weeds, harvested, or hay made.

Keeping farm records is an ongoing activity. Records are kept for sheep, paddocks and machinery. Attendance at field days and seminars is important, to improve knowledge and keep up to date with changing technology. The use of the internet to access financial and technical information is very important. Ken James has been a member of a farm management advisory group where important skills in marketing farm products and management are acquired. Private farm consultants, such as the following, are brought on to the farm.

1. *Agronomist.* Agronomists identify weeds and the most appropriate herbicides to use.
2. *Soil scientist.* Soil samples are taken and sent away for testing to determine the nutrient levels. Fertiliser recommendations are then made.
3. *Stock agents.* Agents look at first-cross ewes and wether lambs to determine the optimum time to sell.

Management practices to address environmental sustainability are listed below.

1. *Whole farm planning.* This is achieved by looking at the whole farm and how the different enterprises interact.
2. *Tree planting.* This is carried out to provide windbreaks, protection from soil erosion, shade and shelter for sheep, tree corridors for wildlife and biodiversity. Biodiversity is provided by maintaining Kurrajong, Cyprus pine, eucalypt and wattle trees. Windbreaks are especially important for shorn sheep.
3. *Crop and pasture rotation.* Rotation increases fertility, improves soil structure and breaks disease cycles.
4. *Growing legumes.* They provide a high-protein feed to sheep and improve the soil fertility. Pasture grasses in the paddock will make use of the nitrogen that legumes add to the soil.
5. *Minimum tillage.* On Rockdale, the approach to tillage is moving towards using direct drilling when possible.
6. *Crop stubble.* The stubble (stems and leaves from the previous crop) is incorporated into the soil after the crop is harvested. The benefit is to reduce the incidence/level of wind and water erosion and evaporation.
7. *Avoidance of overstocking/overgrazing.* Leaving sheep in a paddock for a long time or too many animals in a paddock can result in pasture being eaten out and erosion occurring. The aim is to have a minimum of 70% ground cover (i.e. vegetation covering 70% of the ground) in any paddock.
8. *Stock rotation.* Smaller paddock sizes are maintained and a good lane-way system is in place so that sheep can be easily rotated between paddocks.

connect

More information on stubble retention and zero till practices

12 What farm records are kept on Rockdale?

13 Outline management practices used to address environmental sustainability on Rockdale.

Machinery and plant

Ken James is able to operate Rockdale with some casual labour and contractors (who have invested in larger scale machinery) because of the large amount of modern equipment kept and used. His equipment includes a large tractor, small tractor, disc plough, scarifier, disc harrows, trash seeder, combine seed drill, boom spray unit, header, two trucks, haymaking

 ISBN 9780170265560

equipment, and farm one-tonner plus quad bikes. With increased awareness of sustainable farming, more use is being made of minimum tillage implements and techniques.

The sheep plant includes a shearing shed with yards (Fig. 24.8).

Alamy/Frank Paul

Figure 24.7 Some machinery and equipment used at Rockdale

Figure 24.8 The shearing shed at Rockdale: **a** wool classing **b** shearing

ISBN 9780170265560

Production and marketing

Rockdale produces wheat, wool, prime lambs and smaller quantities of oats, barley and lupins. The output of each varies from one year to the next depending on seasonal conditions and market prices.

1 *Wheat.* Approximately 100 hectares are sown with wheat each year. With an average yield of approximately 2.5 tonnes per hectare; the total harvest exceeds 250 tonnes. The wheat is marketed or sold to Grain Corp or to private buyers, such as Manildra flour mills. Price received varies depending on grain quality (e.g. weight of wheat, grain size, protein level and freedom from disease and weed seeds).
2 *Wool.* The total clip is usually approximately 100 bales. The wool is sold to local wool buyers, who arrange testing and sale at auction.
3 *Prime lambs.* Prime lambs are sold at the Central West Livestock Exchange (located between Parkes and Forbes) by auction. The ewe portion of the first-cross lambs is often sold to a neighbouring farmer for use as first-cross mothers for their second cross prime lambs.

Gross margin for the wheat enterprise

A **gross margin** provides an indication of the profitability of an activity or enterprise on the farm. It includes only variable costs and not fixed or overhead costs. The gross margin is prepared by deducting all variable costs from gross income obtained from a particular farm enterprise or activity.

Gross margin = total income – variable costs

Gross margins for cropping enterprises are usually calculated on a per hectare basis. The following applies for Rockdale (per hectare).

Income	$
Wheat 2.5 t at $250.00/t	625.00
Total income	625.00
Variable costs	
Seed	19.00
Tractor costs (including fuel)	41.00
Fertiliser	66.00
Sprays	35.00
Harvesting (contract)	60.00
Cartage	34.00
Total variable costs	255.00
Gross margin	
Gross margin = 625.00 – 255.00 =	370.00

14 Describe the production and marketing of wheat on Rockdale.

The gross margin for wheat per hectare is $370.00.

Constraints on production and profitability

The levels of production and profitability of Rockdale are determined by environmental and economic constraints.

Environmental constraints

Environmental constraints are mainly physical.

1 *Rainfall.* Rain is needed in May–June when the wheat seed is germinating, and in September–October for flowering and head filling. Rain also affects pasture growth and therefore sheep-carrying capacity.
2 *Frosts.* Severe frosts and cold snaps adversely affect wheat and young lambs. Late frosts cause the wheat flowers not to be fertilised and result in a reduced seed set.

 ISBN 9780170265560

3 *Soil erosion.* Contour ploughing and contour banks have helped to reduce soil erosion on the hilly country (Fig. 24.9).

CSIRO/John Coppi

Figure 24.9 Contour banks

4 *Soil type.* Soil type affects the productivity of the plant production system. The red-brown earths that cover most of the property are well drained and have good structure and a loamy texture, which make them suitable for cultivation. However, these soils tend to be low in phosphorous and nitrogen.

5 *Topography.* Some of the hilly country is not cultivated. This land contains rock outcrops and is used for grazing. The rest of the property is arable.

6 *Pests and diseases.* Wheat is affected by rust, crown rot, take-all fungus and stripe rust. The red-legged earth mite, heliothis moth and weevils also are important pests. Sheep are affected by worms, lice, flystrike and sometimes footrot.

15 List the main environmental constraints that affect level of production on Rockdale.

Economic constraints

Economic constraints are of two types.

1 *Costs of production.* Costs of inputs are continually rising (e.g. drenches, vaccines, dips, seed, fuel, fertiliser, shearing, new machinery). Interest rates are a major cost.

2 *Returns for products.* Returns from the sale of wool, wheat and first-cross lambs have not increased at the same rate as costs of production.

16 List the main economic constraints that affect the profitability of Rockdale.

Farm technology

Technology is used in management and production on the farm. Examples of technologies being used in the management of modern sheep and wheat farms include:

- computers for record keeping, gross margin calculations and comparisons, weather data, flock management, market information and market reports for wheat, wool and lambs
- EBVs for ram selection; for example, fleece weight, diameter (microns), staple strength and body weight
- NLIS ear tags, used when selling sheep (they assist with traceability)
- pregnancy scanners
- improved drenches and vaccines
- improved machinery and equipment for cultivation, sowing, applying chemicals, harvesting and handling sheep. Quad bikes have revolutionised sheep movement. Contractors use GPS guidance systems when they carry out weed spraying, harvesting and hay making. GPS devices can also identify poor yield areas in a paddock during harvesting.

ISBN 9780170265560

Farm safety

On a wheat and sheep farm there are many potential hazards, therefore safety is very important. Table 24.1 summarises the main hazards encountered and the safe work practices that should be used.

connect

Farm hazards

Table 24.1 Safe management of common sheep and wheat farm hazards

Hazard	Example	Safe work practice
Operating machinery	Operating tractors, working near belt or PTO driven machinery. Operating quad bike outside their specifications and capabilities.	PTO covers on all machines. Tractors ROPS in place. Disengage all equipment when cleaning or inspecting. Operate to manufacturer's recommendations. Repair and maintain machinery.
Animal handling	Working with rams, carrying out animal treatments; e.g. drenching and at shearing.	Awareness of animal behaviour and avoid frightening the animal. Also good control of working dogs.
Manual handling	Lifting fertiliser bags, animals, feed and fleeces.	Safe lifting procedures, animal handling devices and raised boards in shearing sheds.
Chemicals	Use and storage of sprays, pour on insecticides, fuel, detergents.	Read labels, use according to directions, wear personal protective clothing, e.g. respirator; store safely and dispose of according to local government requirements.
Electricity	Auger conveyors near overhead wires, machine movement and equipment such as boom sprays and long irrigation pipes being moved near overhead power lines.	Be aware of locations of overhead wires, signage indicating where electrical outlets and cables are.
Work place	Sun exposure (when possible move animals in the early morning or late afternoon), general farm environment and woolshed e.g. wool press and grinder.	Sun aware behaviour; e.g. hat, sunscreen. Have laneways to move sheep. Eliminate damaged flooring, rails. Tidy shed environments. Have guards in place on grinders.
Movement awareness	Movement of vehicles, animals around sheds and farm house especially in relation to children.	Know where young children are at all times. Clear indications when driving machinery or operating powered equipment.
Noise	Tractors, machines.	Use ear muffs and sound-proof cabins on tractors.
Confined spaces	Working in silos.	Have silo lids that can be opened from the ground and observation windows on the side of the silo.
Dust	Milling and mixing feed.	Use a respirator.

17 Describe two ways in which technology is used in the management and production of wheat–sheep farms.

18 Identify three potential safety hazards on a wheat–sheep farm and describe safe working practices that can be used to overcome the problem.

A systems approach

It is important for a farmer to identify goals. Goals are what the farmer is trying to achieve on the farm – usually to make a profit. On a wheat and sheep farm this would be achieved by producing good quality wheat, wool and first-cross lambs, while minimising expenses. To minimise expenses the farmer tries to reduce the cost of inputs.

Ken James's wheat and sheep farm can be examined in terms of a **black box model** (see page 179). The parts of a black box model of a farm are as follows.

1 *Inputs* are the materials that go into the farming system (e.g. seed, fertiliser, fuel, pesticides, drenches, dips, vaccines, labour, new machinery).

 ISBN 9780170265560

2 *Outputs* are the materials produced by the farming system and removed from it (e.g. wheat, wool, prime lambs).
3 *Processes* are the activities that change inputs into the desired output (e.g. shearing, harvesting, culling, lambing).
Figure 24.10 shows the inputs and outputs of a black box model of Rockdale.

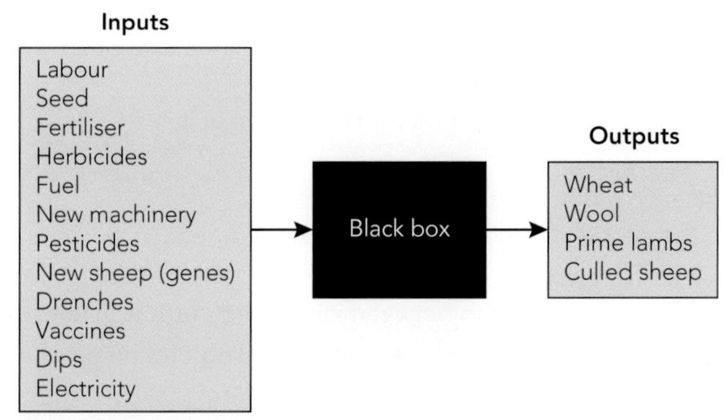

Figure 24.10 A black box model of Rockdale

A **static display model** gives more information about the **subsystems** (see page 3) of the farm and their interaction with one another, and the boundaries (Fig. 24.11). Boundaries are the limitations of the system, which may be physical (e.g. boundary fence), financial (e.g. bank credit), technological (e.g. weather forecasts), entrepreneurial (e.g. managerial skill and knowledge) or other limitations.

19 What does a black box model of a farm show?
20 What does a static display model of a farm show?
21 What is a boundary?

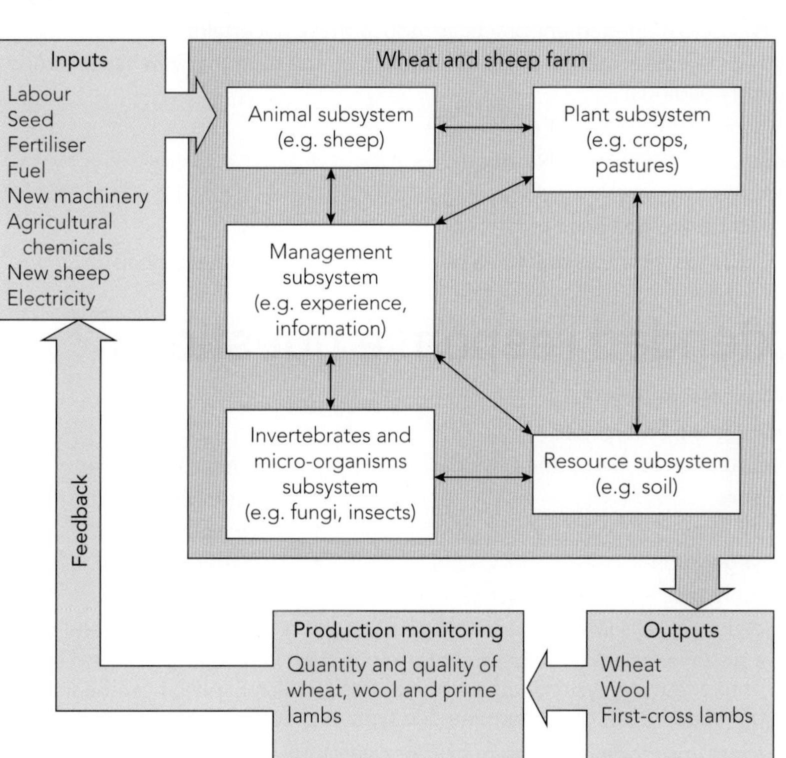

Figure 24.11 A static display model of Rockdale

connect
District agronomist
Information on occupations in the agriculture and rural studies fields.

connect
Agricultural engineering

connect
Conservation earthworks

connect
Agricultural Technical Officer

connect
Irrigation

Careers

In addition to the careers mentioned in this case study, there are many career pathways in agriculture relating to working with plants. Technical officers and agricultural engineering jobs are needed for plant production systems, because of the importance of irrigation and drainage systems and many other technologies related to the growing of crops.

ISBN 9780170265560

Chapter review

Things to do

For the farm you have studied:

1. Describe the factors that a farmer might consider when making farm management decisions.
2. a Identify the subsystems operating within the farm unit.
   b Describe the boundaries.
   c Name the main inputs, outputs and processes.
3. a Describe the annual routine management procedures.
   b Explain why each exists.
4. Examine the physical relationship between inputs and outputs in farm production.
5. Evaluate the economic efficiency of the farm (by finding out the gross margin per hectare and comparing it with that of other farms).

Things to find out

Answer these questions in relation to the farm you have studied.

1. Identify instruments used to measure resources such as weather and soils.
2. What are the main subsystems and the interactions between them?
3. a Name four areas of uncertainty on the farm.
   b How could management practices reduce these uncertainties?
4. Identify problems associated with production on the farm that you have studied. Use the following headings:
   Management
   Environment
   Disease
   Nutrition
   Genetics
5. a Describe the goals of the farmer.
   b Discuss the attempts that have been made to achieve these goals.

Extended response questions

1. For your case study farm:
   a identify the enterprises
   b record information on the physical and biological resources of the farm, including:
      - soils
      - topography
      - climate
      - water sources
      - vegetation
      - infrastructure.
   c identify technologies used on the farm
   d explain the ways in which technology is used in farm management and production.
2. Draw a systems model or diagram that shows the physical, biological and socioeconomic inputs into a farming system with which you are familiar. Explain how the inputs that you have identified interact to determine one type of production occurring in the area.
3. Draw a systems model or diagram of your case study farm. Explain how analysis of the model could enable you to identify:
   a important interactions between subsystems
   b problems with the operations of the farm
   c problems with the efficiency of the farm.
4. a For a farm that you have studied, identify potential farm safety hazards, assess the risk and suggest strategies to reduce or overcome these problems.
   b Outline the Work Health and Safety legislative requirements that affect a farm.

ISBN 9780170265560

UNIT 5
MICRO-ORGANISMS AND INVERTEBRATES

Shutterstock.com/Sinisa Botas

CHAPTER 25

Outcomes

Students will learn about the following topics.

1. The role of nutrient cycles in Australian agricultural systems, including the nitrogen and carbon cycle.
2. The role of microbes and invertebrates in the decomposition of organic matter.
3. Beneficial relationships between microbes and plants, including the fixing of atmospheric nitrogen in legumes.

Students will be able to demonstrate their learning by carrying out these actions.

1. Distinguish between micro-organisms and invertebrates.
2. Describe the role of micro-organisms and invertebrates in mineral or nutrient cycling (the carbon and nitrogen cycles).
3. Describe the role of micro-organisms in the rumen and in legume nodules.
4. Outline the role of micro-organisms and invertebrates in the decomposition of organic matter.
5. Define the term 'disease'.
6. Describe the role of micro-organisms in food production.
7. Describe the role of micro-organisms in silage making.
8. Describe the effects of micro-organisms in food spoilage.
9. Outline the methods used to reduce food spoilage, such as heat, low temperature, dehydration, chemical preservatives, radiation and aseptic packaging.
10. Explain the terms 'food poisoning' and 'food infection'.
11. Describe the industrial uses of micro-organisms.
12. Explain how micro-organisms are used in the treatment of sewage waste.
13. Outline how genetic engineering has been applied in micro-organisms to produce useful products.

ISBN 9780170265560

MICRO-ORGANISMS AND INVERTEBRATES IN AGRICULTURE

Words to know

active immunity antibody production by an organism to combat disease

aerobic in the presence of oxygen

anaerobic without oxygen

antibiotic a chemical substance, produced by certain micro-organisms, that kills or inhibits the growth of other micro-organisms

antibody a protein, made by animals in response to bodily invasion by a pathogen (poison), that combines with the pathogen and renders it harmless

attenuated strains weakened strain

disease any condition that produces an adverse change in the normal functioning of an organism

erosional wearing down

invertebrates animals without backbones

live vaccines vaccines that contain pathogens that are no longer able to cause disease

microbiology the study of micro-organisms and their activities

pasteurisation the process of mild heating to kill particular spoilage organisms or pathogens

passive immunity immunity gained through the use of ready-made antibodies being introduced into the body

psychrophilic (an organism that) grows best at low temperatures (less than 20°C)

silage plant material that is stored with little drying – it slightly ferments and is preserved and used for stock feed

symbiosis an association in which two organisms or populations live together for their mutual benefit

thermophilic (an organism that) grows best at high temperatures (between 50°C and 60°C)

yeast a single-celled fungus

Introduction

Micro-organisms and invertebrates are important in agricultural production systems. The groups of micro-organisms to be discussed include algae, bacteria, protozoa, viruses and fungi. The groups of invertebrates to be discussed include arthropods (e.g. insects, arachnids), helminths, annelids, nematodes and molluscs.

Micro-organisms

Microbiology is the study of micro-organisms and their activities. Micro-organisms, or microbes, are too small to be seen clearly by the unaided human eye. In general, organisms with a diameter of less than 1 mm are studied by microbiologists.

Micro-organisms are important to humans in many ways. Bacteria and fungi are the main decomposers of organic matter in the soil. This breakdown is important as it enables minerals to be recycled.

Soil is also the site of most nitrogen fixation by micro-organisms. In this process nitrogen in the soil atmosphere is converted to ammonia by both symbiotic and free-living micro-organisms.

Ruminant animals rely on the micro-organisms in their rumen. Bacteria, anaerobic fungi and protozoa break down cellulose as well as other plant materials.

Micro-organisms are also important in food manufacture. They are used to make cheese, yoghurt, bread, wine, beer and silage. Some micro-organisms can cause food spoilage.

Micro-organisms are used in industry. Antibiotics and vitamins are products of industrial fermentation. Micro-organisms are also used in the treatment of sewage wastes.

On the negative side, many micro-organisms also cause food spoilage. Food contains large amounts of water, sugar and proteins, used by bacteria and fungi to grow and reproduce. To prevent foods from being destroyed, or people from being poisoned, care must be taken when handling and storing food.

Several methods exist for preserving food: heat, low temperature, dehydration, chemical preservatives, radiation and good aseptic packaging. Food poisoning and food infection are now infrequent due to care in food preservation and strict control methods.

Invertebrates

Invertebrates are all animals without backbones. Many invertebrates have highly specialised structures that allow them to live in a wide range of environments.

Invertebrates involved in the recycling of organic matter include earthworms, millipedes, beetles and slaters.

Invertebrates may also be involved in food production, crop pollination and biological control of pests. For example, bees produce honey and can be involved in crop pollination.

Some invertebrates are important parasites of animals and plants. Ticks and mites are important external parasites of animals, and flatworms and roundworms are important internal parasites. Aphids, nematodes and thrips live on plants.

1 What size are organisms studied by microbiologists?

2 Explain the term 'invertebrate'.

Influence on agricultural production systems

Micro-organisms and invertebrates are important because of their effect on agricultural production systems. They play a role in:

- mineral cycles
- microbial symbiosis
- ruminant digestion
- soil fertility
- plant and animal disease
- food manufacturing
- silage production
- food spoilage
- food preservation
- food poisoning and food infection
- industrial microbiology
- sewage waste treatment
- genetic engineering.

3 List eight ways that micro-organisms and invertebrates affect agricultural production systems.

ISBN 9780170265560

Mineral cycles

Nutrients in the form of minerals are cycled between the living and the non-living aspects of the environment, by means of micro-organisms. Microscopic bacteria and fungi, as well as many invertebrates, decompose the remains and wastes of plants and animals found in the soil, releasing the chemical elements they contain. These elements can then be absorbed by roots and reused by plants. Figure 25.1 illustrates how decomposer-type organisms assist in the release of minerals and how these minerals move through a living system in a nutrient cycle.

All elements in nature, especially nitrogen, carbon and oxygen, can be recycled in this way.

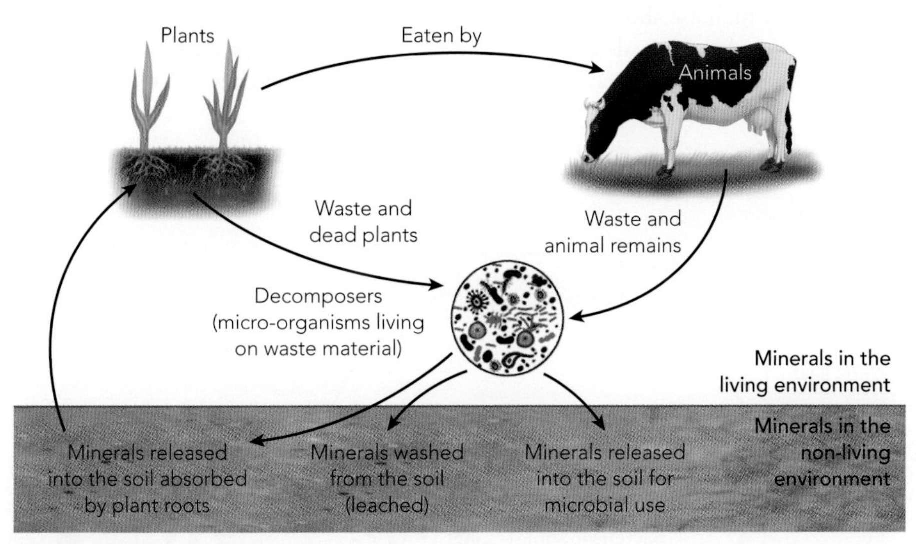

Figure 25.1 A nutrient cycle

Carbon cycle

The cycling of carbon is very important. Almost all the energy needed to drive the activities and processes that take place in living things is derived from photosynthesis or the decomposition of organic matter. Figure 25.2 shows the circulation of carbon in the carbon cycle.

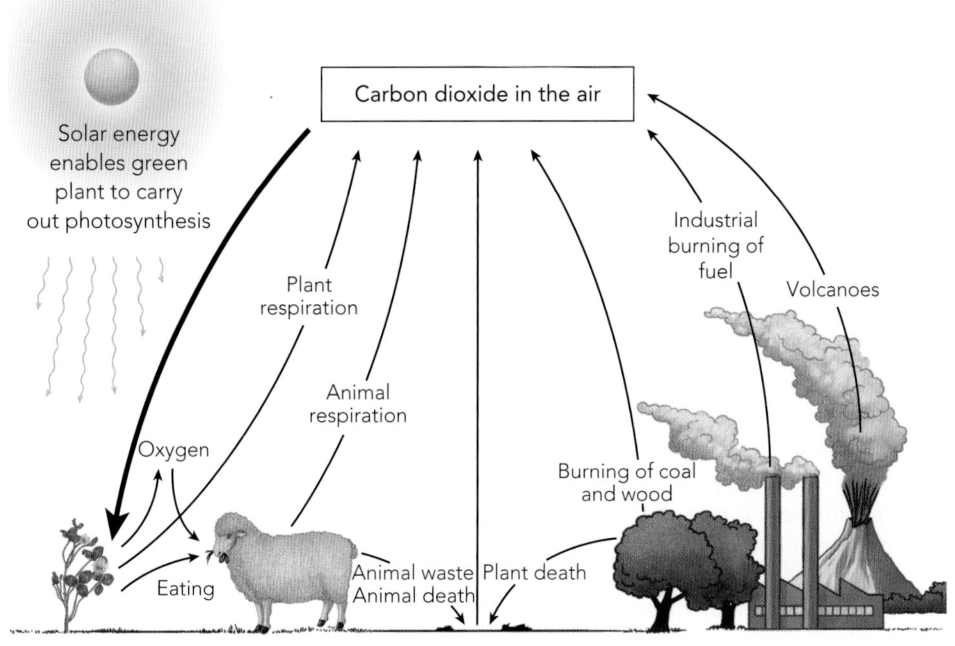

Figure 25.2 The carbon cycle

ISBN 9780170265560

Carbon dioxide is removed from the atmosphere and converted into plant material by the process of photosynthesis. When a plant is eaten, some of this fixed carbon enters the animal. Animal waste, dead animals and dead plants contribute to the organic matter pool. There is more carbon present in organic matter than there is in living organisms. Therefore the microbial decomposition of this material is an important way of returning carbon dioxide to the atmosphere.

Nitrogen cycle

Micro-organisms are responsible for many of the transformations in the nitrogen cycle, the main features of which are shown in Figure 25.3. Plant growth is often limited by low levels of nitrogen in a suitable form, despite the abundance of nitrogen gas in the atmosphere. Micro-organisms make nitrogen available to plants in a form that they can use.

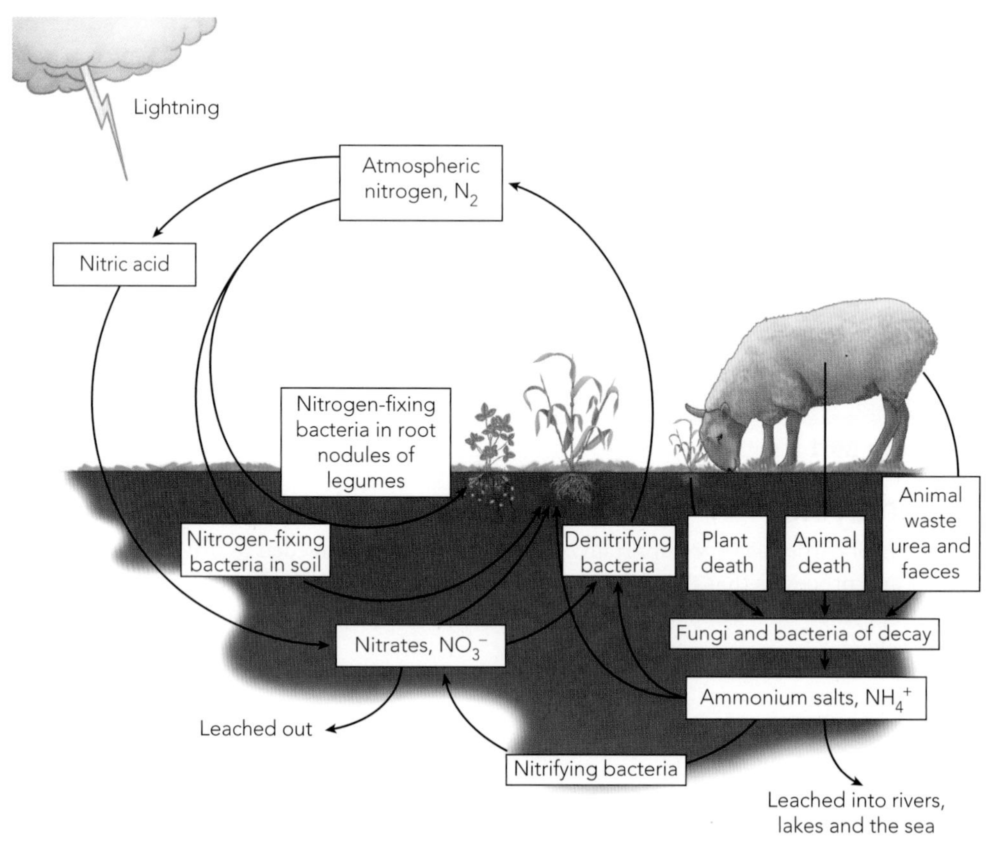

Figure 25.3 The nitrogen cycle

The main features of the nitrogen cycle are as follows.

- *Nitrogen fixation.* Nitrogen gas in the atmosphere is made available to living organisms by being converted to ammonia (NH_3), or 'fixed'. Nitrogen fixation is carried out by bacteria and blue-green algae. The bacteria may be either free living (e.g. *Clostridium*) or symbiotic (e.g. *Rhizobium*, which invades the roots of legumes and forms nodules).
- *Precipitation.* Rainfall contains a certain amount of nitrogen, which has been released by lightning or comes from industrial pollution. It is present mainly as ammonium (NH_4^+) and nitrate (NO_3^-) ions.
- *Ammonification.* Most nitrogen is present in the soil as complex organic compounds. These are decomposed by fungi and bacteria during the process of mineralisation. Organic nitrogen is released as ammonium (NH_4^+) ions.
- *Nitrification.* During nitrification, ammonium ions are oxidised to nitrite (NO_2^-) and then to nitrate (NO_3^-) ions. Bacteria are responsible for nitrification. The most common nitrifying bacteria are *Nitrosomonas* and *Nitrobacter*.

Mineralisation refers to the complete decomposition of organic material, with the release of available elements, including nitrogen compounds:

$$\text{Mineralisation} = \text{ammonification} + \text{nitrification}$$

ISBN 9780170265560

- *Denitrification.* Many soil bacteria are able to use nitrate, nitrite or ammonium ions to produce nitrogen gas (N_2). This process, called denitrification, occurs in waterlogged soils where oxygen levels are limiting. It is desirable to reduce denitrification, as it results in a loss of nitrogen available to plants.
- *Leaching.* This is the loss of nutrients from the soil. The nitrate ion (NO_3^-) is absorbed only weakly by the soil and is easily leached out. Leaching is more likely to occur in high rainfall areas.

4 Why is the microbial decomposition of organic matter an important process?
5 What is nitrogen fixation?
6 Name the two types of bacteria involved in nitrification.
7 What is denitrification?

Microbial symbiosis

A symbiotic association is one in which two organisms or populations live together. The term **symbiosis** usually refers to an association that benefits both organisms or populations. In agriculture there is a number of important microbial symbioses, including those occurring in the root nodules of leguminous plants and in the rumen of ruminant animals.

Legume nodules

Legumes include clovers, lucerne, soybeans, peas and beans. Most legumes have swellings or nodules on their roots. The presence of nodules enables legumes to grow in soils that are low in nitrogen. In these root nodules, formed by the association of *Rhizobium* bacteria and a legume, nitrogen from the atmosphere is converted to ammonia or 'fixed'. This conversion leads to a rapid formation of plant proteins.

The association between *Rhizobium* and the legume is symbiotic, as both plant and bacteria benefit. The bacteria are provided with their food, and synthesise nitrogen compounds made from nitrogen in the air in return. The nitrogen compounds pass from the bacteria into the small nodules, where they are converted into amino acids for use by the plant.

Nodules can be pink or white bumps on the plant roots. Pink nodules are generally more efficient producers of protein. They are larger and fewer in number than the corresponding white nodules.

There are a number of strains of *Rhizobium* bacteria in the soil. Each strain will infect only a particular group of related legume plants. Where there is uncertainty as to the type and number of *Rhizobium* bacteria in the soil, inoculation is recommended; that is, the farmer coats the seed with the correct strain of *Rhizobium* bacteria before the seed is sown. Legume seeds are usually inoculated just before sowing by mixing the seed with water, the inoculum and an adhesive (glue). The seeds are dried before sowing.

8 Explain why *Rhizobium* bacteria in a nodule of a legume plant is regarded as a symbiotic association.
9 Under what circumstances should legume seed be inoculated?

The rumen

A second symbiosis of great importance to agriculture is that between ruminant animals (e.g. cattle, sheep, goats) and the microbial population of the rumen. The rumen can be regarded as a large microbiological fermentation chamber. It contains a large number of bacteria (10^9 per millilitre), protozoa (10^6 per millilitre) and a lesser population of fungi. The rumen is kept slightly acid and at a temperature of 38°C to 42°C.

One of the main purposes of the bacteria, protozoa and fungi in the rumen is to break down cellulose as well as other plant materials (Fig. 25.4). The rumen provides a chamber to allow sufficient time for the decomposition processes to occur. Because the rumen is **anaerobic** (i.e. lacking in oxygen or oxygen-free), microbial fermentation occurs.

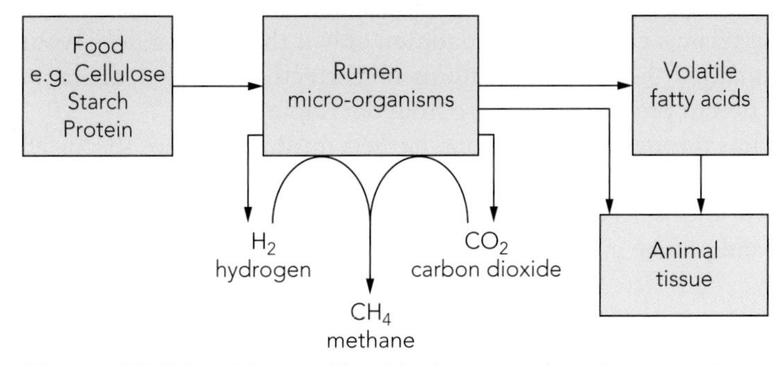

Figure 25.4 Breakdown of food in the rumen by micro-organisms

ISBN 9780170265560

connect

Microbes in bovine rumen fluid

10 What micro-organisms are present in the rumen?

11 How do the micro-organisms in the rumen benefit from their association with the rumen, in ruminant animals?

The food (cellulose, starch and protein) is broken down to produce volatile fatty acids (acetic, propionic and butyric), carbon dioxide and hydrogen. Methane gas is produced and is removed by belching. The fatty acids are absorbed into the bloodstream of the animal through the rumen wall and supply its carbon and energy needs.

Micro-organisms pass down into the fourth stomach of the animal (see Fig. 5.6, page 79), where they are broken down by digestive enzymes to provide the vitamins and essential amino acids that the animal requires for growth.

It is a symbiotic association, as both animal and micro-organism benefits. The micro-organisms gain from a rich supply of nutrients, which allows them to grow rapidly and continuously. In return they produce volatile fatty acids, amino acids and vitamins, which can be used by the host animal. Rumen micro-organisms are particularly important in the synthesis of vitamin B12 (cyanocobalamin), which is made up of complex molecules that contain cobalt.

Soil fertility

Of the micro-organisms that live in the soil, bacteria are present in the greatest number and variety. Bacteria in the soil take part in nutrient cycling. Some are involved in cellulose and protein breakdown; others are involved in nitrogen fixation. Some of the common bacteria genera include *Azotobacter, Bacillus, Clostridium, Nitrobacter, Nitrosomonas, Pseudomonas* and *Rhizobium*.

Soil is an important reservoir for pathogens, particularly those that affect plants.

Fungi are also abundant in the soil. They are important in the decomposition of cellulose and lignin. Fungi also play an important role in improving soil fertility. Their filaments or hyphae help to bind soil particles together to form aggregates or soil crumbs. This improves aeration and drainage of the soil.

The population of algae in soil is generally smaller than that of either bacteria or fungi. The major types present are green algae, blue-green 'algae' (not true algae, actually a type of bacteria) and diatoms. Their photosynthetic nature accounts for their presence on the surface or just below the surface layer. Their growth contributes to soil structure and is beneficial for erosion control. They have a significant role in nitrogen fixation, particularly in paddy soils used for the cultivation of rice.

12 What type of micro-organism is present in the soil in greatest numbers?

13 Outline two roles of fungi in the soil.

14 In what type of farming situation do algae play an important role in nitrogen fixation?

Many invertebrates are involved in recycling organic matter. These include earthworms, millipedes, beetles and slaters. Earthworms burrow into damp soil, helping to aerate the soil by their burrowing action. Earthworms also take in the soil, and in passing the soil through their bodies they help to break it down and release minerals.

Micro-organisms and invertebrates in the soil assist to improve soil fertility and make Australian agricultural systems more sustainable.

Plant and animal disease

A **disease** occurs when there is an adverse change in the normal functioning of a plant or animal. Disease-causing agents are called pathogens. Diseases might cause death, but many reduce production levels over a period and are known as **erosional** diseases. Microbes are one of many causes of disease in plants and animals.

For a disease condition to develop, a special relationship must be created between the disease-causing organism, the host plant or animal and the environment. For example, the bacteria causing tetanus can cause the condition only if they enter a deep wound in an animal. Additionally, if the disease symptoms are correctly diagnosed, the farmer can give the animal an injection to prevent the disease from developing fully.

To prevent loss through viral infections, farmers must plant crops specifically bred to resist the particular disease, or treat animals in advance if a particular disease is known to occur in the local area. This technique is called immunity resistance and takes three forms: natural, active and passive immunity.

 ISBN 9780170265560

Active immunity

Animals that have been exposed to a disease may produce antibodies to fight the pathogens. An **antibody** is a protein produced in response to a toxin produced by a pathogen; it combines with the toxin and renders it harmless. The more antibodies there are, the better the animal's immunity will be.

Protection can be given to an animal through the injection of a suspension of living or dead organisms (i.e. a **vaccine**) into the animal's bloodstream. **Live vaccines** contain pathogens that are no longer able to cause disease. These pathogens are called **attenuated strains**. For example, live vaccines were used in the prevention (treatment) of brucellosis, while dead suspensions are used for polio and cholera vaccinations in humans. Blackleg and tetanus are prevented through the injection of bacteria-free toxins or poisons into the host to encourage antibody production.

Gaining immunity through exposure to a disease or by use of vaccines is called **active immunity**.

Passive immunity

Blood can be removed from animals that have been vaccinated or exposed to disease and the fluid or serum portion can then be separated from the blood sample. This serum will contain antibodies. The serum can be injected into an animal to give short-term protection from the disease. Tetanus is one disease that is treated in this way. For continued immunity, the injections should be repeated at regular intervals (Fig. 25.5). Immunity gained in this manner is called **passive immunity**.

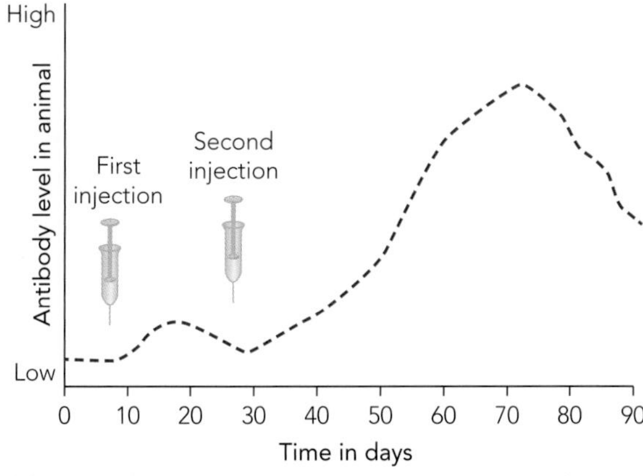

Figure 25.5 Developing immunity in animals by serum injection

Natural immunity

Natural barriers exist against infection, such as an animal's skin or the outer covering of a plant. These create natural immunity.

Plants do not have an antibody system and rely on natural mechanisms to resist disease, such as a waxy cuticle or bark. For a plant to be infected by a fungus or bacterium, the organism must find an opening into the plant from the surrounding environment. For instance, fungi may enter through the pores or stomata in a leaf; insects that suck sap from plants may transmit disease from sick to healthy plants; the farmer may encourage disease through poor pruning techniques and by not planting disease-free plants or seeds. Continuous planting of the same crop on one area of land also results in a build-up of disease organisms in the soil.

If the natural barriers are breached an infected plant may resist the disease by killing all the cells surrounding the infected area and effectively isolating the pathogen. It may also produce chemicals that hinder the growth of the organism.

15 What is passive immunity?

16 What is natural immunity?

17 How do plants resist disease-causing organisms?

ISBN 9780170265560

Food manufacture

Many food items are produced in whole or part by the activities of micro-organisms. They include butter, cheese, yoghurt, bread and alcoholic beverages. Some milk products are shown in Figure 25.6.

iStockphoto.com/brebca

Figure 25.6 Some food items produced by the activities of micro-organisms

Butter

Butter is made by churning cream of a specified butterfat content until the fatty globules of butter separate from the liquid buttermilk. The buttermilk is then drained off.

The typical diacetyl flavour and aroma of butter are developed by adding diacetyl to butter, or by including bacteria able to produce diacetyl in the cream. These include *Streptococcus lactis* and *Leuconostoc citrovorum*.

connect
Butter manufacture

Cheese

Although there are many types of cheese, all require the formation of a curd or solid portion, which can then be separated from the whey or liquid portion.

The curd is made up of a protein, casein, and is usually formed by the action of an enzyme, rennin. Rennin is aided by acidic conditions provided by using a starter culture of lactic acid bacteria. *Streptococcus lactis* and *Streptococcus cremoris* are common starter cultures. They provide the characteristic flavour and aroma of fermented dairy products during the ripening process.

The curd then undergoes a microbial ripening process. The hard cheddar and Swiss cheeses are ripened by lactic acid growing anaerobically (i.e. without oxygen) inside them. Blue vein and Roquefort cheeses are ripened by *Penicillium* mould inoculated into the cheese. The texture of these cheeses allows adequate oxygen to reach the aerobic (i.e. oxygen-using) moulds.

connect
Manufacture of goat cheese

Yoghurt

Commercial yoghurt is made from low fat milk, from which much of the water has been evaporated. The milk is then inoculated with a mixed culture of *Streptococcus thermophilus* and *Lactobacillus bulgaricus*. The temperature of the fermentation is kept at approximately 45°C for several hours. Lactose in the milk is converted to lactic acid.

connect
Making yoghurt

Bread

In the presence of oxygen, yeast ferments the sugars in bread dough into carbon dioxide and water. Aerobic conditions favour carbon dioxide production, which raises the dough and forms the typical bubbles of leavened bread. The yeast organism involved in bread making is called *Saccharomyces cereviseae*.

connect
Bread manufacture

ISBN 9780170265560

Beer

Beer is made from barley grain that is fermented by yeast to form carbon dioxide and alcohol. The yeast involved is *Saccharomyces carlsbergensis.*

The barley grain is allowed to sprout (germinate) and then dried and ground. The product, called malt, contains starch-degrading enzymes (amylases) that convert cereal starches into sugars that can be fermented by yeasts.

Hops are added for flavour; then yeast is added and the beer is fermented at 37–49°C.

Wine

Wine is made from grape sugar that is fermented by yeast to form carbon dioxide and alcohol. The yeast involved is *Saccharomyces cereviseae.* Red wines are fermented at 25°C. White wines are incubated at 10°C to 15°C. Figure 25.7 shows the process for making red wine.

Different strains of grape provide various flavours and sugar concentrations.

18 What gives butter its diacetyl flavour?

19 What gives blue vein cheese its characteristic appearance during ripening?

20 How is yoghurt made?

21 What causes the dough to rise and bubbles to form during bread making?

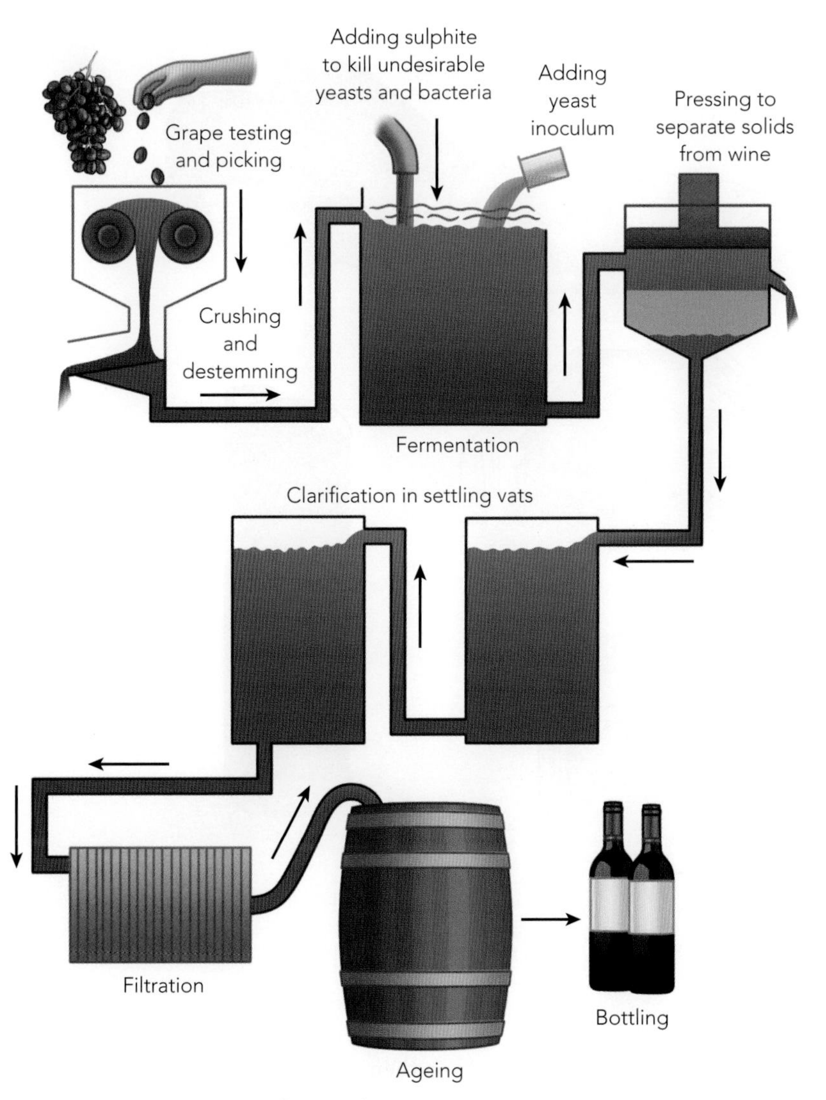

Figure 25.7 Steps in making red wine

Vinegar

Vinegar can be made by allowing wine to go sour under controlled conditions. Wine producers who allowed wine to be exposed to the air found that it turned sour, from the growth of aerobic bacteria that convert the ethanol in the wine to acetic acid. This process is now used deliberately to make vinegar.

22 Why are hops used in beer making?

23 Name the yeast involved in wine making.

24 a Why does wine exposed to the air turn sour?

b What is produced as a result?

ISBN 9780170265560

Ethanol is first produced by anaerobic fermentation of carbohydrates by yeasts. The ethanol is then aerobically oxidised to acetic acid by certain bacteria, such as *Acetobacter* and *Gluconobacter*.

Silage production

Micro-organisms are important on the farm in the manufacture of silage. **Silage** is made when freshly cut green plant material is compressed and stored in oxygen-free (anaerobic) conditions. This promotes desirable anaerobic organisms, such as *Lactobacillus*, but discourages aerobic organisms, which tend to putrefy silage.

The preservation of silage is, in principle, a pickling process. Freshly cut green plant materials are placed in a silage pit, bunker silo, large 'plastic bag' or steel glass-lined tower. Figure 25.8 shows these different storage methods. As much air as possible is excluded by squashing the freshly cut plant material down in some way.

Figure 25.8 Various silos for storing silage

The aerobic bacteria naturally found on the plant surfaces remove oxygen from the air in the silo and release carbon dioxide. Anaerobic bacteria then convert sugars in the plant sap into lactic acid and acetic acid, and the temperature of the plant material rises. The resulting acidity prevents the development of other bacteria that would produce less desirable acids (e.g. butyric acid), decompose plant proteins and putrefy the silage. If this happened, the silage would be unpalatable and indigestible. This will occur if the silage is not properly made. In silage that is made correctly the plant material is pickled in a vinegar-like solution and preserved indefinitely and is a nutritious and palatable stock feed.

 ISBN 9780170265560

Crops for silage are cut or chopped by a forage harvester (Fig. 25.9). Maize, oats and sorghum are ideal plants for making silage. The moisture content of the material for making silage should be between 65% and 75%.

Figure 25.9 A forage harvester

To obtain good silage, the harvested plant material must be packed down tightly to force out as much air as possible. A tractor may be used to assist in packing. Packing down results in a rapid collapse of the plant cells and allows their contents to be fermented by bacteria.

Silage, provided that it is not exposed to the air, can be stored on a farm for many years.

25 Name one desirable anaerobic organism involved in silage making.

26 Why does harvested plant material have to be packed down tightly when making silage?

Spoilage

On the negative side, many microbes also cause food to spoil. Food contains large amounts of water, sugar and proteins used by bacteria and fungi to grow and reproduce. To prevent foods from being destroyed or people from being poisoned, care must be taken when handling and storing food.

Micro-organisms can cause a variety of changes in food. Different kinds of micro-organisms produce different types of changes in food, which include changes in appearance, colour, flavour, odour and the production of gases. Spoilage is caused mainly by bacteria and fungi.

The important bacteria include *Pseudomonas, Escherichia coli, Bacillus* and *Clostridium*.

Pseudomonas species cause souring and putrefaction of meat and discolouring in eggs and milk. Because *Pseudomonas* is aerobic, it is usually involved in surface spoilage of food. *Pseudomonas* bacteria are also **psychrophilic**; that is, they grow best at low temperatures (approximately 15°C) and cause spoilage at these temperatures.

Escherichia coli is the bacterium usually present in human and animal intestinal tracts, and its presence in food is undesirable. These bacteria produce gases and acids. If these bacteria appear in drinking water in large numbers, the water is not fit for human consumption.

Bacillus and *Clostridium* are important spoilage organisms in canned foods – especially preserved vegetables. These two organisms are **thermophilic**; that is, they grow best at high temperatures (between 50°C and 60°C). If canned foods are incubated at high temperatures, such as exist in a truck in the hot sun, the thermophilic bacteria that often survive commercial sterilisation can germinate and grow.

The important fungi involved in food spoilage include *Rhizopus, Penicillium* and *Monilia*. These fungi discolour and cause mould in bread, meat and vegetables.

Spoilage might also occur during the manufacture of microbial products, preventing their sale. *Acetobacter* bacteria can convert alcohol to acetic acid or carbon dioxide, and are therefore important spoilage organisms.

27 List the changes that occur in a food as a result of the growth of micro-organisms.

28 Name two groups of micro-organisms that cause food spoilage.

29 Explain the term 'psychrophilic'.

30 Name two types of bacteria that can cause spoilage in canned foods.

ISBN 9780170265560

Food preservation

Our diet consists very largely of perishable foods (foods that may be spoiled by the growth of bacteria and fungi). Before food can be sold to the public, most of the disease-causing micro-organisms have to be destroyed, and the few that remain must be prevented from growing. The food must also be palatable and have the necessary nutritive value.

connect

Food spoilage and preservation

A number of methods are used to reduce food spoilage, including heat, low temperatures, dehydration, chemical preservatives, radiation and aseptic packaging.

Heat

Heat (or high temperature) treatment is one of the safest and most reliable methods to preserve food. Heat is used to destroy micro-organisms in bottled or canned products. There are two methods: pasteurisation and canning.

Pasteurisation

In **pasteurisation**, mild heat treatment is used to kill some, but not all, of the micro-organisms in a product that would be spoiled by high temperatures. A reduction in the total number of bacteria takes place during pasteurisation, so that the keeping quality is improved.

Milk is pasteurised by heat treatment at 72°C for 15 seconds. This is sufficient to kill such pathogens as *Mycobacterium tuberculosis* and *Brucella abortus,* which cause tuberculosis and brucellosis, two dangerous diseases. Ice cream, yoghurt, beer and certain fruit juices are pasteurised to destroy heat-resistant micro-organisms.

Canning

Foods can be preserved by heating sealed containers. Higher or more severe temperatures are used than in pasteurisation. Research has determined the exact heat treatment necessary to sterilise the food but minimise the reduction in food quality.

The main aim of commercial sterilisation is to destroy the dangerous botulism organism *Clostridium botulinum* and to kill spoilage organisms. If it succeeds in destroying *Clostridium botulinum* bacteria, any other significant spoilage or pathogenic bacterium will also be destroyed. Figure 25.10 shows the industrial canning process.

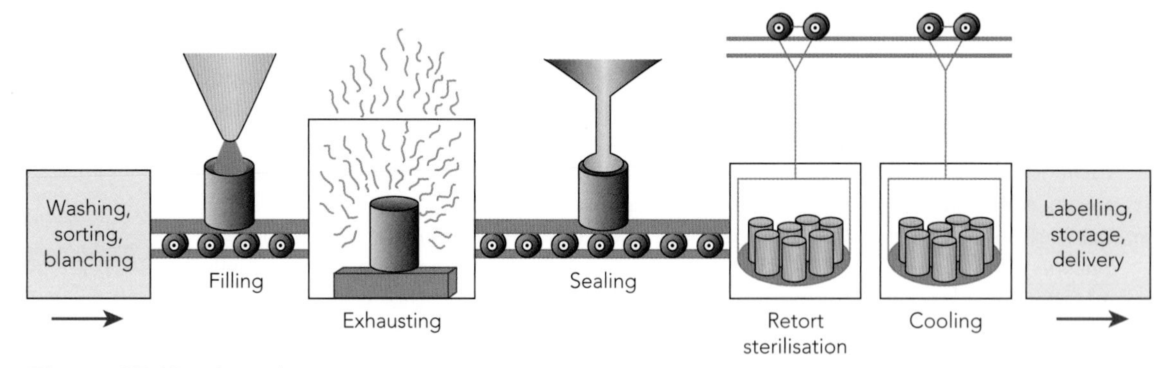

Figure 25.10 Industrial canning

Low temperatures

Temperatures approaching 0°C and lower retard the growth of most micro-organisms. Low temperatures also lengthen the reproduction time of micro-organisms. Modern refrigeration and deep-freezing equipment have made it possible to transport and store perishable foods for long periods. Vegetables, fruit, meats and fish can now be preserved in a near-fresh condition by refrigeration and deep freezing.

Refrigeration

Household refrigerators chill food to approximately 4°C to 7°C. This slows down the growth of micro-organisms to keep most foods in reasonable condition for a few days.

 ISBN 9780170265560

Freezing

When food is frozen, it will keep indefinitely. The food must be frozen to approximately –20°C for long-term storage. Once frozen food defrosts it should never be refrozen, because the few micro-organisms that have survived will grow and multiply.

Dehydration

Dehydration involves the removal of water from food, or a reduction in the availability of water in a food. Dehydration results in insufficient water to support the growth of micro-organisms, but the micro-organism is not killed. A number of dehydration techniques are used, including: sun drying, freeze drying, mechanical drying, syruping and salting.

Chemical preservatives

Chemical preservatives are added to foods to retard spoilage. Most of the preservatives used today are simple organic acids, or salts of organic acids, that the body readily metabolises, and are thought to be safe in foods.

The most effective added preservatives are benzoic, sorbic, acetic, lactic and propionic acids. Sorbic acid and propionic acid are used to inhibit the growth of mould in bread. Nitrates and nitrites are used in curing meats to preserve the pleasing red colour and to inhibit anaerobic bacteria. Benzoic acid is used to prevent mould growth in cheese and soft drink. Foods prepared by fermentation processes (e.g. sauerkraut, pickles, and silage for animals) are preserved mainly by the acetic, lactic and propionic acids produced during microbial fermentation.

Radiation

Ultraviolet radiation is used to reduce the numbers of micro-organisms in food preparation areas. It is used in the surface sterilisation of utensils and meat.

Gamma rays and high energy electron beams are being investigated for their suitability as agents for food preservation. Canned and packaged foods can be sterilised by an appropriate radiation dose.

A major factor affecting the adoption of radiation techniques to preserve food is that the public is concerned about foods exposed to radiation. The effect of radiation on the flavour, colour, odour, texture and nutritional quality of the food has to be determined.

Aseptic packaging

A recent development in food preservation is the increasing use of aseptic packaging. Packages are usually made of some material that cannot tolerate conventional heat treatment (e.g. laminated paper, plastic). The packaging materials come in continuous rolls that are fed into a machine that sterilises the material with hot hydrogen peroxide solution, sometimes aided by ultraviolet light. High-energy electron beams also can be used to sterilise the packaging materials. While still in the sterile environment, the material is formed into packages, which are then filled with liquid foods that have been conventionally sterilised by heat. The filled package is not sterilised after it is sealed.

Food poisoning and food infection

Food poisoning and food infection are now infrequent due to care in food preservation and strict control methods.

Food poisoning

Food poisoning refers to an illness or intoxication caused by toxins present in the food consumed. In this case the organism does not need to invade the body to produce ill effects. *Clostridium botulinum* and *Staphylococcus aureus* (both bacteria) are the main causes of food poisoning. To cause food poisoning, the organisms must be able to grow in the foodstuff and produce a toxin.

Meats, fish, oysters, prawns and canned vegetables all support the growth of *Clostridium botulinum*. The botulinus toxin is very potent; only a small amount is needed to cause death. Ingestion of the toxin causes nausea, fatigue, vomiting, double vision, paralysis and death.

31 Name two types of micro-organisms that spoil food.

32 List the methods used to reduce food spoilage.

33 Name two pathogens killed during pasteurisation of milk.

34 How do low temperatures affect the growth and reproduction of micro-organisms?

35 How does dehydration reduce food spoilage?

36 What chemical preservatives are used to inhibit mould growth in bread?

37 What type of radiation is used in the surface sterilisation of utensils and meat?

38 Give two examples of material used in aseptic packaging that cannot tolerate conventional heat treatment.

connect

Food poisoning and food preservation

Consolidate concepts about food preservation and spoilage

ISBN 9780170265560

39 Explain the term 'food poisoning'.

40 What foods support the growth of *Clostridium botulinum*?

41 Explain the term 'food infection'.

42 What are the symptoms in humans of food poisoning?

Custards, cream products and pig meats support the growth of *Staphylococcus aureus.* The toxin produced is not very potent. It causes nausea, vomiting, abdominal cramps and diarrhoea.

Food infection

Food infection is due to the invasion of the body by pathogenic organisms present in the food consumed. *Salmonella* and *Streptococcus* are the two main causes of food infection. They produce similar symptoms. Symptoms usually develop 5–24 hours after the food is consumed and include nausea, headache, vomiting and diarrhoea.

Industrial microbiology

Many micro-organisms are used in industry. Antibiotics, vitamins, ethanol and citric acid are typical products of industrial fermentation.

Of the microbial products manufactured commercially, probably the most important are **antibiotics**. These are chemical substances produced by certain micro-organisms that kill or inhibit the growth of other micro-organisms. They are produced commercially by microbial fermentation. For example, penicillin is produced by the fungus *Penicillium*.

Vitamins and amino acids are often used pharmaceutically or are added to foods. Several important vitamins and amino acids are produced commercially by microbial processes. Most vitamins are made commercially by chemical synthesis. However, a few are too complicated to be synthesised inexpensively, but fortunately they can be made by microbial fermentation (e.g. vitamin B12, riboflavin).

43 Name three products of industrial fermentation.

44 Name two vitamins produced by microbial fermentation.

Citric acid, which is widely used in foods and beverages, is produced by fungi able to convert sugar to citric acid. *Aspergillus niger* is the fungus species most widely used in the commercial production of citric acid. Molasses is the most common sugar used as a **substrate** (raw material – see Chapter 26) for citric acid production.

Sewage waste treatment

45 Where is sewage sludge stored?

46 Name two gases produced when bacteria feed on sludge in a digestion tank.

Micro-organisms are used in the treatment of sewage waste from intensive animal production systems, such as piggeries. The solid component (sewage sludge) is collected and stored in a digestion tank. Anaerobic conditions exist in this tank, and bacteria feed on the sludge to produce methane gas and carbon dioxide. The small amount of material left is then either fed into sludge lagoons, or dried and buried, or burned.

Genetic engineering

Genetic engineering involves the manipulation of an organism's genetic material. Micro-organisms can be changed so that they make a number of compounds (or gene products) that are useful to humans. Such compounds include insulin, growth hormones, interferons, vaccines, enzymes and other biological substances. Genetic engineering has resulted from scientists' ability to isolate, manipulate and reproduce identified segments of DNA from viral, bacterial and yeast sources.

47 Name five compounds that can be produced by micro-organisms that have been changed.

48 How is it possible to produce cheap and safe vaccines?

Genetic engineering can be used to produce unlimited quantities of gene products from any species. It is now possible to produce cheap and safe vaccines using micro-organisms. This is done when the gene from the pathogenic organism (e.g. foot and mouth virus) is cloned in a safe, easily cultured micro-organism, such as *Escherichia coli*.

Careers

There are many career opportunities in this diverse field related to agricultural studies and the application of scientific study to the needs of industry. Consider the possibilities shown on the Requirements for cheese making and Microbiologist websites.

connect

Requirements for cheese making

connect

Microbiologist

ISBN 9780170265560

Chapter review

Things to do

1. Label a diagram of the nitrogen cycle and the carbon cycle.
2. Collect and classify a range of insects, arachnids, helminths and molluscs on the school farm.
3. Observe and record the symptoms of an invertebrate infestation in plants and animals.
4. Perform husbandry practices designed to reduce disease in:
   a animals; for example, drenching, vaccination
   b plants; for example, dusting, hygiene.

5. **Experiment 25.1** Micro-organisms and bread making

Aim: To use the micro-organism yeast to make bread

Method

1. Put 100 g of plain flour and 1 g of salt into a bowl.
2. Mix in 1.5 g of fresh yeast and 3 g of sugar.
3. Make a well in the mixture, and stir in 75 mm of warm water.
4. With clean hands, knead the dough for approximately 3 minutes.
5. Cover the dough and leave it to stand in a warm place for approximately 20 minutes.
6. Knead the dough again for 1 minute, and place it on a baking tray.
7. Again cover the dough and allow it to rise for approximately 20 minutes.
8. Put the dough in an oven at 200°C, and cook for approximately 15–20 minutes.

Questions

1. What is the name of the gas produced when the yeast reacts with sugar?
2. What is fermentation?

6. **Experiment 25.2** Micro-organisms and silage making

Aim: To make silage on the school farm under aerobic and anaerobic conditions

Method

1. Place grass clippings, maize stalks etc., into two plastic garbage bags, and pack down as tightly as possible.
2. Seal the top of one bag so that it is airtight.
3. Put both bags in a warm place, and leave for several weeks.
4. Open the closed bag, and compare the two bags of silage.

Questions

1. Which bag made the best silage? Why?
2. Describe the conditions needed for making good quality silage.

7 **Experiment 25.3** Micro-organisms and food spoilage

Aim: To look at the environmental conditions that favour food spoilage.

Method

1. Obtain several small pieces of each of the following foods: ham, cheese, potato and apple.
2. Mark four clean test tubes with the name of one food to be tested, and place three small pieces of this food in the bottom of each tube.
3. Put some cotton wool in the top of three test tubes. Place one tube in the freezer, another in the refrigerator and the third in a warm position.
4. Leave the fourth test tube also in a warm position, but without cotton wool in the top.
5. Observe the test tubes each day for 1 week.
6. Record observations.

Questions

1 How does temperature affect spoilage?

2 What two groups of organisms cause most food spoilage?

8 Identify root nodules on the root systems of different legume plants on your school farm or farm case study.

Things to find out

1 Describe the role of the various organisms that are involved in cycling nitrogen.

2 Outline why yeasts are important in food manufacture.

3 What does the word 'vinegar' mean?

4 How are oysters and prawns stored to prevent food poisoning?

5 Pasteurisation does not kill all micro-organisms; so food can still spoil. Why, then, are dairy products pasteurised?

6 Describe the methods by which micro-organisms are used for:
- a industry
- b waste treatment.

7 *Escherichia coli* is the bacterium usually present in human and animal intestinal tracts, and its presence in food is undesirable. How is drinking water tested for the presence of *Escherichia coli*?

8 Explain how genetically engineered micro-organisms can be useful to humans.

9
- a Describe the methods that are used by plants to resist attack by pathogens.
- b How do these methods differ from the production of substances by animals for the same purposes?

10 Explain how and why root nodules are important in the management of soil fertility.

 ISBN 9780170265560

Extended response questions

1. a Describe the role of microbes and invertebrates in the decomposition of organic matter and nutrient cycling. Include nitrogen and carbon.
   b Explain the importance of microbes and invertebrates in the decomposition of organic matter and nutrient cycling.
2. Micro-organisms and invertebrates can have a beneficial effect on agricultural production systems. Discuss their positive value in:
   a mineral cycling
   b food production
   c the rumen.
3. Discuss the harmful and beneficial effects of micro-organisms and invertebrates in either plant production systems or animal production systems.
4. Micro-organisms are involved in food production and food spoilage.
   a Describe the beneficial roles of micro-organisms in the production of food products such as butter, cheese, yoghurt, bread, vinegar and alcohol.
   b Describe the harmful effects of micro-organisms in causing food spoilage.
5. The diet of modern civilised people consists largely of perishable foods. Discuss the methods used to reduce food spoilage, including heat, low temperatures, dehydration, chemical preservatives, radiation and aseptic packaging.

CHAPTER 26

Outcomes

Students will learn about the following topics.

1. The nature and importance of micro-organisms in agriculture.
2. The nature and importance of invertebrates in agriculture.
3. The occurrence, survival and spread of disease.
4. Resistance and immunity.
5. Population dynamics.

Students will be able to demonstrate their learning by carrying out these actions.

1. Distinguish between micro-organisms and invertebrates.
2. Define the term 'disease'.
3. Explain the difference between infectious and non-infectious diseases.
4. List important characteristics of the following micro-organisms and invertebrates: algae, bacteria, protozoa, viruses, fungi, insects, arachnids, platyhelminths, nematodes and molluscs.
5. Give examples of micro-organisms and invertebrates that are important in agricultural production systems.
6. Give examples of beneficial micro-organisms and invertebrates involved in: mineral cycling, food production, ruminant metabolism, energy flow and biological control.
7. Give examples of harmful micro-organisms and invertebrates involved in: competition, disease and food spoilage.

10. Explain how infective agents enter an animal's body.
11. Describe how animal pathogens are spread to new hosts.
13. Discuss resistance to disease in plants
15. Explain how infective agents enter a plant.
16. Describe how plant pathogens are spread to new hosts.
17. Outline the occurrence and distribution of disease in plants.
18. Discuss resistance to disease in plants.
19. Explain the terms 'population' and 'economic threshold'.

ISBN 9780170265560

MICRO-ORGANISMS AND INVERTEBRATES AS DISEASE AGENTS

Words to know

algae simple plant-like organisms that possess chlorophyll and perform photosynthesis

bacteria single-celled microscopic organisms. Some cause organic matter to decay or cause disease, others are important in the formation of cultured dairy products and fixing nitrogen both in the soil and in the roots of legumes

binary fission a type of asexual reproduction in which a parent cell divides into two

economic threshold the population level of a pest when it is necessary to start control measures

endemic diseases that occur in a population year after year

epidemic a rapidly increasing incidence of a disease

exoskeleton the external body covering of an arthropod

flagellum (plural: flagella) a whip-like extension of certain cells; their rotation causes movement

hermaphrodites organisms that have both male and female reproductive organs (e.g. slugs and snails)

heterotroph an organism that needs a supply of food from its surroundings

infectious disease a disease caused by a pathogen transmitted from a diseased individual to a healthy one

metamorphosis change in body form. Some insects have a complete change from larva to adult (e.g. flies)

mycelium mass of hyphae

nematodes roundworms, threadworms and eelworms

non-infectious disease a disease not caused by a pathogen that consequently cannot be transmitted from one individual to another

parasite an organism that lives on or in, and obtains its food from, another living organism

phytoalexin an inhibitory chemical produced by a plant following infection by a micro-organism

platyhelminths flatworms; the simplest of the worm groups; can be free-living or parasitic

population (in biology) a group of one kind of organism living in a particular place at any one time; (in statistics) the entire pool from which a statistical sample is drawn

spores reproductive structures of fungi and some bacteria

substrate the non-living material on which an organism lives or grows

ISBN 9780170265560

Introduction

In agriculture it is important to examine the:

- activities and behaviour of populations of micro-organisms and invertebrates, not of individuals
- role or function of micro-organisms and invertebrates, rather than emphasising the particular group to which they belong.

The groups of micro-organisms to be discussed include algae, bacteria, protozoa, viruses and fungi.

The groups of invertebrates to be discussed include insects, arachnids, platyhelminths, nematodes and molluscs.

One of the major influences of micro-organisms and invertebrates on agriculture is their capacity to cause disease. A disease occurs when there is an adverse change in the normal functioning of an animal or plant. Diseases in both animals and plants may be either infectious or non-infectious.

- **Infectious disease**. The disease-causing agent, or pathogen, is transmitted from a diseased individual to a healthy one. Most infectious diseases are caused by micro-organisms.
- **Non-infectious disease**. The disease cannot be transmitted from one individual to another. These diseases are not caused by pathogens but have a number of causes, including mineral deficiencies, mineral toxicities and various other environmental factors.

Animals and plants affected by the activities of micro-organisms and invertebrates may show various symptoms according to the particular interaction of host, pathogen/parasite and environment.

1. List five groups of micro-organisms.
2. List five groups of invertebrates.
3. Explain the difference between infectious and non-infectious diseases.
4. What is a pathogen?

Micro-organisms

Micro-organisms possess features of both plants and animals. Many types show primitive levels of organisation in their cells. For this reason they are often grouped together and classified as 'halfway houses' for life between the plant and animal kingdoms. For instance, algae (including seaweed) and fungi (e.g. mushrooms and toadstools) are both very plant-like in appearance. However, only algae have the green-coloured chemical called **chlorophyll**, used by real plants to produce food. However, while some algae are green, they do not actually have the internal structures seen in plants. Similarly, animal-like microbes called protozoa (e.g. amoeba) lack features common to animal cells.

Some algae and organisms called bacteria are very primitive in their cellular makeup. Viruses, rickettsias and mycoplasmas are almost non-living in many respects, as they have only a protein coating around a nucleic acid such as DNA or RNA. However, because these organisms can grow and reproduce, under certain conditions, they are classified as living creatures rather than chemicals.

Algae

Algae are simple organisms that possess chlorophyll and perform photosynthesis.

Algae range in size from microscopic single-celled forms to huge kelps that stretch several metres in length and float in the oceans. Single-celled algae are usually found floating on the surface of inland river systems.

Cell walls are not always present. When cell walls are present, they usually contain cellulose. The cell walls of diatoms are impregnated with silica.

Some algae have structures for movement, such as a **flagellum** (whip-like tail; plural flagella).

The simpler forms of algae reproduce by **binary fission** (i.e. divide into two) whereas more advanced forms may reproduce either sexually or asexually.

Importance in agriculture

Blue-green algae (actually a form of bacteria but frequently grouped with algae) are able to fix nitrogen in anaerobic (i.e. no oxygen) situations. In rice paddies this organism supplies the rice plant with nitrogen.

Algae also provide the material called **agar**, used for growing micro-organisms in laboratories.

In many Asian countries a giant alga called kelp is farmed as a food source, mineral source and cheap garden fertiliser.

connect

Some uses for algae

Describe the main products that can be generated through farming algae on a commercial scale.

ISBN 9780170265560

Many algae are at the base of food chains for fish, yabbies and other aquatic organisms harvested for human food.

Unwanted rapid growth of algae (algal blooms) in river systems can result in decreased water quality for stock and domestic use. Algal blooms occur when there is a high concentration of nutrients (in particular, nitrogen and phosphorus) from farm run-off or sewage effluent. Warm weather and favourable sunlight result in rapid reproduction of algae. Some algae produce toxins that are harmful to fish, animals and humans.

5 Describe the size range in algae.

6 Why are algae so important in agriculture?

Bacteria

The several types of **bacteria** are classified according to their shape (Fig. 26.1). Several types of bacteria can be grown or cultured on agar plates. The colour, shape and size of the colonies can be used to identify the bacteria type.

Spherical cells are called cocci (singular: coccus). In the simplest form the cells always divide along the same axis. Often the cells do not separate but remain attached to each other and thus form pairs of cells (as in a diplococcus) or linear chains (as in a streptococcus). When a grape-like cluster of cells form, they are known as a staphylococcus.

Rod-shaped bacteria are called bacilli (singular: bacillus) and can be long or short. There are several groups of spiral or curve-shaped bacteria.

connect

View types of bacteria

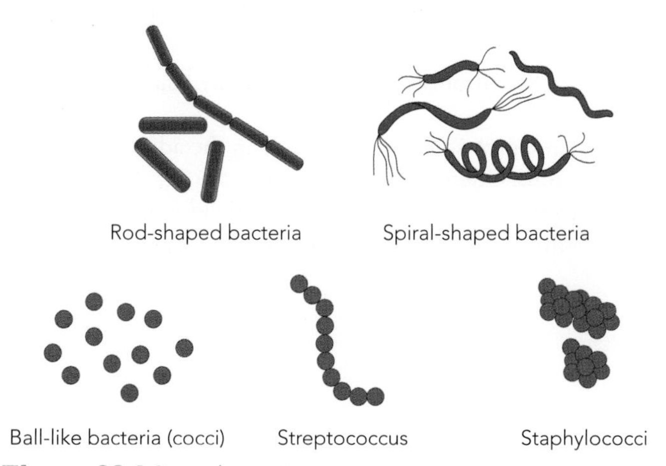

Figure 26.1 Some bacteria types

Bacteria have cell walls that may be rigid or flexible. Differences in the chemical composition of bacterial cell walls have enabled bacteria to be divided into two groups according to their Gram stain (i.e. ability to retain a violet stain after decolonisation, named after Danish physician Hans Gram). Some are Gram-positive; and others are Gram-negative.

Some bacteria are able to move. They do this by either swimming through liquids by means of flagella (Fig. 26.2) or by gliding in contact with a solid surface.

Bacteria reproduce by binary fission.

connect

Bacteria growth

This video shows the process of binary fission.

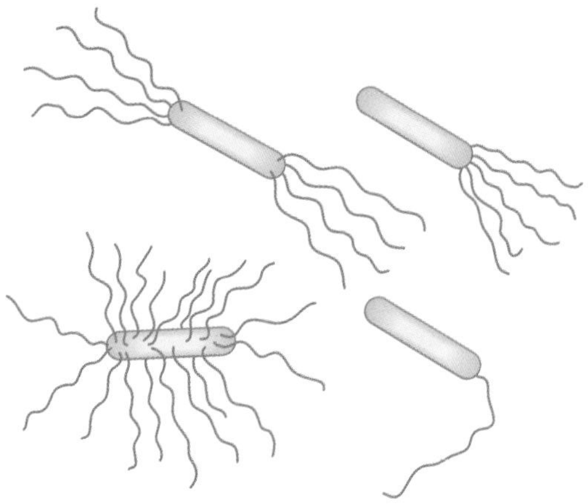

Figure 26.2 The arrangement of flagella in bacteria

ISBN 9780170265560

Importance in agriculture

Bacteria are involved in various mineral cycles in nature, such as the nitrogen and carbon cycles. Bacteria release and recycle minerals and gases that were once part of a living system. Other bacteria, such as the nitrogen-fixing bacteria found on the roots of clover, are responsible for incorporating gases into living systems.

Bacteria are involved in cheese production, silage production and cellulose digestion in ruminants.

Bacteria are the cause of many plant and animal diseases and losses during storage of agricultural products. Plants can die from soft rots and galls caused by bacteria. Dairy cattle can suffer from bacterial diseases such as mastitis or tetanus.

7 Briefly describe the main shapes of bacteria.

8 Why are bacteria important in agriculture?

Protozoa

Protozoa are single-celled **heterotrophs** (i.e. they cannot produce their own food).

Protozoa are 5–250 micrometres long. They can be either free-living or parasitic.

The cell wall is not always present, and if present it does not contain cellulose.

Protozoa are classified according to their method of movement (Fig. 26.3).

- The simplest protozoa move by extending part of their cytoplasm in one or more 'false feet' or pseudopodia. A common example is amoeba.
- In more highly developed species, propulsion is by whip-like flagella. A common example is *Trypanosoma*.
- Other protozoa move by synchronised movement of tiny hairs or cilia. A common example is *Paramecium*.
- Another group of parasitic protozoa are non-motile in the adult stage; that is, they are not capable of spontaneous, independent movement.

Some protozoan groups reproduce by binary fission (dividing into two); one group reproduces by multiple fission (dividing into many parts).

connect

Images of a ciliated protozoan

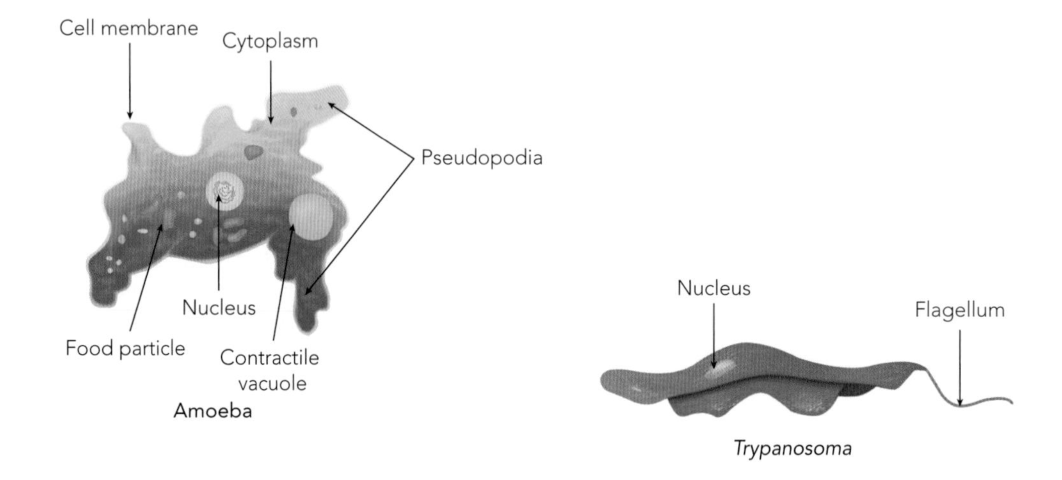

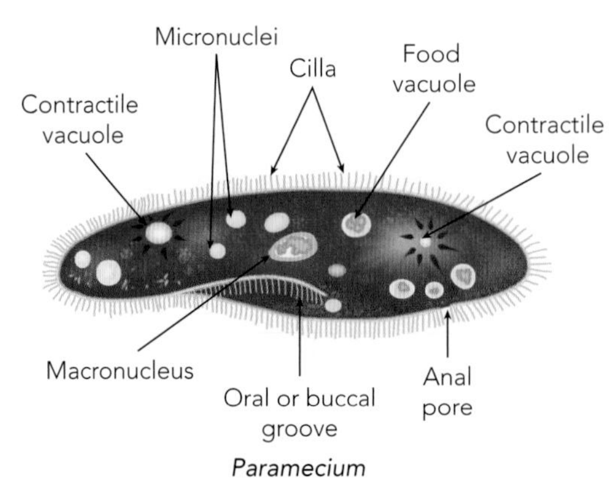

Figure 26.3 Protozoa are classified according to their method of locomotion

ISBN 9780170265560

Importance in agriculture

- Protozoa are important in agriculture and animal nutrition.
- Protozoa occur in the rumen and play an important role in ruminant digestion.
- Protozoa are also involved in energy and nutrient cycles.
- Protozoa can cause a number of animal diseases. Coccidiosis affects poultry. Tick fever affects cattle (cattle ticks are the vector for spreading the protozoa).
- Protozoa are not known to cause any plant diseases.

Protozoa can also cause a number of human diseases (e.g. amoebic dysentery, African sleeping sickness and malaria).

9 How are protozoa classified?

10 Why are protozoa important in agriculture?

Viruses

Viruses can live only as parasites within the cells of a living organism. Even though viruses may survive outside their host's cells, they can reproduce only inside a living host cell. Hence they are called obligate parasites.

The largest viruses are approximately the size of the smallest bacteria, and most are much smaller. Because of their small size, viruses cannot be seen with a light microscope, but they can be observed with an electron microscope.

Virus particles consist of a core of nucleic acid (either DNA or RNA) enclosed in a protein coat. The nucleic acid is the infectious material that, once inside a living cell, makes the cell produce more virus particles. In this process the nucleic acid causes changes, which usually result in disease in the host organism. Figure 26.4 illustrates a general virus lifecycle.

Plant viruses can be transmitted from infected to non-infected plants during pruning, budding or grafting or by insects. Animal viruses can be spread by infected body fluids or through insects that have bitten infected animals. For these reasons, many farmers quarantine diseased animals.

connect

Types of viruses

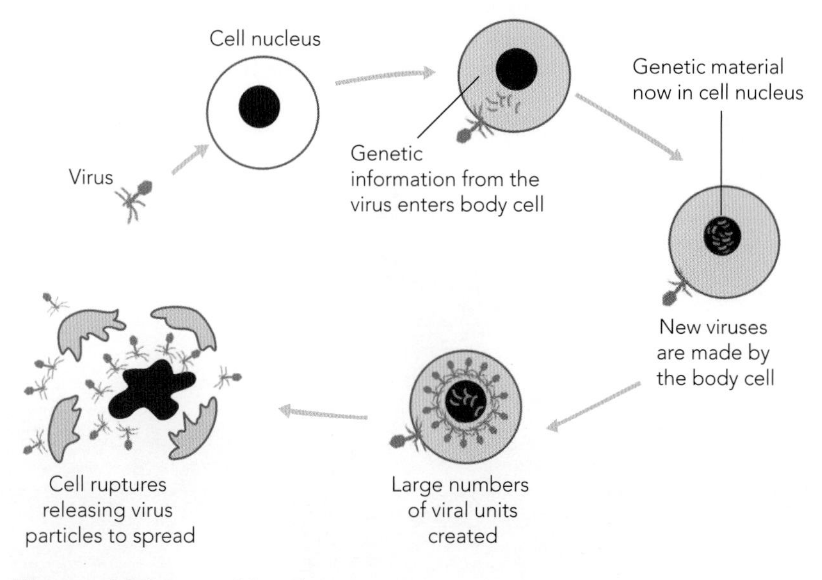

Figure 26.4 A virus lifecycle

Importance in agriculture

Viruses cause a number of diseases in animals. In Australia these include three-day sickness in cattle, scabby mouth in sheep, and Marek's disease in poultry. Due to good quarantine practices, foot-and-mouth disease has not entered Australia.

Viruses cause a number of diseases in plants, including spotted wilt of tomatoes, yellow dwarf virus of wheat, and yellow mosaic of beans.

On the positive side viruses are responsible for the variegated colouring in some flower crops such as tulips.

11 Explain how viruses reproduce.

12 Why are viruses important in agriculture?

ISBN 9780170265560

Fungi

Fungi are plant-like organisms that lack chlorophyll and are heterotrophic. They derive their energy from the decomposition of organic materials.

They range in size from single-celled yeasts, important in the manufacture of alcoholic products and bread, to the multicellular mushrooms. Fungi are composed of fine-branching threads called **hyphae** (see page 295). A mass of these threads is called a **mycelium** (Fig. 26.5).

Fungal cell walls are rigid or semi-rigid and contain cellulose or chitin.

Fungi are able to grow over and through a substrate, such as plant tissue, dead organic matter or a laboratory growth medium (Fig. 26.6). In this process the hyphae form a colony.

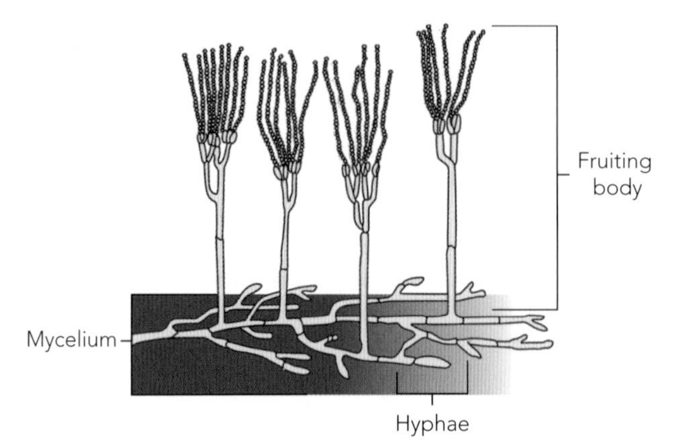

Figure 26.5 A fungal mycelium, showing long chains of conidia (spores)

Fungi reproduce asexually and sexually by forming **spores**. The spores enable the fungus to survive and/or be dispersed. In many instances the fruiting body, which contains the spores, is the only obvious part of a fungus, as the rest of the fungus grows hidden in the

Figure 26.6 Mushrooms

substrate. The main function of most of the spores is to disperse or spread the fungus. The spores are easily carried by rain splash or wind. In some species of fungi, fruiting structures called perithecia act as survival structures by releasing spores when conditions become favourable for spore germination and growth.

Fungi are divided into groups or classes based on their method of reproduction.

ISBN 9780170265560

Importance in agriculture

Fungi are involved in mineral and energy cycles. Fungi and other micro-organisms decompose the remains and wastes of plants and animals found in the soil, releasing the chemical elements found in them.

A number of fungi are used in the brewing, wine and bread industries. Yeasts are important in fermentation and are essential in the production of beer and wine. Yeasts are also used in the production of leavened bread.

Various fungi produce antibiotics, used in human and animal medicine to control bacterial and fungal diseases. Antibiotics are obtained from fungi cultured under controlled conditions.

Mushrooms are an important source of food. Some species of edible fungi are produced commercially on mushroom farms. The part eaten is the spore-forming structure (fruiting body).

Research is being carried out to determine whether fungi can be used for the biological control of weeds and insect pests. A rust fungus has been successfully used to control skeleton weed.

Fungi cause many diseases in plants, including black stem rust of wheat, loose smut of oats and peach leaf curl. Fungi can also cause a few diseases in animals, including ringworm in cattle and lumpy jaw in cattle.

connect

Mycorrhizal applications

View information on the commercial use of mycorrhizae.

13 How are fungi spread?

14 Why are fungi important in agriculture?

Invertebrates

Insects

Insects are one of the most prolific and successful life forms on earth. Insects belong to the Phylum Arthropoda, or 'jointed-feet' organisms.

To be classified as an insect an organism must have the following five characteristics (see Figure 26.7):

1 a body divided into three distinct sections – head, thorax and abdomen
2 three pairs of legs attached to the thorax (which may also have wings attached to it)
3 no legs attached to the abdomen in the adult stage
4 air sacs called tracheae (through which the organism breathes)
5 an **exoskeleton** (i.e. protective or supportive structure covering the outside of the body).

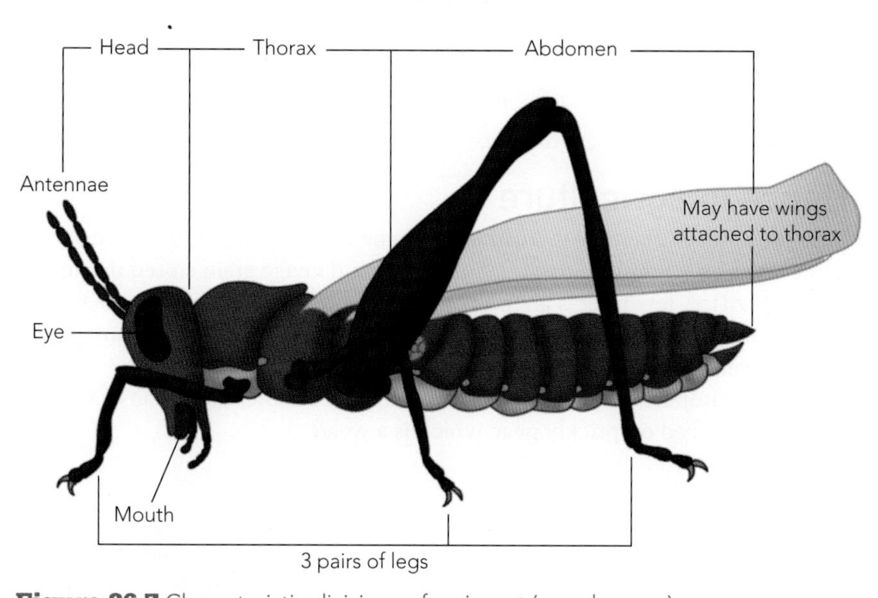

Figure 26.7 Characteristic divisions of an insect (grasshopper)

Insects may be divided into three subgroups based on lifecycle and wing development.

1 Primitive insects, which are wingless and have changed little in form over time (e.g. silverfish).
2 Nymph-type insects that basically follow the lifecycle illustrated in Figure 26.8 (e.g. aphids, locusts). Wings are present, attached externally to the thorax. The young resemble the adult, and insects undergo **incomplete metamorphosis**; that is, there is little change in body form as the insect goes through the different stages.

ISBN 9780170265560

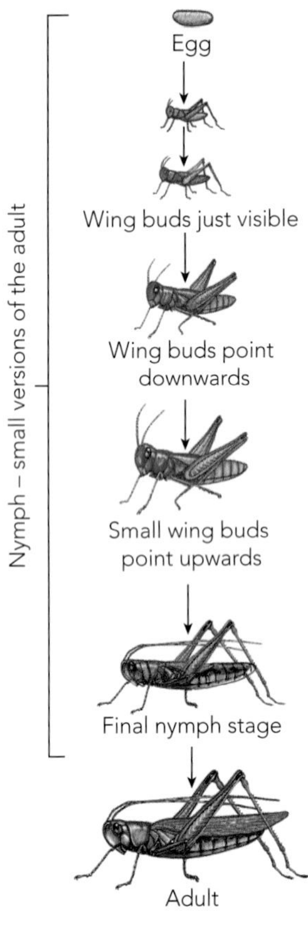

Figure 26.8 Insect lifecycle nymph type (e.g. grasshopper)

3 Metamorphosis-type insects with a lifecycle similar to that shown in Figure 26.9 (e.g. bees, flies, mosquitoes, butterflies). They are able to exploit two completely different environments. The insects pass through entirely different stages (i.e. egg, larva, pupa and adult). They undergo **complete metamorphosis**. That is, there are big differences in the form, size and shape of the four developmental stages.

A number of features of insects have made them one of the most successful life forms. These include: mobility, small body size, high rates of reproduction, an exoskeleton, and adaptability in a wide variety of environments.

Insects feed in various ways. The locust is a chewing insect with several mouthparts designed to strip leaves. The butterfly has a sucking mouthpart; this sucking tube is seen coiled under the insect's head. The housefly has a sponging mouthpart, through which it can dissolve its food then suck it up. Because of this variety of feeding methods, many different forms of insect damage may be inflicted on plants (Fig. 26.10).

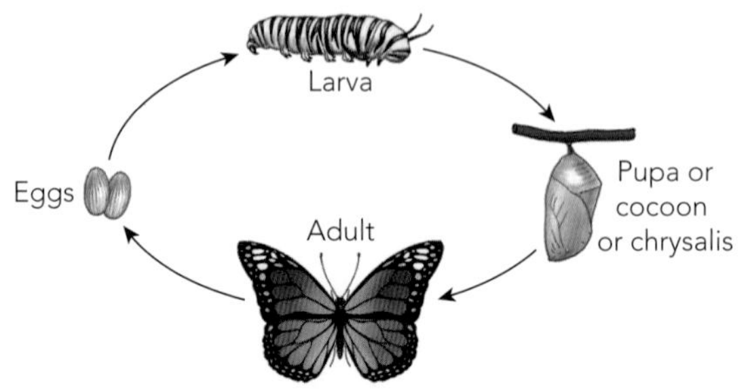

Figure 26.9 Insect lifecycle complete metamorphosis type (e.g. butterfly)

Chewing mouthparts Sucking mouthparts Sponging mouthparts

Locust

Grasshopper Butterfly

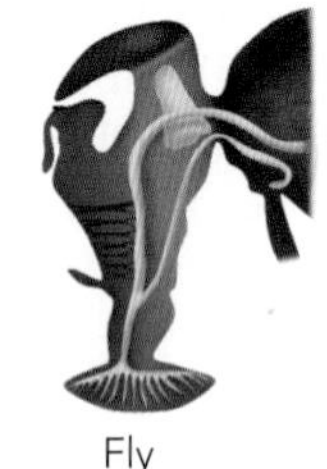

Fly

Figure 26.10 Insect mouthparts

connect

Locust plagues

15 List five features of insects that have made them one of the most successful life forms on Earth.

16 Why are insects important in agriculture?

Importance in agriculture

Some insects are pests of plants (e.g. Australian plague locust, aphid, Queensland fruit fly, scale insects, cabbage white butterfly). Some insects damage grain stored in silos (for example, rice weevil). A few insects are pests of animals (e.g. sheep blowfly, sheep ked).

Insects are beneficial to agricultural production because they pollinate many plants. Bees produce honey, a valuable product.

Some insects are important in biological control. For example, the larvae of the moth *Cactoblastis cactorum* feed on prickly pear, which is a weed.

Arachnids

Arachnids are closely related to insects. Arachnids also belong to Phylum Arthropoda, or 'jointed-feet' organisms. They include ticks, mites, spiders and scorpions.

To be classified as an arachnid, an organism must have the following three characteristics (Fig. 26.11):

1 a body divided into two distinct sections – cephalothorax and abdomen
2 four pairs of legs
3 no antennae.

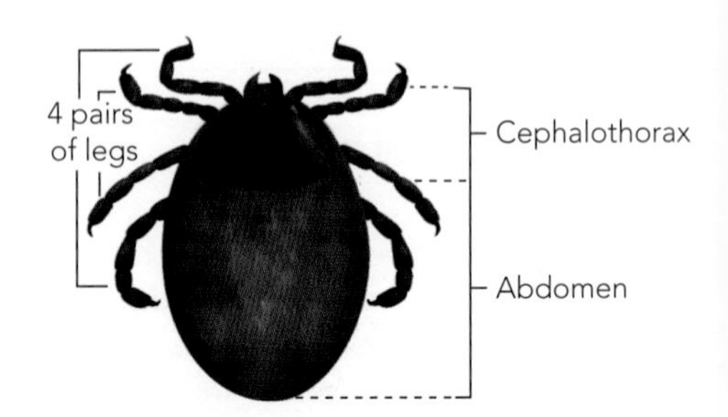

Figure 26.11 Characteristic divisions of an arachnid (tick)

ISBN 9780170265560

The most important arachnids affecting agriculture are ticks and mites. The lifecycle of a tick is shown in Figure 26.12. Ticks have specialised holding structures that enable them to anchor themselves to a host. They also have specialised mouthparts for piercing and sucking. Mites have limbs that have been modified for attachment or burrowing. Some mites also have specialised mouthparts for feeding.

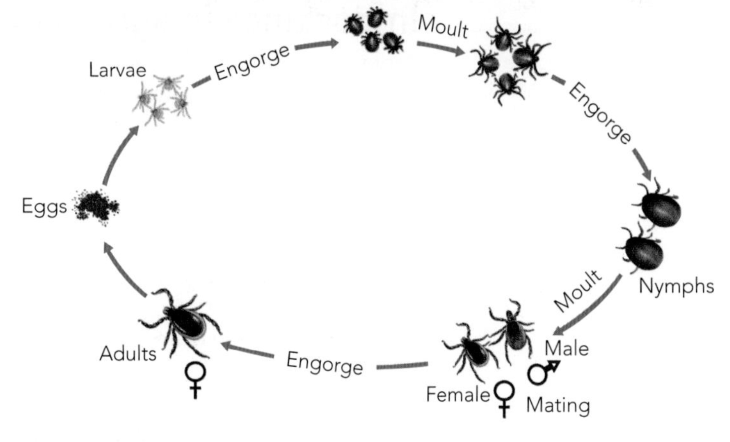

Figure 26.12 Lifecycle of a tick

Importance in agriculture

Ticks and mites are ectoparasites (i.e. external parasites) of farm animals. Cattle ticks cause serious production losses in northern New South Wales and tropical areas. They suck blood from the cattle and cattle lose condition. Mites can cause serious production losses in animals. In poultry, red mites feed on the blood of the host animal. This results in anaemia and irritation.

A number of mites are important pests of pastures, crops and orchards. These include red-legged earth mite, red spider and citrus mite.

17 List three features of an organism that enable it to be classified as an arachnid.

18 Why are arachnids important in agriculture?

Platyhelminths

Platyhelminths are flatworms and belong to Phylum Platyhelminthes ('flatworm'). They can be free-living or parasitic.

To be classified as a platyhelminth, an organism must have the following characteristics:

1 body usually flattened
2 definite reproductive and excretory organs
3 a mouth leading to a simple gut
4 attachment organs, such as suckers and hooks
5 no circulatory or respiratory system.

Two classes of platyhelminths are important parasites of farm animals: Class Trematoda (flukes) and Class Cestoda (tapeworms).

The liver fluke is an **endoparasite** (i.e. internal parasite; see page 143) of sheep and cattle. In sheep, the adult fluke lives in the bile duct. Figure 26.13 shows the lifecycle of a sheep liver fluke.

Figure 26.13 The lifecycle of a sheep liver fluke

ISBN 9780170265560

19 Examine the lifecycle of the liver fluke. Suggest stages in the lifecycle that could be broken to control this parasite.

20 Why are platyhelminths important in agriculture?

Importance in agriculture

Tapeworms and liver flukes are important endoparasites of farm animals. Both cause poor condition. Tapeworms live in the gut of animals and compete with the animals for the food that is ingested, causing a deterioration in weight and general health of an animal. Liver flukes cause liver damage in sheep, which may then lead to a condition called Black disease, caused by the bacterium *Clostridium novyi*. Fluke infections cause anaemia, poor growth, weight loss and possible death.

Nematodes

Nematodes are also called roundworms, threadworms or eelworms and belong to Phylum Nematoda ('thread'). They occur in fresh and salt water and in soil, and are parasites of both plants and animals.

To be classified as a nematode an organism must have the following characteristics:

1 a long, round body
2 a non-segmented body
3 a digestive tube with a mouth and anus.

In farm animals, almost all roundworms have similar lifecycles. Thousands of eggs are laid by adult females in the digestive tract and pass out with faeces. In moist and warm conditions, the eggs hatch and develop into infective larvae. The larvae crawl up blades of grass, are swallowed, and reinfection occurs.

The soil nematode also is a roundworm; it infects the root system of many plants.

connect

View of a nematode in soil

Importance in agriculture

Nematodes are important parasites of sheep and cattle. They include barber's pole worm, nodule worm, black scour worm and small brown stomach worm. The only accurate way to see if animals are infected by worms is to do a worm egg count. Fecal samples are collected and under a microscope worm eggs are identified and counted. Visual signs of infection are only obvious after a significant population of worms has established itself in the animal and extensive damage has been done to the animal. Infestations of the barber's pole worm can cause anaemia, lack of weight gain and even death in sheep. This worm is an intestinal blood sucking parasite. The nodule worm lives in the large intestine and produces nodules in the large intestine and colon of sheep. They cause damage to the lining of the gut. Sheep lose condition and scour frequently. They often have a 'humped' appearance. The black scour worm lives in the small intestine of sheep and damages the gut lining. Weight loss and scouring are symptoms of the infection. The small brown stomach worm lives in the abomasum of sheep and damages the wall of the stomach. This causes weight loss and a reduced appetite.

Some nematodes cause diseases of plants. For example, root knot nematode causes galls (lumps) of different sizes to develop on the root system of tomatoes, cabbages and cauliflowers, interfering with the intake and translocation of water, mineral nutrients and sugars.

21 List the features of an organism that enable it to be classified as a nematode.

22 Why are nematodes important in agriculture?

Molluscs

Molluscs are soft-bodied invertebrates with shells, belonging to Phylum Mollusca ('soft'). Sometimes the shell is internal. The largest group of molluscs is the snail or gastropod group. Their shells are in one piece and are usually coiled in a spiral. Slugs also are gastropods, but they have internal shells. Ocean molluscs include clams, oysters and mussels. The squid and octopus also are molluscs.

To be classified as a mollusc an organism must have the following characteristics (Fig. 26.14):

1 a large, muscular foot used for movement
2 a soft body with a mantle or fold of tissue over the body – with possibly a shell, secreted for protection
3 a central mass above the foot, which contains most of the internal organs.

 ISBN 9780170265560

In agriculture, molluscs are most commonly found in the garden. Snails and slugs require damp surroundings in which to live. Mucus seals the snail in the shell and keeps it moist. The snail moves on a large pad called a foot, leaving a trail of mucus. They usually remain hidden until after dark or after rain.

Slugs and snails are egg-laying **hermaphrodites** (i.e. they have both male and female reproductive organs).

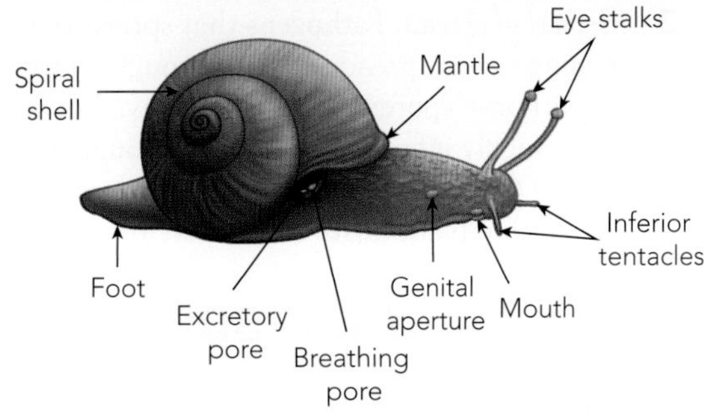

Figure 26.14 A common garden snail

Importance in agriculture

Snails and slugs are agricultural pests. They eat plant leaves and also carry disease; a particular small snail is part of the liver fluke lifecycle, for example. The main control measures for these pests include a combination of cultural, chemical and biological methods. Oysters, mussels and squid are used for human food. Some molluscs such as oysters and mussels are farmed.

23 List three features of an organism that enable it to be classified as a mollusc.

24 Why are molluscs important in agriculture?

Occurrence and distribution of disease in animals

When an animal in a population of highly susceptible individuals becomes diseased, and when environmental conditions are favourable for a rapid build-up and spread of the pathogen, an outbreak of disease may occur. **Epidemics** occur when the incidence of disease outbreaks increases rapidly, usually building up and declining within a few months. **Endemic** diseases are those that occur in a population year after year.

The effect of an infectious disease depends on the environment, method of spread, and susceptibility and size of the host population.

1 *Environment*. The environment has a major influence on the spread of the pathogen. High rainfall increases the number of biting insects, which may act as vectors for viruses. For example, the unusually high rainfall in 1950 increased the numbers of biting insects, with the result that myxomatosis spread rapidly in rabbit populations around the inland river systems of Australia.
2 *Method of spread*. Airborne organisms (for example, some bacteria) can spread rapidly through a flock or herd where large numbers of animals are kept crowded together. However, they may be more limited in their spread to another geographical area. Some pathogens can be carried by wind-borne insects. Ephemeral fever, for example, is caused by an arbovirus that is carried by an insect vector.
3 *Host population*. For an infectious disease to spread there must be a continuous supply of susceptible animal hosts. These may result from the birth or purchase of animals. Outbreaks of disease will end when the supply of susceptible animals is exhausted by death or when animals develop **immunity** (see page 59).

25 What is an epidemic?

26 Briefly describe how the environment, method of spread and host population affect the spread of a pathogen in animals.

Occurrence and distribution of disease in plants

When a plant in a population of highly susceptible individuals becomes diseased, and when environmental conditions are favourable for a rapid build-up and spread of the pathogen, an outbreak may occur.

The effect of an infectious disease depends on the environment, method of spread and susceptibility of the host.

1 *Environment*. The environment has a major influence on the spread of the pathogen. For example, the disease late blight of potatoes requires specific humidity conditions for the fungus to infect healthy plants.

ISBN 9780170265560

27 What factors cause an outbreak of a plant disease?

28 Briefly describe how the effect of an infectious plant disease depends on the environment, method of spread and susceptibility of the host.

2 *Method of spread.* Pathogens that spread rapidly cause the most damage. Airborne organisms can spread quickly through a crop. The fungus that causes stem rust of wheat has airborne spores which can survive between seasons on volunteer wheat plants and very quickly infects and spreads through a new crop.
3 *Hosts.* For an infectious disease to spread, there must be a continuous supply of susceptible plant hosts. These may result from planting new crops.

Survival and dispersal of pathogens to new hosts

To spread disease, pathogens must be transmitted to new and susceptible hosts. Methods by which this is achieved in plants include the following.

1 *Aerial contamination.* Many fungal pathogens are spread as spores by wind over large distances (e.g. stem rust, powdery mildews). They rely on air currents to bring them in contact with a susceptible host. Pathogenic bacteria are spread by rain splash, aided by wind, to infect other leaves.
2 *Crop residues.* Many fungal pathogens survive between seasons as specialised fruiting bodies on crop residues (e.g. crop stubble, tree prunings), and then release spores during the growth of the next crop.
3 *Biting insects.* Sap-sucking insects (e.g. aphids, leafhoppers) are important virus vectors.
4 *Contaminated instruments.* Contaminated instruments can spread pathogens to new hosts. Some plant viruses are spread during pruning.
5 *Soil- and seed-borne infection.* Soil-borne fungal pathogens survive as hyphae in the root or stem residue of diseased plants. For example, fungi that cause crown rot of wheat survive as hyphae on stem residue. Fungi, bacteria and viral pathogens can survive on the seed, in the seed coat, or by infecting the embryo.
6 *Vegetative propagation.* The use of infected plant parts such as cuttings, buds for grafting and tubers is an effective way of spreading pathogens such as viruses.

Knowledge of the ways that pathogens can be transmitted is important when trying to develop methods of disease control.

29 List the methods by which plant pathogens are spread to new plant hosts.

30 Why is it important to know how pathogens are spread to new plant hosts?

Resistance to disease in plants

Plants can resist attack by micro-organisms and invertebrates, but unlike animals they do not have immunity, because they do not produce antibodies.

Plants resist infection by a number of methods.

1 *Physical barriers.* Natural barriers (e.g. leaf cuticle, closed stomata) act as barriers against penetration by micro-organisms and invertebrates.
2 *Chemical barriers.* Chemical inhibitors of fungi and bacteria may be released by the invaded cell or neighbouring cells to resist infection.
3 *Absence of nutrients.* Absence of nutrients excreted onto the surface of the plant prevents micro-organisms from growing or multiplying.
4 *Production of phytoalexins.* Phytoalexins are particular inhibitory chemicals produced by a plant following infection by a micro-organism. Such chemicals are toxic warding-off compounds and inhibit micro-organisms.
5 *Hypersensitivity reaction.* Some plants respond to infection by the very rapid death of the penetrated cell and sometimes neighbouring cells. This hypersensitivity reaction is a form of resistance to infection by some fungi, nematodes and viruses. The hypersensitive collapse of cells prevents further development of the pathogen. The pathogen either dies or fails to reproduce, and so the disease does not develop.
6 *Plant surfaces.* Some plants are resistant to insect attack due to the physical characteristics of the leaf (e.g. hairiness, waxiness, shape). The resistant plant is not selected for egg-laying or feeding by insects.

31 Explain why plants do not have immunity to disease.

32 List the ways that plants resist infection.

ISBN 9780170265560

Population dynamics and ecology

This section looks at the effects of large numbers (populations) of micro-organisms and invertebrates.

Agricultural production systems may be in a balance with micro-organism and invertebrate populations, or there may be the need for continued intervention by humans to regulate or control these populations. The control method used depends on its effectiveness, cost, legal regulations (in relation to pesticides) and environmental considerations.

The growth and development of any organism, from its 'birth' to its maturity and reproduction, can be described by a number of stages. This is called a life history. Insects, for example, have a lifecycle that involves three or four stages; Figure 26.15 shows the lifecycle of the sheep blowfly. All organisms of the same species have the same life history, with the same stages in their lifecycle occurring in the same sequence. The period of the lifecycle may be less than 30 minutes for some micro-organisms.

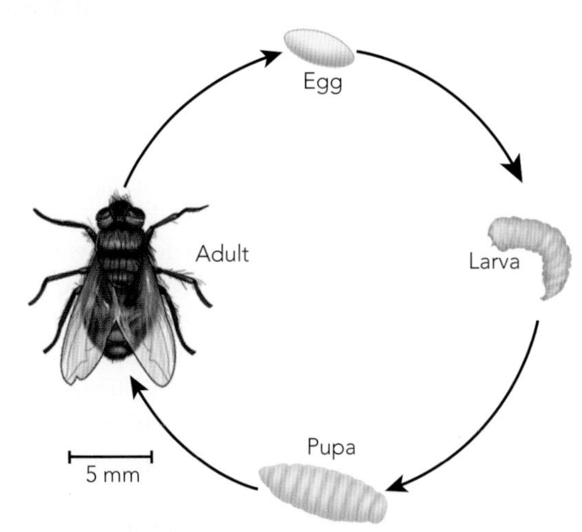

Figure 26.15 The lifecycle of the sheep blowfly (*Lucilia cuprina*)

A **population** is a group of one kind of organism living in a particular place at any one time. The place where the organism lives is called its habitat.

Population density refers to the number of individuals present per unit area. If a population is studied over a long period, and the number of individuals remains relatively constant, the population is said to be in a state of equilibrium with the environment.

Population density will increase if the birth rate in an area exceeds the death rate. Population can increase due to births and immigration, and decrease due to deaths and emigration. The number of pests in a population is thus determined by:

$N = (B + I) - (D + E)$

where

N = number of pests in the population

B = number of births

I = number of immigrants into the area

D = number of deaths

E = number of emigrants out of the area.

The growth rate of a population is determined by change in population size over time.

$$\text{Growth rate} = \frac{\text{Change in population size}}{\text{Time}}$$

33 Explain the term 'population'.

34 Write the equation used to determine the number of pests present in a population, and explain each of the letters.

35 What factors cause fluctuations in insect populations over a long period?

If a population of a particular species is studied over a long period, the number of individuals (or organisms) will oscillate about an equilibrium point. If conditions are favourable, numbers will increase; but if conditions are unfavourable, there will be a fall in numbers. Figure 26.16 shows how a population fluctuates about an equilibrium population. Fluctuations in insect populations, for example, result from changes in availability of food, availability of breeding sites, populations of predators and parasites, and climate. Severe winters and droughts will decrease insect populations.

The potential for a species to reproduce is governed by its genetic makeup and its environment. This potential is greatly modified by characteristics of the organism itself (intrinsic or internal factors) and of the environment (extrinsic or external factors).

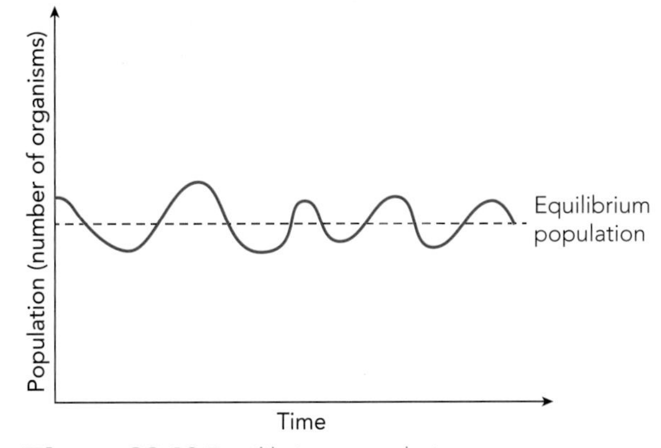

Figure 26.16 Equilibrium population

ISBN 9780170265560

Integrated pest management

Integrated pest management (IPM) involves using several strategies to control pests. It differs from other control methods in that two or more control methods are used together or integrated. A common feature of many IPM programs is the use of chemical control in conjunction with biological control. Examples are the programs used in Australia to control pests in apples and cotton.

IPM aims to keep or maintain pest populations at the lowest level. IPM relies as much as possible on natural factors to restrict the build-up of pests, including predators, parasites, unfavourable weather and host resistance.

Pests need to be controlled only when they are causing a financial loss to the farmer. A farmer will control a pest only if the cost of control is less than the damage caused by the pest (i.e. if the pest is causing economic injury).

There are several components in an IPM program that reflect the fact that control of a pest is only required when pests and diseases start to cause financial loss to the farmer. These seven aspects are outlined below.

1 *Equilibrium population (EP)*. This is the population density that an organism fluctuates near over a long period. Fluctuations result from such causes as droughts, predators and lack of food. This means that control strategies may not be required all the time.
2 *Economic injury level (EIL)*. This is the population of a pest where the costs of control measures are less than the damage caused by the pest.
3 *Economic threshold (ET)*. **Economic threshold** is the population of a pest where damage is starting to occur and control measures need to be started.
4 *Pest ecology*. This involves knowledge of the factors that regulate pest numbers. It includes details of the lifecycle of the pest and the time of egg laying, which would allow control measures to be started at an appropriate time. It also includes an understanding of any 'natural enemies' of the pest, which would help with controlling the pest.
5 *Monitoring populations*. It is important to monitor both pests and natural enemies.
6 *Decision making*. Decisions about pest management should be made only after various factors have been considered. What control methods are available? What are the costs of alternative methods? Will natural enemies be released? What is the timing of control methods?
7 *Environmental considerations*. It is important, for example, to ensure that any insecticides used are toxic to the pest but have little toxic effect on the pest's natural enemies, wildlife and farm animals.

In practice, control measures for most pests should be started before economic injury level is reached, because there is usually a lag before control measures are effective. Figure 26.17 shows the relationship between EP, ET and EIL for a pest population.

36 Explain the term 'integrated pest management'.
37 Explain the term 'economic injury level'.
38 Explain the term 'economic threshold'.

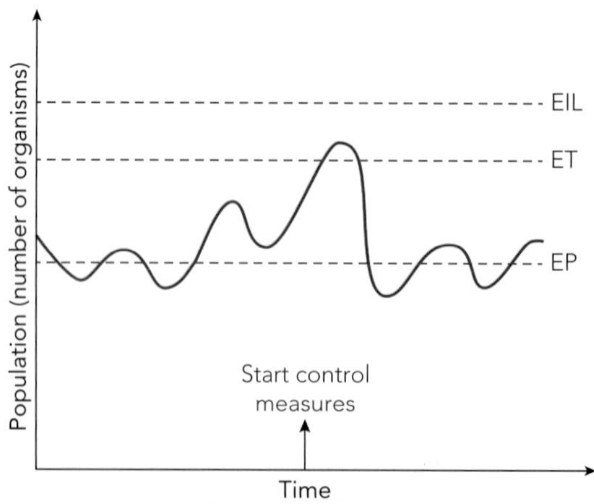

Figure 26.17 The relationship between equilibrium population (EP), economic threshold (ET) and economic injury level (EIL) for a pest population. The arrow indicates when to start control measures

ISBN 9780170265560

Chapter review

Things to do

1. Describe how the mouth, respiratory surfaces and breaks in the skin provide entry for disease-causing organisms to an animal's body.
2. Study the prepared slides of some animal parasites.
   - a Obtain prepared slides of sheep liver fluke, sheep tapeworm and cattle tick.
   - b Draw diagrams of the parasites showing their approximate size and structure.
   - c Compare the mouthparts of the liver fluke, tapeworm and tick.
   - d List any features that suit them to a parasitic way of life.
3. Consider varieties of sheep.
   - a Examine a plain-bodied sheep.
   - b Examine a sheep with wrinkles.
   - c Explain why the plain-bodied sheep is more likely to be resistant to blowfly strike.
4. Study the spread of plant viral diseases.
   - a Obtain a tomato plant infected with spotted wilt.
   - b Use a mortar and pestle to crush some of the infected leaf tissue. You might need to add some sand as an abrasive.
   - c Prepare several healthy plants by scratching the leaves.
   - d Transfer some of the juice from the infected tissue onto the leaves of the healthy plants.
   - e Observe the inoculated plants, and compare them with a control plant.
5. Further investigate disease symptoms in plants.
   - a Collect diseased plant material on the school farm.
   - b Examine the plant material using a hand lens.
   - c Draw and describe the symptoms of disease for each piece of plant material.
6. Examine fungi under a microscope.
   - a Make a wet mount of a small segment of tissue containing the fungus (e.g. wheat leaves with rust, or citrus with mould).
   - b Observe under high power and low power, and record your observations.
   - c Stain a sample of the specimen with lactophenol blue. Allow several hours for the stain to penetrate the tissue. Apply a coverslip.
   - d Observe the stained specimen. Identify spores, hyphae, mycelium, cell walls and cell contents.
7. Visit a poultry farm. Draw a plan of the layout of the farm. Describe how disease is prevented and controlled on the farm.
8. Examine a range of insecticides used on the school farm. Draw up a table, and do the following for each one.
   - a List the insects controlled.
   - b Identify the active ingredient (chemical).
   - c Calculate the dilution rate.
   - d List any special precautions needed when using the chemical.
   - e Give the withholding period for each chemical (i.e. the time required between the last spray and harvest).

Things to find out

1. List the reasons why micro-organisms and invertebrates are so biologically successful.
2. Describe how algal blooms can be controlled in river systems.
3. Explain the term 'obligate parasite'.
4. Explain the terms 'economic threshold' and 'economic injury level' for a pest that you have studied.
5. Discuss how knowledge of the lifecycle of the liver fluke helps when developing a program to control this parasite.

Extended response questions

1. Describe the biological characteristics of micro-organisms and invertebrates that cause them to be important in agricultural production systems. Use examples to illustrate your answer.
2. Resistance and immunity of plants and animals to attack by micro-organisms have similar mechanisms, except that higher animals have a well-developed ability to produce antibodies.
   - a Outline the mechanisms of disease resistance in plants and animals.
   - b Explain how higher animals have immunity to disease.
3. Agricultural production systems may not be in balance with the invertebrate and microbial populations.
   - a Suggest some reasons why this situation occurs.
   - b Outline strategies that can be used by farmers attempting to regulate or control these populations.
4. 'Many micro-organisms and invertebrates have beneficial roles; others are harmful; and some have little effect on agricultural production systems.' Discuss this statement, using examples.

ISBN 9780170265560

UNIT 6
PRODUCING AND MARKETING FARM PRODUCTS

Shutterstock.com/Flaxphotos

CHAPTER 27

Outcomes

Students will learn about the following topics.

1 The farm as a business.
2 The place of the farm in the wider agribusiness sector.
3 Decision-making processes and management strategies.
4 Factors affecting farm decision making.
5 The impact of financial pressures on farmers.

Students will be able to demonstrate their learning by carrying out these actions.

1 Discuss the nature of farming in Australia.
2 Describe a wide variety of farm business structures.
3 Outline the key steps in managing a farm.
4 Discuss why risk is an element in any process relating to farm management.
5 Describe strategies used to reduce uncertainty relating to farm management.

ISBN 9780170265560

GENERAL ASPECTS OF FARM MANAGEMENT

Words to know

corporate farms farms owned by private businesses

cost–price squeeze the situation of falling prices for farm output at the farm gate and increasing farm production costs

diversification producing more than one output on the farm from unrelated farm subsystems (e.g. wheat and wool)

farm problem the apparent difference in yearly incomes between people in rural activities and city workers caused by rising production costs and low market prices for farm outputs. This then leads to depressed rural income, which limits the ability of farmers to obtain finance for further farm development and forces labour to move out of farming and into other areas of employment

Introduction

Australian farming is characterised by a high ratio of farm size in relation to the labour employed to run the farm. This is a reflection of the poor quality of Australian soils, harsh climate and the consequently low carrying capacity of the land. Despite this relationship, large farms mostly remain family farms. The majority of broadacre farms are run by families, where the owners actually operate and manage the farm. The family farm has a future, as the family can be thought of as a stable source of labour, knowledge and capital for the business. However, in some cases where children of farming families do not take up a rural vocation, lack of skilled and reliable labour to take over from ageing owners causes the family farm to be sold. Family-based businesses depend on the needs of larger society. The farm must compete for scarce resources (e.g. capital, labour, knowledge) with the needs of secondary and tertiary industry. Government regulation and the buying policies of firms determine in many instances how the family farm is operated.

In this time of rapidly developing technology and access to worldwide communication, financial and marketing systems through computers, fax, digital and satellite facilities, international market trends and anomalies impact on the local marketing and farming practices in more direct ways than in the past. Individuals can interact with financial, marketing and educational institutions worldwide, and conduct trade in traditional ways or use e-commerce facilities to arrange contracts and to negotiate trading arrangements. Many industries have services available to farmers to assist them to develop international markets and to export farm products overseas.

Farm management

The farmer (i.e. farm manager), as a decision-maker, seeks to maximise profit from a farming system in a sustainable manner. To do this, the farmer must plan labour requirements and the level of monetary and physical inputs required to obtain the greatest income from the farm. Farm management involves making a series of decisions. Effective farm managers need to review existing situations with an emphasis on increasing returns. The farmer becomes a 'jack (or jill) of all trades', a decision-maker, a supervisor, an innovator, and a person who can maintain accurate records of the farm's physical and financial situation. Another major consideration is ensuring the environmental, economic and social sustainability of the farming system.

Key steps

Management of a farm involves several key steps:

1 collecting facts by researching and making accurate observations
2 carefully analysing information
3 formulating a plan of action
4 specifying clearly the required outcomes
5 putting a plan effectively into action, using skills related to producing, marketing and renewing resources
6 effectively countering adverse conditions that may lead to production losses, such as drought, flood, disease, market gluts or changes in consumer preferences
7 taking responsibility for the outcomes of management decisions.

1 What are the main characteristics of Australian farms?
2 Outline the key steps involved in managing a farm.

Farming as a business

A farm business system is significantly different from most other small businesses in that the farm and home are much more closely interrelated. Farm planning can be influenced not only by economic factors but also by social factors, family dynamics, personal preferences and perceived family needs.

Use of family labour in such tasks as the establishment of fence lines, the maintenance of irrigation systems or the construction of haysheds always means that family members are active participants in the farm as a business. Hard work is rewarded by visible achievements in the form of sheds, land ploughed or other activities that give a sense of accomplishment.

 ISBN 9780170265560

With large projects, such as land clearance, harvesting and transport of farm outputs (e.g. grain), local community members often share each others' skills, labour and equipment to meet deadlines. This relationship leads to the formation of strong social bonds within the local community. Friendly neighbours are both social contacts, support in terms of a willingness to assist in times of need, and security in times of trouble associated with accident and disaster.

For some farming families, children no longer take up a rural vocation. So as the parents grow old, the workload associated with operating a farm increases, making it difficult for the family as a unit to continue on the land. As owners age, they need to employ experienced people to assist with routines such as tail docking, vaccinating or dehorning. In some instances, due to lack of skilled and reliable labour, large family farms are subdivided and sold to realise the capital invested in them.

The farm is part of a production chain. Numerous service businesses provide materials and support to the farmer in order to run the farm. Processing factories receive farm products and various government agencies licence and regulate production through a variety of market mechanisms. Corporations promote, market and carry out research for many of our major agricultural products.

Within the dairy industry there are many examples of these linkages between the farm and the broader agribusiness sector, such as Dairy Australia (see the associated websites). Agribusiness refers to those activities involved in the production, processing and distribution of food and fibre products. The farm is only one part of this system. The types of farm products developed for particular markets, along with the range of processing, distribution and marketing arrangements, are also part of the agribusiness system.

Dairy Australia (funded by a levy collected from farmers and manufacturers) is a research organisation with a focus on animal health and nutrition, sustainable farm production systems, disease control, herd fertility and efficient milking practices.

Another agribusiness that assists farmers to improve the genetics of their herd and therefore production is the Australian Dairy Herd Improvement Scheme (ADHIS) that calculates Australian Breeding Values (ABVs) for dairy traits.

The New South Wales Food Authority (New South Wales Department of Health) is a government body that licenses dairy farmers. To be licensed the dairy farm requires a HACCP (Hazard Analysis Critical Control Point) food safety program. The dairy is inspected annually for registration purposes. Companies such as Dairy Farmers Pty Ltd provide product specifications to farmers. These companies also promote dairy products in the marketplace, such as various types of milk. Production records, such as monthly production reports, provide the farmer with feedback. These are provided by companies that carry out herd recording on individual farms.

Service businesses (e.g. veterinarians, farm supply companies and accountants) support the farmers with products, advice and finance to run the farm.

connect

Dairy Australia

connect

Australian Dairy Herd Improvement Scheme

3 How does a farm business differ from a small retail business?

4 Explain two family or personal considerations that might influence what a farmer produces on a farm.

5 Define the term 'agribusiness'.

6 Use examples to show how a farm is supported by a network of service businesses and industry specific organisations.

Farm business structures

There is a wide variety of farm business structures in Australia from the family farm to the **corporate farm** enterprise. Table 27.1 on page 390 outlines the advantages and disadvantages relating to the most common farm business structures, namely family owned farms, farming corporations (both public and private) and cooperatives.

Agriculture and financial pressures on farmers

The irregular nature of income to a farmer varies from one enterprise to another. Irregular income can arise from one or more of the following reasons:

- market instability
- production lag
- rural–urban wage differential
- consumer preferences
- seasonality of production
- the farm problem
- interest rates.

Table 27.1 Farm business structures

Business structure	Advantages	Disadvantages
Family farm Owned and operated by a family.	Intimate knowledge of the farm. Motivated to pass a sustainable and successful farm to the children. Labour needs supplied by family.	Tradition discourages innovation and change. Setbacks limit reinvestment. Farmer has little bargaining power.
Farming corporations Public or private company; e.g. Kidman and Vestys beef.	Have the resources to draw on. Can negotiate prices and reduce input costs by buying in bulk. Corporation properties can share resources and machinery.	Board of Directors and shareholders. All property managers are employees, some may lack motivation. Large staff turnover, new staff need to be trained.
Cooperatives Many family farms are brought together, to sell their produce; e.g. Riverina rice.	Monopoly can set prices. Processing and packaging plants and produces value-added products. Scientists make new breeds/products. Buys inputs in bulk.	Farmers lose independence in terms of how to manage their farms. Can't sell outside cooperative; set prices.

Market instability

Agriculture is characteristically made up of a large number of small producers (farmers), who as individuals do not produce sufficient levels of output to control market price totally. At the beginning of any production phase, the farmer is unable to predict just how many other farmers will be producing the same product to be marketed at approximately the same time. This is especially true in Australia, where distances are so great. As a result of this imperfect knowledge of market and production situations, periods of product oversupply (glut) or undersupply (shortage) occur. In addition, disease, pests or climatic factors can induce market instability by affecting production.

Production lag

Farmers, once committed to a particular production cycle – whether this is animal based (e.g. wool production) or plant based (e.g. cotton production) – are unable to increase or decrease production levels rapidly in the short term as a result of increases or falls in market prices. Orchardists, for instance, cannot hold fruit on the trees waiting for an improvement in market price, as natural conditions will cause the fruit to ripen and quickly rot. Farmers are unable to adjust supply levels over short periods in relation to market price fluctuations, because they manage a natural system.

Rural–urban wage differential

In developed economies, labour earnings in agriculture are lower relative to the earnings of city or urban workers. Reasons for this include lack of recognised skills to allow the determination of salary scales, limitations (as outlined below) on the demand for agricultural goods, which limit wage increases, high production costs relative to market prices and a high level of casual labour within the workforce. Demand for farm products will be lower, and consequently market prices lower, because of the following limitations.

- People can eat only so much food. As incomes increase, food consumption remains reasonably constant. As the level of wealth increases in the community, substitution of low-grade foods with higher quality forms or more refined foods occurs. However, in wealthy communities a basic need for staple foods (e.g. bread, milk) remains.

ISBN 9780170265560

- Population increases obviously generate an increased demand for food, but as the population of a country ages – as in the case of many Western nations – the type of food consumed changes, and the quantity required declines.
- Synthetic and highly refined materials continue to place pressure on the market share of more basic agricultural products. Wool and other such materials (e.g. cotton) must mount varied and aggressive advertising campaigns to attract buyers away from synthetic materials.

7 Why are farmers unable to adjust farm output levels over short time intervals?

8 Give reasons why rural earnings might become depressed relative to wages paid to city workers.

Consumer preferences

Consumer preference is influenced by advertising, health concerns, market price and fads causing consumer demand to vary over time; for example, the increasing demand for low-fat milk rather than whole milk, and for high-protein milk. Seasonality of consumer demand can also affect farm incomes, such as the increase in consumption of ice cream and other frozen dairy products during the summer period. Nutritional awareness programs and an increase in general community awareness in regard to diet and healthy lifestyles impact on the range of foods people are eating. Foods classified as contributing to increasing levels of body fats and cholesterol are being consumed in lower amounts, while interest in organic foods is spreading among consumers. Concerns about animal welfare, genetically modified plants and animals and the role of technology in production all influence consumer demand and lead to the development of specialised or niche markets.

Seasonality of production

Seasonality can result in irregular farm income unless other enterprises are operational on a farm. A crop is usually only harvested and marketed once a year so income only comes once a year. The combination of sheep and wheat enterprises spreads labour resources across the year rather than in a particular season and provides income from two markets that are not linked to each other. In dairying there is a production decline in winter due to poor pasture growth, which can impact on income. Seasons can also affect the quality of a product. For instance, hot summers can lower milk protein levels.

The farm problem

Farming produces many outputs that require further refining or processing before sale to the public or consumers. Prices paid to farmers for the raw output produced from the farm remain low in comparison with prices paid by consumers for the final refined product. Wool attracts a low price at farm level in comparison to a woollen garment sold in shops to consumers. Similar examples can be given for many types of farm outputs.

Production costs on farms are subject to increase, especially due to government manipulation of the economic system through such means as the value of the Australian dollar, inflation effects, sales tax or tariffs. As a result, farm income is squeezed between forces resulting in rapid increases in the cost of inputs and generally low market prices obtained for farm outputs at the farm gate, as is illustrated in Figure 27.1. Over time the increase in the cost of inputs rises at a much faster rate than the increase in the price of outputs. This situation is referred to as the farm **cost–price squeeze**.

The long-term result in most growing economies is that rural incomes are depressed relative to city incomes. This situation is known as the **farm problem**.

Interest rates

In the broader economy interest rates are always a factor in determining the profitability of an enterprise. By eroding available cash through interest rate charges the cash flow levels available to a business decrease restricting the farmer's ability to pay bills or expand operations. Structural adjustments in the broader economy due to the mining boom causing fluctuations in the value of the Australian dollar, which has resulted in a high price for Australian exports on the world market and associated high costs of production.

ISBN 9780170265560

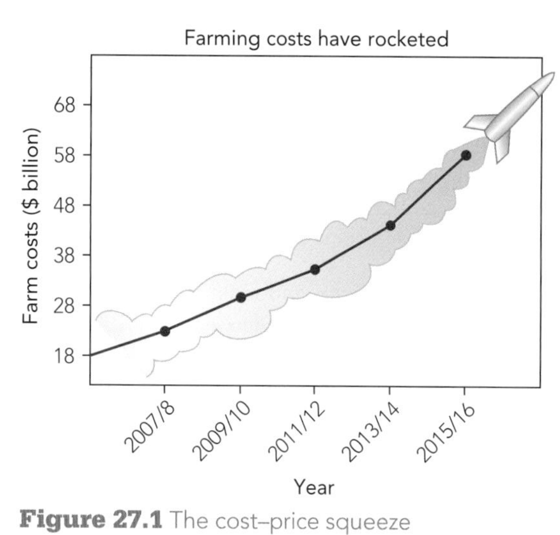

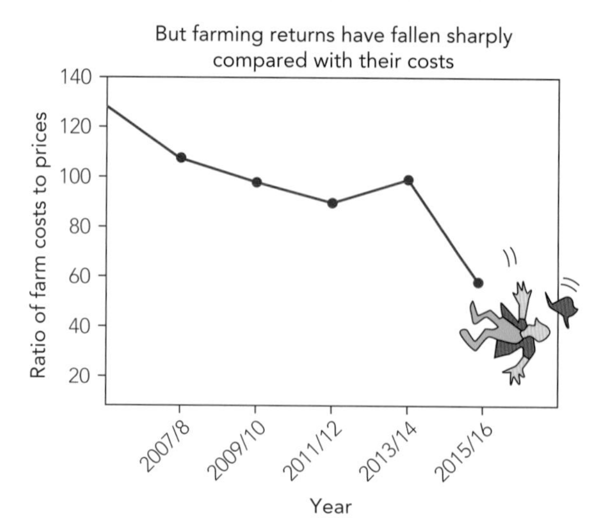

Figure 27.1 The cost–price squeeze

9 Explain how seasonality can impact on farm finances.

10 What does the 'cost–price squeeze' mean?

11 What is the farm problem?

12 Describe an example of how a change in consumer preferences can impact on farm finances.

connect

Financing the farm

This video outlines alternative strategies to securing finance to access overseas markets and adopt new technologies.

13 Describe at least six steps that farmers can follow to reduce uncertainty relating to making decisions on farms.

Risk management

Any farm decision involves a certain amount of risk because farmers make long-term decisions and commit themselves to a plan of action with incomplete knowledge of future possibilities. Drastic weather changes (e.g. floods, droughts), insect or disease problems, and fluctuating market conditions often cause loss and hardship in farming.

Risk taking can be hedged against by taking precautions. Common risk avoidance mechanisms include the following.

- Farmers should thoroughly research market outlets, processing options and processing costs. The product produced on a farm must be able to be sold or processed rapidly and cheaply, so that a competitive market price can be obtained.
- Farmers need to determine for themselves whether the market is big enough to make production worthwhile. This may involve the possibility of fostering overseas markets for products.
- Farmers should consider whether a reasonable profit margin would still exist if currency fluctuations or government intervention occurs, based on current knowledge.
- **Diversification** of farm production should be encouraged; that is, more than one commercial subsystem should be developed on the farm. Provided that the various farm enterprises are not linked to the same market, this is an effective farm management technique to avoid risk, as market prices for all types of farm outputs are not likely to fall at the same time.
- Where possible, farmers may invest extra capital in off-farm ventures, such as purchasing shares on the stock market, buying property, or becoming part owners in a business venture in the retail industry.
- Farmers may attempt to secure a reliable market outlet prior to, or early in, the production process. This usually involves obtaining contracts to supply large retail chains with perishable goods.
- Farmers may investigate the possibility of insurance against disaster, although this is an expensive option.
- All farmers should maintain sufficient reserves of money, fodder and water to allow survival during unfavourable times.
- Farmers should not overinvest in capital equipment, especially expensive items that are used for only small periods during the year (e.g. harvesters). Instead, farmers can use contractors to perform these specialist tasks, or jointly buy and share expensive equipment.

ISBN 9780170265560

Chapter review

Things to do

1. Survey farmers in your local area. Construct a survey that allows you to determine the following trends:
   a. the type of agricultural systems located in your area
   b. how farmers are coping with the problems of rising costs and falling market prices
   c. how farmers offset risk and uncertainty in their operations
   d. future options for the farmers.
2. Examine newspapers, rural publications and specialist magazines for information on the effects of other countries (e.g. United States, Japan, European Community) on the level of Australian farm exports.

Things to find out

1. For the farm that you have studied, outline the financial pressures that might affect the farmer.
2. For the farm that you have studied, discuss the supporting services on which it relies. Include businesses, government organisations, processing and marketing companies.
3. Explain how interest rate trends over time can affect decision making by a farmer.

Extended response questions

1. Analyse the structure of the Australian agribusiness sector, indicating the importance of the family farm within this sector.
2. Farmers' incomes have stabilised or fallen, while at the same time incomes of people employed in other industries have increased.
   a. Is it farmers' fault that this situation has occurred?
   b. Can such a situation be resolved?

CHAPTER 28

Outcomes

Students will learn about the following topics.

1. The functions of a market.
2. The various markets that exist for agricultural products.
3. The law of demand.
4. The law of supply.
5. How market price is a reflection of the forces of supply and demand.
6. Factors that influence consumer demand.
7. Factors that influence product supply.
8. The concept of derived demand.
9. The purpose of various types of farm budgeting procedures.
10. The advantages and disadvantages of gross margins.

Students will be able to demonstrate their learning by carrying out these actions.

1. Describe strategies available to farmers to market their products.
2. Use techniques to analyse the financial situation of a farm enterprise.

ISBN 9780170265560

MARKETS AND PRODUCTION

Words to know

AuctionsPlus a nationwide electronic system for buying and selling livestock; formerly known as Computer Aided Livestock Marketing (CALM)

budget a planning tool used to assess the profitability of alternative farming plans

commodity goods of trade

derived demand demand generated by the processed goods rather than the raw materials used in production

dse dry sheep equivalent; 1 dse equals one dry sheep or wether per hectare; it is a measure of the stocking rate of the land

equilibrium market price the market price at which the quantity of goods demanded is exactly balanced by the quantity of goods supplied

gross margin a planning tool used to compare enterprises of a similar nature, determined as gross income minus variable costs; gross margins must be expressed in terms of per hectare, per animal, or a similar quantitative measure

law of demand at low prices consumers buy more of a product and as market prices increase, less of the product is purchased

law of supply producers are willing to supply more of a product onto the market place as market prices increase

partial budget a budget used to estimate the effect on farm profits of a change that will affect only farm subsystems or enterprises, not the entire farm operation

total production costs the sum of the costs of variable inputs and the costs of fixed inputs at any given level of production

variable costs costs which vary as the size or level of an enterprise varies

whole farm budget a budget used to assess the operation of the entire farm business

ISBN 9780170265560

Introduction

If you visit any city, town or location where a crowd of people gathers, the chances of locating an area where goods are marketed is fairly high. A market exists where buying or selling occurs for a particular **commodity**.

Due to a number of dynamic forces, the prices offered for similar goods tend to equalise as competition between buyers and sellers occurs. For a truly competitive market to evolve, the products for sale must be more or less the same, so that the people who buy commodities (consumers) show no preference towards which products they buy. Competitive markets should have a large number of buyers and sellers, and not be dominated by any single group of buyers or sellers. Neither can any restriction be placed on the number of buyers or sellers entering or leaving the market. Such a market is also known as an open market or free market. In many situations, the conditions for a free market operation to exist are not met; so various unique market situations have evolved for many agricultural products.

Functions of a market

Farmers are usually concerned in the marketing of products they produce. They can often increase returns by carefully timing the sale of produce. However, the process of marketing produce is not simple. It is not only buying and selling, it is also the associated activities of transport, quality assessment, packaging, storage and promotion. In addition there is research and development associated with both the production of raw materials and its consequent processing, and the development of new markets and marketing strategies. The combined effect of these processes associated with buying and selling is to add value to the raw farm product or, in the case of fresh produce, to extend the shelf life of these products.

As a consequence of these additional features of a successful marketing process, many specialised business areas have become established. Four major business areas are outlined below.

1 *Exchange and merchandising.* Functions associated with the exchange of goods include the finding and gathering of material to sell and the maintenance of quality standards. The merchandising functions of advertising, display and general public promotion also are part of this business area.
2 *Transport and storage.* Functions associated with the transport, storage and general control of supply of material onto the market ensure a continual supply of the product to consumers. They attempt to even out the market and maintain a guaranteed level of income for producers (suppliers).
3 *Market research.* Extensive consumer survey activities and promotional activities may be conducted to provide a product in the form that consumers demand. Financing of ventures to ensure success of projects is included in this process.
4 *Product development.* Research may be carried out, to develop new forms of the product to improve its appeal to the consumer. It may be directed towards providing information that assists in the development of new markets, or improved handling or storage facilities for products.

As a result of the above processes, a margin exists between the price paid to the farmer and the price paid by the consumer. This margin covers the costs of various marketing processes and additional processing of the product so it can be sold at a higher price, and is known as **value adding** (see page 3) to the product.

Many marketing operations have now been deregulated (they are no longer controlled by government) and farmers deal directly with processors. In some instances farmers have formed groups called cooperatives to market their products. In other cases, governments have set up marketing boards.

A highly developed marketing system is required in order to distribute farm products to the place, at the time, and in the form required by consumers.

The process is illustrated in Figure 28.1.

1 Describe the main features/functions of a free market.

2 What does the term 'value adding' mean?

ISBN 9780170265560

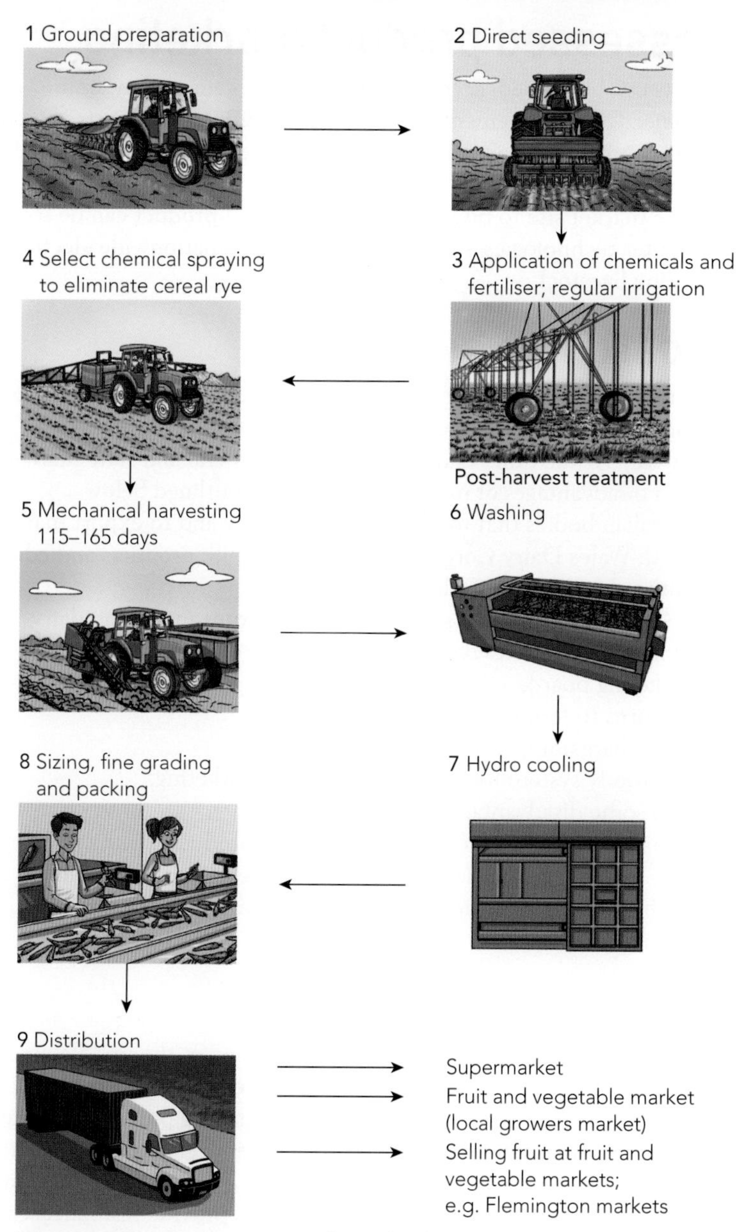

Figure 28.1 Getting carrots from the farm to the consumer

Agricultural markets

Direct marketing/selling

For many products there are no formalised marketing systems. The farmer must negotiate either directly with buyers or sell to agents at recognised market locations. Vegetable and flower growers take their produce to the wholesale markets where prices are negotiated between buyers and growers or they are determined through auction.

Direct marketing has several advantages:

- it is an inexpensive method of sale
- it allows immediate feedback from buyers (which could encourage the farmer to change management).

There are also some disadvantages:

- unsold produce returns to the farm or is placed into cold store (an expense)
- the farmer may be encouraged to lower price through lack of demand (interest) from buyers.

ISBN 9780170265560

Objectively assessed product marketing

Many agricultural products can be **graded**, according to purity, protein quality, frame scores or other types of objective measurement. Certain commodities, due to this objective assessment, can be sold by description or on the basis of sample (e.g. wool). When goods are sold by description, buyers simply need to have the information in front of them; they do not have to travel to the market site to buy. Objectively assessed product can be sold via telecommunication/computer technologies, such as **AuctionsPlus**, a nationwide electronic system for buying and selling livestock.

Marketing boards

Several types of agricultural marketing boards exist. At one end are the business organisations that are operated on a non-profit basis and are under democratic control by members (e.g. agricultural cooperatives). At the other end are compulsory federal and state marketing boards. The advantages and disadvantages of marketing boards are outlined below.

These are state or Australian bodies that market produce locally and to export markets. For instance, the New South Wales Dairy Corporation promotes milk products and educates the consumer on the health value of milk products.

The Australian Dairy Corporation finds export markets, promotes Australian dairy products internationally and also promotes dairy products on the domestic market.

The advantages of marketing boards are that they can:

- maintain or increase returns to farmers
- provide increased market bargaining power
- above all, maintain an orderly system for regulated produce marketing.

However, there are also some disadvantages:

- marketing boards ultimately cater for research, regulation of produce quantity and quality, and promotion, rather than the development of marketing opportunities
- farmers may perceive that marketing boards don't always obtain the best market price for the product.

Agricultural cooperatives

Agricultural cooperatives are set up to allow farmers to bypass some of the middle operators in marketing, to cut costs and gain some guaranteed return for their goods. Members of a cooperative are able to trade as a group, so obtaining greater bargaining power. Cooperatives may provide a single service or combinations of services for farmers, relating to marketing produce, lowering transport costs or purchasing some inputs (e.g. seed, fertiliser, fuel) in bulk and passing the savings on to member farmers. Some cooperatives provide insurance, credit and specialist services; a few can process and package members' produce.

There are three types of cooperative.

1. *Market-focused cooperatives* are owned by primary producers and provide assistance to farmers in meeting market quality and quantity needs. Consequently, processes that add value to the raw farm product are important. These cooperatives assist farmers with transport availability and standards, grading, packing, storage, promotion, wholesale, processing, distribution and export information and facilities. These services are usually funded through a levy imposed on the producers.
2. *Supply cooperatives*, again, are owned by their members and use economies of scale to purchase feed, fuel, fertiliser, chemicals, seed and hardware for sale to farmers at lower prices.
3. *Agriculture service cooperatives and jointly owned operations*: provide specific and economic services for farmers, such as herd improvement, supply of plant or equipment to process farm outputs, and a range of marketing services.

Examples of Australian cooperatives

Australian Cooperative Foods (ACF) (food company) owns many dairy factories, including Dairy Farmers. Any dairy farm that supplies milk to this factory is a shareholder.

ISBN 9780170265560

ACF collects milk from the farm. Milk is tested, processed, packaged, promoted, marketed and distributed to depots or large retailers, such as Woolworths.

The Rice Growers Cooperative (for the Leeton, Murray and Murrumbidgee irrigation areas) has been set up, owned and managed by rice farmers. All rice farmers are suppliers of rice to the cooperative's mills and all farmers are shareholders in the cooperative. This cooperative bulk buys inputs, such as fertilisers and machinery, allowing savings to be passed onto farmers. It dictates quantities and varieties of rice to be grown each year. It stores, processes, packages, promotes, markets and distributes the products and also pays researchers to develop new varieties of rice and to develop new products.

The advantages of cooperatives include:

- increased presence in the market and the consequent ability to set the price for produce
- being more able to value add (leading to greater profits), as cooperatives have capital to set up processing and packaging plants.

The disadvantages of cooperatives are:

- farmers may lose some independence; for example, they may be told (under contract) what to produce and how much
- often, cooperative shareholders/farmers can only sell their produce to the cooperative.

Statutory marketing boards

State marketing boards were set up when voluntary cooperatives had problems in effectively marketing products because of their relatively small size. With statutory marketing boards, producers can seek the assistance of government so that legal powers can be used to force producers to market their produce through the board. The board is usually made up of producers and a small number of government officials. For example, the Rice Marketing Board, under a formal agreement with the Rice Growers Cooperative, coordinates the purchase and marketing of rice in New South Wales. Statutory marketing boards are of decreasing importance as agricultural industries become more deregulated, as has occurred in the dairy industry, where milk is now sold by farmers directly to various milk processors.

Export control boards

In addition to the statutory marketing boards in the various states, there is a series of marketing boards set up under Commonwealth legislation. Originally, boards such as the Australian Meat Board (now Meat and Livestock Australia, MLA) were set up to check that the products being exported from Australia met minimum standards of quality, and to regulate shipments to overseas markets at a steady rate. Firms had to have a licence from the boards before they could be involved in export trade.

However, a number of these boards have been given actual trading powers instead of being restricted to the supervision of export transactions. More recently, the efficiency of marketing has become determined by the amount of competition between all sectors of the marketing process. This has led to the deregulation of markets for many farm products, such as wool and wheat.

3 Explain the difference between direct selling and objectively assessed selling of produce.
4 Why did marketing boards come into existence for many products?
5 Outline the functions of an agricultural cooperative.

Marketing strategies

Vertical and horizontal integration

Two styles of management that affect the financial returns of Australian farmers are vertical and horizontal integration.

Vertical integration

Vertical integration refers to the coordination between a number of stages in the production and marketing processes. One way of achieving control over another stage of production or marketing is to use contracts. This occurs, for instance, in the broiler industry. Here a processor may have a contract with a farmer to supply a certain number of birds at a

ISBN 9780170265560

certain age and an agreed price. The processor pays for the day-old chickens, the feed and the veterinary bills. Such a contract gives the farmer greater security. Along similar lines, a large firm operating in the meat industry in the Northern Territory has its own cattle station, coastal killing centres and meatworks, refrigerated shipping lines, wholesale distribution facilities and retail butcher shops.

The advantages of vertical integration are:

- a stronger market position can be secured by being able to control input or market linkages
- food production and distribution systems can be streamlined, reducing the cost of food production (more efficient)
- the single company also promotes, markets and distributes marketable products to retailers.

However, there can also be some disadvantages:

- The economic independence of family farms can be destroyed as farmers have to buy their inputs from a few large firms that have the power to ask high prices
- they can lead to the monopolisation of markets.

Horizontal integration

Horizontal integration involves the linking together of firms at the same level of marketing. In a department store supermarket, for instance, groceries, meat and fruit are sold together.

The advantages of horizontal integration include:

- companies working together create more services or products
- a larger turnover of products at relatively low mark-ups can be created
- advertising is more effective.

The disadvantages can include:

- poor performance by one of the businesses can have detrimental effects on the other businesses
- large broadly-based companies can be harder to manage
- some governments restrict this type of monopoly.

6 Define the term 'vertical integration'.

7 Define the term 'horizontal integration'.

Contracts

Under contract, farmers supply their produce to processing/retailing or retailing companies.

These companies do the marketing; for example, beef cattle feedlots are contracted to supply meat to McDonald's, vegetable growers are contracted to Edgell–Birds Eye.

The contract specifies a guaranteed price and expects the farmer to supply a certain quantity and quality.

The advantages of contracts are that farmers gains for their produce:

- a guaranteed market
- a guaranteed price

The disadvantages of contracts are that the farmer:

- must maintain inputs (irrigation, fertiliser) to meet quality specification and yields
- may miss out on better prices, if open market prices increase

8 Compare the use of contract marketing with direct marketing strategies.

Demand and supply

Demand

Market demand for any single commodity refers to the collective buying power of consumers. A relationship exists between different market prices of a commodity and the quantity of product that consumers are willing to purchase at these varying prices. This relationship is shown in Figure 28.2 on page 401.

The straight line illustrated is a theoretical representation of what occurs. True demand relationships are curves, but Figure 28.2 demonstrates a fundamental law applied to markets, called the **law of demand**. Basically, at low prices consumers buy more of the product; and as market prices increase, less of the product is purchased. This is in part because at higher market prices low income earners cease to purchase the product and other consumers buy less of it.

 ISBN 9780170265560

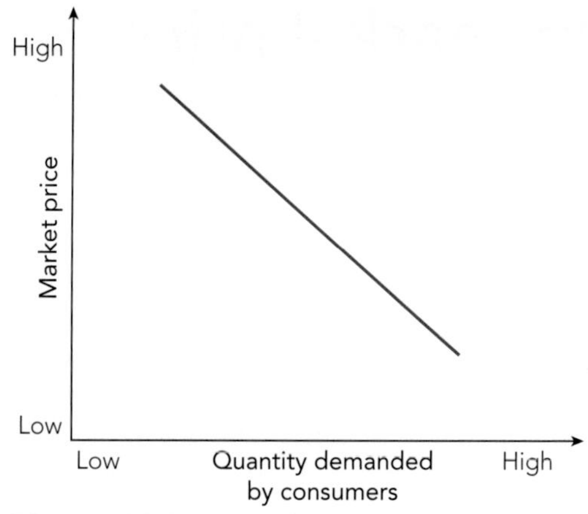

Figure 28.2 Demand curve

Factors affecting demand

Many factors influence demand, including:

- market price of the commodity
- consumer tastes as determined by advertising, culture, education and habit
- disposable income of consumers at any particular time
- number of people wanting the commodity
- prices of alternative products available to consumers.

Supply

Market supply for any single commodity refers to the collective amount brought to market by producers. The relationship between the amount of a product supplied to the market by producers, and different market prices, is shown in Figure 28.3.

Again only a basic trend is illustrated, which holds true for most open markets, namely that producers are willing to supply more of the product onto the market as market prices increase. This is known as the law of supply.

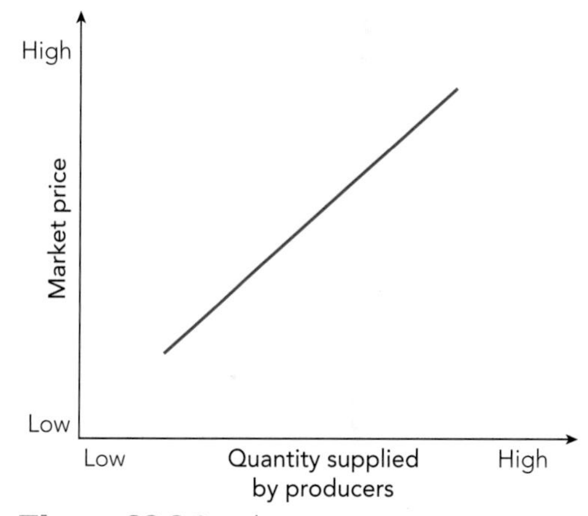

Figure 28.3 Supply curve

Factors affecting supply

The following factors have a dominating influence on the supply of goods to a market.

- The market price offered for the product will affect the amount of the product released onto the market. There is an upward limit to this in agriculture before unripe or immature goods are encountered in the market place.
- The degree of technological innovation involved in the production of agricultural material will affect the efficiency of operations (e.g. harvesting) or the amount of land that can be effectively managed. As production efficiency increases, more material is supplied to the market, lowering market price if demand fails to increase as the supply of goods increases.
- The number of people producing and marketing the product at any one time will affect the quantity offered for sale.
- The cost of production will influence the scale of production.
- Prices offered for alternative commodities may influence the management directions of farmers, such as the varying market prices offered for wool versus wheat grain.
- Environmental factors (e.g. weather conditions, disease and pests) may increase or decrease yield, or encourage or discourage production.

9 What is 'the law of demand'?

10 List the factors that influence the demand for a product.

11 What is 'the law of supply'?

12 List the factors that affect the supply of a product.

ISBN 9780170265560

Equilibrium market price

As a result of the mutually antagonistic forces of demand and supply, a situation rapidly establishes itself in any open market place, where consumers will not buy highly priced commodities and, as more of a commodity enters the market, the lower its market price becomes. Figure 28.4 illustrates the opposing nature of the laws of supply and demand.

The point in the diagram where the supply and demand lines intersect is the **equilibrium market price**, which represents the price at which the product is traded. This is the market price where the quantity demanded is exactly balanced by the quantity supplied. At points above and below this equilibrium market price, the forces of supply and demand push the price back to an equilibrium position. This is the case whenever products are for sale where no particular preferences are displayed by consumers, where the market is not dominated by one particular producer, and where there are no restrictions on new buyers or sellers entering or leaving the market place.

13 Explain the concept of an equilibrium market price.

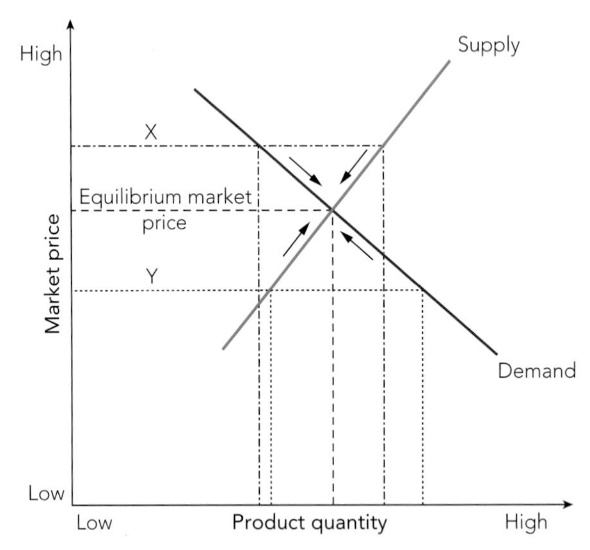

At X, the quantity supplied exceeds the quantity demanded. Suppliers will compete with each other to sell more of the product, forcing market price down.

At Y, the quantity demanded exceeds the quantity supplied. Consumers will compete with each other to get more of the product, forcing market price up.

Figure 28.4 Equilibrium market price determination

Derived demand

For many agricultural products, demand is not directly related to the raw material released at the farm gate; it is more closely related to the processed materials available in shops. Producers, in other words, are more interested in the prices of woollen garments, biscuits, or tinned fruit or fruit juices rather than the prices of wool, wheat grain or fresh fruit paid to the farmer directly. **Derived demand** is the term used to describe the demand for the product at the farm gate that results from the demand for the final refined or value-added product the consumer buys.

There is a long chain of intermediate products between the farmer and the final shopper. Long production chains may mean that increasing or decreasing levels of consumer demand for the refined final product result in very large reactions to production levels of raw materials from the farm.

 ISBN 9780170265560

Types of costs

Costs can be considered as variable, fixed or total.

Variable costs

A **variable cost** is one that changes or can be easily changed during a period of production. Since such costs vary according to the time span applied to the production system, variable costs are often known as short-run costs. They include such aspects as varying the seeding rates or fertiliser levels, or controlling the amount of chemicals or veterinary products used on the farm. Many of these types of decisions are related to the intensity of land use required by a farmer.

Fixed costs

A fixed cost is one that must be paid for irrespective of whether any productive enterprise is entered into. Fixed costs consequently do not vary with the level of production on a farm. Common fixed costs include rates, depreciation rates on equipment, and wages of permanent employees. When considered in terms of a time span, these costs can be referred to as long-run costs. A general rule in the management of a farm is to ignore fixed costs when making decisions, as these must be met even if there is no production.

Total costs

Over a long period, to stay in production, both the fixed and variable costs associated with farm operations must be covered. **Total production costs** are the sum of both fixed and variable costs on the farm.

14 Explain the term 'derived demand'.

15 List the types of costs that exist in farm record keeping.

Farm budgets

Agricultural production involves the use and management of scarce resources (e.g. land, water, electricity, people) to produce goods or services for the community. The farm is the basic unit of production. How well the farmer utilises the available capital, land and labour to achieve desired outcomes depends on careful planning and management. This is done through the effective use of management tools (e.g. budgets) and the manipulation of farm inputs and output levels. Forces that are external to the farm may act on the farm manager and consequently exert some effect on farm output.

Budgets are summaries of probable financial outlays and incomes over a specified period. They help farmers in making decisions for a particular farm at a set point in time. The financial information used in the construction of a budget is subject to uncertainty; so alternative investment strategies also should be investigated. Budgets can assist farmers in a number of ways, namely:

- determining optimal rates for production factors, such as the level of fertiliser usage on a farm
- calculating the profitability of various farm practices
- determining the advantages of adding extra enterprises or changing the combination of enterprises on a farm
- assessing the need for hiring additional labour or machinery rather than purchasing or employing permanent additions to the farm
- assessing the profitability of expanding the entire farm or improving various farm subsystems, such as pastures, by installing irrigation systems.

Whole farm budgets

A **whole farm budget** is constructed to estimate entire farm expenditure and income over the financial year. Whole farm budgets are useful in calculating taxation requirements, in assessing the financial situation for the entire farm when applying for a bank loan, and in assessing the impact of a future radical change to whole farm organisation. Such changes might be the introduction of new technology across the farm, alteration of whole farm labour levels, assessment of future risk in established marketing situations, or adoption of alternative production systems.

Components

There are a number of components to a whole farm budget.

1 *Capital value of the property.* The farmer establishes the value of the land, equipment and improvements on the farm in order to establish the amount of capital invested in the operation. Values are determined for land, water supplies, fencing, buildings including the farmhouse, machinery, and livestock or other products.

2 *Assumed annual returns.* Based on farm records of production and existing market prices (with allowance for inflation effects over the next 12 months), the farmer makes a series of assumptions to estimate income for the upcoming 12 months. Income is determined through the sale of outputs, such as livestock, crop and animal products.

3 *Assumed annual costs.* Again the farmer estimates the costs for the next 12 months on the basis of farm records, experience and professional advice. Costs are both short-run or variable costs, which relate to levels of production, and long-run or fixed overhead costs, which relate to just having a farm. It is prudent to include a charge for labour in this area of the budget, even if it is family labour, to gain an accurate view of long-term profitability.

4 *Profit/loss.* Profit or loss for an enterprise is calculated from:

Profit/loss = total returns – total costs

The figure obtained is only as accurate as the figures used in calculating estimated farm income and costs. Should unforeseen circumstances arise (e.g. a shortage of product being supplied to the market because of too few suppliers), market prices and consequent income levels will rise.

5 *Return on capital invested.* Return on capital invested in the farm enterprise is calculated to allow the farmer to compare this return with other scenarios regarding investment of money. Return on capital investment is calculated from:

$$\text{Return} = \frac{\text{net income}}{\text{capital invested}} \times 100\%$$

Example: Beef cattle

The following is an example of a whole farm budget for a beef cattle property, which has diversified to include an interest in a stud dairy herd. The owner also runs some sheep.

1 Capital invested in the property	**$**
Land	970 000
Buildings	560 000
Stock value	412 500
Total	1 942 500
2 Anticipated income	
Sale of sheep	6 000
Sale of cattle	87 500
Sale of wool and skins	2 750
Milk sales	118 500
Contract work in the area	14 500
Total	229 250
3 Anticipated expenses	
a Fixed or overhead costs	
Equipment depreciation	15 750
Rates, land tax	22 000
Salary for permanent worker	57 500
Insurance, interest	3 000

 ISBN 9780170265560

b Variable expenses	
Fodder, seed, fertiliser	37 500
Casual wages	37 500
Cartage	5000
Fuel	7500
Veterinary fees	6250
Light, power	1875
Total	193 875

4 Profit/loss statement

Profit = total return – total costs
= \$229 250 – \$193 875
= \$35 375

5 Return on capital invested

$$\text{Return} = \frac{\text{profit}}{\text{capital invested}} \times 100\%$$

$$= \frac{\$35\,375}{\$1\,942\,500} \times 100\%$$

= 1.82%

At this point the farmer would compare the interest return on capital with that from bank, building society or other forms of investment.

Partial budget

A **partial budget** is used to examine changes in income and expenditure when a change in one section of the farm occurs. This is achieved by comparing a new situation with the existing setup. When a subsystem is altered, usually with one subsystem replacing another, it is not always necessary to do a whole farm budget. The farmer is interested in determining what additional revenue is obtained, what costs may be incurred, and whether the proposed change would leave the farmer in a better or worse financial situation. This method does not assist the farmer in determining the optimum combination of resources to use in the production system. It simply compares one alternative situation with another.

If the farmer wishes to invest in a new technology relative to the existing operation, a partial budget is useful in comparing costs for adopting the new technology with anticipated increased returns. For example, where a farmer wishes to change from producing square bales of hay, which are stacked in sheds, to producing round bales, to be covered in plastic and left in the paddock, a partial budget is used to assess the profitability in changing management systems.

Example: Cattle versus prime lamb

The sample partial budget (Table 28.1, page 406) shows the effect of changing from a current situation of running 60 cattle to a possible option of developing a first-cross prime lamb operation based on 600 Merino ewes that are crossbred with Border Leicester rams. Prior to developing the partial budget, the farmer has calculated that it will cost \$56 100 to purchase the sheep and upgrade existing farm structures to accommodate the new stock. This cost is offset by the sale of cattle anticipated to fetch \$32 000. Capital outlay for the change is therefore \$24 100. The following questions must then be answered.

1 Which enterprise generates the better income?
2 Does the income obtained from the new enterprise justify the cost of converting from cattle production?

ISBN 9780170265560

Table 28.1 Calculating a farm budget

Present activity: 60 cows	$	Proposed activity: 600 ewes	$
Annual income		**Estimated annual income**	
CFA* cows	3375.00	Wool	19312.50
CFA bull	500.00	CFA ewes	4500.00
Vealers	29700.00	CFA rams	112.50
		Lambs	26400.00
Total	**33575.00**	**Total**	**50325.00**
Annual running costs		Estimated annual running costs	
Husbandry	**900.00**	**Shearing**	**2875.00**
Cow replacements	6250.00	Husbandry	3200.00
Bull replacements	1250.00	Ewe replacements	8910.00
		Ram replacements	900.00
Total	**8400.00**	**Total**	**15885.00**
Estimated annual profit	**25175.00**	**Estimated annual profit**	**34440.00**
Expected difference in profit = $9265.00			

*CFA means cast for age, or sold because of old age.

16 What are three reasons for using farm budgets?

17 Distinguish between whole farm and partial farm budgets.

From this information it can be seen that, when the plan is in full operation, additional income of $9265.00 per year will be obtained.

18 Discuss when long-term budgets would be used.

Long-term, developmental or capital budgeting

Long-term, developmental or capital budgeting considers the impact of time. It is basically a series of partial budgets over time. This type of budget is useful when evaluating long-term farm operations where returns on initial investment would not be forthcoming for a number of years (e.g. pasture improvement, woodlot schemes and soil conservation systems).

Gross margins

Gross margin calculation is a simplified method of showing relative returns for enterprises or activities so that direct comparisons can be made. Gross margin is determined as follows.

Gross margin = total revenue – variable costs

Table 28.2 indicates several common variable costs. Fixed costs are ignored because it is assumed that a farm is being used. Table 28.3 indicates many common types of fixed costs.

Table 28.2 Typical variable cost items

Crop related	Livestock related
Seed	Seed and fertiliser for fodder crops
Fertiliser	Haymaking, purchased feed
Sprays	Drench, vaccine, other veterinary costs
Fuels and lubricants	Shearing and crutching
Repairs and maintenance to machinery and equipment	Other contract livestock services
Contract harvesting	Casual labour
Casual labour	Animal insurance
Cartage	Cartage
Irrigation running costs	Shed and yard repairs

ISBN 9780170265560

Table 28.3 Typical overhead cost items

- Administration – accounting, telephone, postage
- Rates and rents
- Depreciation of machinery and equipment
- General farm insurance and workers' compensation interest payments
- Wages of permanent employees
- Repairs to water supplies, farm roads, buildings
- Taxation payments
- Lease payments

Gross margins must be expressed in comparable terms; so units on a per-hectare, per-animal, per-tractor or per-labour-hour basis are commonly used. Units must be stated to allow farmers to make direct comparisons.

Farmers are able to determine the income forgone by developing certain combinations of enterprises on farms, through the use of gross margins. While gross margins are useful for farm analysis, they should not be seen as a measure of farm profit.

Gross margins per hectare calculated for a variety of enterprises on farms in similar environments are often used in forward planning. However, limitations do exist with these direct comparisons. Cash crops often illustrate high gross margins, but to be economically viable over time the farmer needs to be aware of long-term physical and financial limits to expansion. Comparison of the gross margins of long-term crops (e.g. woodlots) with annual crops may give misleading results that do not allow for the many years of little or no income obtained from woodlots or other long-term farming systems (e.g. orcharding).

Example: Wheat

Gross margin for a cash crop such as wheat, grown on 50 hectares, is calculated as follows.

	$
Total return	6750.00
Variable costs	
Equipment repair	375.00
Seed	200.00
Fertiliser	250.00
Sprays	300.00
Fuel	187.50
Insurance	250.00
Harvest labour	375.00
Rail cartage	125.00
Total	**2062.50**
Therefore: Gross margin = 6750.00 – $2062.50 = 4687.50, or $93.75/ha	

Once calculated, this gross margin for wheat can be compared with other annual crop enterprises for a similar locality. In the most commonly used situation, wheat, barley and canola, which all require similar equipment, labour levels and environmental conditions, can be reasonably compared.

ISBN 9780170265560

Example: Sheep versus cattle

The gross margin analysis in Table 28.4 indicates how enterprises can be compared. Note the units used.

Table 28.4 Sheep and cattle operations: comparison of gross margins, and assumptions used

Sheep	Cattle
Merino wethers $18.02/dse	**Steers $16.67/dse**
• Wethers purchased at $30/head and replaced every 4 years • CFAs sold at $8/head o/s • Wool cuts 6 kg/head at $5/kg net	• Heifers retained • 2-year-old calving • 90% weaning • Cows retained for 6 calvings • Steers sold at 16 months (450 kg) at $1.20/kg • Cull heifers sold at 14 months (320 kg) at $1.20/kg • Cull cows and bulls sold at 90c/kg
Merino breeding $17.67/dse	**Vealers $15.43/dse**
• 80% weaning • Cull ewes and wether hoggets sold at $25/head • Ewes retained for first 4 lambings • CFA ewes sold at $8/head • Wool cuts: – Ewes 5.0 kg/head at $5.00/kg net – Hoggets 4.5 kg/head at $5.20/kg net – Weaners 1.5 kg/head at $3.50/kg net • Replacement rams purchased at $400/head	• Heifers retained • 2-year-old calving • 90% weaning • Cows retained for 6 calvings • Steer vealers (300 kg) and heifer vealers (250 kg) sold at 10 months at $1.20/kg • Cull cows and bulls sold at 90 c/kg

19 Define the term 'gross margin'.

20 Discuss why gross margins are useful.

Gross margins are given on an equivalent unit of measurement – dse or dry sheep equivalent – allowing some degree of comparison. When looking at these gross margins, also bear in mind differences in stocking rates, and capital required to develop yards, fencing and watering systems. The demands of cattle are higher than those of sheep. Note also the assumptions used to calculate the gross margins.

There are many factors to consider when evaluating alternative enterprises for farms or assessing farms in general. Budgets allow an economic assessment of the situation; gross margins are useful for enterprise comparison. The farmer also needs to assess physical and technological risks involved in the operation of the farm.

ISBN 9780170265560

Chapter review

Things to do

1. With the aid of agricultural magazines, newspapers or internet sites, obtain current market prices for the major livestock and crop enterprises in your local area. Continue this for several weeks. What trends do you notice?
2. For a continuous period no shorter than 6 weeks, establish market price trends for a major local product. What effect will the market price have on farmers' anticipated levels of production and future production levels?
3. Talk to a local farmer and identify what factors influence the farmer's main production goals. Are these factors related to the farm environment (e.g. soil type, local climate), or are they off-farm forces (e.g. government policy, social preferences, family preferences)? Make a table identifying what forces are affecting the farmer's decision-making process.

Things to find out

1. Describe the factors that affect demand for the product that you have studied.
2. Develop a gross margin for two major crop or livestock industries in your district.
3. Research the role of information technology (telephone, computer systems, objective assessment of products) in livestock marketing.
4. For a major local agricultural product, find out how this product is marketed. What mechanisms exist to:
   - a guarantee the supply to the market?
   - b provide a marketing outlet for farmers?
   - c promote the product?
   - d add value to the product?
5. Why is the concept of 'adding value' to an agricultural product considered important by marketing organisations?

Extended response questions

1. Discuss the marketing strategies available to farmers for products from one plant and one named animal production system.
2. How do the laws of supply and demand influence marketing strategies for agricultural products?
3. Discuss strategies available to farmers to market farm products, including vertical integration, contract selling, direct marketing, cooperatives and marketing boards.
4. Discuss, with examples, the ways that farmers are able to assess and plan future farm operations.
5. Discuss how calculating a gross margin and return to capital can be used to analyse the financial situation of a farm enterprise.
6. 'Farming is less a business and more a way of life.' Discuss the implications of this statement in relation to farm situations that you have studied.

CHAPTER 29

Outcomes

Students will learn about the following topics.

1 Government influence on production and marketing.

Students will be able to demonstrate their learning by carrying out these actions.

1 Outline reasons why governments might want to intervene in the market place.
2 List the main mechanisms by which governments intervene.
3 Discuss methods used to control the supply of products onto the market.
4 Discuss methods used to influence the demand for farm products.
5 Discuss methods used by governments to compensate farmers directly.

ISBN 9780170265560

GOVERNMENTS AND MARKETS

Words to know

bounty money paid by the government to farmers to offset production costs

buffer stock scheme a scheme designed to maintain a stable market price by withholding excess quantities of product in times of oversupply and progressively releasing this when market prices are more favourable

contract a regulated quantity of output allowed to be produced (replaces the term 'quota')

dumping flooding the market with a product

multiple-price scheme a scheme designed to maximise returns by obtaining several prices for essentially the same output, by splitting the market on the basis of distance, consumer income levels or varieties of possible manufactured goods (e.g. flavoured milk, buttermilk, skim milk, low-fat milk)

subsidy aid provided by governments in the form of money to stabilise and maintain incomes and market price

Introduction

There are several reasons why governments seek to intervene in the market. As the agricultural sector of the economy develops and generates more income for a country, government interest in the long-term stability and earning potential of agricultural industries tends to increase. In many instances, government intervention is directed towards stabilising the prices received by farmers for their farm outputs. There is also a developing need to address the inequality of incomes between rural and city workers. Balance-of-payment considerations may also prompt government intervention in the market.

Governments seek to increase or stabilise prices by:

- controlling or diverting supply
- influencing demand for a product
- directly compensating farmers for movements in price.

Usually governments do not rely on just one method; rather, they adopt a series of measures.

Stabilisation of prices received by farmers enables farm budgets and farm improvements to be planned and developed with greater certainty of outcome. The need for intervention by governments is brought about by unpredictable environmental factors that expose farmers to widely varying income levels from year to year and the impact of fluctuations in overseas markets.

Although price support mechanisms are not the most desirable way to encourage equity in income levels between farmers and city workers, they have been widely used. The main problem is that price support mechanisms help all farming operators, including corporate farms.

1 Why do governments intervene in open markets?
2 Outline the principal factors that influence farm income levels.

Controlling supply

The main methods used to control supply are:

- input restrictions
- market quotas
- product destruction
- import restrictions.

The main mechanisms used often require legislation to restrict inputs. Governments may restrict the input of production factors such as land or water. A land-licensing system limits farmers in the sugar industry by restricting the amount of land that they can devote to the production of sugar cane. In irrigation areas, water quotas that must be purchased restrict the amount of land devoted to irrigation. Both of these methods affect total production levels on the farm. This method of control is often ineffective, for example because farmers substitute limited water with more fertiliser to increase production levels.

Market **contracts** restrict the amount of product that can be sold. Limitations are commonly found in the liquid milk industry, where dairy farmers supply milk on a contract basis.

In extreme circumstances, farmers may be recompensed for destroying a certain percentage of the farm subsystem. Orchardists who supply canneries are occasionally faced with periods of drastic oversupply. To maintain a satisfactory market price for all producers, farmers may be paid to destroy a certain percentage of mature fruit-bearing trees. Although the practice is discouraged, refined agricultural products such as coffee may be simply buried in times of oversupply.

Diverting supply

The aim of diversionary techniques is to raise prices or stabilise prices received by producers, by diverting part of supply from one period to another or from one market to another.

Buffer stock schemes

In **buffer stock schemes**, a certain quantity of goods is removed from the market at a period of peak supply and stored, to be returned to the market place at a time when demand for the product is more buoyant. Such operations are usually carried out by some cooperative

 ISBN 9780170265560

arrangement between producers or via special government agencies. An example is the use of cold storage rooms to store apples, so avoiding periods of seasonal oversupply.

Success depends on the buffer stock agency's having enough capital to sustain a weak market (i.e. to buy up all excess stock). It also depends on the same agency's having a storage area of sufficient size to store excess product and to prevent it from deteriorating when stored for long periods. The coordinating body must release stored stocks at a time when producers gain the most benefit.

Two-price or multiple-price schemes

In two-price or **multiple-price schemes**, total returns for a product can be maximised by dividing the market into separate sections, and then charging different prices on these markets (a form of price discrimination). In this way as much of the product as possible is sold for as high a price as possible on one market, while the rest is sold for whatever it will bring on the alternative market. The different markets must be kept separate, for example by geographical distance (an export market and a domestic market) or by developing different end-uses for the product (e.g. modified milks such as low fat and Omega 3 milk). Consumer incomes may also allow the development of split markets, so similar products can be supplied at different prices, depending on the socioeconomic standing of the communities concerned; for example, using internet marketing.

One common example is the use of a domestic market where a portion of the product is sold at a price that the government feels consumers can pay, while the remainder of the product is sold cheaply overseas at whatever the world market price is at the time of sale. The returns from both markets are usually pooled and returned to the producers. Care must be taken when using this method, as claims of product **dumping** are quickly levelled at countries that dump low-priced material on the world market.

3 Through what mechanisms do governments seek to control or divert supply of agricultural products?
4 Distinguish between buffer stock and two-price schemes.
5 Why is it of some benefit to split the market for some products? Give examples.

Influencing demand

Demand for a product can be influenced by a number of direct government intervention schemes:

- restrictions on substitutes
- mixing regulations
- preferred market agreements
- product promotion.

Governments may place a restriction on the sale or manufacture of products that can easily substitute for natural agricultural products. In Australia, for example, restrictions were placed on the manufacturing levels for margarine for many years, because it competes with butter.

Mixing regulations may be used, requiring that a certain percentage of a domestically grown product be used in the manufacture of the final marketable product.

Trade agreements with overseas countries (e.g. Japan) may be negotiated in which Australian products have preference or guaranteed access to markets in return for equivalent concessions.

Government-financed promotions and advertising campaigns are occasionally developed to encourage consumers to purchase large quantities of a product. Marketing boards are entrusted to carry out many of these marketing functions. Meat and Livestock Australia promotes the sale of lamb through funds gained from both government sources and levies on producers.

connect

Lamb campaigns

Compensating for price movements

Governments may directly increase prices received by farmers by paying subsidies of various types.

Deficiency payments

A deficiency payment is a form of variable **subsidy**, where the government determines the level at which it believes the price of a commodity should be supported. Farmers sell their produce on the open market, and if the price obtained is less than the guaranteed price,

ISBN 9780170265560

the government makes good the deficiency in the form of supplementary payments. The advantages of these schemes are that market prices are not raised, consumption is encouraged, the inflationary effect on the economy is less, and the burden is largely borne by the taxpayer.

Flat rate schemes

An amount of money, commonly known as a **bounty**, is paid by the government to offset costs to farmers when they buy inputs such as fuel or fertiliser. This form of subsidy often leads to inefficiency in production and adds to inflationary pressures in the economy.

Price stabilisation funds

Governments often find it risky to provide guaranteed prices for products, especially when dealing with export markets. Stabilisation funds are analogous to buffer stock schemes, except in this case money rather than physical stocks of a product are set aside.

In the operation of a price stabilisation fund, a tax or retention levy is imposed on producers when the market price exceeds a predetermined 'fair market price' – usually determined by the government. When the market price for the product falls below market price expectations, money from the tax fund is used to improve returns to producers. In many cases, limits are set on the size of the stabilisation fund that can accumulate within the industry.

Fixed amount

Industries can be allocated set amounts of money for assistance in their development by governments.

In the past, these mechanisms often resulted in:

- larger producers benefiting the most from various interventions
- other agribusinesses, such as feed suppliers and transport industries, benefiting from some of the market support systems
- an increase in domestic prices for goods, causing reduced domestic consumption.

Many inefficient industries were kept viable because of the various forms of government intervention, many expanding because of these mechanisms, not because of production levels.

6 How do governments seek to influence consumer demand or producer demand for materials?

7 List ways of directly compensating farmers for price fluctuations.

Rationalising government intervention into the market

The above mechanisms have been used by successive Australian governments to maintain and stabilise farmer returns. However, these measures for assistance actually distort the use of scarce resources in agriculture, such as water, fertiliser and land. In consequence, productivity becomes limited by the very mechanisms designed to sustain farmers, which work against structural change and the adoption of more efficient methods of production. In the past 10 years all levels of government have worked toward reforming the nature of market intervention to a point where the level of farmer support is the second lowest in the OECD. This has enabled more efficient production systems to evolve, especially in the dairying and the extensive grain and livestock production systems.

Reforms to agricultural marketing arrangements

Since the 1970s, competition has been gradually introduced into most agriculture industries where compulsory agricultural marketing arrangements had governed processes between the farm and domestic and/or export markets. Key reforms include the following.

- *Wheat.* In the 1970s, move from guaranteed to stabilised prices; provision for 'grower to buyer' sales outside the pooling arrangements; home consumption price limited to wheat for human consumption and determined by a formula to take account of export prices.
- *Dried vine fruits.* End of price stabilisation arrangements in 1980.

 ISBN 9780170265560

- *Citrus.* Decade-long phase down of tariffs from 30% to 5%, beginning in 1986; state marketing boards amalgamated, reducing geographical barriers to competition.
- *Cotton.* Queensland Cotton Board deregulated in 1989.
- *Eggs.* State-based production and pricing controls progressively withdrawn from 1989.
- *Sugar.* Domestic administered price arrangements and export controls terminated by the Commonwealth in the late 1980s.
- *Wheat.* Domestic market deregulated in 1989; grower levy fund introduced to replace the Commonwealth guarantee of Australian Wheat Board borrowing.
- *Barley.* Competition gradually introduced into domestic feed and malting barley marketing in South Australia and Victoria from 1998.
- *Dairy.* Phased reductions in market support payments on export of dairy products.
- *Dried vine fruits.* Commonwealth price equalisation levy and statutory equalisation of domestic sales removed in the early 1990s, as was the industry's exemption from section 45 of the *Trade Practices Act 1974* (which reduced the scope for collusive price discrimination).
- *Horticulture.* Underwriting scheme for apples and pears terminated in 1990.
- *Tobacco.* Local Leaf Content Scheme and the Tobacco Industry Stabilisation plan ceased in 1995; withdrawal of vesting powers in 1995.
- *Sugar.* Import tariffs and domestic price supports removed in mid-1997.
- *Wheat.* Australian Wheat Board converted from statutory authority to a grower-owned company in 1999.
- *Wool.* Reserve Price Scheme ceased in 1991.
- *Dairy products.* Dairy state-based controls over sourcing and pricing of market milk ceased in 2000; 9-year Dairy Industry Adjustment Package (DIAP) concluded in 2009.
- *Barley.* South Australian single-desk arrangements terminated in 2007; Western Australian market deregulated in 2009 (allowing any number of licensed entities to export barley).
- *Canola.* Exports of canola and lupins deregulated in Western Australia in 2009 (traders no longer required to apply for licenses to export).
- *Sugar.* Queensland Sugar Limited lost its compulsory acquisition powers in 2006 and lost exemption from the *Trade Practices Act* in 2009.
- *Wheat.* Bulk exports deregulated in 2008, meaning proposals to export bulk wheat no longer needed approval from the single-desk seller (Australian Wheat Board).
- *Rice.* New South Wales Rice Marketing Board still retains powers to vest, process and market all rice produced in New South Wales (approximately 99% of Australian rice is produced in New South Wales).

8 Describe the marketing reforms that have occurred in the wheat and dairy industries since the 1970s.

E. M. Gray, M. Oss-Emer and Y. Sheng (2014), *Australian agricultural productivity growth: Past reforms and future opportunities*, ABARES report 14.2. Adapted from Industry Commission (1998), *Microeconomic reforms in Australia: a compendium from the 1970s to 1997*, research paper, Australian Government Publishing Service, Canberra; Productivity Commission (1999), *Impact of competition policy reforms on rural and regional Australia*, report no. 8, AusInfo, Canberra; Productivity Commission (2005), *Trends in Australian agriculture*, Productivity Commission, Canberra; World Trade Organization (2007), *Trade policy review: Australia — report by the Secretariat (revision)*, Trade Policy Review Body, WT/TPR/S/178/Rev.1, 1 May, World Trade Organization and WTO (2011), *Trade policy review: Australia — report by the Secretariat (revision)*, Trade Policy Review Body, WT/TPR/S/244/Rev.1, 18 May, World Trade Organization.

Key Australian Government agricultural programs

Reforms, such as the removal of tariffs, and a more diverse approach to managing risk on farms have seen lower prices for many farm inputs, including machinery, fertiliser, water and labour. In addition, farmers have better access to a deregulated financial system to access credit and a range of financial products and information. Funding programs now seek to support the agricultural sector more broadly by providing funding for research and development and biosecurity initiatives, and also by providing services associated with hardship in the farming community, such as financial counselling, family payments, financial and taxation assistance. Particular programs are outlined in Table 29.1 on page 416.

ISBN 9780170265560

Table 29.1 Australian Government agricultural programs

Program	Elements
Funding for rural research and development (R&D)	The Australian Government has a range of programs spread across several departments that provide funding for rural R&D (R&D for the agricultural, fishery and forestry industries). The Australian Government invests approximately $715 million in rural R&D annually.
Biosecurity	The Australian Government Department of Agriculture primarily manages biosecurity risk at the border and offshore. This involves inspecting vessels, goods and passengers as they enter Australia, and assessing risks posed by proposed import of goods, including plants, animals and their products. While the *Quarantine Act 1908* does not provide powers for the Australian Government to manage post-border pests and diseases in general, it does allow the Australian Government to play a role during emergency situations.
Drought-related programs	Assistance provided to farmers under drought programs aims to help farmers prepare for and manage the effects of drought and other challenges.
Rural Financial Counselling Service	Free financial advice is provided for primary producers, fishers and small rural business experiencing financial hardship.
Transitional Farm Family Payment	Payments are provided to farmers experiencing significant financial hardship, paid at a fortnightly rate equivalent to the Newstart Allowance.
Taxation assistance	A number of special tax measures and concessions are available to primary producers, including: • tax averaging across years • Farm Management Deposits (allowing farmers to set aside pre-tax income to smooth income across years) • ability to access a range of other offsets, deductions and concessions to reduce their assessable income.
Farm finance initiative	Announced in April 2013, this program aims to support farmers currently struggling with high levels of debt, who nevertheless demonstrate long-term viability. Eligible farmers are able to access short-term (up to 5-year) concessional loans.
Carbon Farming Futures and the Carbon Farming Initiative	Programs aim to create opportunities for land managers to enhance productivity, gain economic benefits and help the environment by reducing greenhouse gas emissions. Through the Carbon Farming Futures program, funds are available for research, on-farm demonstration, extension and outreach activities. The Carbon Farming Initiative operates as a voluntary offset scheme to facilitate the sale of carbon credits generated from eligible activities within the land sector to international and domestic carbon markets. It funds eligible on-farm activities that generate carbon credits.
Caring for our Country	The program aims to protect Australia's natural environment and sustainability. Farmers and other land managers can apply for funding to undertake projects that improve biodiversity and sustainable farm practices. This includes funding for Landcare, a community-based organisation that has worked to raise awareness and influence farming and land management practices since the 1980s.
Disaster income recovery subsidy	This subsidy is provided to assist farmers (and other businesses) who experience a loss of income as a result of disasters such as bushfires and flooding.

Source: E. M. Gray, M. Oss-Emer and Y. Sheng (2014), *Australian agricultural productivity growth: Past reforms and future opportunities*, ABARES report 14.2

9 Outline and discuss the value of two Australian Government programs for funding agriculture. Note: When discussing the value of the programs give your own opinion.

Careers

There are a variety of careers associated with agricultural economics. View the Agricultural resource economist website for an example.

connect

Agricultural resource economist

ISBN 9780170265560

Chapter review

Things to do

1. For your local area, compile a table of the main products produced on farms, and outline the main methods used to market these products.
2. a Talk to a local farmer and write down how the main farm product is harvested, processed and transported to the storage or marketing facility in your area.
   b Outline any mechanisms used by the government to maintain market prices and regulate production levels in the market.
3. Collect articles to maintain a scrapbook on ways in which governments seek to stabilise farm prices in various industries.

Things to find out

1. Describe one way that government influences the production or marketing of a named product.
2. For a product that you have studied, outline the mechanisms used to control the supply and demand for this product.
3. Outline reasons why there has been an ongoing strategy to reduce support to the farming sector.

Extended response questions

1. The types of agricultural systems found in a region depend on both economic and environmental factors. Briefly describe why this is the case. Illustrate your answer by outlining the situation for a named farming system in your area.
2. Government has an influence on the production and marketing of agricultural products.
   a Outline how two government bodies influence the production of a named product.
   b Outline how two government bodies influence the marketing of a named product.
3. What do you consider is an appropriate role for governments to play in agricultural marketing? Outline reasons for your arguments.

CHAPTER 30

Outcomes

Students will learn about the following topics.

1. The marketing chain for a product.
2. The influence of government on production and marketing strategies.
3. Quantity and quality criteria for a product.
4. The importance of product specification in the marketing of a product.
5. The problems in meeting market specifications of a product.
6. Processing raw agricultural commodities.
7. The nature and potential of value adding to a product.
8. The role of advertising and promotion in the marketing of a product.
9. Supply and demand relationships for a product.
10. The impact of research and technology on milk production and marketing.

Students will be able to demonstrate their learning by carrying out these actions.

1. Determine the marketing chain for the product.
2. Explain various marketing options for the product.
3. Outline government influence on the production and marketing of the product.
4. Assess the quantity and quality of the product.
5. Analyse market specifications for the product.
6. Evaluate the management strategies used to assess and meet market specifications.
7. Schedule the timing of operations in a production cycle to meet market specifications.
8. Analyse marketing information for the product.
9. Construct a flow chart of steps involved in processing the raw agricultural commodity into its various forms.
10. Evaluate ways in which the product can be value added.
11. Outline strategies for advertising and promotion of the product.
12. Assess a current advertising campaign for a product.
13. Describe factors affecting supply and demand for the product.
14. Interpret supply and demand information for a product.
15. Outline the importance of ongoing research and describe recent technologies which have impacted on milk production and marketing.

 ISBN 9780170265560

PRODUCT CASE STUDY: MILK

Words to know

homogenisation dispersal of fat globules within milk to stop the cream from separating out

pasteurisation killing bacteria that can cause disease by heating milk to 72°C for 15 seconds; the milk is not sterilised; it will still spoil after pasteurisation even if the container is not opened

somatic cell count a measure of the white blood cell count in a sample of milk; used as an indication of mastitis

ISBN 9780170265560

Introduction

Australia produces, on average, 9 102 million litres of milk per year, valued at $4 billion. Australia exports 45% of its annual milk production and the country produces 10% of world trade in dairy products. The main export markets are Japan, China, Singapore, Indonesia and Malaysia. There is likely to be future opportunities for increased sales into Asian countries.

There are two main markets into which milk is sold. The first is called liquid (fresh) milk, where the milk is processed for immediate fresh consumption. Farmers usually receive a higher price for this milk, although in many areas farmers receive a single price for milk delivered to the processing factory. The second is called manufacturing milk. This milk is 'manufactured' or made into dairy products, such as cheese, butter, yoghurt and milk powders. Farmers usually receive a lower price for this milk.

The major consumer dairy products on the Australian market are drinking milk (fresh and UHT, white and flavoured), cheese, butter and dairy blends, and yoghurt.

Prior to July 2000, milk production on farms was regulated by a quota system supervised by a marketing authority. This has been replaced by a contract system administered by individual processing companies. Individual milk factories control the pricing and distribution of milk for the liquid milk market. The price paid to farmers depends on the quality of milk, as determined by factors such as high fat and protein levels, and low levels of micro-organism activity. Dairy farmers now operate in a completely deregulated environment, where international prices are the major factor in determining the price received by the farmer for milk. Australian dairy farmers receive a low price by world standards; consequently they must efficiently manage their production system.

Milk is processed both by farmer owned cooperatives and by public and private companies.

Substantial change has occurred in the dairy industry, with many farmers leaving and others accessing loans in order to stay in production. The problem is twofold – the low price paid to the farmer for the raw product; and the need to market milk and milk products in forms that encourage consumer demand and arrest the decline in consumption. Farmers are also assessing ways to add value to their raw product; for example, developing niche marketing opportunities with speciality Australian cheeses that are successfully expanding both their domestic and their export markets.

Dairy Australia is an organisation funded by farmer-paid levies, which are imposed on the fat and protein content of all milk produced in Australia. It provides money to improve farm productivity and farmers' management skills.

Recently, due to competition between the major supermarket chains, milk pricing arrangements have changed and farmers generally receive a lower price per litre.

Farm incomes are under pressure from milk pricing competition and increasing input costs, and productivity growth appears to have slowed in recent times.

connect

Nature of markets and the dairy industry

1 Identify the five main export markets for Australian milk products.
2 Name the major dairy products consumed in Australia.

The marketing chain for dairy cow whole milk

Milk collected from dairy cows by a milking machine that uses a pulsating vacuum system is at a temperature of 37°C, and must be cooled and stored on farm at a temperature of 4°C until the milk tanker collects it. Milk going into the tanker is assessed for taste, smell and appearance by the operator of the tanker. The milk tanker has multiple chambers within the trailer to separate milk from the individual farms and prevent cross-contamination.

The refrigerated tanker then transports the milk to the milk factory where it is initially stored in refrigerated vats and then tested in a laboratory for the presence of antibiotics and sediment. In the laboratory, tests for microbial activity that would indicate infection are also carried out. A **somatic cell count** measures the number of white blood cells in the milk, which is an indicator of mastitis. A high number indicates infection, often caused by poor milking practices (such as leaving the milking cups on the cow for too long) or unhygienic practices (such as poor washing of udders to remove manure, soil or grass matter, or not dipping the teats after milking with an iodine solution to kill any bacteria

ISBN 9780170265560

inside the teat canal). A total plate count for bacteria is performed in the laboratory along with testing for butterfat percentage and protein percentage. These quality checks are made prior to the milk being processed and determine the price paid to the farmer. If the milk does not meet predetermined standards it may be rejected.

Drinking milk is then **pasteurised** (heated to 72°C for 15 seconds) to kill pathogenic bacteria and homogenised (fat globules are dispersed through the milk to stop cream separating out from the milk) to make it creamier. The milk is then packaged into cartons and plastic bottles and distributed to retail outlets by factory-owned trucks. Because milk is a perishable product the movement of milk to retail outlets needs to occur quickly. Milk used to make products other than drinking milk is diverted to a separator prior to pasteurisation.

A variety of value-added products are also produced at the processing factory, such as cheese, cream, yoghurt, ice cream, skim milk, butterfat and many types of milk (Fig. 30.1). Milk and the many value-added products are transported to retail outlets and milk vendors from processing factories daily.

connect

Marketing milk and its products

3 Briefly describe the marketing chain for dairy cow whole milk.

Figure 30.1 Processing factories can produce many types of milk.

Government intervention within the market place

Government intervention into the milk market can affect either the production of whole milk or the marketing of whole milk and milk products. Three government bodies that influence the production of whole milk are described below.

1 *New South Wales Food Authority.* This government body licenses farms to produce milk. This ensures that dairy farmers produce uncontaminated milk that is safe for consumption. Hygiene management practices, including stock management, must meet predetermined standards. Compulsory maintenance of health standards in the milking sheds includes serviceable paint work, flyscreens, gutters, and cleanliness of floors, tiles, milk lines and machines. These areas must meet the requirements of the Code of Practice for Dairy Buildings.

   Farmers must also develop a food safety program. One way to do this is to use a Hazard Analysis Critical Control Point (HACCP) food safety program. It includes things like keeping milk cold (less than 4°C) and away from contaminants. Food safety programs, such as HACCP (quality control procedures), ensure consumers obtain a safe, high-quality product.

2 *Animal Welfare Code of Practice (administered by the Department of Primary Industries).* The farmer is responsible for the welfare of the animals on the farm. The Animal Welfare Code of Practice – Cattle sets the minimum standards that must be met on the dairy farm. Animal welfare regulations regarding the treatment and husbandry operations permitted to be carried out by the farmer may result in changes to the husbandry and routine of animal management; for example, tail-docking of dairy cows for udder health.

3 *Environmental Protection Authority (EPA).* The EPA determines the disposal method for effluent from dairy premises. Considerations would include drainage of effluent to waterways, water movement during prolonged wet weather and waterway pollution (eutrophication).

Two government bodies that influence the marketing of milk and other dairy products are as follows.

1 *Milk Marketing New South Wales.* An entity of the New South Wales Food Authority, this body promotes and encourages the production, supply, use, sale and consumption of milk and dairy products in New South Wales; for example, marketing dairy products at agricultural shows and through the media.

2 *Department of Foreign Affairs and Trade.* This organisation promotes and markets dairy products at food shows overseas and offers contacts and advice to develop export markets.

connect

New South Wales Food Authority and the dairy industry

4 Outline how two government bodies influence the production of whole milk.

5 Outline how government bodies influence the marketing of milk and other dairy products.

Product quality and quantity criteria

The product specifications that need to be met by the farmers when supplying milk to their processing factory are shown in Table 30.1. The market specifications as shown on milk cartons are listed in Table 30.2.

Table 30.1 Product specifications for one particular factory

Product component	Specification
Butterfat (milk fat)	3.95%
Milk protein	3.15%
TPC (total plate count)	<15 000/mL
BMCC (bulk milk cell count – somatic cell count)	<500 000/mL
Antibiotics	<0.003 μg/mL

Table 30.2 Market specifications

Product component	Specification
Butterfat	3.6%
Milk protein	3.4%
Pasteurised	Yes
Homogenised	Yes
Milk sugar	4.8%

6 List the quality criteria for milk that must be met by farmers when supplying milk to a processing factory.

ISBN 9780170265560

Role of product specification in marketing of milk

The value of product specification to the consumer lies with the dairy farmer having to achieve the minimum standards that are specified by the processing factory. If these standards are regularly not met, the milk:

- is rejected and not collected, meaning a loss of income (e.g. this could happen if total plate counts (TPC) are too high TPCs (total plate counts) indicating excessive bacterial contamination)
- receives a lower payment per litre (e.g. if the milk has lower butterfat or protein percentage than specified)
- may be used to produce manufactured products rather than whole milk (such as cheese, which has a lower butterfat percentage).

Farm management strategies to meet product specifications

Monthly production statements are sent to farmers for feedback so they can maintain quality control measures for the product. Many dairy farmers keep records on all of the cows in their herd. One day per month milk from each cow is sampled from the evening and morning milking. From the sampling vessel the quantity of milk produced per cow can be measured on the farm. These samples are then sent to be tested to work out the percentage of protein and fat in the milk and for other tests. This procedure is known as herd recording.

The disadvantages of such reporting procedures include:

- the time taken for milking is longer when taking samples
- the expense or cost, in terms of labour and time.

Herd testing gives the farmer information about the fat and protein percentages of the milk samples delivered to the factory and the presence of any contaminants or elevated bacterial counts. High somatic cell counts would indicate high levels of the bacteria that cause mastitis, indicating the farmer needs to teat dip animals at milking and use a Rapid Mastitis Test on milk from each cow to track the infection levels. The presence of antibiotics in the milk can present a health danger to the consumer and also makes the milk unsuitable for cheese making. If iodine (a common disinfectant used to clean milking machines and pipes) is present, the milk is rejected after a warning. If coliform bacterial counts, such as *E. coli*, are high then faecal contamination of the milk is indicated. Better farm hygiene procedures are required.

The following are three common problems farmers encounter, and the strategies they may use to overcome these problems.

1 If milk protein levels are too low:
   - feed a protein supplement, such as by-pass protein pellets, dairy meal protein pellets
   - grow quality legume/grass pastures for grazing, such as white clover and kikuyu
   - selectively breed using high milk protein bulls, artificial insemination (AI) and embryo transfer (ET).

2 *If milk fat levels are too low:*
   - protect cows from cold/strong winds
   - feed good quality lucerne hay as a supplement (high fibre)
   - grow the fodder crops oats (winter) and sorghum (summer) for grazing (high fibre)
   - selectively breed using high milk fat bulls, AI and ET.

3 If total plate counts (TPC) are too high (coliform or bacteria from faeces):
- wash and dry udders well before applying teat cups
- keep the milking shed as clean as possible while milking (wash down if too much faeces)
- thoroughly wash down milking shed after milking, giving time for the area to dry before being used again.

connect

Robot milking

connect

Electronic identification system

Quantity measurements received in a production report from the factory include lifetime milk (in litres, L), days dry and calving interval. The production report also shows the highest milk producing cow (L/cow/305 days).

On the basis of this information the farmer can also determine:
- the cows that could be used as donor cows in embryo transfer programs
- the cows that should be culled, such as poor producers; that is, those that produce low litres of milk/day, or have low protein in their milk
- the need for increased feeding and supplements, such as concentrates, at milking to get a higher protein percentage in the milk
- the effectiveness of treatments performed
- if calving interval is too long.

7 Describe the role of product specification in the marketing of whole milk and modified milks.

8 Describe farm management strategies used by farmers to meet product specifications.

In some dairies the cows are fitted with computerised ear tags or microchip collars that register each cow as they enter the dairy and collect data on the quantity of milk produced at each milking. The Electronic identification system website illustrates this use of electronic collars in dairies.

Consumer expectations of the product

The consumer has certain product expectations that need to be met for companies to maintain a demand for the product in the market place. For milk these demands include that it is:
- always available at a reasonable price
- of consistent quality; that is, the same taste, smell, freshness, colour, thickness, cleanliness (bacteria-free)
- always looking, smelling and tasting good; for example, not lumpy and creamy white (not past use-by date)
- conveniently and well packaged; that is, the container is easy to fit in fridge, reseals easily, strong, easy to pour, does not leak, can be recycled
- available as a variety of products; for example, whole milk, flavoured, low fat, high calcium, low lactose
- an endorsement logo; for example, the Heart Foundation tick.

To help meet these expectations, market specifications are shown on the milk carton or plastic bottle, including nutritional information. This information is useful because it:
- allows the consumer to compare different drinking milks
- keeps the consumer informed
- allows the consumer to choose the milk product that matches their dietary and health needs
- is a guarantee of certain quality standards
- encourages consumer confidence in the product.

9 List six expectations that consumers of milk demand.

Production cycles and market specifications

There are several farm operations that need to be specifically timed to meet market specifications in terms of both quality and quantity of milk production.
- *Vaccinations*. Ephemeral fever (three-day sickness) vaccination is performed in early summer because biting mosquitoes that are prevalent in summer and into early autumn spread the disease. The vaccine provides protection over summer and into early autumn. Cows suffering from ephemeral fever are very sick, rarely eat, become listless and may dry off if lactating, which reduces total milk production.
- *Pasture sowing routines*. Winter pastures and fodder crops are established toward the end of summer and into early autumn. This is to allow good growth of the pasture before it is slowed by the low temperatures of winter. Commonly a winter oats crop or rye grass is planted on many dairy properties on the coastal areas of New South Wales to provide food over winter and so maintain a constant supply of milk to the factory.

ISBN 9780170265560

- *Drenching*. Warm, moist seasons (spring and autumn) are associated with worm infestations. Drenching is performed to reduce worm numbers and prevent a fall in milk production (caused by the cow's energy being used to combat the disease rather than to produce milk).

The production cycle of the cow needs to be managed as well to maintain both milk quality and quantity. This includes management operations, such as:

- *mating*. Mating occurs generally 3 months after the cow has calved, allowing one calf to be produced per year. Cows need time to come back on heat. The cow is dried off (production of milk is stopped) 2 months before producing the next calf. This is because the cow needs to conserve energy and protein for the rapidly growing calf (growth is most rapid in last 3 months of pregnancy). The farmer aims to re-mate the cow as soon as possible so that the cow calves and returns to lactation quickly. A tight mating program means that in its lifetime a cow will produce seven calves and have regular lactation periods (instead of fewer lactations).
- *pregnancy testing*. Testing approximately 42 days after mating enables the farmer to move pregnant animals quickly onto better pastures or to provide them with supplementary feeding. This is to promote foetal growth. Non-pregnant animals can also be quickly re-mated or culled.
- *feeding supplements during pregnancy*. Growth of the calf is most rapid in the last three months of pregnancy. It is necessary to increase the level of nutrition (food quality and quantity) to sustain the growth of the calf and to meet the nutritional needs of the mother.

Figure 30.2 indicates the timing of various management decisions based on the production cycle of the cow. A 5-in-1 vaccination is given to the cow 4 weeks before parturition or birth. It is an annual booster for the cow (boosts immune system). It also increases the antibodies provided from the cow to the calf in colostrum (first milk).

10 Name six farm operations that need to be timed to meet market specifications in terms of both milk quality and quantity.

11 For two of the operations listed in question 10, outline why timing is important.

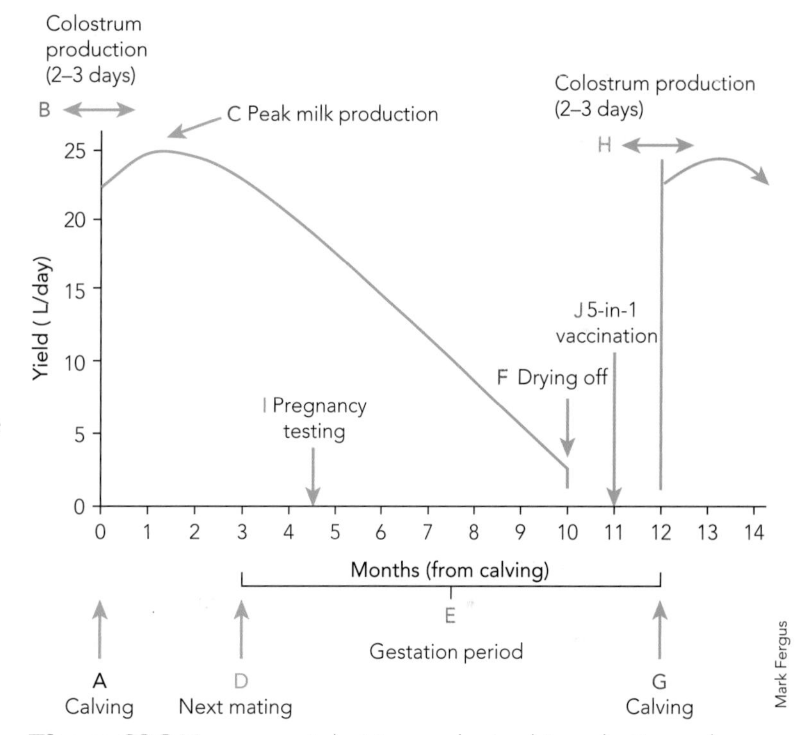

Figure 30.2 Management decisions and animals' production cycle

The processing of dairy products and value adding

Figure 30.3 illustrates the processes involved in the manufacture of butter and cheese, two value-added products derived from whole milk.

Value adding refers to creating new products, modifying existing products and providing new packaging methods (e.g. for convenience and shelf life) with the result that the new product can be sold for a higher price than raw whole milk. Examples include:

- creating new products from the raw product; for example, making cheese or butter from whole milk
- modifying an existing product to meet new markets; for example, producing different milks from whole milk, such as 'Lite White', 'Shape', 'Hi-Lite' and flavoured milks
- creating convenience packaging; for example, 3 L plastic milk bottles.

The advantages of value adding include that products:

- are more profitable; with each additional processing step, the profit margin tends to increase
- are often more easily sold
- may expand markets (both domestic and export) by catering for a new niche or specialised market
- are often more likely to capture new export markets; for example, ice cream to Japan, cheese to Hong Kong and Indonesia, and butter to Singapore and Thailand.

12 Explain the term 'value adding'.

13 Name three advantages of value adding.

14 Name three value-added dairy products.

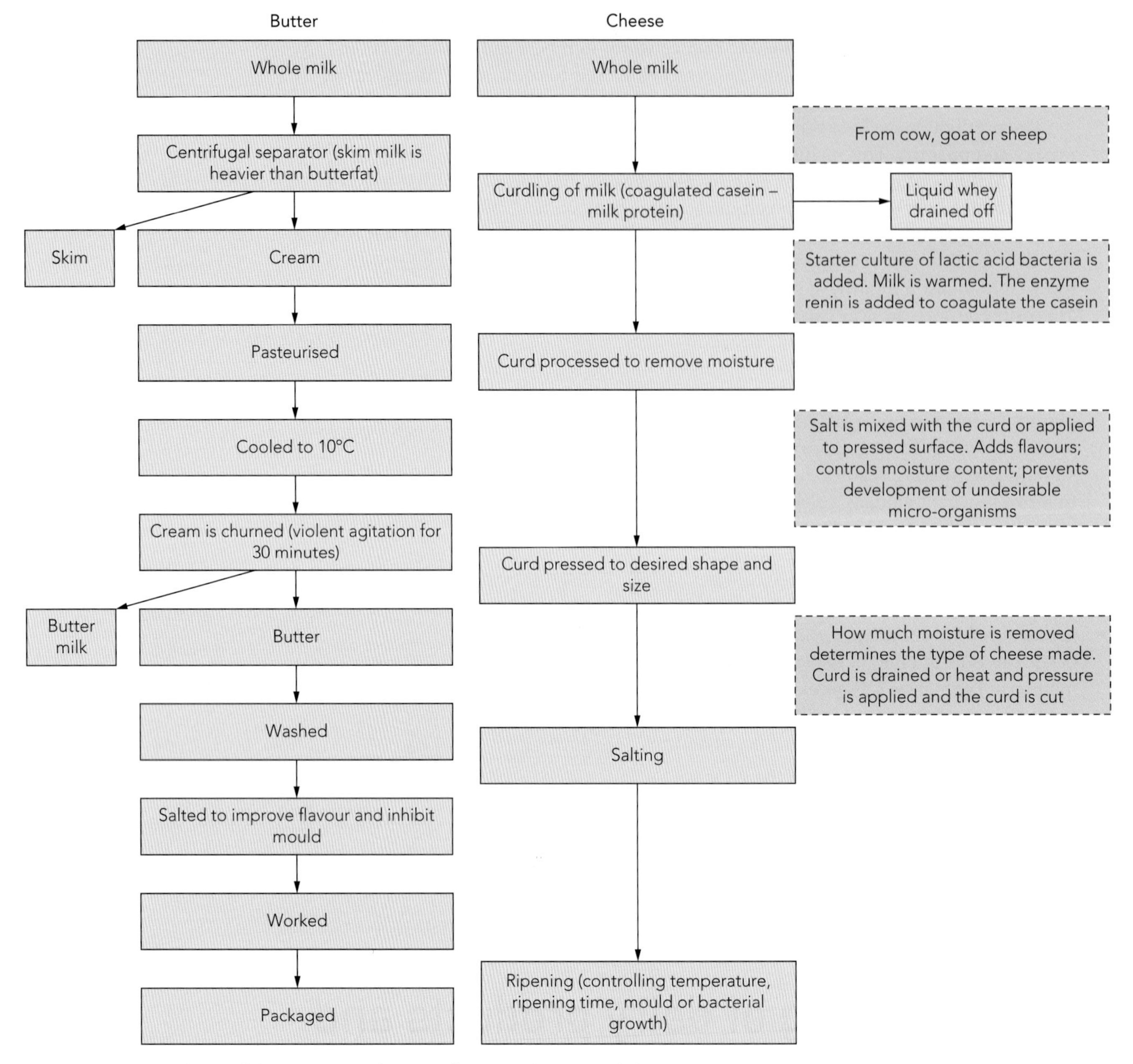

Figure 30.3 Outlines of the processes for manufacturing butter and cheese

connect

Butter manufacture

connect

Making yoghurt

connect

Manufacture of goat cheese

15 Explain how dairy farmers can add value.

Dairy factories, such as Dairy Farmers, produce a wide variety of products from raw whole milk (e.g. pasteurised and homogenised milk, flavoured milks, skim milk, ice cream, long-life milk, cheese and yoghurt). These products are the result of processing that value adds to the raw whole milk. Other factories produce specialised products for the export market, often adding ingredients such as egg yolk and vanilla essence to the milk product for the Asian market.

Farmers can also value add. Farmers may elect to run a herd of Jersey cattle instead of Friesian cattle. Jersey cows produce milk with a very high butterfat content, making it a creamy milk. Farmers can set up their own milk processing and bottling factory where pasteurisation, **homogenisation**, bottling and selling all occurs from the farm. On this basis a number of localised niche markets can be developed. Other farmers can make specific types of cheeses to sell into specialised market places, such as local farmers' markets.

ISBN 9780170265560

Strategies for the advertising and promotion of milk

Strategies used to advertise and promote whole milk include:

- promotion on the milk carton
- tasting at agricultural shows
- health and nutritional sheets used in primary schools
- internet promotion, showing the variety of milks available
- television commercials.

Dairy Farmers has developed different drinking milks for different consumers to expand their share of the drinking milk market. The company has targeted different consumers as follows.

- Whole milk (good levels of fat for energy, protein for growth, sugar for energy, calcium for growth) is marketed toward families with growing young children.
- Shape (very high protein and calcium, no fat) targets consumers such as middle-aged females to reduce risk of osteoporosis, elderly that suffer from osteoporosis, athletes.
- Skim milk (high protein, no fat, riboflavin) targets consumers such as people at risk of heart disease or who are weight conscious.

The expected outcomes of such market strategies include:

- increased sales and profits
- raised awareness of a wide product range
- product or brand is well known and recognisable
- improved consumer knowledge about the health benefits of consuming a particular product
- consumers aware that this is an Australian product
- increased consumer awareness of new products.

connect

Dairy farmer products

16 Name five strategies used for the advertising and promotion of milk.

17 Name three advantages of market strategies used to advertise and promote whole milk.

Demand for drinking milk

Consumer demand for drinking milk is said to be inelastic, which means that they will buy milk even if there is fairly large changes in price and therefore demand is fairly constant throughout the year. Demand drops slightly over the colder months and increases marginally in summer. These changes are easily overcome by:

- decreasing milk production in winter slightly
- increasing milk production in summer slightly.

Essentially, demand is driven by the size of the population and will only increase when the population grows. Once this occurs the farmer will then be offered to supply more milk on contract.

Supply of drinking milk

Farmers are under contract to supply milk for each week of the year at contract level plus 10%. The contract specifies the quantity of milk to be supplied weekly and a base return in cents per litre is given.

Farmers may seek to increase the amount of milk supplied by:

- increasing the level of supplementary feeding
- buying more cows
- genetic improvement of the livestock.

18 Describe the demand for drinking milk over 12 months of a year.

19 Describe how farmers can increase the amount of milk they supply.

ISBN 9780170265560

The impact of research and technology on the production and marketing of milk

The importance of research in improving production includes methods to:
- improve product quality
- improve yield
- increase efficiency of production
- produce sustainability.

Technological impacts on milk production include such innovations as the use of robotic milking systems, sexed semen and embryos, and the use of tetraploid ryegrass in pasture mixes.

Technological impact on the marketing of milk includes:
- modified milks, such as 'Lite White' or 'Omega 3'
- internet marketing
- on-farm pasteurisation, homogenisation and bottling of a niche milk type then subsequent marketing of the product.

connect

Country Valley

Niche market: on-farm pasteurisation, homogenisation and bottling of milk

20 Outline the impact of research and technology on the:

a production of milk

b marketing of milk.

 ISBN 9780170265560

Chapter review

Things to do

1. Outline the features of a major market for a product that you have studied.
2. Outline a problem a farmer might experience in meeting a market specification for the product that you have studied.
3. Collect articles from magazines and newspapers that outline the changes to the marketing system for the product that you have studied.
4. Collect market price information for 8 weeks for the product that you have studied. Examine any trends in price that occur, and list reasons for these trends.
5. Evaluate a current advertising or promotional campaign for a product that you have studied.

Things to find out

1. Name one method used to assess the quality of a named farm product.
2. Outline the steps involved in harvesting, storing and manufacturing products for an industry you have studied.
3. Outline the importance of ongoing research in the production or marketing of a product that you have studied.
4. Describe ways in which the farm product that you have studied can be value added.
5. Describe recent technologies and their impact on agricultural production and/or marketing.
6. Outline a problem a farmer might experience in meeting a market specification for a named farm product.
7. Investigate future proposals for the marketing of a product you have studied.

Extended response questions

1. For the named product that you have studied:
   a determine the marketing chain
   b explain various marketing options
   c outline government influence on the production and marketing
   d assess the quantity and quality
   e analyse market specifications
   f evaluate the management strategies used to assess and meet market specifications
   g schedule the timing of operations in a production cycle to meet market specifications
   h analyse marketing information for the product
   i construct a flow chart of steps involved in processing the raw agricultural commodity into its various forms
   j evaluate ways in which the product can be value added
   k outline strategies for advertising and promotion
   l assess a current advertising campaign
   m describe factors affecting supply and demand
   n interpret supply and demand information.
2. Explain how raw farm outputs are stored, processed and marketed for the product that you have studied.
3. Government intervention in the marketing system is designed to guarantee an income for farmers and to ensure the optimum use of scarce resources by farmers. Outline how governments achieve these ends with reference to an agricultural industry that you have studied.

CHAPTER 31

Outcomes

Students will learn about the following topics.

1. Innovation, ethics and current issues.
2. Aspects of biotechnology and the meaning of the term.
3. Current areas of development in biotechnology.
4. The wide range of potential applications of gene technology in agriculture.
5. The wide range of potential applications of biotechnology in agriculture.
6. Ethical concerns and debate around the use of biotechnology in agricultural production.
7. The benefits and problems of biotechnology and genetic engineering in agricultural industries.
8. The benefits and problems of gene technology in agricultural markets.
9. Biofuel production.
10. Potential applications for biotechnology in agriculture.
11. Potential applications for genetic engineering in agriculture.

Students will be able to demonstrate their learning by carrying out these actions.

1. Define the terms 'biotechnology', 'DNA', 'genetically modified organism' (GMO), 'gene markers', 'genetic engineering' and 'protein synthesis'.
2. Describe current biotechnology applications.
3. Investigate the uses of biotechnology in agriculture, such as genetic modification in plants and animals.
4. Describe the implications of biotechnology in the agrifood, fibre and fuel industries.
5. Outline the importance of food safety and labelling of genetically modified products.
6. Discuss the issues relating to food production using GMOs.
7. Examine the regulations that surround the development and use of GMOs and biotechnology.
8. Explain the role of biosecurity.
9. Analyse the conflict between increased production and ethical concerns around biotechnology innovation.
10. Describe ways biofuel is produced.
11. Identify and describe industries that consume biofuel products.
12. Evaluate biofuel production.
13. Research one agricultural biotechnology.

ISBN 9780170265560

AGRIFOOD, FIBRE AND FUEL TECHNOLOGY

Words to know

agrifood food obtained from agriculture

biosecurity the protection of Australia's animal and plant industries and the natural environment from exotic pests and diseases

biotechnology any technique that changes living organisms at the molecular level (genes and chromosomes) to produce useful products, such as medicines (e.g. insulin) or an insect-resistant plant, or offspring through embryo transfer

deoxyribonucleic acid (DNA) the biological material that stores genetic information for an organism; one chromosome is one DNA molecule, a double stranded molecule in the shape of a helix; along each strand is a sequence of bases, comprising many genes

gene marker a gene or DNA sequence with a known location on a chromosome that can be used to identify individuals or species, or a gene used to determine if a nucleic acid sequence has been successfully inserted into an organism's DNA

genetically modified organism (GMO) an organism whose genetic material has been altered using genetic engineering techniques; genes from different species are combined

protein synthesis the process of creating proteins, starting with amino acid synthesis and ending with the assembling of amino acids into protein molecules

transgene a gene or genetic material that has been transferred from one organism to another

ISBN 9780170265560

Introduction

The term **biotechnology** refers to the use of living things in industry, technology, medicine and **agrifood** (food obtained from agriculture).

Biotechnologies such as Bt corn, Roundup Ready® soybean in food, Bt cotton, Roundup Ready Flex® cotton in fibre, ethanol production from sugar cane waste, methane production from piggery effluent and fermentation in fuel industries are all examples of the various uses of biotechnology in agriculture.

Table 31.1 outlines the advantages and disadvantages of some biotechnologies in use within agricultural production systems.

Table 31.1 Biotechnologies in agricultural production systems

Biotechnology	Positive contributions	Consequences
Biopesticides; e.g. NPV, viral pesticide against heliothis caterpillars in cotton and lettuce crops. Dipel, a bacterial insecticide used against caterpillars such as cabbage white butterfly in brassica crops.	• Less toxic than conventional pesticides. • Very specific to target pest. • Environmentally friendly.	• Takes time to kill the insect pest. • Farmer needs to know what to do. • Costly and takes time to develop. • Relatively expensive to buy. • Shorter shelf life.
Genetically modified goats: human gene for lysozyme inserted into dairy goats; human gene for anti-blood clotting factor inserted into dairy goats.	• Rapid way of producing a gene determined product. • Greatly reduces diarrhoea incidence in human infants.	• Initial cost of research. • Difficult to supply developing countries. • Ethical issue of taking human genes and inserting into a goat.
Bioethanol production from crops such as sugar cane, corn and wheat or from green wastes or straw (waste from harvested crops).	• Produced from renewable resources. • Reduced crude oil dependency. • Less pollution.	• Large areas of farmland would need to be devoted to growing crops, therefore there would be less land available for cropping. • Ethanol and biodiesel production contribute to greenhouse emissions.
Bt crops, such as cotton and soybeans, where genes from *Bacillus thuringiensis* have been inserted into the cotton plant. The genes produce proteins that are toxic to heliothis caterpillars.	• Reduce the need for use of pesticides. • Less pesticide run-off going into creeks and rivers means eco-friendly. • Insect pests do not become resistant to the chemical. • No withholding period.	• Technologies are all in the hands of a few large companies that may monopolise prices. • Super weeds that have the Bt gene may develop if pollen escapes. • Eliminates the choices of farmers should genetically modified cotton pollen fertilise non-Bt cotton crops and grow in crops certified GMO-free or organically produced.

1 Name two biotechnologies used in agricultural production systems.

2 For one named biotechnology used in Question 1, describe the positive contributions and consequences.

The Agricultural Biotechnology Council of Australia (ABCA) is the national coordinating organisation for the Australian Biotechnology sector. This group was established to pursue the recognition of the current and potential benefits of agricultural technology. Further information can be found at the Australian Biotechnology Council of Australia website.

connect

Australian Biotechnology Council of Australia

 ISBN 9780170265560

Innovation: genetically modified foods and biotechnology in agricultural production

So far, there has been no sustainable increase in food production or decline in world hunger through the use of genetically modified (GM) crops. Golden rice is a rice variety produced through **genetic engineering** (see page 131) to make a chemical precursor of vitamin A in the endosperm of the rice. Rice can produce this chemical in the green parts of the plant but not in the seed. Scientists discovered that by inserting three genes into the plant this could be overcome. The aim was to produce a product that could be consumed to reduce the incidence of vitamin A deficiency. Many problems were encountered when the variety was introduced; for example, balanced indigenous diets were sacrificed to grow the rice. The advantages obtained were not as significant as the advantages of providing populations with a varied diet.

In developing countries, monoculture and industrial-scale farming are the threats to food security. The Bill Gates Foundation is assisting local farmers in Third World countries to improve food security with simple irrigation schemes. There is insufficient money in developing economies to support the introduction of GM crops.

Ethical concerns

Genetically modified organism (GMO) technology, its development, ownership and decision-making apparatus are all owned and therefore controlled by a small number of powerful corporations, such as Monsanto and Ciba-Geigy. Essentially, these companies have control over the farmers because they hold the seed stocks. Poor farmers can't afford to buy genetically modified seed.

A number of dangers exist in unregulated environments where policy and supervision standards are inadequate, such as:

- potential for the development of
  - super-weeds
  - new and harmful viruses
  - targeted pest resistance
  - transgenic allergens
- patenting of GMOs could stop other research
- use of terminator genes to reduce the risk of pollen escaping could stop seed being saved to sow the next crop (saving seed is a common practice in developing countries).
- there could be a resultant loss of biodiversity.

3 Explain the positive contribution that genetically modified Golden rice could bring to consumers.

4 Outline the ethical concerns about using GMO technology.

Regulations and the development of genetically modified foods

The use of genetically modified materials in Australia is controlled by various state and territory regulations and the *Gene Technology Act 2000*. The Act seeks to protect the safety of both the environment and people. Other regulatory agencies include:

- the Veterinary Medicines Authority
- Australian Pesticides
- Therapeutic Goods Administration
- Food Standards Australia New Zealand.

ISBN 9780170265560

Food Standards Australia New Zealand (FSANZ) assesses the safety of all GM foods. Food Standards Australia and New Zealand states the following goal:

> The safety assessment is not to establish the absolute safety of the GM food but rather to consider whether the GM food is comparable to the conventional counterpart food; i.e. that the GM food has all the benefits and risks normally associated with the conventional food.

For further information about Food Standards Australia New Zealand, view the Food safety website.

connect

Food safety

Food assessment includes:

- consideration of the intended and unintended effects of the genetic modification; for example, along with insect protection, compositional changes in the food might arise that could affect the health and safety of the population
- comparisons with conventional foods that have an acceptable standard of safety. This involves identifying similarities and differences between the genetically modified food and a comparable conventional food. Differences are checked for safety and nutritional issues.

Problems and potential problems arising from the use of food produced from GM plants may include:

- GM involves the expression of new **protein synthesis** in the GM organism, which might be allergens; for example, peas were genetically modified for protection against weevils, but when mice were fed the GM peas, they suffered an allergic reaction
- antibiotic marker genes are used when developing GM plants to identify the cells that have the inserted gene, and these marker genes could be transferred to microbes in the gut
- there are currently no official programs for monitoring the long-term effects of GM foods.

connect

Food labelling

For more information of GM labelling

5 Identify the organisation that regulates the development of genetically modified foods.

6 Outline the legal requirements in Australia concerning the labelling of foods that contain GM plant products.

Several legal requirements exist regarding the labelling of foods that contain GM plant products. These arose because consumers are concerned with the safety of GM foods and the ethical issues related to the use of gene technology. They want to be informed and have a choice.

Labelling requirements include:

- GM food and food ingredients where new DNA and/or new protein is present must be labelled; the label must use the words 'genetically modified'
- when genetic modification has changed the nutritional components of the food, labelling is also required; for example, high oleic acid soybeans and high lysine corn.

connect

FLAVR-SAVER tomato

The success of the FLAVR-SAVER was fleeting due to other factors. Read about the history of this tomato.

connect

Drought tolerance

7 Identify two examples of genetic manipulation in plants.

8 For one plant example used in Question 7:

a describe the genetic modification

b outline the advantage(s) to the farmer and/or consumer.

Genetically modifying plants

The following are examples of genetic modification in plants.

- *Non-browning potato.* The PPO gene (a gene associated with the browning of products after cutting) was identified, isolated, modified and inserted back into the potato. This resulted in a significant reduction in the number of potatoes rejected for processing into crisps or fries because of browning or bruising, due to a reduction in oxidation rate in the potato.
- *FLAVR-SAVER tomato.* The gene for softening in tomatoes was isolated, copied and inserted backwards in the tomato. This slowed down the natural softening process, so the tomato can be left longer on the vine to develop extra flavour.
- *Wheat.* The gene for resistance to nematode, fungal rust and bacterial diseases was identified, isolated and transferred to wheat, resulting in increased resistance to pests. Characteristics such as protein and starch content may be transferred in this way, resulting in wheat specifically for bread, pasta or noodles. There are many other examples of genomic research. Visit the Drought tolerance website to read about research involving the genetics of wheat and the interplay of the environment and an organism's genetic makeup.

ISBN 9780170265560

- *Enhanced flowers*. An extra gene that blocks production of gaseous ethylene has been inserted into carnations so that vase life can be extended two to three times longer than normal.

Genetically modifying animals

There are many issues arising from the genetic manipulation of animals.

The Enviropig and the Aquadvantage salmon are examples of genetic modification in animals.

connect

Genetic engineering of animals

View the website and answer the questions.

Enviropig

Process of genetic modification

Bred from a line of Large White pigs (a species of pig), these transgenic pigs are designed to digest phosphorus in their food using their genetically modified salivary glands. These pigs were developed in a process where a segment of DNA containing a gene sequence isolated from another organism was introduced into their genetic makeup. Genes were introduced into fertilised embryos by injection then surgically implanted into an oestrous synchronised sow. Piglets were checked for the desired enzyme, phytase, in their saliva. After the seventh generation, pigs bred true for the transgene.

connect

Enviropig

Expected benefits

In cereal grains, 50–75% of phosphorus is found as phytic acid, and this can be digested in the stomach of these transgenic pigs. This results in reduced feed costs as diet supplements of phosphorus are not required. The phosphorus content of manure is reduced by 20–60%, depending on the stage of growth and diet. Soil build up of phosphorus is reduced, leading to less phosphate in run-off. Consequently, there is less chance of eutrophication and excessive algal growth in waterways (problems which cause reduced oxygen content in water and the resulting death of fish).

Aquadvantage salmon

Process of genetic modification

A growth hormone regulating gene from a Pacific Chinook salmon and a promoter gene from an ocean pout (an eel-like fish) were inserted into the Atlantic salmon. The gene transfer was through the physical insertion of genes into the nucleus using a micro syringe and a gene gun.

connect

Salmon

11 Identify two examples of genetic manipulation in animals.

12 For one animal example used in Question 11:

a describe the genetic modification

b outline the advantage(s) to the farmer and/or consumer.

Expected benefits

The modified fish can grow year-round, gaining marketable size after 600 days (16 months) rather than 800 days (2 years). The salmon can be grown in ponds, with no risk of disease spreading to wild stocks or interbreeding. As they are more efficient at converting food to fish meat, feed costs are reduced.

Visit the Salmon website and examine the reasons why there is a delay in the commercial release of this species of fish.

Advantages and disadvantages of GM plants and animals

The production and use of GM plants and animals appears to offer many advantages to the modern world we live in. However, as with most technological advances, there is the possibility of adverse and unforeseen negative effects.

Advantages

Potential advantages of GM plants include:

- better quality and quantity of food
- vaccines can be produced in plants
- produce may reach the consumer in better condition with less spoilage

ISBN 9780170265560

- reduced maturation time of crops
- increased nutrients, yields, and stress tolerance; improved resistance to disease, pests, and herbicides of crops
- new products and growing techniques of crops
- environmentally friendly bio-insecticides
- conservation of soil, water, and energy
- better natural waste management
- increased food security for increasing populations
- benefit from genetic engineering in the mass production of insulin and human growth hormone
- possible future use of genetically engineered animals to produce substances in their milk or blood that could be of benefit, such as certain antigens or antibiotics
- reduced pesticide use saves money and reduces environmental pollution and risks to human health; for example, genetically engineered nematode-resistant bananas and Bt crops.

connect

Genetically modified bananas

Answer the questions.

connect

Gene technology and farming

An outline of the CSIRO's work in gene technology in plant and animal farming systems

Disadvantages

There are also many potential disadvantages of genetic modification of plants:

- potential human health effects, including increasing human exposure to allergens, allowing the transfer of antibiotic resistance markers, and unknown long-term effects
- potential environmental impacts, such as the unintended transfer of **transgenes** through cross-pollination, and the loss of flora and fauna biodiversity
- intellectual property rights being used as a way of manipulating markets – the organisation holding the rights has exclusive use of the technology or discovery
- domination of world food production by a few companies
- mixing genes among species,such as introducing pig genes into another food, is problematic for some religious and cultural groups
- consumer objections to consuming animal genes in plants and vice versa
- labelling not mandatory in some countries, resulting in the consumer unwittingly eating GM food
- new advances may be skewed to interests of rich countries
- the use of genetically identical plants leads to a loss of biodiversity
- transgenic species might escape into wild populations and super weeds could be formed by the cross-fertilisation
- business has control of the GM products and their motive is profit
- genetically engineered animals are not approved for commercial use in Australia.

connect

Risks of gene technology

Answer the questions.

9 Identify four advantages relating to the production and use of GM plants for food production.

10 Identify four disadvantages relating to the production and use of GM plants for food production.

Biosecurity

Biosecurity involves the protection of Australia's animal and plant industries and the natural environment from exotic pests and diseases. An important part of biosecurity is plant and animal quarantine. Biosecurity is needed when carrying out research using GM plants, growing GM crops on the farm and importing GM crop seeds.

Biosecurity and the use of GM plants and their products

A number of requirements are placed on institutions and individuals who use GM plants and their products. These regulations include the following.

- *Research.* Any research group must obtain approval and a licence from the Office of the Gene Technology Regulator. The *Gene Technology Act 2000* prescribes conditions for conduct, management and containment of work involving GMOs. Research groups must supply a submission that includes the aim of the trial, details of the genetically modified plant, the research trial plan, risk assessment and a risk management plan.
- *Growing GM crops on the farm.* Many farmers are already growing approved and licensed crops. The risk is greatest at planting and harvest, where seed or pollen can easily spread to a nearby GM-free farm. The farmer is advised to grow his crop using risk-prevention strategies, such as:
  - locating the GM crop to reduce the risk of seed or pollen drift to a neighbouring farm
  - understanding the crop (e.g. pollen and seed spread varies between crops)

connect

Biosecurity

ISBN 9780170265560

- seeking advice from grower organisations on 'best practice'
- growing buffer crops or pastures around GM crops
- following the seed suppliers' instructions, which will include any special conditions that are required on the licence from the gene technology regulator.

- *Importing genetically modified seed*. The importer must obtain approval and a licence from the Office of the Gene Technology Regulator and the Australian Quarantine and Inspection Service by:
  1 submitting an application
  2 preparing a risk assessment and risk management plan (RARMP), which considers legal implications, risk posed, advice received from experts and any licence conditions imposed; this is developed by the Office of Gene Technology
  3 consulting the RARMP; submissions are sought from authorities and public
  4 licence decision and conditions; the decision will consider the RARMP, health risks, applicant suitability as well as monitoring, auditing and reporting to ensure compliance with licence conditions.

13 Define and outline the role of biosecurity in Australia.

14 Outline the three requirements/regulations placed on institutions and individuals using GM plants and their products.

Biotechnology applications

Biopesticides

- Biopesticides are pesticides where the active ingredient comes from natural materials, such as plants, bacteria and viruses. Two examples are described below.
  - Microbial pesticides contain a microbe such as a bacterium, fungus or virus that is the active ingredient. Each pesticide is specific for its target pest; for example, Bt (*Bacillus thuringiensis* bacteria) pesticides, where proteins from the bacteria kill one or a few related species of caterpillar.
  - Biochemical pesticides, where a naturally occurring and non-toxic substance is used; for example, sex pheromones interfere with insect mating.

  Advantages of biopesticides include:
  - they are less toxic than conventional pesticides and largely avoid the associated pollution problems
  - they are very specific – they only affect targeted pest and related species, and they avoid causing the death of beneficial organisms (e.g. pollinator bees, soil earthworms).
  - as part of IPM, they can reduce use of synthetic pesticides, and so reduce associated hazards (e.g. polluted waterways and algal blooms)
  - pest resistance problems are reduced.

  Disadvantages of biopesticides include:
  - it takes time to kill targeted pests, so there is some crop damage in the meantime
  - the farmer needs to have a good knowledge of the pest (e.g. must know the most vulnerable stage of the insect lifecycle).

Rumen modification (micro-organisms)

Research aims to improve digestion of fibre (cellulose) and reduce methane production in animals. Researchers have concentrated on three areas – modifying diet; searching for different micro-organisms that could be inoculated into the rumen of cattle (South African cattle have been tested); and genetically engineering more efficient microbes.

Gene markers

Gene markers are used to test for the presence of desired genes. A radioactive or fluorescent probe attaches to a very small base sequence that lies alongside a targeted gene. When the DNA profile chart is made for a particular bull the presence or absence of the targeted gene will be picked up. So it can be determined if a particular bull is carrying a favourable gene (e.g. a gene that improves milk production). There are gene markers for milk fat yield, milk protein yield and meat tenderness.

Vaccine production

Clostridia vaccines protect against bacterial diseases, such as tetanus, pulpy kidney and blackleg in sheep and cattle. Vaccine production involves growing the clostridia bacteria in a fermentation vat that contains proteins, salts, vitamins and buffers (to control pH) with the aim of producing large quantities of toxins.

The process is stopped with the addition of formalin. The toxin is inactivated and is now called a toxoid, which is not damaging to a host but stimulates the host's immune system to produce antibodies. The toxoid is filtered off, diluted and an adjuvant is added to form the vaccine. An adjuvant contains aluminium salts and acts as an irritant to attract immune system cells. It presents the vaccine antigens to the immune system cells.

Embryo sexing

After embryos are flushed from the reproductive tract of a donor cow, they can be sexed. A small number of cells are removed from each embryo and a probe is used to detect part of the Y chromosome. A male embryo will have both X and Y sex chromosomes; a female embryo has two X chromosomes.

Semen sexing

Each sperm cell will contain an X or Y chromosome. Sperm containing a Y would produce a male offspring. The sperm cells are stained with a fluorescent dye and a flow cytometer machine identifies and separates the sperm cells based on DNA density. The technique is 95% accurate and straws are filled with either X or Y sperm cells. Embryo and semen sexing has applications in animal industries where the product comes from one sex; for example, in the egg laying and dairy industries.

15 Identify six current developments in biotechnology.

16 For two current developments in biotechnology mentioned in Question 15, describe the technology and its application.

Embryo splitting

Collected embryos are placed into a culture medium that causes them to sink to the bottom of the plastic dish. A fine surgical blade is then used to split the embryo in half. This is done under a microscope. The inner cell mass must be split as these are the cells that give rise to a foetus. Each 'demi-embryo' can be transferred immediately into recipient cows where they are able to grow into an entire individual.

Managing processes in agricultural systems

Biofuel production

Biofuel is produced from grain, sugar, vegetable oils, algae and green waste/straw. Biofuels are a potential alternative to petroleum; however, further research (Fig 31.1) should be conducted in the field of biofuels for more sustainable and efficient use of carbon to reduce the release of greenhouse gases into the atmosphere.

Further research to enhance current biofuel technologies is aimed at:

- improving efficiency of solar powered distillation processes
- genetic engineering of bacteria to increase the concentration of alcohol produced in fermentation to higher than 15%
- developing mechanisms for the decomposition of cellulose to produce glucose economically (e.g. looking for alternative pathways of producing ethanol directly from cellulose).

ISBN 9780170265560

Alamy/© Jim West

Figure 31.1 Biofuel research. Biofuels provide cleaner burning fuels and produce less harmful greenhouse fases in their combustion than fossil fuels.

Advantages

The advantages of biofuels are many.

- Biofuels are produced from renewable resources (plant crops and wastes; e.g. molasses from sugar refining).
- The production process is 'greenhouse gas neutral'. Carbon dioxide produced during combustion is used up during photosynthesis by the next crop so the net amount of carbon dioxide released into the atmosphere is zero.
- Biofuels reduce our petroleum dependency and greenhouse gas emissions from transportation; they make more efficient use of carbon.
- Toxic carbon monoxide emissions are reduced when ethanol is used as a fuel as ethanol undergoes complete combustion more readily than octane (the major component of petrol).

Disadvantages

There are also disadvantages to biofuel production.

- Large areas of farmland would need to be devoted to growing suitable biofuel crops, with subsequent land degradation and pest problems associated with large scale monoculture systems.
- Carbon dioxide would be produced in preparing land for biofuel crops (e.g. ploughing paddocks using fuel-powered tractors), and in the fermentation and distillation processes, reducing the overall sustainability of production
- Disposal of large amounts of smelly fermentation waste when producing ethanol could present major environmental problems.
- Farmland use has changed, because crops once grown as a source of grain production for food and livestock feed are now being grown for biofuel production. As a consequence grain prices have increased. This is a particular concern for developing countries, which are more vulnerable to increasing prices.
- Aid organisations have limited budgets and if biofuel production means that prices increase, less food can be bought and distributed to people in poverty, resulting in continuation of starvation.
- World food demands continue to increase, and it will be harder to meet demand if land is given over to biofuel production.
- Large quantities of water are required for ethanol production.

connect

Biofuels

17 Biofuels are a potential alternative to petroleum. Outline areas for further research for more sustainable and efficient use of carbon.

18 Describe how bioethanol and biodiesel are produced.

Uses for biofuels

- *Bioethanol.* Bioethanol is a solvent in cosmetics (e.g. perfumes and deodorants), food colourings and flavourings, antiseptics and some household cleaning agents. It is used in fuel blends for petrol engines (e.g. e10 is a blend of 10% ethanol and 90% petrol). Figure 31.2 illustrates the production of bioethanol.

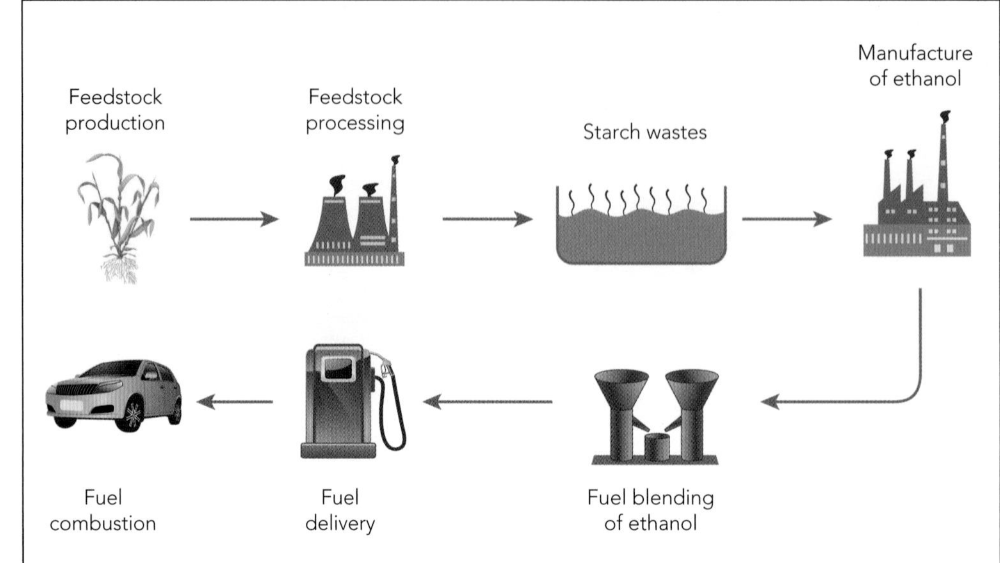

Figure 31.2 Bioethanol manufacture

connect

Ethanol

Detailed information on the production of bioethanol

connect

Use of ethanol and its production by an Australian company

- *Biodiesel.* Biodiesel is used in diesel engines for land/rail/ocean transport, road and rail haulage of goods and merchant marine shipping (e.g. container transport, fishing). The manufacture of biodiesel in shown in Figure 31.3.

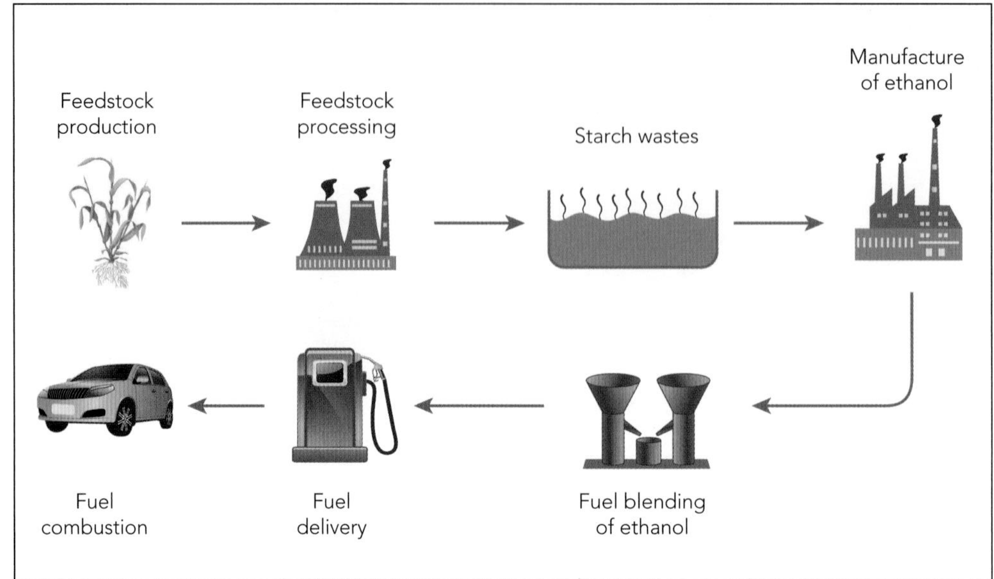

Figure 31.3 Biodiesel manufacture

connect

Biodiesel

Detailed information on the production of biodiesel

connect

Biogas

Detailed information on the biogas industry

- *Biogas.* This is a gas that is produced from the biological breakdown of organic matter when little or no oxygen is present. The gases produced are methane, carbon monoxide and hydrogen. These can be used as fuels for the generation of energy and heating purposes, such as cooking. Detailed information on this industry can be obtained from the Biogas website.

19 Identify three biofuels.

20 Identify and describe industries or activities that consume the following biofuels:

a bioethanol

b biodiesel

c biogas.

ISBN 9780170265560

Chapter review

Mandatory analysis of a research study for NSW HSC Agriculture course

Students choose only **one** of the research study electives given in Chapters 31, 32 and 33.

Research into technological developments

a Analyse a research study of the development and/or implementation of one agricultural biotechnology in terms of:
- design of the study
- methodology of the study
- collection of data for the study
- presentation of data
- analysis of data
- conclusions and recommendations.

b Explain the need for research in the development of agricultural technologies.

Things to do

1. Define the following terms:
   a DNA
   b GMO
   c genetic engineering.
2. Identify two ways of producing biofuel from agricultural crops.
3. Outline the importance of food safety and labelling of GMOs.
4. Explain how biosecurity helps protect Australia's agricultural industries.

Things to find out

1. Outline the issues relating to food production using GMOs.
2. Examine regulations that surround development and use of GMOs and biotechnology.
3. Identify and describe industries or activities that consume biofuel products.
4. Outline a method used to analyse data in a research study on the development and/or implementation of an agricultural biotechnology.
5. Outline the implications of using biotechnology in agriculture.

Extended response questions

1. Discuss a current biotechnology development.
2. Analyse the conflict between increased production and ethical concerns in biotechnology innovation. Outline specific examples to help illustrate your answer.
3. Evaluate biofuel production with respect to world food demands and sustainable and efficient use of carbon.

CHAPTER 32

Outcomes

Students will learn about the following topics.

1. Australian climate and its variability.
2. Causes of climate variability.
3. Changes in climate that may be attributed to human activity.
4. Management of resources such as water, greenhouse gases and carbon levels.
5. Management techniques available to the farmer to minimise risk by managing climate variability.
6. Research into climate variability.

Students will be able to demonstrate their learning by carrying out these actions.

1. Examine local climate and contrast it with another region.
2. Calculate rainfall and temperature statistics.
3. Analyse data to determine the frequency of weather events.
4. Explain the implications of climate variability for agricultural production.
5. Compare the variability of climate in different geographical regions of Australia.
6. Extrapolate from climate variability data to predict the effects of climate change on production.
7. Investigate research evidence in relation to long-term climate variation.
8. Outline the effect of sea surface temperatures on the Southern Oscillation Index (SOI) and subsequent effects on climate.
9. Describe the processes causing La Niña and El Niño.
10. Identify the main greenhouse gases and their sources of emission.
11. Recognise the effect of greenhouse gases on atmospheric temperature and climate change.
12. Explain how land clearing and vegetation changes can affect local climate.
13. Explain the contribution of nitrogen fertilisers and ruminant farm animals to greenhouse gas production.
14. Identify methods farmers can use to reduce the concentration of greenhouse gases (principally methane) in the atmosphere.
15. Outline methods used to sequester carbon in agricultural soils.
16. Describe the methods to store and trade water resources.
17. Analyse issues related to water storage and trading.
18. Evaluate a range of management options available to farmers to manage climatic variability.
19. Analyse a research study into climate variability or management strategies related to climate.

ISBN 9780170265560

CLIMATE CHALLENGE AND AGRICULTURE

Words to know

carbon credit a certificate or permit to allow production of 1 metric tonne of carbon dioxide or another greenhouse gas

carbon sequestration a natural or artificial process that removes and traps carbon dioxide from the atmosphere

carbon trading the basis of an emissions trading approach that relies on the buying and selling of carbon credits

effective rainfall where mean monthly rainfall exceeds the monthly evaporation rate

El Niño a period marked by the development of a warm ocean current off the South American coast and a major shift in weather patterns across the Pacific with cooler than normal surface sea temperatures across the north of Australia

greenhouse effect the overall increase in temperature when solar radiation is unable to escape from Earth's atmosphere because of the build-up of atmospheric gases, particularly carbon dioxide

La Niña a period marked by the extensive cooling of the central and eastern tropical Pacific Ocean, often accompanied by warmer than normal sea surface temperatures (SST) in the western Pacific, and to the north of Australia.

rainfall variability a measure of the likelihood of rainfall

Southern Oscillation Index (SOI) a calculation of the monthly or seasonal fluctuations in the air pressure difference between Tahiti and Darwin

water trading trading of water in Australia through the buying and selling of water access entitlements and allocations

ISBN 9780170265560

Introduction

Climate (see page 3) is a term used to describe average weather conditions over a long period of time. The climate of a district includes the average rainfall and temperature. These averages are calculated from daily recordings over a long period (generally in excess of 30 years). The success of most forms of agriculture in Australia is largely dependent on the reliability of rainfall and a suitable existing temperature range.

Weather (see page 19) is a term describing the short-term changes in the atmosphere from day to day. It is this variation that has an immediate effect on the farm production system. The weather is the result of a number of individual factors acting together to produce certain environmental conditions.

When studying aspects of weather, a meteorologist considers mainly solar radiation, temperature, rainfall, evaporation rates, humidity and wind.

Farmers can get information on weather and climate from:

- the weather bureau (Bureau of Meteorology)
- records kept on the farm or in the local area
- mobile phone apps
- the internet.

Some people view daily weather forecasts to decide what to wear or if activities would be better indoors or out, but the weather has far more significant influences on farming operations from day, to day as extreme weather conditions can affect the growth of both plants and animals. Seasons also vary around Australia. In the south of Australia there are usually short, cold wet winters, followed by long, hot summers that are often dry. In the tropical north of Australia it is a different picture. Day length does not vary greatly from season to season and there is a hot, wet season corresponding to the southern summer period and a mild dry season corresponding to the southern winter-spring-autumn period.

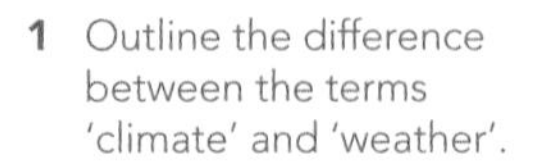

1 Outline the difference between the terms 'climate' and 'weather'.

2 Describe the variation in seasons between northern and southern Australia.

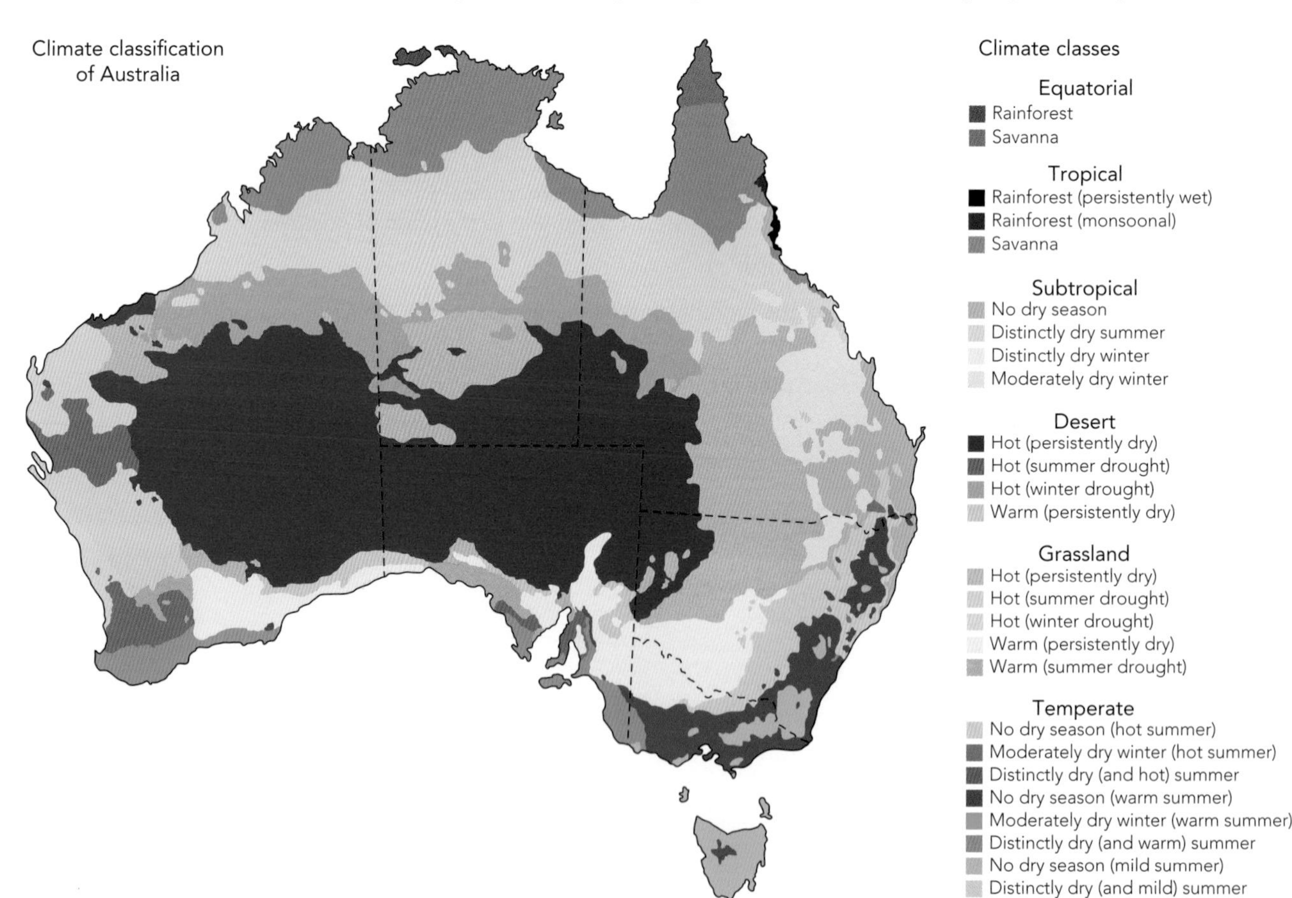

Figure 32.1 Australia's climatic regions

ISBN 9780170265560

It is the climate (daily and seasonal weather patterns over time) that determines which agricultural industries can be efficiently run in any area, although soils and topography also have important effects. Australia can be divided into six main climatic regions:

1 equatorial
2 tropical
3 subtropical
4 desert (arid and semi-arid)
5 grassland
6 temperate.

They are shown in Figure 32.1 along with their main characteristics.

3 Use Figure 32.1 to describe the different climatic zones across Australia.

Australia's variable climate

Climate is described mainly in terms of temperature and rainfall; however, wind velocity, humidity levels, day length and evaporation rates are also important. Refer to Chapter 14 of this book for extensive information on climate and its effect.

Temperature

An understanding of mean monthly temperatures and when either frosts or extremely high temperatures are likely to occur can be useful when deciding which crops or pasture plants are suitable for a locality. For example, wheat, oats and barley crops and pasture species, such as subterranean clover, are only slightly affected by frosts. Maize (corn), sugar cane, and legumes, such as glycine and lotononis, are killed by frosts and are mostly tropical or subtropical in their distribution.

Cattle can also be affected by heat, diverting energy to maintain body warmth in the cold or in the case of European breeds of cattle, suffering heat stress in periods of high temperature. Tropical cattle, such as the Santa Gertrudis, can maintain body temperatures in humid, hot environments. Figure 32.2 illustrates the average daily maximum temperatures for Australia and Figure 32.3, the average daily minimum temperatures. The trend indicated is that northern areas are warmer than southern areas. Topography also has an effect with higher elevations having lower temperatures. Figure 32.4 (page 446) indicates the mean temperature across Australia.

4 Using the information shown in Figures 32.2, 32.3 and 32.4, describe the temperature pattern across Australia.

5 Outline how this pattern could affect agricultural activity.

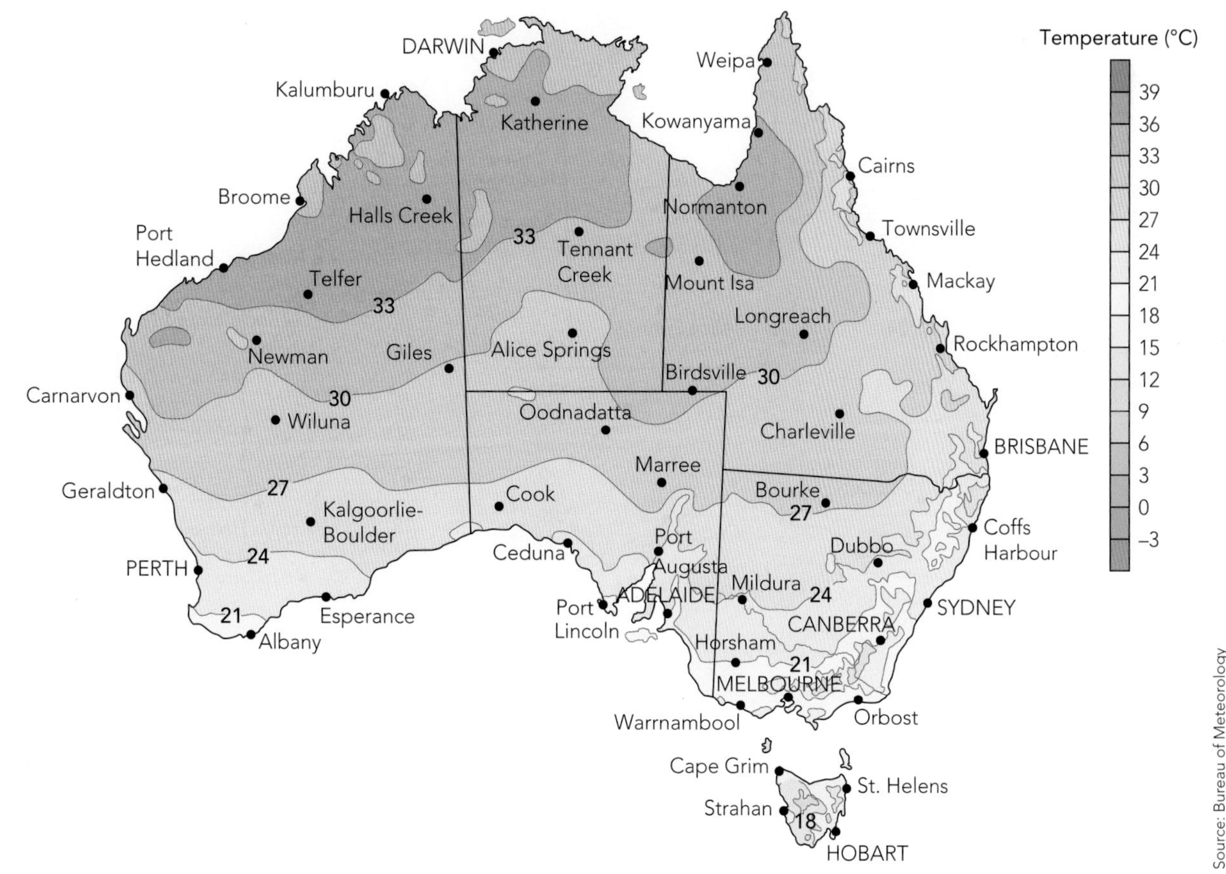

Figure 32.2 Average daily maximum temperatures across Australia

Figure 32.3 Average daily minimum temperatures across Australia

Figure 32.4 Average daily mean temperatures across Australia

ISBN 9780170265560

Table 32.1 Minimum and maximum temperatures (°C) and rainfall (mm) at specific locations

Mildura, Victoria

Temperature	Jan	Feb	Mar	Apr	May	Jun	Jul	Aug	Sep	Oct	Nov	Dec	Annual
Mean	32.3	31.6	28.3	23.6	19.1	16.0	15.4	17.3	20.5	23.9	27.5	30.1	23.8
Lowest	29.0	27.5	25.8	20.1	16.2	13.7	13.8	14.8	16.6	20.0	23.7	26.6	22.3
Highest	37.1	35.2	31.7	27.5	22.3	19.7	18.0	21.4	25.5	27.0	32.5	33.5	25.0
Rainfall	**Jan**	**Feb**	**Mar**	**Apr**	**May**	**Jun**	**Jul**	**Aug**	**Sep**	**Oct**	**Nov**	**Dec**	
Mean	21.4	23.1	20.6	18.4	25.1	22.3	26.0	25.7	26.8	29.8	25.8	25.8	

Broome, Western Australia

Temperature	Jan	Feb	Mar	Apr	May	Jun	Jul	Aug	Sep	Oct	Nov	Dec
Mean	33.3	32.9	33.9	34.3	31.5	29.1	28.8	30.3	31.8	32.9	33.6	33.9
Highest	35.5	35.9	36.1	37.0	34.3	32.0	31.3	32.1	34.2	35.4	35.5	36.3
Lowest	30.6	30.9	31.5	30.5	27.6	26.5	25.8	27.8	29.2	28.9	31.2	32.1
Rainfall	**Jan**	**Feb**	**Mar**	**Apr**	**May**	**Jun**	**Jul**	**Aug**	**Sep**	**Oct**	**Nov**	**Dec**
Mean	181.5	179.7	100.7	26.2	27.1	19.6	7.1	1.7	1.4	1.5	9.2	57.6

Camden, New South Wales

Temperature	Jan	Feb	Mar	Apr	May	Jun	Jul	Aug	Sep	Oct	Nov	Dec
Mean	29.5	28.5	26.8	23.8	20.5	17.7	17.2	19.0	21.9	24.1	26.1	28.4
Highest	32.5	32.0	29.9	26.4	22.3	19.2	18.9	22.0	25.5	28.0	30.1	32.4
Lowest	26.1	25.4	24.7	21.5	19.4	16.2	15.6	17.5	19.3	20.2	23.0	23.9
Rainfall	**Jan**	**Feb**	**Mar**	**Apr**	**May**	**Jun**	**Jul**	**Aug**	**Sep**	**Oct**	**Nov**	**Dec**
Mean	76.6	100.2	88.8	66.8	56.4	60.8	37.9	41.4	39.5	63.8	78.9	54.4

Mackay, Queensland

Temperature	Jan	Feb	Mar	Apr	May	Jun	Jul	Aug	Sep	Oct	Nov	Dec
Mean	30.1	30.0	29.2	27.6	25.1	23.0	22.5	23.6	25.7	28.3	29.4	30.7
Highest	31.8	31.6	30.8	28.8	26.5	24.5	23.8	26.3	28.1	30.5	31.5	33.8
Lowest	28.7	28.0	28.2	26.3	23.7	20.7	20.7	21.4	23.5	26.0	26.8	28.4
Rainfall	**Jan**	**Feb**	**Mar**	**Apr**	**May**	**Jun**	**Jul**	**Aug**	**Sep**	**Oct**	**Nov**	**Dec**
Highest daily	219.0	121.0	174.0	94.0	29.0	3.0	7.0	0.0	2.0	12.0	46.0	13.0
Monthly Total	359.0	586.0	417.0	309.0	124.0	12.0	15.0	0.0	5.0	12.0	147.0	26.0

Heat stress and milk production

Heat stress affects milk production in dairy cattle, as shown in Table 32.2.

Table 32.2 Heat stress and milk production

Maximum temperature for 6 hours (°C)	Dry matter intake (kg/day)	Milk yield (L/day)	Water intake (L/day)
25	17.7	25	71
30	16.9	23	76
35	16.5	17	115
40	10.2	12	102

6 Graph the mean temperatures for Mildura, Broome, Camden and Mackay.

7 Graph the rainfall for the above centres.

8 List the trends that the data indicates for these regions.

9 What trends emerge from the data in Table 32.2?

10 What could be occurring as temperatures exceed 35°C?

Extreme temperature data

Figure 32.5 illustrates the general trend of warming since 1910 to the early 1990s. Greater warming occurred during the second half of the 20th century, particularly for night temperatures. Increases in both annual mean minimums and in annual mean maximum temperatures have resulted in an increase in the area of Australia experiencing maximum temperatures and a decrease in the area experiencing below minimum temperatures. Put simply, this means that the maximum and minimum temperatures were higher than expected over more of Australia. For areas affected by frosts, the annual number of frost days and the length of the frost season have declined. However, this trend has weakened in 2010 to 2011 with temperatures actually cooling due to the effects of sea surface temperatures.

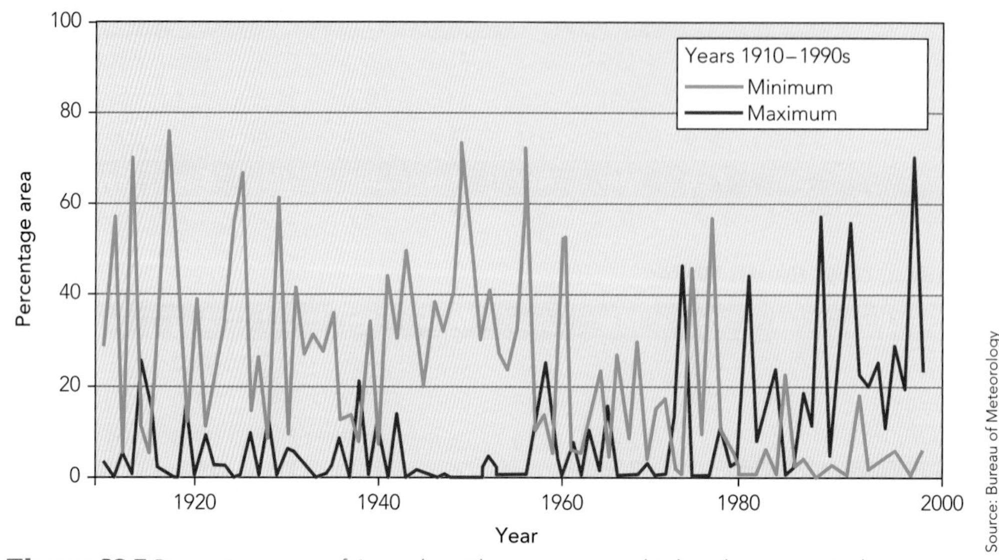

Figure 32.5 Percentage area of Australia with temperatures higher than expected means 1910–1990s

Rainfall

Average annual rainfall is a broad measure of the amount of rain an area will receive. Figure 32.6 illustrates the average annual rainfall across Australia. Average annual rainfall is calculated by adding rainfall totals over a specified period and dividing by the number of years. The map indicates that large areas of Australia are dry.

However, farmers also need to know the seasonal distribution of rainfall and how variable this might be. Based on seasonal rainfall patterns in the north of Australia, crops are required that grow in the wet summer months, but grow very little, if at all, in the dry season. In southern Australia crops of importance would grow in the late autumn, winter and spring months when water is available for growth. Figure 32.7 illustrates the percentage of the average annual rainfall that falls in January and Figure 32.8 (page 450) the percentage of the average annual rainfall that falls in July. The majority of rain for northern Australian falls during the wet season (October to April), while in southern Australia most rain falls in winter (June to August). In Australia most rain falls on the coastal areas and rainfall reduces quickly as you move inland.

11 Compare the pattern of rainfall across Australia for the January and July months.

12 Look at Figure 32.9 (page 451) and determine the areas of Australia that have the most reliable rainfall.

13 Look at Figure 32.9 (page 451) and determine the areas of Australia with the least reliable rainfall.

14 How do these patterns affect agriculture across the country?

ISBN 9780170265560

Figure 32.6 Average annual rainfall across Australia

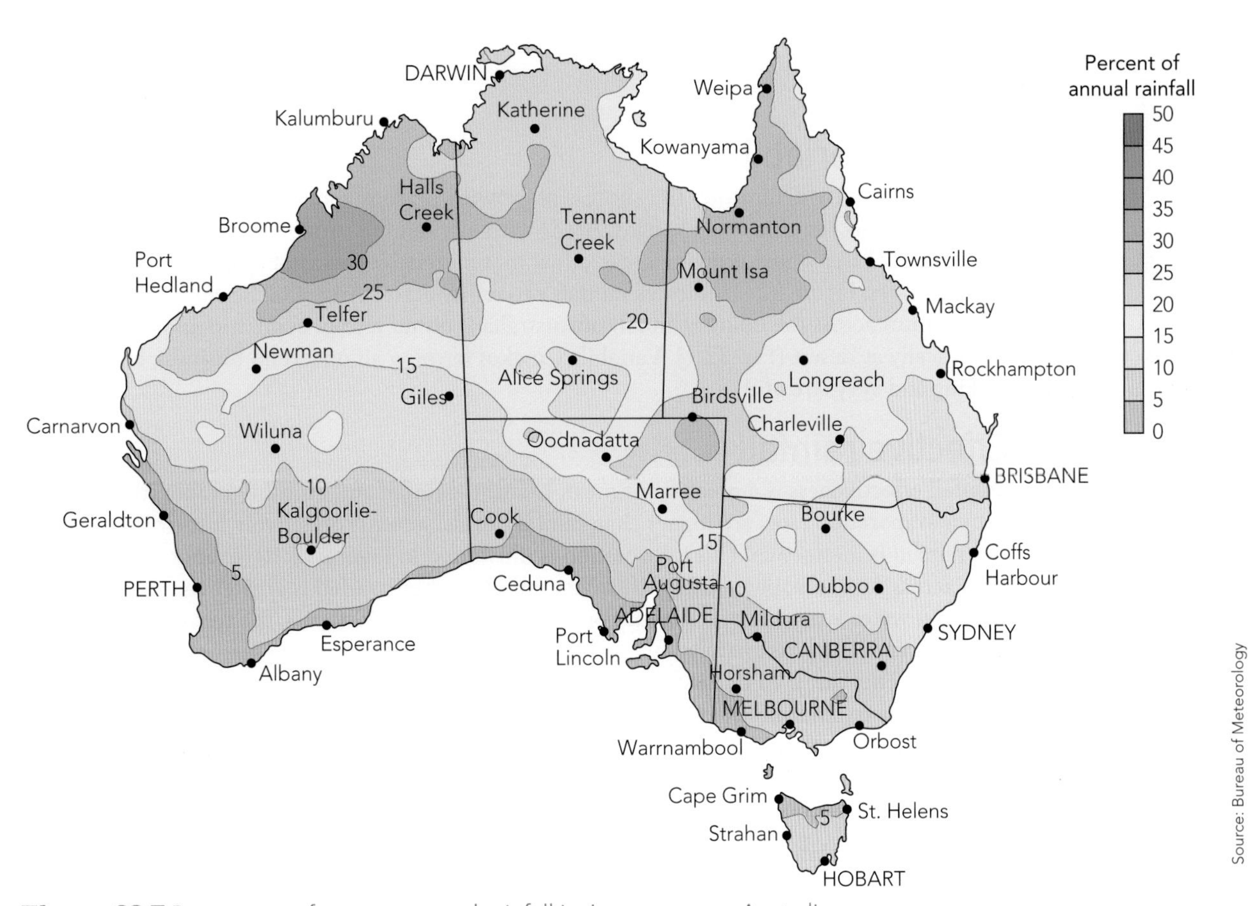

Figure 32.7 Percentage of average annual rainfall in January across Australia

ISBN 9780170265560

Figure 32.8 Percentage of average annual rainfall in July across Australia

Rainfall variability

Rainfall variability determines the success of growing seasons and, consequently, how quickly animals can gain body weight. Areas with low rainfall variability, which correspond with much of the wheat growing areas, allow more intensive forms of cropping and higher stocking rates. Figure 32.9 indicates rainfall variability across Australia. Because rainfall is not uniform across the country, much of Australia is either arid or semi-arid and classified climatically as desert (Fig. 32.1). Variability makes farming an uncertain business, often loss making and very stressful.

Effective rainfall

Rainfall effectiveness is a measure of mean monthly rainfall and monthly evaporation rates. When evaporation rates are higher than rainfall levels, water is effectively lost from the soil, and rainfall is ineffective. This information is illustrated in Figure 32.10 (evapotranspiration rates across Australia, page 452). Effective rainfall is when rainfall is greater than evaporation rates and water is available to plants. For the months that this occurs, the term 'growing season' is used. Approximately 75% to 80% of Australia has a growing season of less than 5 months. This vast area of Australia has a cover of native vegetation, or is used by the grazing industry. Merino sheep and beef cattle enterprises dominate the agricultural systems in this area where crops cannot be grown. The cereal growing, wool and prime lamb areas of Australia have a growing season of 5–9 months but are subject to droughts. Intensive agricultural production systems dominate the coastal areas where there is a growing season

ISBN 9780170265560

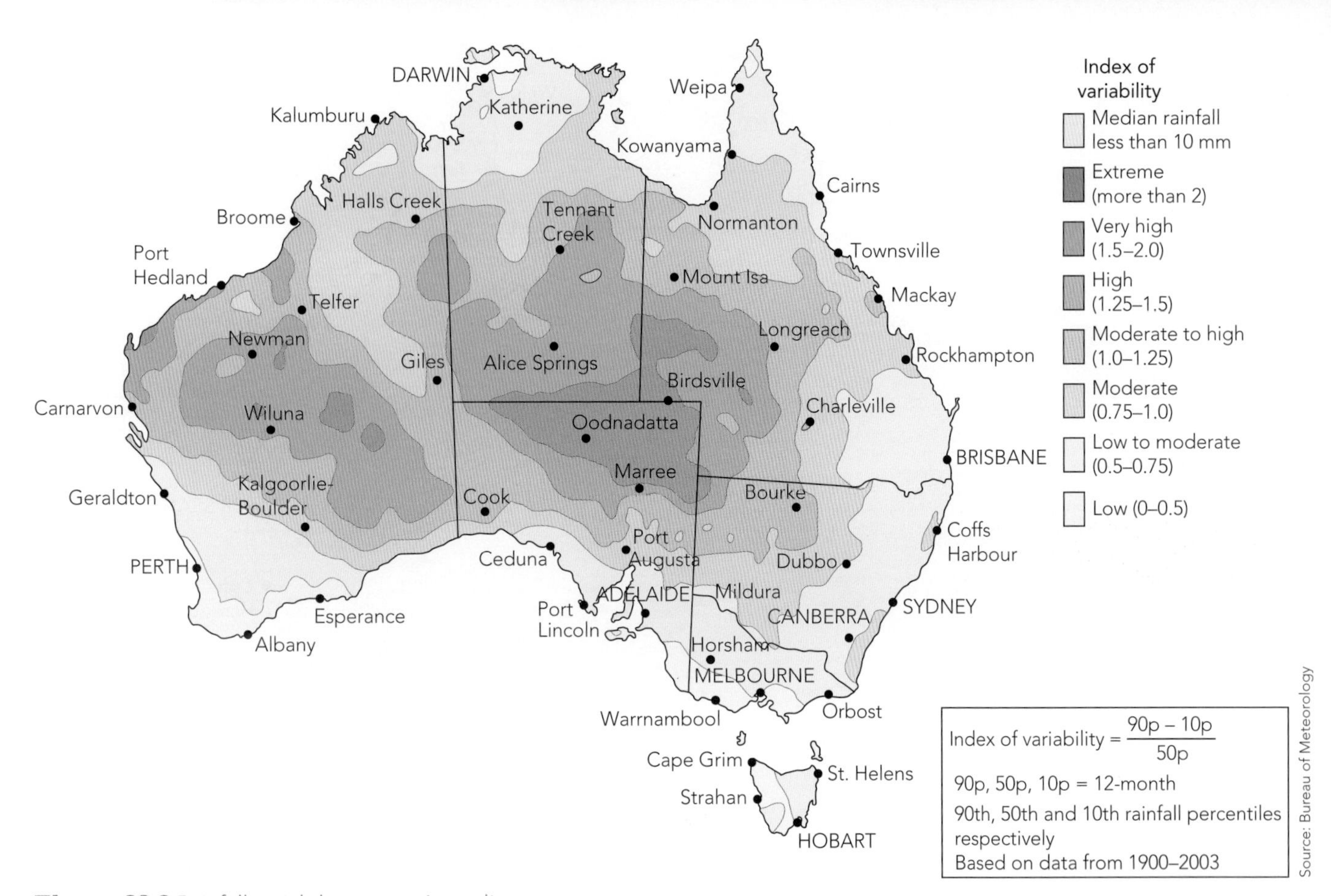

Figure 32.9 Rainfall variability across Australia

of 9–12 months. Effective rainfall areas are largely in the south of Australia where rainfall occurs over the cooler winter months whereas in the north of Australia, rainfall occurs during the hot months and high rates of evaporation make much of this rainfall ineffective.

The most important factor of climate in Australia is the number of consecutive months in a year that water is available for the growth of crops.

In 2010–11, irrigated agriculture used less than 1% of agricultural land in Australia but made up nearly 30% of the gross value of agricultural production. The major irrigated industries, by value, are vegetables, fruit (excluding grapes) and dairy (Australian Bureau of Statistics, 2012).

15 Outline what effective rainfall means.

16 Refer to Figures 32.6 and 32.10 to determine the areas of the country that receive effective rainfall annually.

Extreme rainfall events

Extreme rainfall events have become more common in Australia during the 20th century. Increasing trends in extreme rainfall (95th percentile) and total rainfall have occurred for the period November to April. For the months May to October, increases in extreme rainfall have also occurred. This has led to an increase in the frequency of floods followed by droughts and has affected stream flows. Figure 32.11 (page 452) illustrates the percentage change in extreme daily rainfall in the period 1910–1995.

Figure 32.10 Evapotranspiration rates across Australia

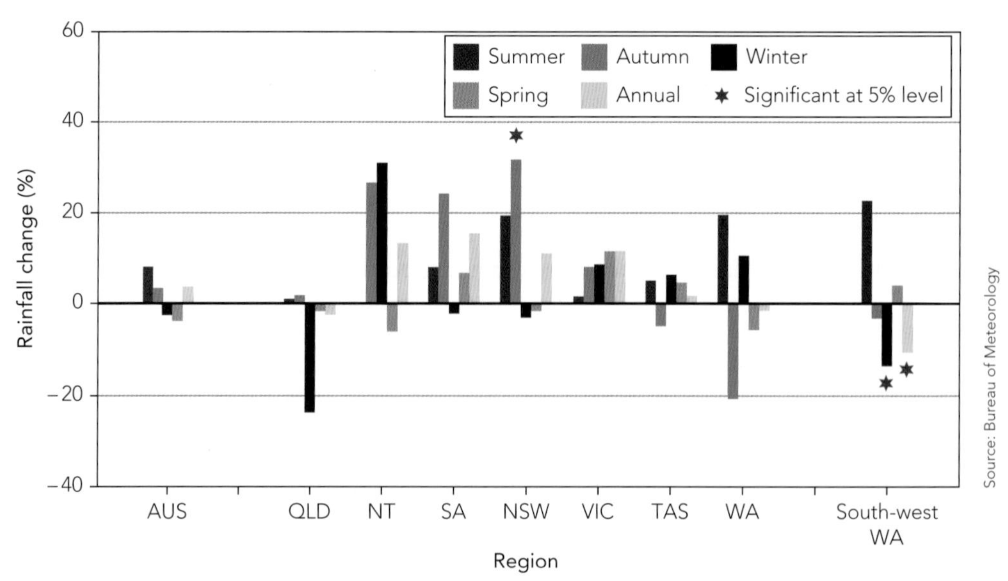

Figure 32.11 Extreme rainfall events

ISBN 9780170265560

Agricultural production zones

Figure 32.12 indicates the major broadacre areas in Australia, determined not only by climate (principally rainfall and its effectiveness) but also soil type and topography.

17 Discuss the importance of climatic factors in the location and extent of the agricultural production zones across Australia, as illustrated in Figure 32.12.

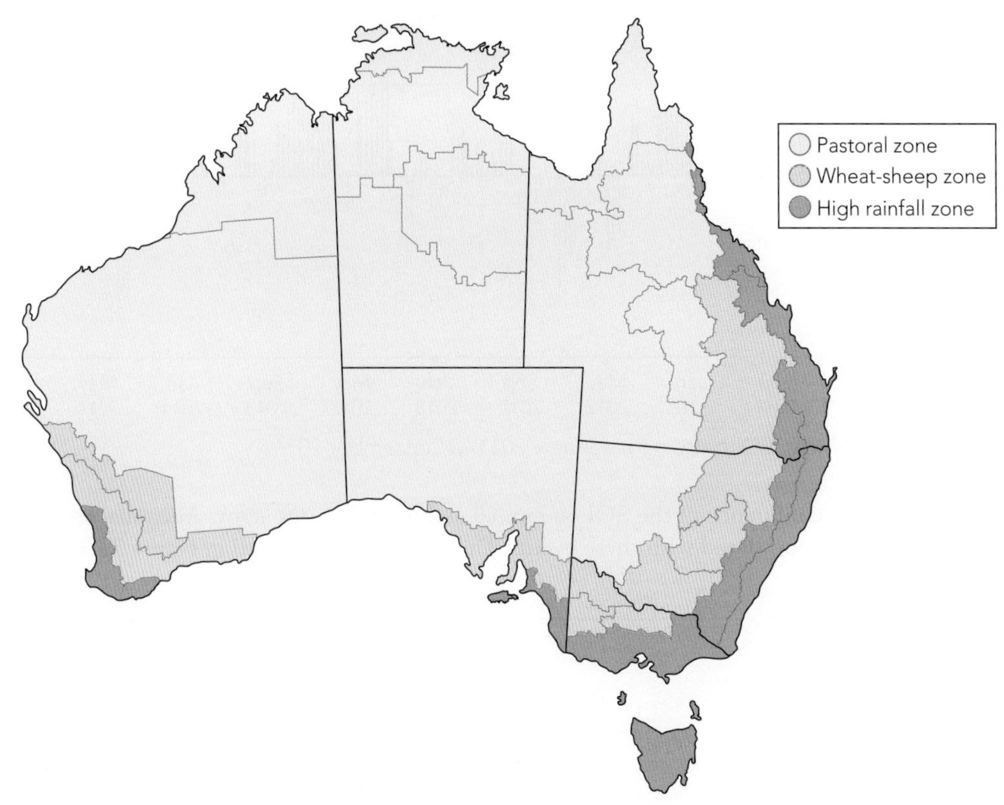

Figure 32.12 Agricultural production zones across Australia

connect

Climate change

Extensive information on the Australian climate, and climate variability and its causes

connect

Future climate

Extensive information on climate, the greenhouse effects, and the impact and management of climate change

Southern Oscillation Index

The **Southern Oscillation Index (SOI)** is calculated from the monthly or seasonal fluctuations in the air pressure difference between Tahiti and Darwin. A strongly and consistently positive SOI pattern (e.g. consistently above +6 over a 2-month period) is related to a high probability of above the long-term average (median) rainfall for many areas of Australia (areas of eastern Australia become wetter than normal). This is known as **La Niña**. La Niña refers to the extensive cooling of the central and eastern tropical Pacific Ocean, often accompanied by warmer than normal sea surface temperatures (SST) in the western Pacific, and to the north of Australia. La Niña events have been correlated with higher numbers of tropical cyclones during the cyclone season (November to April).

Conversely, a 'deep' and consistently negative SOI pattern (less than approximately −6 over a 2-month period, with little change over that period) is related to a high probability of below median rainfall (drier conditions) for many areas of Australia (particularly eastern Australia) at certain times of the year – this is known as **El Niño**. This period is marked by the development of a warm ocean current off the South American coast and a major shift in weather patterns across the Pacific with cooler than normal surface sea temperatures across the north of Australia.

However, it is important to remember that the pattern of relationship between SOI and rainfall (and temperature) can vary depending on the particular season and region.

connect

The Southern Oscillation Index

ISBN 9780170265560

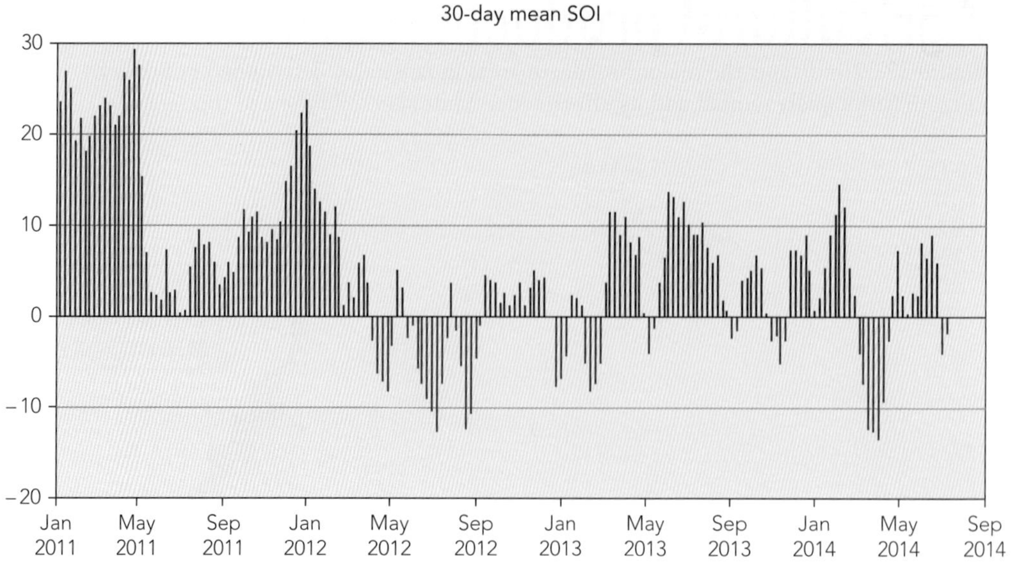

Figure 32.13 Mean SOI pattern for January 2011 to September 2014

SOI graph is provided courtesy of www.weatherzone.com.au and is

connect

Model for the development of La Niña and El Niño events

Because of the influence of the SOI on rainfall and temperatures across Australia, crop yields are directly correlated with these events. This pattern is illustrated in Figure 32.14.

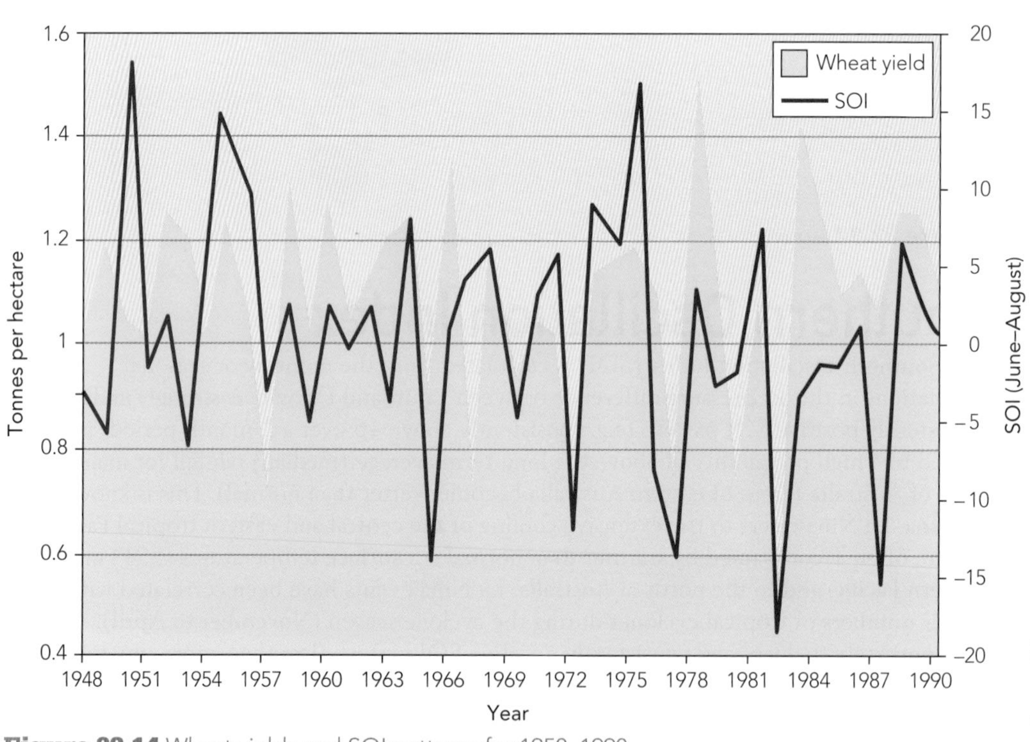

Figure 32.14 Wheat yields and SOI patterns for 1950–1990

Soruce: ABS

connect

Measures of climate

connect

Southern Annular Mode

Southern Annular Mode

Southern annular mode (SAM) also affects Australian weather. When SAM is in a positive phase there are weaker westerly winds in southern Australia and stronger westerly winds at higher altitudes. High-pressure systems develop across southern Australia. This affects the variability of winter rainfall in south-west Australia and of winter, spring and summer rainfall in south-east Australia.

ISBN 9780170265560

In general, a positive phase relates to stable, dry conditions. In a negative phase, the band of westerly winds across the globe expands towards the equator, and more or stronger low-pressure systems develop across southern Australia, which can lead to increased rainfall.

Many other measures exist to determine what is happening to our climate – the SOI and SAM are just two.

18 What conclusions can you draw from the information in Figures 32.13 and 32.14?

19 From the working model, describe how La Niña and El Niño events occur and what effect these events have on Australian weather.

Evidence of climate change

Over time Earth's climate has changed greatly, from extreme cold to extreme heat and dry conditions. Records of these changes are being found in the analysis of ice cores from the Antarctic ice sheet and Greenland. Many changes have been widespread, extreme and fairly rapid. Further information on data obtained from ice cores can be found at the Ice core evidence, Ice cores reveal climate change and Temperature websites.

Over time other events can also be used to indicate the changing nature of the world's climate over time. These events include:

- variation in the distribution of vegetation and wildlife around the world over time (e.g. extensive remains of tropical rainforests under the Antarctic ice cap).
- variations in the thickness of tree rings, reflecting harsh or favourable times for growth
- the movement of human populations displaced by climatic events in history
- recorded changes in sea level
- fossil and pollen evidence, indicating the presence or absence of life forms in areas as climate changed.

Extensive work by the CSIRO has also gathered evidence indicating that, currently, climate patterns are changing. The evidence includes the following facts.

- Australia's mean surface air temperature has warmed by 0.9°C since 1910.
- Seven of the 10 warmest years on record have occurred since 1998.
- Over the past 15 years, the frequency of very warm months has increased five-fold and the frequency of very cool months has declined by approximately a third, compared with 1951–1980.
- Sea-surface temperatures in the Australian region have warmed by 0.9°C since 1900.
- Rainfall averaged across Australia has slightly increased since 1900, with a large increase in north-west Australia since 1970.
- A declining trend in winter rainfall persists in south-west Australia.
- Autumn and early winter rainfall has mostly been below average in the south-east since 1990.
- Extreme fire weather has increased, and the fire season has lengthened across large parts of Australia since the 1970s.
- Global mean temperature has risen by 0.85°C from 1880 to 2012.
- The amount of heat stored in the oceans has increased, and global mean sea level has risen by 225 mm from 1880 to 2012.
- Annual average global atmospheric carbon dioxide concentrations reached 395 parts per million (ppm) in 2013 and concentrations of the other major greenhouse gases are at their highest levels for at least 800 000 years.
- Australian temperatures are projected to continue to increase, with more extremely hot days and fewer extremely cool days.
- Average rainfall in southern Australia is projected to decrease, and heavy rainfall is projected to increase over most parts of northern Australia.
- Sea-level rise and ocean acidification are projected to continue.

connect

Ice core evidence

connect

Ice cores reveal climate change

connect

Temperature

An outline of the CSIRO's work gathering evidence of changes to current climate patterns

20 Outline the evidence that climate change has always been occurring.

21 Outline the evidence for recent climate change.

22 Compare Figures 32.15a and b with Figure 32.16a and b. What conclusions can you make from the information shown on the graphs for 2013 compared with 2007?

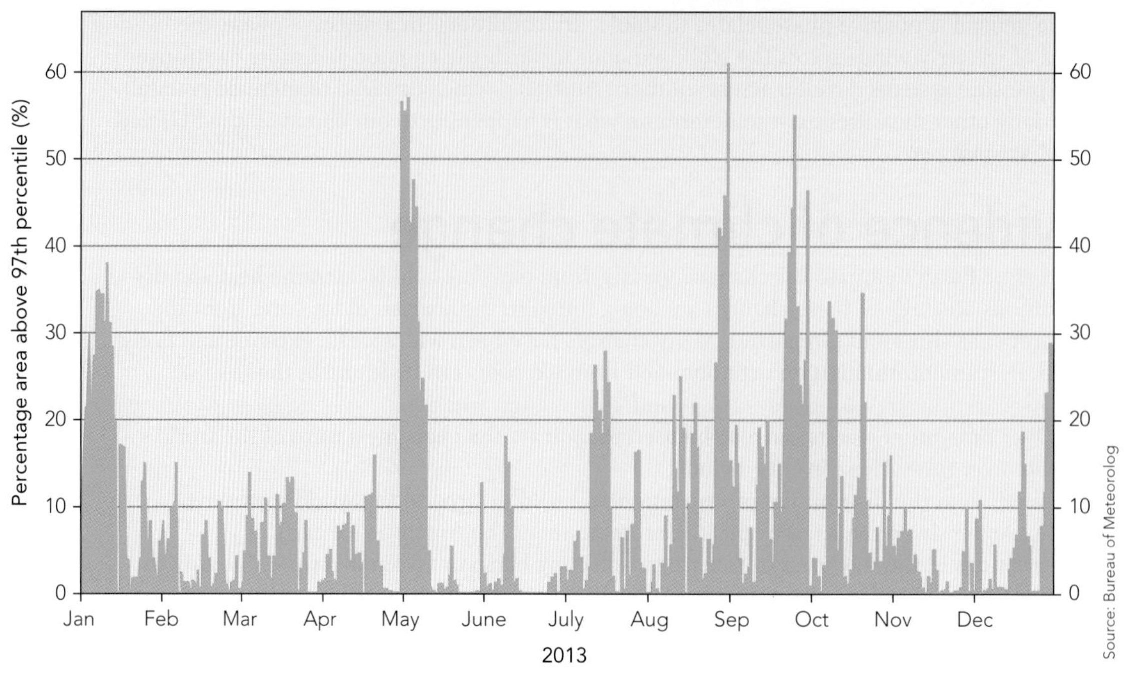

Figure 32.15 a Daily extreme maximum temperature in Australia – 2013

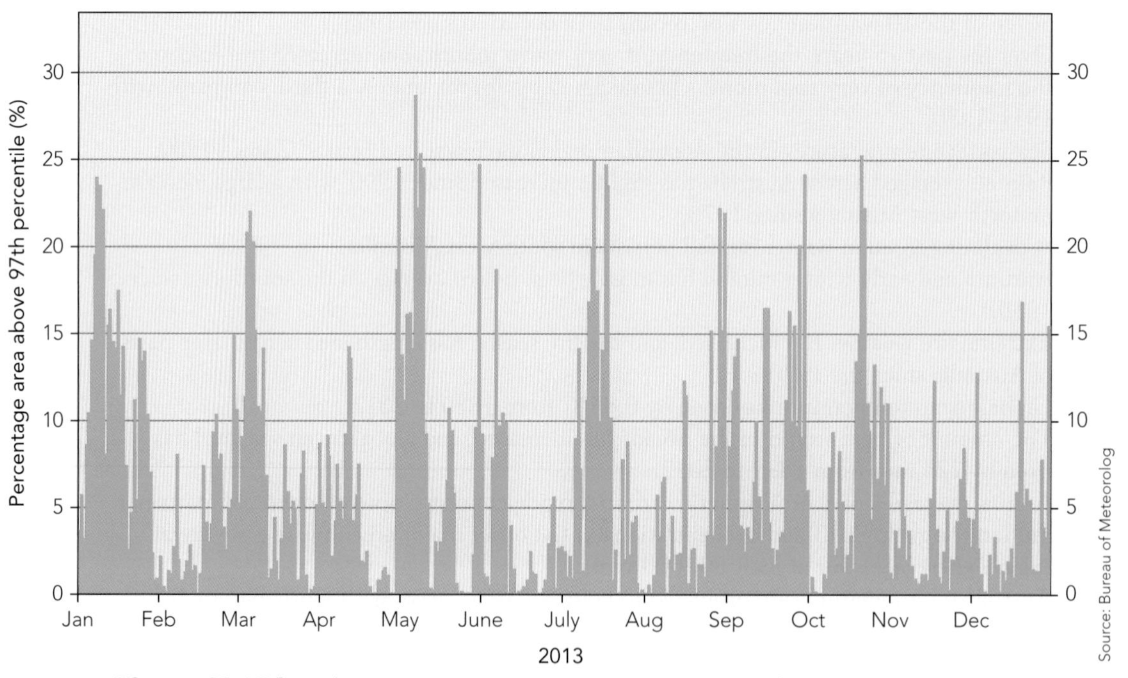

Figure 32.15 b Daily extreme minimum temperature in Australia – 2013

ISBN 9780170265560

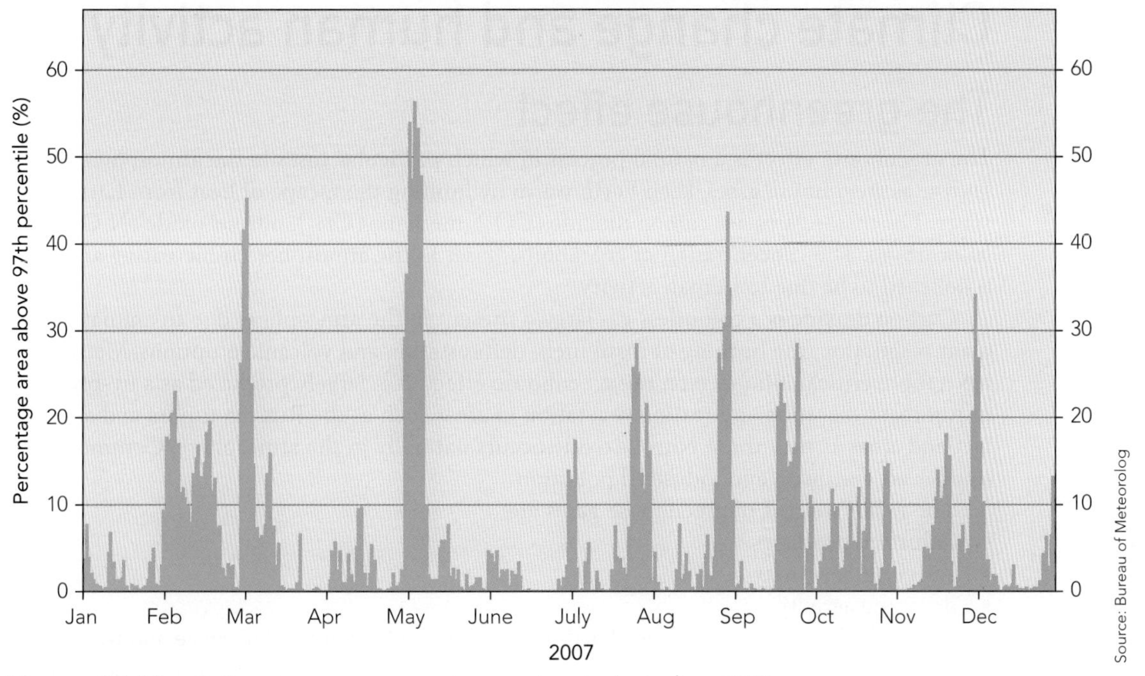

Figure 32.16 a Daily extreme maximum temperature in Australia – 2007

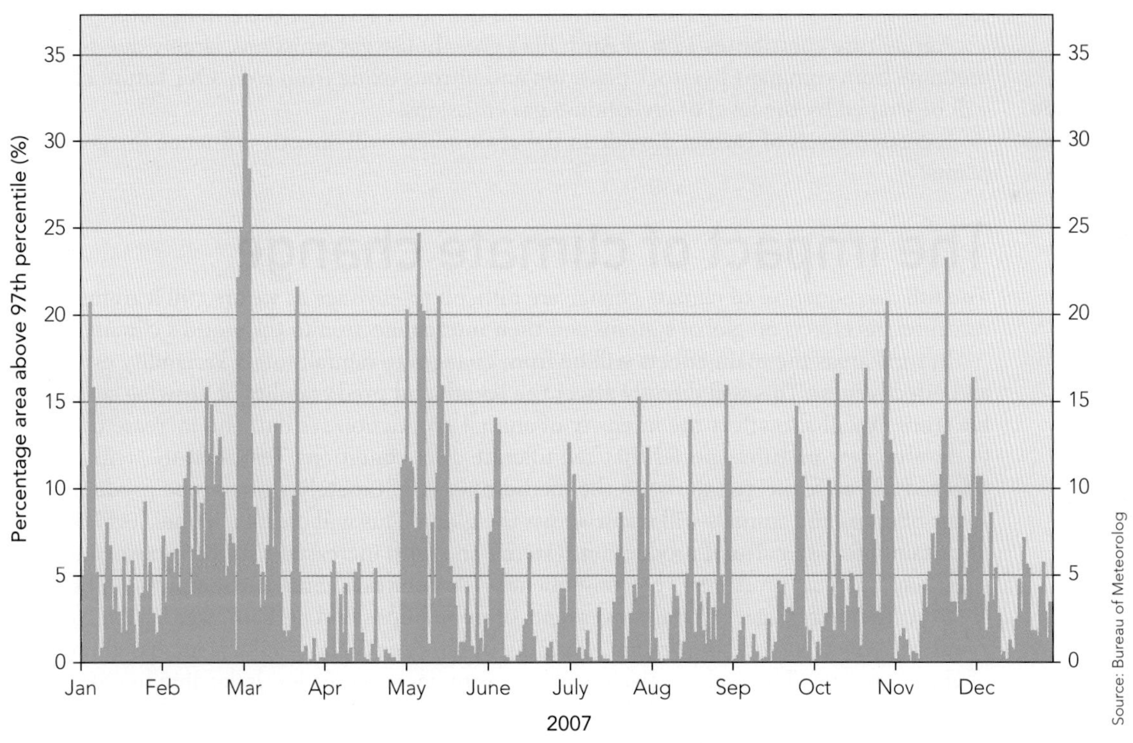

Figure 32.16 b Daily extreme minimum temperature in Australia – 2007

Climate change and human activity

The greenhouse effect

The greenhouse effect is a natural process in which particular gases in the atmosphere, known as greenhouse gases, keep Earth warm by limiting the escape of heat from Earth into space. The major gases are carbon dioxide (CO_2), methane (CH_4), nitrous oxide (N_2O) and halocarbons. The concentration of greenhouse gases in the atmosphere is increasing and this is believed to be due to human activity.

Carbon dioxide is a colourless gas largely present in the atmosphere due to animal and plant respiration, the burning of fossil fuels, deforestation and volcanic eruptions. Methane has a more critical influence on the greenhouse effect. It is largely produced as a by-product of ruminant digestion and rotting vegetation in areas such as the Russian tundra and similar wetland areas as they thaw. Nitrous oxide occurs naturally in the atmosphere. Common sources are fertilisers and industrial pollution.

The major sources of greenhouse gases are:

- industrial processes
- agricultural activity
- domestic activity.

When the Sun's energy reaches Earth some of it is reflected back to space and the rest is absorbed. The absorbed energy warms the atmosphere and Earth's surface, both of which then emit heat back toward space as long-wave radiation (refer to Chapter 14). This outgoing long-wave radiation is partially trapped in the atmosphere by greenhouse gases, such as carbon dioxide, methane and water vapour, which then radiate the energy in all directions, further warming Earth's surface and the lower levels of the atmosphere.

Agriculture contributes 12% of Australia's greenhouse gas emissions, largely due to methane from ruminant livestock digestion and nitrous oxide from soils. Our future climate will be shaped by the level of greenhouse gas emissions.

Increased levels of carbon dioxide in the atmosphere will directly influence the growth of plants.

connect

What is the greenhouse effect?

This video illustrates the cause and outcomes of the greenhouse effect.

connect

Understanding the science

Information on how the greenhouse effect occurs and the various gases that contribute to it

connect

Future climate

Possible scenarios for climate change in the future

connect

Levels of carbon dioxide

View the video and answer the associated questions.

23 Explain the greenhouse effect.

24 Outline the role of the gases responsible for increasing the greenhouse effect.

25 Explain some of the possible scenarios arising from the greenhouse effect.

The impact of climate change

The full consequences of climate change are difficult to envisage as we are still learning about its causes, its effects on global systems and their interconnection to the world's climate.

In rural areas the main effects will be from changes to rainfall totals, variability estimates and effectiveness. This will directly affect the length and predictability of growing seasons for crops. It will also affect the intensity of animal production systems and increase the need to develop new pastures species that can tolerate drier conditions. Temperatures will also be affected and are expected to rise in the medium term. Ultimately, shifts in the production zones of crops and animals will occur across the globe. This will create a higher level of uncertainty for agricultural production systems, and with increased risk comes associated stress and financial pressures. The consequences for poor rural-based economies will be the greatest. This may trigger global movements of populations and perhaps, wars.

Other effects include the gradual warming of the oceans and resultant increased rate of glacier melting. The melting glaciers will, in turn, increase sea levels, leading to coastal flooding of low areas; many small Pacific islands will be completely submerged.

connect

Agriculture

More information about the impact of climate change on Australian farms

26 List the impact of climate change on one of the following, using information on the Department of Primary Industry website listed above: horticulture, pastoral industries, cropping or irrigated crops.

Managing climate change

Carbon trading

Agricultural industries contribute to the production of greenhouse gases. Examples of farming practices that lead to this situation include land clearing, stubble burning, grain harvesting and intensive livestock enterprises. The major greenhouse gases involved are carbon dioxide, methane and nitrous oxide.

ISBN 9780170265560

The pricing of carbon was part of an Australian Government incentive scheme called the Clean Energy Plan, which aimed to reduce greenhouse gas emissions in Australia by 5% below year 2000 levels by 2020. However, the Liberal government that was elected in 2013 abolished the carbon tax in favour of a 'Direct Action Plan' for reduction of carbon emissions. The *Carbon Farming Initiative Amendment Bill 2014* will expand the *Carbon Credits (Carbon Farming Initiative) Act 2011* to allow the Clean Energy Regulator to conduct auctions and enter into contracts to purchase emissions reductions. It would enable a broader range of emissions reduction projects to be approved and amend the project eligibility criteria and processes for approving projects and crediting **carbon credit** units.

Another method of carbon emissions trading is 'cap and trade' – a market-based approach to control emissions. A limit, or cap, is set on the amount of pollutants emitted. This limit is allocated to companies in the form of emission permits or carbon credits. Companies must purchase and hold enough permits to cover their carbon-based emissions. Should more permits be required, companies must trade with other, more-efficient companies to buy permits. Hence, heavy polluting industries in relation to carbon dioxide production must buy additional permits. High costs associated with high levels of pollution should force industries to lower pollution levels.

Australia generates 1.5% of global emissions but on a per capita basis is a very high emitter. Australian primary industries have reduced greenhouse gas emissions by 40% in the past 10 years.

Carbon sequestration, which refers to processes and practices that remove and trap carbon dioxide from the atmosphere, are widely reflected in many farming practices today. It has been found that 94% of Australian farms actively undertake some form of natural resource management, such as:

- increasing biodiversity by planting native plant species, which also act as carbon sinks
- increasing total vegetative cover
- increasing agroforestry activity
- protecting creek banks through revegetation practices.

These changes also provide shelter for livestock, protect waterways and represent a source of financial return based on a **carbon trading** scheme. Farmers gain carbon offsets, which provide a financial return for land set aside for plantings to capture carbon dioxide.

There has also been an increase of 70% in the number of farmers using direct drill and minimum tillage practices in the last 30 years. Many farms are also adopting zero tillage practices. These changes in whole farm management and environmental management have assisted in the reduction of greenhouse gas emissions from agricultural industries.

Research is being undertaken on how large areas of forests in different regions of Australia – agroforestry sinks composed of fast-growing commercial trees, such as eucalyptus and plantation pines – can reduce carbon dioxide levels in the atmosphere. Companies that produce high levels of carbon dioxide can finance the development of these agroforest areas, thereby trading their carbon production for carbon credits and assisting Australia to meet the lower emissions of carbon dioxide required under international agreements. There are a number of other agricultural activities that can be regarded as providing carbon sinks in nature. These include Landcare revegetation projects and pasture improvement, especially using perennial pastures, such as lucerne and saltbush, as well as soil structure improvement.

In addition, the leading international body for the assessment of climate change, the Intergovernmental Panel on Climate Change (IPCC) has made a number of recommendations, including that there should be:

- reduction of deforestation
- better crop and livestock management systems
- efforts to increase the organic matter in soils
- improved biodiversity
- better water management policies
- reduced loss of food from pests, especially in the storage period
- investigation of the use of biofuels, with consideration to the type of crops and the extent of land used to grow them.

27 Outline the concept behind carbon trading.

28 Describe the concept of carbon sequestration.

connect

Carbon capture and storage

For more information on carbon capture and methods of reducing greenhouse gas emissions on farms

connect

Climate change and productivity

For more information on various strategies used in adapting to climate change

connect

Farming and climate adaptation

Excellent information from the CSIRO on farm adaptation to climate change

29 List various strategies to allow farmers to adapt to climate change.

ISBN 9780170265560

Carbon emissions and livestock production

On-farm greenhouse gas emissions can be reduced in livestock production systems by farmers adopting the following management practices:

- improving the quality of feed available to animals
- ensuring breeding stock are managed according to their nutritional needs
- improved genetic selection programs to increase growth rates and reduce finishing times
- managing livestock waste.

connect

Livestock emissions

Water management

In Chapter 15 we discussed aspects of water management, which is essential for efficient production in both plant and animal enterprises. If water is extracted from underground sources or surface sources faster than it can be replenished from the water cycle then a serious imbalance occurs. Stream flows are affected, overuse of water can affect salt or pH levels in the soil and economies downstream from the extraction area are affected as farmers struggle to supply their crops and animals with water.

In 2012–13:

- 43% of Australia's agricultural water (5.1 million megalitres) came from irrigation channels
- 25% (2.9 million megalitres) was sourced from rivers, creeks and lakes
- groundwater made up 16% (1.9 million megalitres)
- on-farm dams and tanks accounted for 15% of agricultural water (1.8 million megalitres)
- nationally, the volume taken from irrigation channels rose by 50% to 5.1 million megalitres
- water supplied by irrigation channels was the major source of water for agriculture in the Murray–Darling Basin, accounting for 49% of agricultural water sourced in the region (4.2 million megalitres)
- New South Wales was the biggest user of water for agriculture, with 5.2 million megalitres (or 44% of Australia's agricultural water use), followed by Queensland and Victoria, using 2.6 million megalitres each
- outside of the Murray–Darling Basin, groundwater was the major source of water for agriculture, at 35% (1.2 million megalitres)
- there was an increase of 32% over the 2011–12 period in the volume of water used for agriculture.

Source: Australian Bureau of Statistics (2014), *Sources of Agricultural Water*, cat. no. 4618.0

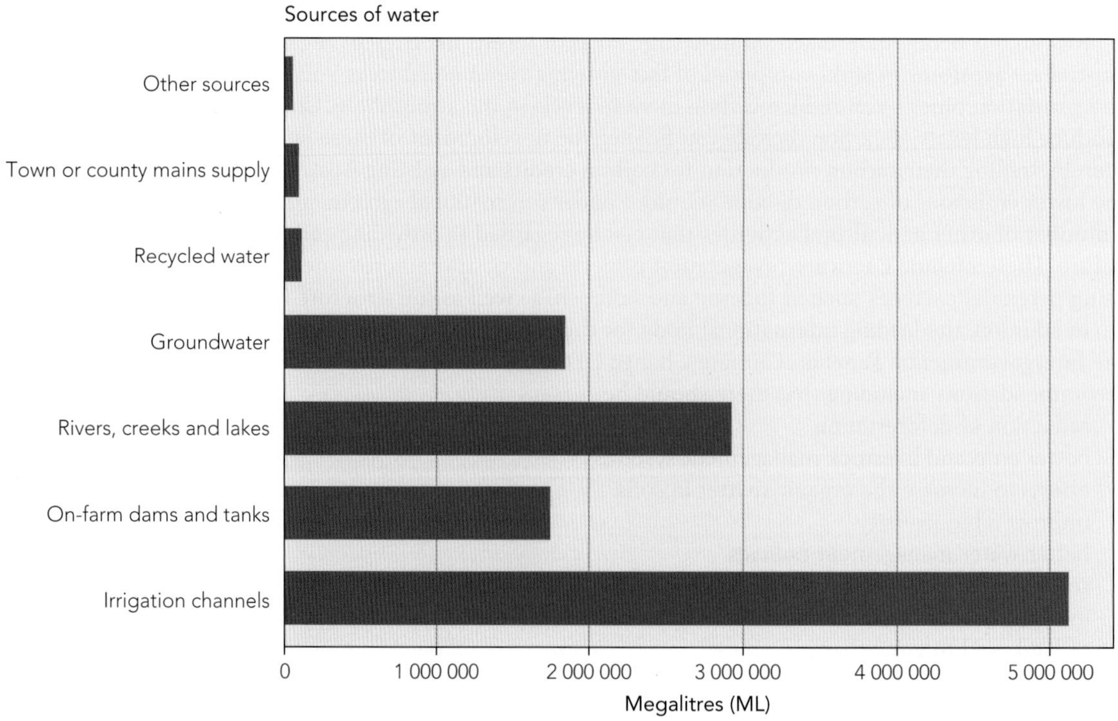

connect

Water use on Australian farms

View ABS statistics for 2012–13

Sources of water

Other sources
Town or county mains supply
Recycled water
Groundwater
Rivers, creeks and lakes
On-farm dams and tanks
Irrigation channels

0 1 000 000 2 000 000 3 000 000 4 000 000 5 000 000

Megalitres (ML)

Figure 32.17 The sources of water used for agricultural purposes – 2012–13

ISBN 9780170265560

Water trading

Figure 32.18 illustrates the proportion of agricultural water entitlements used by the various sectors where the farms are wholly Australian owned.

Water trading can either be on a lease basis or through a permanent trading of water access entitlements. Water licences are an example of water entitlements and these allow farmers to a share of the available water from nearby water sources. Water allocations allow the farmer to use a given volume of water in a particular season. Water allocations will vary from season to season.

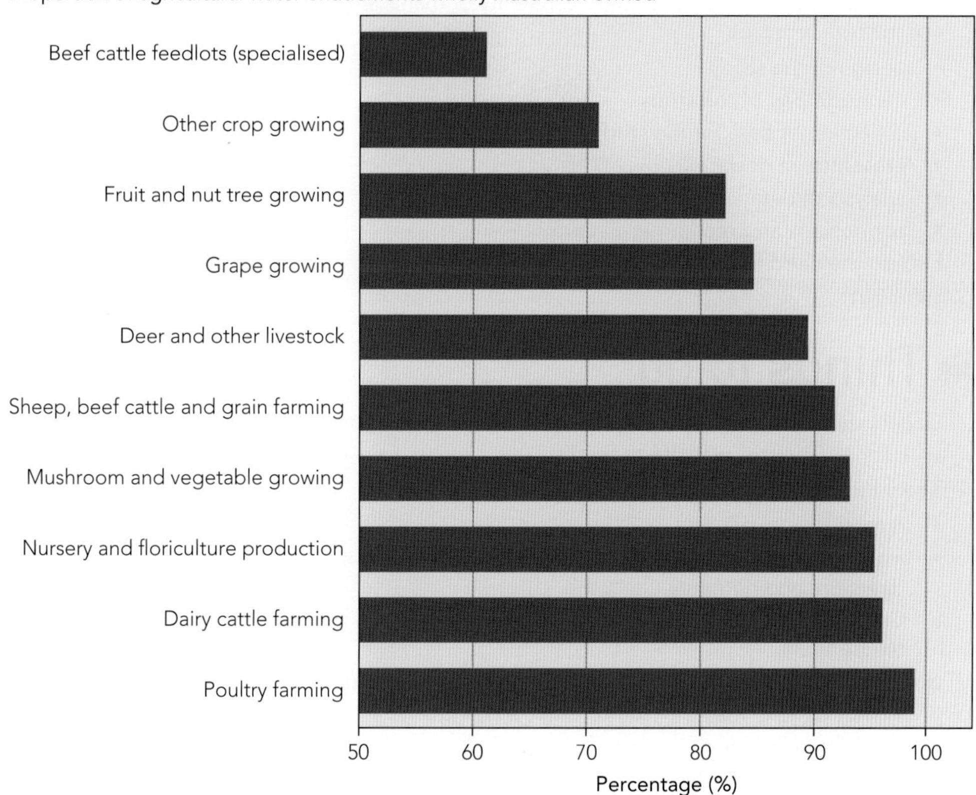

Figure 32.18 Australian water entitlements

Chapter review

Mandatory analysis of a research study for NSW HSC Agriculture course

Students choose only **one** of the research study electives given in Chapters 31, 32 and 33.

Research into climate variability

a Analyse a research study of climate variability or management strategies related to climate variability in terms of:
- design of the study
- methodology of the study
- collection of data for the study
- presentation of data
- analysis of data
- conclusions and recommendations.

b Explain the need for research in climate variability or management strategies for climate variability.

Things to do

connect

Climate information

1 Collect climate information about your local area and another contrasting area. Use the Climate information website to help you.

The information should include:
- rainfall
- maximum/minimum temperatures, and evaporation rates, frost days, wind speed and direction, and humidity.

a Calculate mean and standard deviation of rainfall and maximum and minimum temperature.

b Construct graphs and annotate the graphs with information about:
- hot or cold years
- dry or wet years
- drought or flood events
- 'normal' years.

2 Examine a map of agricultural distribution and compare the agricultural production of different areas (e.g. areas suitable for wheat and barley production).

Compare similar production enterprises in contrasting areas; for example, sowing time of wheat, time of lambing, harvest and sale of product. Use the diagrams showing the Australian climate in this chapter to assist you.

connect

Koshland Science Museum

3 Explore some simulation games. Visit the Koshland Science Museum website to investigate the various questions posed.

Things to find out

1 Outline the effect of nitrogen fertilisers on plant growth. Describe the losses of nitrogen from soils by denitrifying bacteria and outline which gases are released into the atmosphere.

2 Describe the processes causing the climate events of El Niño and La Niña.

3 Describe methods farmers can use to reduce methane emissions from ruminant livestock.

4 a Explain why research is important in assessing the effect of changes in temperature or the impact of a changing climate on farming in Australia.

b Identify and comment on the role of the various organisations involved in the research, such as the CSIRO, government, university, private companies and Meat and Livestock Australia.

5 Compare the methods available for farmers to store water on individual farms with the methods available for communities to store water.

 ISBN 9780170265560

6. Review and describe the water storage facilities in your local area (e.g. dam, river, aquifer).
7. Explain the implications of climate variability for agricultural production.
8. Extrapolate from climate variability data to determine the effects of climate change on production.
9. Evaluate national legislation on vegetation clearing in Australia and the impact of vegetation clearing on climate and on farm management systems.

Extended response questions

1. Explain how nitrogen fertiliser and intensive ruminant animal production systems contribute to greenhouse gas emissions.
2. Discuss locally available water trading systems.
3. Review the current Australian position on an approach to managing greenhouse gas production (e.g. carbon trading scheme).
4. Analyse issues related to water storage and trading, including riverflows, aquifer depletion and enterprise change.
5. Select an appropriate research study on climate variability or management strategies related to climate variability.
   - Comment on aims of the research study and the hypothesis to be tested.
   - Analyse the role of control, treatment, randomisation, replication and standardisation.
   - Discuss the effectiveness of the methodology.
   - Analyse the design validity, appropriateness, ethics and suitability of the statistical analysis.
   - Discuss alternative methods available for conducting research.
   - Analyse the data collection, for example, timeframe, breadth of information, use of second-hand data, accuracy of collection, qualitative and quantitative methods.
   - Discuss the data presentation.
   - Analyse the appropriateness of the presentation.
   - Evaluate alternative forms of presentation, depending on the audience.
   - Evaluate the validity of the conclusion and recommendations made and whether they reflect the aim.
6. Evaluate a range of strategies farmers might use to cope with climate variability in plant production, such as:
   - plant breeding
   - drought-tolerant wheat or other crops.
   - improved irrigation practices (e.g. microjet/drip instead of overhead/flood irrigation)
   - timing of planting
   - soil moisture conservation (e.g. mulches, stubble retention)
   - extended fallows
   - retaining residues
   - moisture monitoring (e.g. use of moisture probes)
   - crop density (e.g. altering up and down according to season).
7. Evaluate a range of strategies farmers might use to cope with climate variability in animal production, such as:
   - grazing strategies (e.g. cell grazing)
   - stocking rates
   - reduction of shelter or shade areas
   - tree planting/effect on water intake
   - fodder conservation, including silage, hay and grain
   - new varieties or breeds, including fat tail sheep/Dorper, Dohnes sheep, Afrikander cattle
   - enterprise changes (e.g. native animals and plants)
   - financial analysis, such as gross margins
   - whole farm and enterprise change.
8. Conduct a debate on the topic: 'Farmers should pay more for irrigation water'.

ISBN 9780170265560

CHAPTER 33

Outcomes

Students will learn about the following topics.

1 Innovations, ethics and current issues relating to research and development.
2 Developments in agricultural technologies.
3 Marketing of technology developments.
4 Reasons for adopting technologies.
5 The adoption of a technological development.
6 Research into technological developments.

Students will be able to demonstrate their learning by carrying out these actions.

1 Discuss issues related to the research and development of technologies.
2 Evaluate a range of new technological developments.
3 Evaluate methods used by companies to market new technological developments.
4 Explain reasons why new technologies are adopted.
5 Analyse a research study of the development and/or implementation of one recent agricultural technology.

ISBN 9780170265560

FARMING FOR THE 21ST CENTURY

Words to know

plant breeder's rights exclusive commercial rights that protect the registrant of a registered variety of plant, which are administered under the *Plant Breeder's Rights Act 1994*

registrant usually the person or company that developed the variety of plant

Introduction

Australian agriculture has become reliant on the application of new technologies and **innovations** (see page 165) as a means of expanding market share in an increasingly interconnected global economy. With the impact of climate change and the need to efficiently use increasingly expensive inputs such as water, land and fertiliser, the use of technology and innovative ways of incorporating new technologies into farming has gained increasing attention and importance.

Throughout this book we have mentioned many applications of technology into farming systems and we have discussed several innovations which indicate the direction of agriculture into the 21st century.

In Chapter 12, the need for adoption of technology into agriculture was discussed along with the pattern of adoption that occurs when a new technology or an innovation is introduced into farming. Examples of new technologies used to produce animal products were discussed, including robot milking, robotic shearing, sheep fat scans, ultrasound pregnancy testing in dairy cattle, breed plan, sexed embryos and semen, and rumen by-pass pellets. New technologies involved in marketing products discussed included sale by description, embryo video sales, marketing standards, internet marketing and modified milks. In Chapter12 we also discussed the role of research and outlined some future directions, including the development of certain vaccines, genetic engineering, the development of more sustainable agricultural systems and environmental considerations. In Chapter 20 we discussed the application of new technologies in plant breeding, including wide crosses, tissue culture and genetic engineering. In Chapter 23 we discussed technologies to improve production in plant industries, including use of plant hormones, production of hybrid cultivars, genetic engineering, hydroponics, increased computerisation to deliver inputs and monitor environments, and improved potting mixes. We also outlined technologies that improved the marketing of plant products.

Factors affecting the adoption rate of technologies

Adoption of technological advances by farmers is crucial to the long-term productivity of Australian agriculture. Technological changes in the more biological aspects of farm systems are rapidly adopted by farmers, especially where such changes can be easily integrated into existing farm management structures. Examples include using improved varieties of crops, pasture plants or animal species, or more effective drenches or fertilisers. In these cases, only those who adopt these newer technologies (the **innovators**) (see page 165) receive any benefit. Biological innovations are less disruptive than mechanical innovations. Some of the factors influencing the take up of new technologies are listed below.

1 *Efficiency.* Increased efficiency gained by the adoption of new technology usually results in lower operational costs.
2 *Lower selling prices.* Lower selling prices for farm goods may assist where natural products compete with synthetic products.
3 *Export.* Where a product is exported, increased production levels may increase returns (depending on the exchange rates).
4 *Sources of funding.* It is very difficult to obtain the required finance to develop or apply a new technology due to the high risk associated with agricultural markets.
5 *Research costs.* The cost of research is high and there can be insecure funding for the time it takes to develop and market a concept or new product.
6 *Legislation.* There are many requirements relating to using new technology on a farm: for example, federal, state and local regulations relating to biosecurity, animal welfare, waste product disposal, work place health and safety, and the use of genetically modified plants or animals.

ISBN 9780170265560

7 *Lack of developed market.* As many farmers have found with innovations in the past, if there is no mature, developed market for the product or processing plant nearby to convert the raw product into a consumer product then, in time, the market for the product collapses.
8 *Set-up costs.* High start-up and establishment costs associated with new technology or innovation sometimes mean the final product is not price competitive on the open market with traditionally produced products. This can cause a slowing down or even a termination in the adoption of the technology or innovation unless consumer demand is extremely strong. Stimulating consumer demand usually requires the use of extensive and effective marketing campaigns.
9 *Advances in technology that make farmers poorer.* In some instances, farmers may be worse off from advances in technology. Production levels may increase to such an extent that product prices fall, and the money saved from acquiring and using the machinery may be less than the fall in market prices.
10 *Social issues.* The community acceptance or non-acceptance of the technology can influence the success of an innovation (e.g. genetically modified foods).
11 *Assessing product quality.* When a new technology is introduced, considerable time and money is invested in determining the standards for product quality and educating the consumer and the producer about these standards.

The slow rate of adoption of ostrich and emu farm products, kangaroo meat, angora rabbits and milking sheep reflect the roles played by many of the above factors.

connect

Sprouting up

A video about how farmers respond to the challenges of creating products using new technologies

1 List and discuss the factors that affect the adoption rate of new technologies.

Issues related to the research and development of technologies

The future development and expansion of agricultural industries relies on farmers being aware of the emerging issues associated with new technologies and innovative practices. It also relies on managers at various levels in the marketing chain, research and development companies, and associated agribusinesses. These issues provide a challenging and dynamic environment in which the farmer continues to convert raw input materials into saleable products that are in demand by consumers.

Plant breeder's rights

Plant breeder's rights (PBR) are used to protect the breeders of new varieties of plants that are deemed unique, stable and uniform in production. These grant exclusive commercial rights for a registered variety of plant and are administered under the *Plant Breeder's Rights Act 1994*.

The protection applies to the **registrant** of the variety – usually this is the person or company that developed the variety. It is legally enforceable and gives the owner exclusive rights to commercially use the variety, sell it, direct the production, sale and distribution of it, and receive royalties from the sale of plants derived from it.

There are several other ways of protecting intellectual property rights.

- *Trademarks*. A trademark is usually used as a brand name under which particular varieties can be sold. It cannot be used as a variety name.
- *Patents*. It is possible to gain a patent for a unique plant gene which can be added to plants of a particular variety. These new plants could be patented, giving the patent-holder exclusive right to the innovation.

connect

Gene wars

Arguments for and against gene patents

connect

Plant breeder's rights

PBR give exclusive rights to:

- produce or reproduce the plant material
- condition the plant material for the purpose of propagation (conditioning includes cleaning, coating, sorting, packaging and grading)
- offer the plant material for sale
- sell the plant material
- import and export the plant material
- stock the plant material for any of the purposes described above.

The exceptions to PBR are the use of the variety:

- privately and for non-commercial purposes
- for experimental purposes
- for breeding other plant varieties.

New varieties of all plant species can, potentially, be registered.

The PBR scheme protects breeders for a period of time and gives them a commercial monopoly, while encouraging plant breeding and innovation. Other provisions enable the use of overseas germplasm for breeding and the use of farm-saved seed.

The general public benefits from PBR by having access to a large, growing pool of new varieties as they become freely available when the protection periods lapse.

PBR protection allows you to exclude others from:

- producing or reproducing the material
- conditioning the material for the purpose of propagation
- offering the material for sale
- selling the material
- importing the material
- exporting the material
- stocking the material for any of the purposes described above.

PBR protection applies for 25 years from the date of granting for grapevines (*Vitis vinifera*) and trees, and 20 years for all other species.

PBR also protects the registered name and synonym of the variety from use in relation to other similar plants.

Plant quarantine regulations and biosecurity risks can affect the issuing of PBR.

The Australian Centre for Intellectual Property in Agriculture (ACIPA) has been established to conduct research and training in intellectual property rights. This is funded by the Grains Research and Development Corporation (GRDC).

2 Explain the concept of plant breeder's rights.

3 List two other methods of protecting property rights to new plant varieties.

4 List the plant breeder's rights gained by a breeder in registering a new variety of plant.

connect

Top 5 trends in agriculture technology

5 Outline the role of biosecurity in the development of new plant varieties.

6 List the main trends in technology development into the future.

Reasons for adopting new technologies

Reasons for adopting new technologies relate to profit, development of disease resistance, weed control and improved marketability and product quality.

The main future technology trends identified in the Top 5 trends in agriculture technology website relate to:

- education
- data management
- robotics
- evaluative metrics – assessment of efficiency and risk
- market feedback based on product quantity and quality.

connect

National Livestock Identification System

connect

NLIS

7 Describe the purpose of the NLIS, using information from the websites.

ISBN 9780170265560

Developments in agricultural technologies

A range of computer-based technologies is currently available for use in agriculture, including computer-based record-keeping systems.

Some examples of these technologies include the following.

- *Electronic identification of livestock.* The National Livestock Identification System (NLIS) is an example of the application of this technology. It gives farmers the ability to track livestock from birth to slaughter. This system is outlined in the National Livestock Identification System and NLIS websites.
- *Marketing systems.* For example, the AuctionsPlus website offers real-time internet auctions of livestock. Farmers do not have to travel many kilometres to trade.
- *Satellite imaging and global positioning systems.* Satellite imaging and global positioning systems are now used to map agricultural land, identify the fertility levels of a soil and deliver inputs, such as fertiliser, efficiently. Refer to Figure 33.1, which illustrates the use of GPS equipment to monitor crop yield.
- *Laser technology in levelling and seedbed land preparation.* Equipment called a laser leveller is usually a self-levelling device that uses a laser beam in conjunction with a sensor device to determine elevations. A mounted laser level is used in combination with a sensor drawn by a tractor to achieve level land or develop a drainage gradient for irrigation.
- *Robotics*
  - FutureDairy is an R&D development program to help Australian dairy farmers manage the challenges they are likely to face during the next 20 years. As one of the big challenges is the availability of labour and the associated lifestyle issues, FutureDairy's focus is on automatic milking systems, or 'robotic milking'. Robotic milking technology is in wide use overseas.

connect

Livestock auction

Use this website to examine the functions of a livestock auction website.

connect

GPS systems in agriculture

connect

Satellite imagery

connect

Geoimage in agriculture

connect

Laser technology

connect

Automatic milking systems

Figure 33.1 GPS technology monitoring crop yield

ISBN 9780170265560

connect

Robotic shearing

- Technology for robotic shearing of sheep has also developed in response to cost and efficiency requirements in the industry. Figure 33.2 illustrates a robotic shearing machine. While robots are not used commercially in shearing sheds at present, they represent an alternative to manual labour for the future.

James Trevelyan, The University of Western Australia

Figure 33.2 Robotic shearing

Shutterstock.com/risteski goce

Figure 33.3 Drone with camera

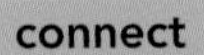

connect

3D robotic drones and agriculture

connect

Climate

An important website about climate data

connect

Creating a digital homestead

8 Discuss how computer-based auctioning systems work for selling livestock.

9 Describe the role of satellite technology and GPS-based systems in agriculture.

10 Describe the use of lasers in farm development.

- Robotic drones can be used assess crop health and productivity, spray hard-to-reach weeds and round up stray livestock. They may be more precise and therefore useful to agriculture (Fig. 33.3). They also allow the development of more profitable and sustainable farming systems.

ISBN 9780170265560

- *Climate data*. Extensive use of computer databases for recording climate data and modelling climate trends continues to evolve. Weather predictions, warnings and analysis of climate changes over time are readily available from many websites.
- *Farm management*. Current research by the CSIRO is attempting to develop technology that will allow a farmer to assess cattle health, feed supply and movement of stock using solar-powered collars on animals. This is called the Digital Homestead project.

Marketing of technology developments

Methods used by companies to market new technologies include:

- advertising on traditional media and internet sites
- presentations and demonstrations at field days
- use of consultants and trial sites, where experiments are often conducted by agricultural organisations
- small demonstrations at agricultural shows.

Gene technology

Genetic engineering is the manipulation of an organism's genetic material. Gene technology, as defined by the CSIRO, means understanding the expression of genes, taking advantage of natural genetic variation, modifying genes and transferring genes to new hosts. For genes to work they must have an on-switch, an off-switch and a section that codes for a particular protein. However, while all these parts need to be present they do not have to come from the same organism.

Spliceosome-Mediated RNA Trans-splicing (SMaRT™) technology gene therapy refers to techniques such as repair of a faulty gene by adding a normal copy of a gene to cells that only have a non-working copy and to 'turning off' a gene that is producing a faulty protein.

There is a need to regulate any technique used to modify genes or other genetic material. This function is carried out by the Office of the Gene Technology Regulator.

The marketing of genetically modified products needs to address many consumer concerns. The GMOs website clarifies some of the concerns to be addressed before this technology can gain further acceptance.

connect

GMOs

connect

The Office of the Gene Technology Regulator

connect

Genetically modified products

View the video and answer the questions.

connect

Gene technology

In-depth information about the processes involved in gene technology and its benefits and regulation, and genetic modification of organisms

connect

Genetic engineering and gene therapy

The advantages of genetic engineering over traditional methods, such as cross-breeding and artificial selection

connect

SMaRT™ gene technology

connect

Genetically modified foods

View the video and answer the questions.

11 Outline three examples of the use of robotic technologies in agriculture.

12 Outline the role of computer technology in predicting, tracking and reporting weather and climate.

13 What is genetic engineering?

14 List some of the methods used in gene technology.

15 List three advantages of gene technology.

16 Define the term 'SMaRT technology gene therapy'.

17 Discuss two concerns around the use of gene technology in agriculture.

18 Outline how the use of gene technologies is regulated.

19 Discuss marketing approaches to improve consumer acceptance of genetically modified foods.

Chapter review

Mandatory analysis of a research study for NSW HSC Agriculture course

Students choose only **one** of the research study electives given in chapters 31, 32 and 33.

Research into technological developments

a **Analyse a research study of the development and/or implementation of one recent agricultural technology in terms of:**
- **design of the study**
- **methodology of the study**
- **collection of data for the study**
- **presentation of data**
- **analysis of data**
- **conclusions and recommendations.**

b **Explain the need for research in the development of agricultural technologies.**

Things to do

1. Describe two methods that companies may use to market new technological developments.
2. Describe issues relating to the research and development of future farming technologies.
3. Compare and contrast two marketing strategies companies may use to promote new technological developments.
4. Identify impediments to the rapid uptake of new technology in agriculture, such as level of exposure, cost, availability of transport, education and training, conservatism and changing markets.
5. Research methods available for marketing new technologies, such as field days, rural newspapers and magazines, industry-based newsletters, websites, direct marketing, local TV and radio.

Things to find out

1. Outline reasons for adopting new technologies in agriculture.
2. Propose reasons why a newly developed agricultural technology may not be widely adopted.
3. Discuss why there is a continuing need for research into agricultural technologies for farming.

ISBN 9780170265560

Extended response questions

1. For one recent technological development:
   - a explain the reason for the development of the technology
   - b outline the historical development of the technology
   - c describe in detail the technological development
   - d evaluate the impact of the technological development in terms of economic, environmental, social, legal and managerial factors.
2. Summarise the benefits of adopting six new agricultural technologies; for example, increased production, disease resistance, efficient weed control, quality and marketability.

   Examples to choose from include:
   - ultrasound scanners
   - NLIS reader in cattle
   - GM crops – herbicide tolerant OR pest resistant
   - semen or embryo sexing
   - marker genes in livestock
   - GPS or precision agriculture – fertiliser application OR herbicide spraying OR monitoring of yield
   - robotic shearing
   - robotic dairy
   - remote sensing to monitor vegetation conservation
   - remote sensing and estimation of crop yields
   - laser land levelling
   - use of drones.
3. Explain the benefits of recent developments in biotechnology and robotics that may assist agricultural production. Include examples in your answer.
4. Discuss, using examples, the benefits of computer-related technologies used to monitor and manage factors associated with agricultural production.
5. Analyse issues related to the research and development of technologies. Outline specific examples to illustrate your answer.
6. During your study of this elective you were required to analyse a research study of the development and/or implementation of one recent innovation in agricultural technology. Analyse the design and methodology used in a named study.
7. Evaluate new technological developments that may assist agricultural industries.
8. Evaluate methods that companies use to market new technological developments.

CHAPTER 34

Outcomes

Students will learn about the following topics.

1 Post-school opportunities.
2 Career profiles.
3 A variety of careers in agriculture.

Students will be able to demonstrate their learning by carrying out these actions.

1 Consider their post-school opportunities.
2 Develop a career profile.
3 Examine career opportunities in agriculture.

ISBN 9780170265560

CAREERS IN AGRICULTURE AND RELATED STUDIES

Words to know

agronomist a person who studies and experiments with various crop plants and pastures, assessing their growth, stainability and economic potential in agriculture

aquaculture the commercial farming of fish, molluscs (e.g. oysters), crustaceans (e.g. prawns), and aquatic plants (seaweeds) in natural or controlled marine or freshwater environments

farrier a person who shoes a horse and maintains the health of the horse's foot

floriculturist a person concerned with the cultivation and marketing of flowering and ornamental plants

horticulturist a person who oversees operations involving the cultivation of plants for research or commercial use

saw doctor a person who maintains and services saws

Introduction

The broad-ranging career opportunities in agriculture presented to you throughout *Dynamic Agriculture Years 11–12* may help you decide whether you want to enrol in further study or seek immediate entry into a given occupation after secondary school. In this chapter, future educational and career pathways are outlined so you can explore, test your perceptions of and develop a critical awareness of the exciting possibilities in the world of agriculture.

Whatever you choose, remember, education is a lifelong process and is not confined to the concepts presented in a textbook!

connect

Industry training packages

Find which skills are recognised by industry training packages through the study of Agriculture Stage 6. For your skills to be recognised, clear documentation is required.

connect

TAFE New South Wales

Find which skills are recognised by TAFE through the study of Agriculture Stage 6. For your skills to be recognised, clear documentation is required.

Post-school opportunities

Studying Agriculture Stage 6 during HSC in high school will enable you to apply for and study courses at universities and other higher education institutions. Over the two years that you study Agriculture Stage 6, you will gain both a theoretical knowledge of current agricultural areas of importance as well as an understanding of the operations of a farm in a physical, scientific, managerial and economic sense, along with associated practical skills. There is an opportunity for recognition of some of these skills in vocational and educational training courses offered by TAFE, industry and private training operators. Recognition of prior learning (RPL) depends on how aligned the outcomes that you achieved in the HSC Agriculture course are with industry training packages.

Where to start

- What are your interests?
- What are your skills?
- What level of education, training or work skills have you acquired?

Use the websites opposite to help you to develop your career profile.

There are many career pathways in the field of rural studies. Look at the options shown in Figure 34.1. Visit the websites My career profile and My future.

connect

My future

connect

Careers in agriculture, natural resource management and food production

Career decisions

The Rural studies and Careers in agriculture, natural resource management and food production websites detail a variety of further study options in the rural studies and agriculture fields, and outline the required qualifications and level of education required to commence the courses.

Alamy/Josie Elias

CSIRO/Willem van Aken

Figure 34.1 Studying agriculture in HSC prepares you for study and training in many related occupations. **a** A Parks and Wildlife Ranger handles a malnourished green turtle rescued from a beach on Magnetic Island, Queensland **b** An agronomist examines dryland salinity-induced hillslope gully and sheet erosion in Northern Queensland

ISBN 9780170265560

Australian Government

LEVEL 4

Agricultural and resource economist
Forest technical officer
Agricultural engineer
Forester
Geographer
Agricultural scientist
Horse manager
Agricultural technical officer
Horticulturist
Agronomist
Park ranger
Natural resource manager
Ecologist
Park ranger
Environmental engineer
Viticulturist
Environmental scientist
Winemaker
Farm manager
Zoologist
Farmer
Fisheries officer

LEVEL 3

Agricultural technical officer
Horse manager
Horse trainer
Animal technician
Natural resource manager
Aquaculture technician
Park ranger
Dairy technician
Primary products inspector
Farm manager
Stock and station agent
Farmer
Fisheries officer
Forest technical officer

LEVEL 2

Animal attendant
Nursery worker
Animal technician
Park ranger
Aquaculture technician
Pest and weed controller
Artificial insemination technical officer
Primary products inspector
Cheesemaker
Saw doctor
Farm manager
Shearer
Farmer
Slaughterer
Farrier
Stock and station agent
Fish farm hand
Timber and wood production worker
Floriculturist
Tree faller
Forest worker
Tree surgeon
Gardener
Winery worker
Harvesting operator
Wool classer
Horse manager
Horse trainer
Horticultural tradesperson
Jackaroo/Jillaroo

LEVEL 1

Animal attendant
Horse manager
Artificial insemination technical officer
Horse trainer
Jackaroo/Jillaroo
Beekeeper
Nursery assistant
Dairy farm hand
Park ranger
Farm hand
Pest and weed controller
Farm manager
Saw doctor
Farmer
Shearer
Fish farm hand
Slaughterer
Floriculturist
Stablehand
Forest worker
Timber and wood production worker
Fruit and vegetable picker
Tree faller
Gardener
Winery worker
Harvesting operator

RURAL STUDIES

Do you enjoy or are you good at

RURAL STUDIES?

Have you considered the occupations above?

Usual training requirements

LEVEL 1 Usually has a skill level equal to the completion of Year 10, the Senior Secondary Certificate of Education, Certificate I or Certificate II qualification. Australian Apprenticeships may be offered at this level.

LEVEL 2 Usually has a skill level equal to a Certificate III or IV or at least three years relevant experience. Australian Apprenticeships may be offered at this level.

LEVEL 3 Usually requires a level of skill equal to a Diploma or Advanced Diploma. Study is often undertaken through TAFEs or Registered Training Organisations. Some universities offer studies at this level.

LEVEL 4 Usually requires the completion of a Bachelor Degree or higher qualification. Study is often undertaken at university.

This chart shows a selection of jobs that have some relation to the subject of RURAL STUDIES.

The four education and training levels are to be used as a guide only. These levels indicate the most common education and/or entry requirements for these jobs.

For further information visit www.jobguide.education.gov.au and www.myfuture.edu.au

ED13-0050

Figure 34.1 Careers in rural studies

connect

Green-Life careers

Working with plants

If you are interested in working with plants or in the landscape, parks and gardens industry then read the information on the Green-Life careers website.

Shutterstock.com/withGod

Figure 34.2 Working with plants: an agronomist charts the results of an experiment in the wheat field

Shutterstock.com/OLEKSANDR SHEVCHENKO

Figure 34.3 Working with plants: A nursery assistant seeding plants in a nursery

Working with animals

If you are interested in working with animals or in the animal agriculture industry, read the information on the Career harvest website.

Alamy/K Philip Harrison

Figure 34.4 Working with animals: farmers check a newborn lamb.

connect

Career harvest

connect

AgCareers

A whole world of choice

For a broader range of career pathways, visit the Job guide and AgCareers websites.

ISBN 9780170265560

Chapter review

Things to do

1. Develop a personal career plan.
2. Look at the variety of jobs available within agriculture from the above websites or from rural newspapers, such as *The Land* in New South Wales or *Queensland Country Life*.
3. Read the job descriptions and consider how these match your interests, knowledge, skills and experiences.
4. Subscribe to AgBizCareers. You will receive regular emails about jobs that are available and a list of the skills and qualifications you need to gain that employment. It is free to subscribe.
5. Subscribe to NRMjobs. You will receive regular emails about jobs that are available and a list of the skills and qualifications you need to gain that employment. It is free to subscribe.

connect

AgBizCareers

connect

NRMjobs

Things to find out

1. Explain the phrase 'being work ready'.
2. List the further training and qualifications will you require.
3. Attend career markets and opportunities to further your knowledge of post-education options.
4. Critically assess the courses and training opportunities offered by universities, TAFE, private education providers and industry.
5. Talk to people involved in the careers that interest you.

Extended response questions

1. Develop your career profile.
2. List a series of questions that you would like to ask a professional in the career field in which you are interested.

Glossary

abscisic acid	a plant growth substance that functions chiefly as a growth inhibitor
absorption	the passage of digested nutrients through the membrane of the alimentary canal into the bloodstream
acidity of the soil	the level of acid in the soil measured by the pH scale, which ranges from 1 (very acidic) to 14 (very alkaline), with a pH of 7 being neutral
active immunity	antibody production by an organism to combat disease
adrenalin	a hormone synthesised by the adrenal glands that produces a fight-or-flight response in animals
adventitious roots	roots that arise from stems, usually at nodes
aerobic	in the presence of oxygen
agribusiness	the many industries that directly or indirectly support the production, processing (value adding), distribution, retailing and marketing of agricultural products
agrifood	food obtained from agriculture
agronomist	a person who studies and experiments with various crop plants and pastures, assessing their growth, sustainability and economic potential in agriculture
albedo	radiation reflected from surfaces
algae	simple plant-like organisms that possess chlorophyll and perform photosynthesis
allele	different form of one gene (e.g. the two forms of the gene for plant height in peas, 'T' tall and 't' dwarf)
allelopathy	where one organism produces one or more biological chemicals that influences the growth, survival or reproductive capacity of another organism
alveoli	the small glands where the milk is made
amino acids	organic compounds that combine to make proteins
anaerobic	without oxygen
analyse	to examine information relating to a problem posed by an experimenter
anatomy	the study by dissection of the structure of the body of an organism
animal ethics	the study of human–nonhuman relations; includes animal rights, animal welfare and animal conservation
animal welfare	the physical and psychological wellbeing of animals
annual plant	a plant that completes its lifecycle within 1 year
anther or pollen sac	the terminal portion of the stamen, containing pollen in sac-like structures
anthesis	the stage of development of a plant when the anthers become prominent and pollen is shed – particularly applies to cereals (e.g. wheat)
antibiosis	the adverse effects of the plant on insects or mites after they have consumed at least some of the plant
antibiotic	a chemical substance, produced by certain micro-organisms, which kills or inhibits the growth of other micro-organisms
antibody	a chemical substance (a protein) made by animals in response to bodily invasion by a pathogen which combines with the pathogen and renders it harmless
antiserum or antitoxin	the preserved serum of an animal that has previously had a specific disease or has been injected with a vaccine. The serum contains high concentrations of antibodies
apical dominance	a state where the growth of all the axillary buds on a stem is suppressed, apparently by the growing terminal bud
aquaculture	the commercial farming of fish, molluscs (e.g. oysters), crustaceans (e.g. prawns), and aquatic plants (seaweeds) in natural or controlled marine or freshwater environments
attenuated strains	weakened strain
AuctionsPlus	a nationwide electronic system for buying and selling livestock; formerly known as Computer Aided Livestock Marketing (CALM)
autotroph	an organism that produces its own food
auxins	plant hormones causing cell elongation, secondary thickening of stems and roots, fruit development and apical dominance
available water	the amount of water available to a plant in the soil as measured between the point of field capacity and the permanent wilting point

ISBN 9780170265560

backlining	pouring a chemical in a line along the sheep's back to kill lice
bacteria	single-celled microscopic organisms. Some cause organic matter to decay or cause disease, others are important in the formation of cultured dairy products and fixing nitrogen both in the soil and in the roots of legumes
balanced ration	a ration containing a balance of nutrients; small quantities of mineral salts and vitamins are often added to give the correct balance of nutrients
bias	a form of prejudice, or slant, in obtaining a result
bibliography	a list of source materials used or consulted in the preparation of a work
biennial plant	a plant that completes its lifecycle within 2 years
bilateral trade	trade between two countries
binary fission	a type of asexual reproduction in which a parent cell divides into two
biological efficiency	how well living processes function
biosecurity	the protection of Australia's animal and plant industries and the natural environment from exotic pests and diseases
biotechnology	any technique that changes living organisms at the molecular level (genes and chromosomes) to produce useful products, such as medicines (e.g. insulin), an insect-resistant plant or offspring through embryo transfer
biotic	related to living things; biotic factors in the environment are the effects of living organisms
black box model	a model showing only inputs and outputs in a system, farm or enterprise
bloat	an unusual build-up of gas in the rumen, which can result in the death of an animal; occurs in cattle grazing pastures containing predominantly lucerne or clover
bone	a hard connective tissue that makes up most of the skeleton of vertebrates
boundary	a limitation of a (farming) system (e.g. fence, money, managerial skills and so on)
bounty	money paid by the government to farmers to offset production costs
breeding	the process of crossing and selecting plants to produce new types or varieties
brucellosis	a bacterial disease in cattle that causes abortion
budget	a planning tool used to assess the profitability of alternative farming plans
buffer stock scheme	a scheme designed to maintain a stable market price by withholding excess quantities of product in times of oversupply and progressively releasing this when market prices are more favourable
bulk density	the mass of soil solid material relative to the total volume of soil material
C/N ratio	the ratio of carbon to nitrogen material in the soil, which determines the effectiveness of micro-organisms in the nitrogen cycle
cambium	a layer of meristematic tissue found in stems and roots between xylem and phloem
capital	money that has been invested or is available for investment
carbon credit	a certificate or permit to allow production of 1 metric tonne of carbon dioxide or another greenhouse gas
carbon sequestration	a natural or artificial process that removes and traps carbon dioxide from the atmosphere
carbon trading	the basis of an emissions trading approach that relies on the buying and selling of carbon credits
carcase	what remains of the animal body after the head, feet, hide, tail and internal organs of the abdomen have been removed
carrier	an animal that can transmit a hereditary characteristic or disease to its offspring, without itself showing the characteristics or any symptoms of the disease
catchment area	the area of land that collects water for a particular waterway; the term may apply to a small stream in a paddock that only flows when it rains or to a very large river system such as the Murray–Darling
cation exchange capacity	a measure of the amount of exchangeable ions that can be held by a clay particle
characteristic	a distinguishing feature of a plant, resulting from the interaction of a gene or genes with the environment
chemical ploughing	the use of Roundup® or similar herbicides to eliminate weeds and plant competition prior to sowing a pasture or crop
chloroplast	a discrete membrane-bound part, within a cell, that contains chlorophyll and is capable of photosynthesis
chlorosis	the loss of chlorophyll in a plant
chromosome	a thread-like structure within the nucleus of a cell, which carries genes

clay double layer	the layer of tightly and loosely bound ions that surrounds a clay particle
climate	the average weather conditions over a long period of time (at least 30 years)
clones	organisms that are genetically identical
coalescence	the degree of moulding or coherence that occurs to a block of soil when subjected to a force at a particular moisture content
coccidiosis	a disease caused by an organism that can damage the lining of the small intestine and affect absorption of nutrients
coefficient of variation	an absolute measure of dispersion of the data
coleoptile	a protective cone of tissue over the terminal bud in the monocot embryo, which gives the bud protection until it reaches the soil surface
coleorhiza	a protective cone of tissue over the radicle in the monocot embryo
collateral security	property pledged as guarantee for the repayment of a loan (usually the farm in agriculture)
colloids	an organic or inorganic particle in a soil that is extremely small, but has a very large surface area. Soil colloids may hold plant nutrients on their surface
colostrum	a protective substance containing antibodies, which is passed on in the first milk
commodity	goods of trade
community	a group of organisms that live together, sharing the same environment
companion planting	growing different crops in rows next to each other usually one type of plant has factors to ward off pests that would normally attack the other plant
concentrates	feeds with high concentrations of major nutrients
conclusion	a series of judgements or inferences made on information gained through an experiment
consumer	person who buys and uses a product to satisfy their needs or wants
contact or stomach poisons	poisons that act by coming into contact with an insect; that is, be absorbed by some part of the insect's body
contour bank	a small bank constructed along the contours of the land where it slows running water and prevents it building up speed, and thus prevents erosion
contract	a regulated quantity of output allowed to be produced (replaces the term 'quota')
control	the standard with which a new technique or variety is compared, or a part of an experiment that does not receive any treatment
corporate farms	farms owned by private businesses
cost–price squeeze	the situation of falling prices for farm output at the farm gate and increasing farm production costs
cover crop	a crop that is grown to protect another tender species until it is established
crop	one kind of plant cultivated to produce some particular product (e.g. wheat is grown to produce grain for bread)
crop rotation	the process of growing different crops, including pasture, from one year to the next on the same piece of land
crossbreeding	the mating of unrelated plants or animals of different breeds or the crossing of unrelated plants (i.e. plants with different genotypes)
cross-pollination	pollination between flowers of different plants of the same species; in fruit trees, of one variety with another
cultivar	a particular strain, or variety, of a plant (e.g. hartog is a cultivar of wheat)
cuticle	a waxy layer covering the epidermis of plants, particularly leaves and herbaceous stems
cytokinins	plant hormones that promote fruit ripening and cell reproduction, and initiate the production of roots and shoots
damping off	where young seedlings wilt and die due to fungal attack
data	information gained through measurement or the collection of observations
deficiency	the lack of a substance important for normal growth, especially mineral nutrients such as nitrogen and phosphorus
density	(of plants) the number of plants per hectare
deoxyribonucleic acid (DNA)	the biological material that stores genetic information for an organism; one chromosome is one DNA molecule, a double stranded molecule in the shape of a helix; along each strand is a sequence of bases comprising many genes
derived demand	demand generated by the processed goods rather than the raw materials used in production

ISBN 9780170265560

development	the changes in the proportions of various parts of an animal's body as the animal gets older
diatoms	microscopic, unicellular, marine or freshwater algae having cell walls impregnated with silica
dicotyledon	a plant that has two seed leaves or cotyledons in its seed
dieback	applies to trees and woody plants and is characterised by the progressive death from younger branches to the base of the plant; causal agent is root pathogens
diet	a general description of the types of feeds eaten by an animal
digestibility	the proportion of the food that is not excreted in faeces, and is assumed to be absorbed by the animal
digestible energy	energy absorbed by an animal after the digestion of food (allowing for energy loss in faeces)
digestion	the breakdown of large insoluble particles into simpler, soluble substances within the digestive (or alimentary) tract so that they can be absorbed
direct drilling	sowing technique in which seed is placed into the previously undisturbed surface of an existing pasture
discussion	a detailed examination of the findings of an experiment
disease	any condition that produces an adverse change in the normal functioning of an organism
disease triangle	a conceptual model that shows the interactions between the environment, the host and an infectious (or abiotic) agent
dispersion	the spread of particles in several directions, from areas of high concentration to areas of lower concentration
dissection	the act of cutting an organism into parts to show its structure
diurnal rhythm	an activity lasting for a day
diversification	producing more than one output on the farm from unrelated farm subsystems (e.g. wheat and wool)
dormancy	a state of a seed or plant where it remains living but does not grow or germinate
dry cow	a cow that has completed her lactation and is not producing milk
dse	dry sheep equivalent. 1 dse equals one dry sheep or wether per hectare; it is a measure of the stocking rate of the land
dumping	flooding the market with a product
dystochia	a difficult birth
economic efficiency	how much money is returned by a particular system in relation to the cost of establishing, running and maintaining the system
economic threshold	population level of a pest when it is necessary to start control measures
economic viability	the ability of a farm to produce sufficient profit to provide a comfortable living for the farming family
ecosystem	the relationship between an interacting community of organisms and its physical environment
ectoparasite	an organism that lives permanently or temporarily on the surface of a host's body
effective rainfall	where mean monthly rainfall exceeds the monthly evaporation rate
efficiency	the extent and rate of conversion of input to output
El Niño	a period marked by the development of a warm ocean current off the South American coast and a major shift in weather patterns across the Pacific with cooler than normal surface sea temperatures across the north of Australia
embryo	the developing organism in early pregnancy or in plants, the small immature plant found in the seed
embryonic mortality	the death of an embryo
endemic	diseases that occur in a population year after year
endodermis	a layer of cells in the root between the cortex and the vascular tissue, that controls the movement of water into the vascular tissue
endoparasite	an organism that lives in the internal organs of the host
endosperm	the food reserve in the monocot seed
environment	any non-genetic factors that affect plant growth, including the climate, the soil, other plants, pests, diseases and management practices; i.e. the immediate surroundings of an organism
epidemic	a rapidly increasing incidence of a disease
epidermis	a layer of cells covering the outside of young roots, stems and leaves
equilibrium market price	the market price at which the quantity of goods demanded is exactly balanced by the quantity of goods supplied
erosional	wearing down

essential amino acid	an amino acid that the animal is not able to make (or synthesise) in quantities sufficient for growth and development
ethylene	a plant hormone that promotes fruit ripening
evapotranspiration	part of the water cycle in which liquid water is removed from an area with vegetation and moves into the atmosphere by the processes of both transpiration and evaporation
exoskeleton	the external body covering of an arthropod
exotic	not native
experimental error	a deviation from the true measurement expressed as a percentage
extensive farming	the production of plants or animals over large areas of land
fallowing	a farming system in which land is left without a crop for extended periods to accumulate soil moisture
farming family	a family operating a farm as a business and a lifestyle
farm problem	the apparent difference in yearly incomes between people in rural activities and city workers due to rising production costs and low market prices for farm outputs. This then leads to depressed rural income, which limits the ability of farmers to obtain finance for further farm development and forces labour to move out of farming and into other areas of employment
farm system	a group of parts (subsystems) that interact to achieve a purpose
farrier	a person who shoes a horse and maintains the health of the horse's foot
fat	a chemical substance made from glycerol and fatty acids
feedback	the information received by the farmer on the performance of the system
fermentation	the breakdown of starches and cellulose to sugars under anaerobic conditions; that is, in the absence of oxygen
fertilisation	the union of male and female sex cells; the union of the male and female sex cells in the ovule
fibre	a feed material consisting mainly of cellulose (from plant cell walls); feeds high in fibre are called roughages
fibrous roots	roots of monocot plants, all about the same size
field capacity	the amount of water retained in a soil profile that was saturated after drainage has occurred over a 24- to 48-hour period
finance	money
first-cross	first generation hybrid offspring produced from mating of parents with different characteristics
flag leaf	the last leaf of a monocot
flagellum	(plural: flagella) a whip-like extension of certain cells; their rotation causes movement
floriculturist	a person concerned with the the cultivation and marketing of flowering and ornamental plants
fodder crop	a crop that is grown to produce feed for grazing animals (e.g. lucerne)
foetus	the developing organism in later pregnancy (the embryo becomes a foetus)
fungicide	chemicals used to control fungal diseases
galls	abnormal swellings and outgrowths found on plants caused by some pathogens and insect activity
gene	a part of a chromosome that determines what a particular characteristic will be. It is a sequence of bases on a DNA strand. One gene produces one polypeptide or protein
gene marker	a gene or DNA sequence with a known location on a chromosome that can be used to identify individuals or species, or a gene used to determine if a nucleic acid sequence has been successfully inserted into an organism's DNA.
gene patents	the ability to register and charge for the use of newly created genes
genetically modified organism (GMO)	an organism whose genetic material has been altered using genetic engineering techniques; genes from different species are combined
genetic engineering	the manipulation of an organism's genetic material or alteration of a chromosome in a laboratory by removing, adding or reversing a gene or genes to produce a new type of organism
genetics	the study of inheritance
genotype	the genetic makeup of an individual organism, consisting of its chromosomes and genes, half of which were inherited from each parent
germination	the process whereby a seed changes from a dormant state to active growth
gestation period	the period of pregnancy
gibberellins	plant hormones causing stem growth, flowering and breaking seed dormancy

ISBN 9780170265560

grassed waterway a wide channel that is permanently grassed and rarely grazed, which slows the velocity of the water and moves it away safely

greenhouse effect the overall increase in temperature caused when solar radiation is unable to escape from Earth's atmosphere because of the presence of atmospheric gases, particularly carbon dioxide

gross domestic product (GDP) the monetary measure that assesses the value of all the finished goods and services that are produced within a country, usually determined on a yearly basis; often used as a measure of the standard of living for a country

gross energy energy content of a food before digestion

gross margin a planning tool used to compare enterprises of a similar nature, determined as gross income minus variable costs; gross margins must be expressed in terms of per hectare, per animal, or a similar quantitative measure

growing season occurs where precipitation exceeds evaporation for a period of time

growth the increase in size and weight of an animal as it gets older

growth habit the way a species of plant grows (e.g. kikuyu grass sends out stems horizontal to the ground surface and is said to have a creeping habit)

gully erosion a type of erosion (which often starts as rill erosion) that is caused by running water, and is characterised by the formation of deep gullies in the soil

harvesting gathering a product from the plants being grown

haustoria specialised hyphae found in fungi for extracting nutrients from inside cells

herbaceous plants or plant parts that are fleshy as opposed to woody

herbicide a chemical used to kill unwanted plants, especially weeds

hereditary disease disease passed on to the offspring by one of the parent's genes

heredity the transfer of genetic traits or factors from parents to offspring

hermaphrodites organisms that have both male and female reproductive organs (e.g. slugs and snails)

heterosis or hybrid vigour the increased vigour of crossbred progeny, or hybrids

heterotrophs an organism that needs a supply of food from its surroundings

heterozygous having two different alleles for any one gene; that is, having one dominant and one recessive gene for a particular character or trait

hogget a shorn or unshorn sheep from approximately 12 to 15 months of age

homogenisation dispersal of fat globules within milk to stop the cream from separating out

homozygous an organism that contains two identical alleles for any one gene

hormone a chemical substance secreted by the ductless, or endocrine, glands directly into the bloodstream to control body actions or processes

hormone growth promotants compounds similar to sex hormones, which increase growth in farm animals; also called anabolic steroids

horticulturist a person who oversees operations involving the cultivation of plants for research or commercial use

hybrid the result of crossing different varieties of plants or different animals of the same species to create new varieties or breeds

hydroponics growing plants without soil in carefully balanced nutrient solutions

hyphae thread-like structures that are part of the structure of fungi

hypothesis a concept or idea to be assessed or tested

immunity the body's ability to resist attacks of pathogens by producing antibodies

improved pasture introduced species of pasture grasses or legumes, or a combination of both, to enhance pasture production

induced sterility sterilisation of male pests (e.g. flies) by chemical or radiation means; the flies are then released into the wild population to breed and no offspring are produced from these matings

infectious disease a disease caused by a pathogen transmitted from a diseased individual to a healthy one

inflorescence the flowers of a plant borne in a particular way on one stalk (the arrangement of flowers on a plant)

innate behaviour behaviour that exists in the animal from birth – it is not learned

innovation a new concept which has not existed previously, or a new application of an idea or form of technology

innovators those who are early adopters of new technologies

inoculums the infective units of a pathogen that start the infection process

ISBN 9780170265560

inorganic fraction	the non-living fraction of a soil, including minerals, soil air and soil water
input	the items or materials that go into a (farming) system (e.g. fertiliser)
insectidicde	a chemical substance used to kill insects
intake	the type and amount of food eaten or water drunk in a period of time
integrated	a combination of practices
integrated pest management (IPM)	the use of two or more methods to control pests or diseases
intensive farming	the production of a large number of plants or animals on a small area of land
intravenous injection	the injection of fluid into a vein
invertebrates	animals without backbones
jetting	forcing an insecticide under pressure into a fleece to prevent flystrike
keds	sheep keds (or ticks); blood-sucking parasites
La Niña	a period marked by the extensive cooling of the central and eastern tropical Pacific ocean, often accompanied by warmer than normal sea surface temperatures (SST) in the western Pacific, and to the north of Australia
lactating cow	a cow that is producing milk
lactation	the secretion of milk
land degradation	the adverse alteration of the land surface and the lowering of the land's capacity to produce; characterised by one or more of the following – soil erosion, increased soil acidity, soil salinity, tree decline and reduced biodiversity
laser levelling	earthmoving technique, guided by lasers, that enables large areas of land to be levelled and giving a gentle slope for irrigation
law of demand	at low prices consumers buy more of a product and as market prices increase, less of the product is purchased
law of supply	producers are willing to supply more of a product onto the market place as market prices increase
limiting factor	any factor that lowers the production potential of a system
line breeding	type of inbreeding based on a single common ancestor (a sire or dam) used over several generations of mating
live vaccines	vaccines that contain pathogens that are no longer able to cause disease
lodge	a term applied to crop plants that fall or are blown over, by wind
lodging	occurs when wind causes tall plants to fall over, making harvesting difficult
long fallowing	the practice of ploughing the land for a crop several months before planting (to accumulate soil moisture) and then cultivating it after rain
long-day plant	a plant that appears to flower in response to a long photoperiod, but in fact responds to the short dark period, (i.e. spring or summer)
maintenance energy	amount of energy derived from food that is needed to keep an animal alive and healthy, but does not allow for growth or production
major minerals	minerals present in an animal's body in large amounts (e.g. calcium, phosphorus, potassium)
mammal	the class of animal that nourishes its young with milk from mammary glands
markets	places where farm produce is sold
maturity	the state of being fully developed
mean	(in statistics) the average value
median	(in statistics) the middle value in a group of values
meristem or meristematic tissue	tissue consisting of cells capable of cell division and thus of producing new cells
metabolic disease	a disease that occurs when one section of the body is not working normally
metabolisable energy	energy available to an animal for use by the body after energy loss in urine and methane production in ruminants
metamorphosis	change in body form. Some insects have a complete change from larva to adult (e.g. flies)
metazoal disease	a disease caused by a metazoan, which can be seen with the naked eye (e.g flatworm or roundworm)
microbial disease	a disease that occurs when a **pathogen** enters an animal
microbiology	the study of micro-organisms and their activities

ISBN 9780170265560

microclimate	the atmospheric conditions near the surface of the ground or adjacent to the crop canopy
micro-organism	an organism that cannot be seen with the naked eye but can be seen with the aid of a microscope; in general, organisms with a diameter of less than 1 mm
milk contract	a quantity of milk that the dairy farmer has agreed to produce
mineral nutrients	elements that the plant requires in ionic form for healthy growth (e.g. potassium (K^+) and nitrate (NO_3^-))
minimum tillage or reduced tillage	methods of farming that minimise soil damage by reducing cultivation and substituting chemicals for machinery use
miticide	chemical used to control mites
mode	(in statistics) the most commonly occurring value
moisture characteristic	the relationship between soil moisture potential and soil moisture content
monocotyledon	a plant that has only one seed leaf or cotyledon in its seed
monogastric	an animal with only one stomach (without a rumen)
morphology	the form and structure of a plant
mulch	material spread over the soil surface to reduce loss of water by evaporation (e.g. crop residues, black plastic, wood chips, compost)
mulesing	surgically removing skin from the breech/rear area of the sheep so wool will not grow there
multiple-price scheme	a scheme designed to maximise returns by obtaining several prices for essentially the same output, by splitting the market on the basis of distance, consumer income levels or varieties of possible manufactured goods (e.g. flavoured milk, buttermilk, skim milk, low-fat milk)
muscle	tissue consisting of elongated cells (muscle fibres)
mutations	a change in the amount or structure of DNA in the chromosomes of an organism
mycelium	mass of hyphae
native pasture	contains plants originating in that particular country (e.g. Australian native plants originating in Australia)
natural pasture	contains plants of both native and naturalised origins
natural vegetation	any vegetation that has not been cleared
necrosis	the death of tissue
nematodes	roundworms, threadworms and eelworms
neonatal mortality	the death of a young animal soon after it has been born
net energy	energy used by an animal for maintenance and production
net radiation	the sum of all radiation energies gained or lost by Earth's surface
non-infectious disease	a disease not caused by a pathogen that consequently cannot be transmitted from one individual to another
nutrition	the study of foods and the food needs of animals
oestrogen	a hormone produced by the ovary and responsible for the development of female sexual characteristics; also responsible for the signs of 'heat'
optimum temperature	ideal temperature
output	the items or materials produced by a system (e.g. farming) and removed from it (e.g. wheat)
oxytocin	a hormone produced by the posterior pituitary, which is responsible for milk letdown
ozone	molecule produced by the breakdown of ordinary molecular oxygen by sunlight, allowing oxygen atoms to combine with other oxygen molecules to form O_3
parameter	the measurements gained from total populations
parasite	an organism that lives on or in, and obtains its food from, another living organism
partial budget	a budget used to estimate the effect on farm profits of a change that will affect only farm subsystems or enterprises, not the entire farm operation
parturition	the act of giving birth
passive immunity	immunity gained through the use of ready-made antibodies being introduced into the body
pasteurisation	killing bacteria that can cause disease by heating milk to 72°C for 15 seconds; the milk is not sterilised; it will still spoil after pasteurisation even if the container is not opened
pasture	a balanced community of plants (generally grasses and legumes) that provides grazing animals with their food requirements
pathogen	a disease-causing organism (can be a micro-organism or an invertebrate)
ped	the basic unit of soil structure

perennial	a plant that continues to grow year after year
permanent wilting point	the point beyond which a plant cannot recover from water loss, even if water is applied to it
pest	any organism that injures, irritates or damages livestock, livestock products or plant products, and can adversely affect productivity
pesticide	any chemical substance, usually dust or spray, used for the destruction of any pest; usually an insecticide
phagocyte	a specialist cell that engulfs bacteria and foreign particles and digests them
phenotype	(in farming) the appearance of a plant or its yield, brought about by the interaction of the genotype and environment
pheromones	a chemical substance released by an animal that influences the development or behaviour of other animals of the same species
phloem	living conductive tissue consisting of sieve tubes and companion cells, through which the products of photosynthesis translocate throughout the plant
photoperiod	the length of time in a 24-hour period that is light
photoperiodism	the response of organisms to changes in day length, as a result of changes in the ratio of daylight to darkness
photosynthesis	the process where carbon dioxide and water are combined using light energy to form high-energy sugar compounds and oxygen
physiology	the way in which organisms, or parts of organisms, function
phytoalexin	an inhibitory chemical produced by a plant following infection by a micro-organism
plant breeder's rights	exclusive commercial rights that protect the registrant of a registered variety of plant, which are administered under the *Plant Breeder's Rights Act 1994*
platyhelminths	flatworms; the simplest of the worm groups; can be free-living or parasitic
pollination	the transfer of pollen grains from the anther to the stigma
polyoestrous	in some non-pregnant animals, oestrus occurs again and again throughout the year (e.g. pigs, cattle)
population	(in biology) a group of one kind of organism living in a particular place at any one time; (in statistics) the entire pool from which a statistical sample is drawn
porosity	the percentage of spaces, or voids, in a soil
post-emergent herbicide	a herbicide that is applied to the crop after it has emerged from the soil
postnatal	the period after birth
pre-emergent herbicide	a herbicide that is applied before or at sowing, before the sown crop emerges from the soil
prenatal	the period before birth, which includes the embryonic and foetal stages
prime lamb	lamb produced for consumption as meat
process	an action that changes input to output
producer	a person who produces goods or services
production energy	additional energy needed by an animal for forms of production, such as growth, pregnancy and lactation
productivity	a measure of the efficiency of a production system that is concerned with the rate at which conversions of inputs to outputs occur
prolactin	a hormone that stimulates the alveolar cells to secrete milk
protein synthesis	the process of creating proteins, starting with amino acid synthesis and ending with the assembling of amino acids into protein molecules
protozoa	single-celled organism that does not produce its own food (e.g. amoeba)
psychrophilic	(an organism that) grows best at low temperatures (less than 20°C)
puberty	the age at which a young animal's reproductive organs are functional (sexual maturity)
pulverescence	the degree of shattering that occurs when a block of soil is subjected to a force at a particular moisture content
quantitative characteristic	a characteristic that can be measured in some way (e.g. yield of wheat in tonnes per hectare)
rainfall variability	a measure of the likelihood of rainfall
randomisation	method used to ensure that all members of a population have an equal chance of being involved in an experiment
range	(in statistics) a measure of spread of the values
rate	a measure of production over time

ISBN 9780170265560

ration	a quantitative measure of the feeds being eaten
receptor	the part of the nervous system that is responsible for detecting changes in an animal's environment
registrant	usually the person or company that developed the variety of plant
replication	repeating the same experiment a number of times
reproduction	the formation of new individuals by the fusion of two sex cells, or gametes, to form a zygote
reproductive phase	the stage of development of a plant where the structures to do with reproduction grow (i.e. flowers, seed, fruit)
respiration	a process that goes on in all living cells where complex organic molecules are oxidised, releasing energy necessary for life
rhizosphere	the area of soil in contact with, and under the influence of, the root system of a plant
rill erosion	the removal of soil caused by small channels of running water
root cap	protective tissue that surrounds the tip of the root, protecting it as it grows through the soil
root hair	a finger-like projection of the epidermal cells of the root, responsible for absorbing water and mineral nutrients from the soil
roughages	bulky feeds that are high in fibre and vary in protein, depending on the source
ruminant	a cud-chewing animal with four stomachs, including a rumen that is inhabited by millions of microbes
salinity	the level of salt in the soil; usually refers to unacceptably high levels
sample	a representative section of a population
saturated soil	a soil in which all pores contain water
saw doctor	a person who maintains and services saws
scientific method	a method of problem-solving that involves the testing of an idea though the use of experiments and the analysis of data
scutellum	the cotyledon in monocot seeds, which absorbs food from the endosperm during germination
seasonally polyoestrous	describes the situation where animals usually breed during particular months of the year (e.g. goats, sheep, horses)
selection	choosing which plants will be used as parents to produce the next generation
selection differential	a measure of the superiority of the selected group over the group before selection was made
selective herbicides	herbicides that target particular weed species but have little or no effect on the desired crop plants
self-pollination	movement of pollen from the anther to the stigma of the same flower
sexed embryos	embryos that have been separated in a laboratory according to their sex
sexed semen	semen that has been processed in a laboratory into two groups; one group is likely to produce male offspring and the other group only female offspring
sheep off shears	sheep that have just been shorn
sheep year	a yearly roster of routine husbandry operations or activities for sheep
sheet erosion	the removal of soil more or less evenly over a wide area of land by wind or running water, usually resulting from heavy downpours
short-day plant	a plant that appears to flower in response to a short **photoperiod**, but in fact flowers in response to the long dark period (i.e. autumn or winter)
silage	plant material that is stored with little drying – it slightly ferments and is preserved and used for stock feed
sodseeding	sowing seed directly into the soil of an existing pasture or crop stubble without any prior ploughing, or cultivation and using a specially designed sodseeding machine
soil consistence	a measure of the mechanical strength of a soil, based on the force necessary to break a block of soil
soil erosion	the removal of soil by running water or wind
soil fertility	a soil's ability to support plant growth
soil moisture potential	a measure of the energy required by the plant to remove water from the soil at a particular moisture content
soil solution	the liquid fraction contained in a soil
soil structure	the arrangement of soil particles in a soil
soil texture	the percentage of sand, silt and clay in a soil, as determined by particle size
solar constant	the average amount of radiation received by a surface at right angles to incoming radiation from the Sun
solar radiation	all forms of radiation received by Earth

somatic cell count	a measure of the white blood cell count in a sample of milk; used as an indication of mastitis
somatotrophin	growth hormone
Southern Oscillation Index (SOI)	a calculation of the monthly or seasonal fluctuations in the air pressure difference between Tahiti and Darwin
species	plants of similar kind that can be crossed and produce fertile offspring
spores	reproductive structures of fungi and some bacteria
squatter's run	an area of land taken up for grazing without the consent of the government in the early days of colonisation of Australia
stamen	the male part of the flower, consisting of the filament (stalk) and anther (pollen sac)
standard deviation	(*s*) the square root of the variance
standardisation	to make all conditions in the different trial of an experiment (e.g. climate, soil, slope) as similar as possible
static display model	indicates inputs, outputs, boundaries, subsystems and the interactions between subsystems on a farm
statistic	the measurements gained from population samples
stigma	the receptive tip of the female part of the plant
stimuli	(singular: stimulus) a change in the external or internal environment of an organism that excites a receptor
stocking rate	the number of animals of a particular type per hectare of land (e.g. 10 sheep per hectare)
stomata	(singular: stoma) small pores or holes in the epidermis of a leaf that allow the diffusion of oxygen and water vapour out of the leaf and the diffusion of carbon dioxide into the leaf; most often occur in the lower epidermis but are found in the upper epidermis of some species
stubble	crop remains (stalks) left after the grain has been harvested
style	tube connecting stigma with ovary
subsidy	aid provided by governments in the form of money to stabilise and maintain incomes and market price
substrate	the non-living material on which an organism lives or grows
subsystem	a system that forms part of a larger system (e.g. a sheep subsystem)
succession	a sequence of different communities in a particular area that form over a period of time
surface energy balance	the balance between radiation gained and lost from the surface of Earth. In daylight hours most of the net radiation energy gained at Earth's surface is used or released as heat
surface scalding	the formation of a surface crust on a soil, which is caused by the build-up of salts in the soil
survey	the collection of sample opinions, facts, figures, etc. in order to estimate the total overall situation
sustainable	able to maintain production levels over the long term
symbiosis	an association in which two organisms or populations live together for their mutual benefit
technology	the practical application of knowledge, such as the use of machinery, computers and/or techniques for undertaking agricultural practices (e.g. the machinery and the method used to grow a wheat crop)
terminal bud	the bud at the end of a stem, responsible for growth in length
terms of trade	for farms, the ratio of the cost of inputs to the price received for products
testa	the coat of a seed
theory	a generally accepted explanation of a principle or observation
thermoperiod	the period of exposure of a plant to a particular temperature
thermophilic	(an organism that) grows best at high temperatures (between 50°C and 60°C)
tiller	a secondary stem in a monocot plant
tissue culture	a method of propagating plants by asexual means
top making	taking wool part-way through processing to the point where it is ready for spinning
total production costs	the sum of the costs of variable inputs and the costs of fixed inputs at any given level of production
toxicity	the effects of the oversupply of a substance, usually a mineral nutrient
toxin	a chemical produced by a micro-organism that has an adverse effect on the life or function of another organism
trace minerals	minerals present in an animal's body in very small amounts (e.g. iron, zinc, copper)
tramlining techniques	controlled traffic farming when working soil
transgene	a gene or genetic material that has been transferred from one organism to another

ISBN 9780170265560

transgenic	animals or plants that have been genetically modified by the addition of a gene or genes from another species into their cells
transpiration	loss of water from the leaves of a plant by evaporation through the stomata
tuber	an underground stem swollen with materials stored by the plant (e.g. a potato)
tunnel erosion	a type of erosion where the saturated subsoil moves and washes away, leaving tunnels
turgid	describes a plant cell that is fully expanded owing to the absorption of water; the cell membrane is pressed against the cell wall
vaccination	an injection of a substance that produces immunity or resistance to a specific disease
vaccine	a biological preparation that improves immunity to a particular disease; a vaccine typically contains an agent that resembles a disease-causing micro-organism, and is often made from weakened or killed forms of the microbe, its toxins or one of its surface proteins
value adding	processing a product in some manner, which enables the product to be sold at a higher price
variable	the particular feature being investigated or measured in an experiment; all other factors are kept constant
variable costs	costs that vary as the size or level of an enterprise varies
variance	(s^2) a measure of how closely values cluster around the mean
variety	types of plants or animals with a particular genotype within a species (e.g. wheat varieties eagle, gatcher and timgalen); In plants, also known as a **cultivar**
vascular bundle	a bundle of tissue in roots and stems, consisting of phloem and xylem
vegetative phase	the stage of development of a plant where roots, stems and leaves grow
vernalisation	the requirement of a plant for a period of exposure to a particular temperature or range of temperatures (usually low), which stimulates the plant to flower
viable	(seed) living and capable of germination
volunteer	(plants) that are not intentionally planted (e.g. wheat plants growing by the side of the road or around silos)
water table	the top of the ground water in the soil
water trading	trading of water in Australia through the buying and selling of water access entitlements and allocations
weather	daily changes in Earth's atmosphere in precipitation, temperature, wind, pressure, cloud cover and other factors
weathering	the process of rock breakdown to form soil
weed	a plant growing where it interferes with other plants or that, because of its characteristics, causes harm to humans or grazing livestock, or degrades the value of animal or plant products
whole farm budget	a budget used to assess the operation of the entire farm business
wilt	a condition where the structure of a plant collapses because of lack of water
wilting point	the moisture level in the soil when wilt occurs
withholding time	the period of time after the application of the pesticide before the product can be harvested and used for human consumption or processing
xylem	dead conductive tissue consisting of tracheids and vessels, through which water and dissolved mineral salts move from the roots throughout the plant
yeast	a single-celled fungus
zero tillage	no tilling takes place to preserve plant cover during the soil preparation phase prior to planting

Index

ISBN 9780170265560

ISBN 9780170265560

ISBN 9780170265560

 ISBN 9780170265560

ISBN 9780170265560

 ISBN 9780170265560

 ISBN 9780170265560

ISBN 9780170265560

ISBN 9780170265560